THE BOOK

OF

THE BUMPING SERIES

A BOOK OF RESEARCH ON NUMBERS

BETWEEN INTERPLAY OF NUMBERS AND FRACTAL GEOMETRY

*B*Y

AITZAZ IMTIAZ

BELGIUM EXCLUSIVE EDITION

NEUROSTOL
GIVE IN ALL YOUR BEST

Printed first in the France.

Project started in 2022

Sold by Amazon International and designed by NeuroStol.

SLAVA UKRAINE

ABSTRACT

The BS Series remains one key concept in the remains of the BS Consortium, which is going to be deprecated in the visible coming time, reminding me of the most nostalgic yet sad times to witness. In that time instance, the BS Consortium witnessed changes after renaming from the AG Consortium, separating the AitSad organization, leaving it to public terms, and now itself presenting the world an attempt of randomness.

It is given by

$$a_n = \begin{cases} (n+1) \cdot 2^{n - \lceil \sqrt{n-1} \rceil^2 + \lceil \sqrt{n-1} \rceil - 2} & \text{if } \lceil \sqrt{n-1} \rceil \leqslant \left\lfloor \dfrac{1+\sqrt{4n-3}}{2} \right\rfloor \\ \\ (n+1) \cdot 2^{\lfloor \sqrt{n-1} \rfloor^2 + \lfloor \sqrt{n-1} \rfloor - n} & \text{otherwise} \end{cases}$$

Proof :

Let us call $a_m, a_{m+1}, \cdots, a_{m+j}$ "increasing part" if, for $i = m, m+1, \cdots, m+j-1$, $a_{i+1} - a_i 0$ hold. For example, the numbers in the "increasing part" of the second line are $a_2 = 3, a_4 = 5, a_8 = 9, a_{14} = 15, \cdots$. Then, we can say that the numbers in the "increasing part" of the M-th line are

$$a_{N^2-N+M} = (N^2 - N + M + 1) \cdot 2^{M-2} \qquad (N \geqslant M-1)$$

In the following, let us prove (1) by induction on M. The numbers in the first line are $a_1 = 1, a_3 = 2, a_7 = 4, a_{13} = 7, \cdots$. Let $b_1 = 1, b_2 = 3, b_3 = 7, b_4 = 13, \cdots$. Since $b_{N+1} - b_N = 2N$, we have, for $N \geqslant 2$, $b_N = 1 + \sum_{k=1}^{N-1}(2k) = N^2 - N + 1$ which holds for $N = 1$. Let $c_1 = 1, c_2 = 2, c_3 = 4, c_4 = 7, \cdots$. Since $c_{N+1} - c_N = N$, we have, for $N \geqslant 2$, $c_N = 1 + \sum_{k=1}^{N-1} k - (N^2 - N + 2)/2$ which holds for $N = 1$. So, The numbers in the first line are $a_{N^2-N+1} = (N^2 - N + 2)/2$, which means that, for $M = 1$, (1) holds. Suppose that (1) holds for M. Then, since the difference of the numbers of the M-th line is 2^{M-1}, we have

$$a_{N^2-N+M+1} = a_{N^2-N+M} + (a_{N^2-N+M} + 2^{M-1})$$

$$= 2 \times (N^2 - N + M + 1) \cdot 2^{M-2} + 2^{M-1}$$

$$= (N^2 - N + M + 2) \cdot 2^{M-1}$$

as long as $N \geqslant M$.

Let us call $a_m, a_{m+1}, \cdots, a_{m+j}$ "decreasing part" if, for $i = m, m+1, \cdots, m+j-1$, $a_{i+1} - a_i 0$ hold. Also, let us call $a_2 = 3, a_5 = 12, a_{10} = 44, a_{17} = 144, \cdots$ "top of triangles".

For example, the numbers in the "decreasing part" of the second line (counted from the top of triangles) are $a_3 = 2, a_6 = 7, a_{11} = 24, a_{18} = 76, \cdots$. Then, we can say that the numbers in the "decreasing part" of the M-th line (counted from the top of triangles) are

$$a_{N^2+M} = (N^2 + M + 1) \cdot 2^{N-M} \qquad (N \geqslant M - 1)$$

In the following, let us prove (2) by induction on M. Substituting $M = N + 1$ into (1) gives $a_{N^2+1} = (N^2 + 2) \cdot 2^{N-1}$, which means that, for $M = 1$, (2) holds. Suppose that (2) holds for M. Then, since the difference of the numbers of the $(M + 1)$-th line (counted from the top of triangles) is 2^{N-M}, we have

$$a_{N^2+M+1} = (a_{N^2+M} + 2^{N-M}) \div 2$$

$$= \left((N^2 + M + 1) \cdot 2^{N-M} + 2^{N-M}\right) \div 2$$

$$= (N^2 + M + 2) \cdot 2^{N-M-1}$$

as long as $N \geqslant M$.

Finally, let us try to represent a_n by n using (1)(2).

For a given n, solving $n = N^2 - N + M$, we get $M = n - N^2 + N$. Since $n - N^2 + N - 1 = M - 1 \leqslant N$, solving $N^2 \geqslant n - 1$ gives $N \geqslant \sqrt{n-1}$. Since $1 \leqslant M = n - N^2 + N$, solving $N^2 - N + 1 - n \leqslant 0$ gives $N \leqslant \frac{1+\sqrt{4n-3}}{2}$. Therefore, for a given n, we have $M = n - N^2 + N$ and $\sqrt{n-1} \leqslant N \leqslant \frac{1+\sqrt{4n-3}}{2}$, so we can take $N = \lceil \sqrt{n-1} \rceil$ if there is such an N.

+ For a given n, solving $n = N^2 + M$, we get $M = n - N^2$. Since $n - N^2 - 1 = M - 1 \leqslant N$, solving $N^2 + N - n + 1 \geqslant 0$ gives $N \geqslant \frac{-1+\sqrt{4n-3}}{2}$. Since $1 \leqslant M = n - N^2$, solving $N^2 \leqslant n - 1$ gives $N \leqslant \sqrt{n-1}$. Therefore, for a given n, we have $M = n - N^2$ and $\frac{-1+\sqrt{4n-3}}{2} \leqslant N \leqslant \sqrt{n-1}$, so we can take $N = \lfloor \sqrt{n-1} \rfloor$ if there is such an N.

In conclusion, we have

$$a_n = \begin{cases} (n+1) \cdot 2^{n - \lceil \sqrt{n-1} \rceil^2 + \lceil \sqrt{n-1} \rceil - 2} & \text{if } \lceil \sqrt{n-1} \rceil \leqslant \left\lfloor \dfrac{1+\sqrt{4n-3}}{2} \right\rfloor \\ (n+1) \cdot 2^{\lfloor \sqrt{n-1} \rfloor^2 + \lfloor \sqrt{n-1} \rfloor - n} & \text{otherwise.} \end{cases} \blacksquare$$

---

a_n can be written as

$$a_n = (n+1) \cdot 2^{n - \lceil \sqrt{n-1} \rceil^2 + \lceil \sqrt{n-1} \rceil - 2} \max\left(\min\left(\left\lfloor \frac{1+\sqrt{4n-3}}{2} \right\rfloor - \lceil \sqrt{n-1} \rceil + 1, 1\right), 0\right)$$

$$+ (n+1) \cdot 2^{\lfloor \sqrt{n-1} \rfloor^2 + \lfloor \sqrt{n-1} \rfloor - n} \left(1 - \max\left(\min\left(\left\lfloor \frac{1+\sqrt{4n-3}}{2} \right\rfloor - \lceil \sqrt{n-1} \rceil + 1, 1\right), 0\right)\right)$$

[signed]

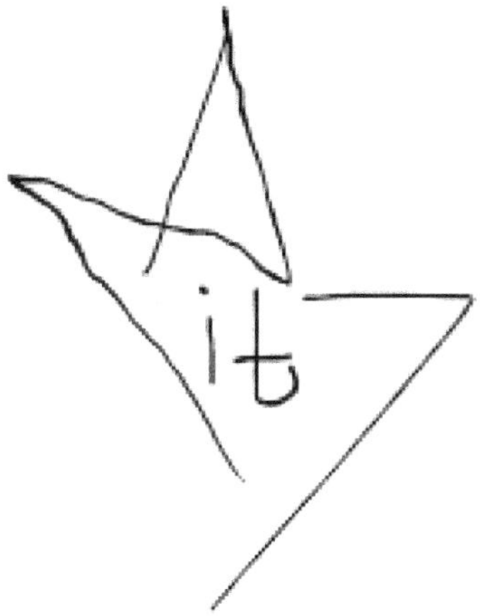

Aitzaz Imtiaz,
December. 22th. 2022.

1 3 2 5 12 7 4 9 20 44 24 13 7 15 32 68 144 76 40 21 11 23 48 100 208 432 224 116 60 31 16 33 68 140 288 592 1216 624 320 164 84 43 22 45 92 188 384 784 1600 3264 1664 848 432 220 112 57 29 59 120 244 496 1008 2048 4160 8448 4288 2176 1104 560 284 144 73 37 75 152 308 624 1264 2560 5184 10496 21248 10752 5440 2752 1392 704 356 180 91 46 93 188 380 768 1552 3136 6336 12800 25856 52224 26368 13312 6720 3392 1712 864 436 220 111 56 113 228 460 928 1872 3776 7616 15360 30976 62464 125952 63488 32000 16128 8128 4096 2064 1040 524 264 133 67 135 272 548 1104 2224 4480 9024 18176 36608 73728 148480 299008 150528 75776 38144 19200 9664 4864 2448 1232 620 312 157 79 159 320 644 1296 2608 5248 10560 21248 42752 86016 173056 348160 700416 352256 177152 89088 44800 22528 11328 5696 2864 1440 724 364 183 92 185 372 748 1504 3024 6080 12224 24576 49408 99328 199680 401408 806912 1622016 815104 409600 205824 103424 51968 26112 13120 6592 3312 1664 836 420 211 106 213 428 860 1728 3472 6976 14016 28160 56576 113664 228352 458752 921600 1851392 3719168 1867776 937984 471040 236544 118784 59648 29952 15040 7552 3792 1904 956 480 241 121 243 488 980 1968 3952 7936 15936 32000 64256 129024 259072 520192 1044480 2097152 4210688 8454144 4243456 2129920 1069056 536576 269312 135168 67840 34048 17088 8576 4304 2160 1084 544 273 137 275 552 1108 2224 4464 8960 17984 36096 72448 145408 291840 585728 1175552 2359296 4734976 9502720 19070976 9568256 4800512 2408448 1208320 606208 304128 152576 76544 38400 19264 9664 4848 2432 1220 612 307 154 309 620 1244 2496 5008 10048 20160 40448 81152 162816 326656 655360 1314816 2637824 5292032 10616832 21299200 42729472 21430272 10747904 5390336 2703360 1355776 679936 340992 171008 85760 43008 21568 10816 5424 2720 1364 684 343 172 345 692 1388 2784 5584 11200 22464 45056 90368 181248 363520 729088 1462272 2932736 5881856 11796480 23658496 47448064 95158272 47710208 23920640 11993088 6012928 3014656 1511424 757760 379904 190464 95488 47872 24000 12032 6032 3024 1516 760 381 191 383 768 1540 3088 6192 12416 24896 49920 100096 200704 402432 806912 1617920 3244032 6504448 13041664 26148864 52428800 105119744 210763776 105644032 52953088 26542080 13303808 6668288 3342336 1675264 839680 420864 210944 105728 52992 26560 13312 6672 3344 1676 840 421 211 423 848 1700 3408 6832 13696 27456 55040 110336 221184 443392 888832 1781760 3571712 7159808 14352384 28770304 57671680 115605504 231735296 464519168 232783872 116654080 58458112 29294592 14680064 7356416 3686400 1847296 925696 463872 232448 116480 58368 29248 14656 7344 3680 1844 924 463 232 465 932 1868 3744 7504 15040 30144 60416 121088 242688 486400 974848 1953792 3915776 7847936 15728640 31522816 63176704 126615552 253755392 508559360 1019215872 510656512 255852544 128188416 64225280 32178176 16121856 8077312 4046848 2027520 1015808 508928 254976 127744 64000 32064 16064 8048 4032 2020 1012 507 254 509 1020 2044 4096 8208 16448 32960 66048 132352 265216 531456 1064960 2134016 4276224 8568832 17170432 34406400 68943872 138149888 276824064 554696704 1111490560 2227175424 1115684864 558891008 279969792 140247040 70254592 35192832 17629184 8830976 4423680 2215936 1110016 556032 278528 139520 69888 35008 17536 8784 4400 2204 1104 553 277 555 1112 2228 4464 8944 17920 35904 71936 144128 288768 578560 1159168 2322432 4653056 9322496 18677760 37421056 74973184 150208512 300941312 602931200 1207959552 2420113408 4848615424 2428502016 1216348160 609222656 305135616 152829952 76546048 38338560 19202048 9617408 4816896 2412544 1208320 605184 303104 151808 76032 38080 19072 9552 4784 2396 1200 601 301 603 1208 2420 4848 9712 19456 38976 78080 156416 313344 627712 1257472 2519040 5046272 10108928 20250624 40566784 81264640 162791424 326107136 653262848 1308622848 2621440000 5251268608 10519314432 5268045824 2638217216 1321205760 661651456 331350016 165937152 83099648 41615360 20840448 10436608 5226496 2617344 1310720 656384 328704 164608 82432 41280 20672 10352 5184 2596 1300 651 326 653 1308 2620 5248 10512 21056 42176 84480 169216 338944 678912 1359872 2723840 5455872 10928128 21889024 43843584 87818240 175898624 352321536 705691648 1413480448 2831155200 5670699008 11358175232 22749904896 11391729664 5704253440 2856321024 1430257664 716177408 358612992 179568640 89915392 45023232 22544384 11288576 5652480 2830336 1417216 709632 355328 177920 89088 44608 22336 11184 5600 2804 1404 703 352 705 1412 2828 5664 11344 22720 45504 91136 182528 365568 732160 1466368 2936832 5881856 11780096 23592960 47251456 94633984 189530112 379584512 760217600 1522532352 3049259008 6106906624 12230590464 24494735360 49056579584 24561844224 12297699328 6157238272 3082813440 1543503872 772800512 386924544 193724416 96993280 48562176 24313856 12173312 6094848 3051520 1527808 764928 382976 191744 96000 48064 24064 12048 6032 3020 1512 757 379 759 1520 3044 6096 12208 24448 48960 98048 196352 393216 787456 1576960 3158016 6324224 12664832 25362432 50790400 101711872 203685888 407896064 816840704 1635778560

3275751424 6559891456 13136560128 26306674688 52680458240 105495134208 52814675968 26440892416
13237223424 6627000320 3317694464 1660944384 831520768 416284672 208404480 104333312 52232192
26148864 13090816 6553600 3280896 1642496 822272 411648 206080 103168 51648 25856 12944 6480 3244
1624 813 407 815 1632 3268 6544 13104 26240 52544 105216 210688 421888 844800 1691648 3387392
6782976 13582336 27197440 54460416 109051904 218365952 437256192 875560960 1753219072 3510632448
7029653504 14076084224 28185722880 56438554624 113011326976 226291089408 113279762432
56706990080 28387049472 14210301952 7113539584 3560964096 1782579200 892338176 446693376
223608832 111935488 56033280 28049408 14041088 7028736 3518464 1761280 881664 441344 220928 110592
55360 27712 13872 6944 3476 1740 871 436 873 1748 3500 7008 14032 28096 56256 112640 225536 451584
904192 1810432 3624960 7258112 14532608 29097984 58261504 116654080 233570304 467664896 936378368
1874853888 3753902080 7516192768 15049162752 30131879936 60330868736 120795955200 241860345856
484257562624 242397216768 121332826112 60733521920 30400315392 15216934912 7616856064 3812622336
1908408320 955252736 478150656 239337472 119799808 59965440 30015488 15024128 7520256 3764224
1884160 943104 472064 236288 118272 59200 29632 14832 7424 3716 1860 931 466 933 1868 3740 7488
14992 30016 60096 120320 240896 482304 965632 1933312 3870720 7749632 15515648 31064064 62193664
124518400 249298944 499122176 999292928 2000683008 4005560320 8019509248 16055795712 32145145856
64357400576 128849018880 257966473216 516469817344 1034013376512 517543559168 259040215040
129654325248 64894271488 32480690176 16257122304 8136949760 4072669184 2038431744 1020264448
510656512 255590400 127926272 64028672 32047104 16039936 8028160 4018176 2011136 1006592 503808
252160 126208 63168 31616 15824 7920 3964 1984 993 497 995 1992 3988 7984 15984 32000 64064 128256
256768 514048 1029120 2060288 4124672 8257536 16531456 33095680 66256896 132644864 265551872
531628032 1064304640 2130706432 4265607168 8539602944 17095983104 34225520640 68518150144
137170518016 274609471488 549755813888 1100585369600 2203318222848 1102732853248 551903297536
276220084224 138244259840 69189238784 34628173824 17330864128 8673820672 4341104640 2172649472
1087373312 544210944 272367616 136314880 68222976 34144256 17088512 8552448 4280320 2142208
1072128 536576 268544 134400 67264 33664 16848 8432 4220 2112 1057 529 1059 2120 4244 8496 17008
34048 68160 136448 273152 546816 1094656 2191360 4386816 8781824 17580032 35192832 70451200
141033472 282329088 565182464 1131413504 2264924160 4534042624 9076473856 18169724928
36373004288 72813117440 145760452608 291789340672 584115552256 1169304846336 2340757176320
4685809319936 2345052143616 1173599813632 587336777728 293936824320 147102629888 73618423808
36842766336 18438160384 9227468800 4617928704 2311061504 1156579328 578813952 289669120
144965632 72548352 36306944 18169856 9093120 4550656 2277376 1139712 570368 285440 142848 71488
35776 17904 8960 4484 2244 1123 562 1125 2252 4508 9024 18064 36160 72384 144896 290048 580608
1162240 2326528 4657152 9322496 18661376 37355520 74776576 149684224 299630592 599785472
1200619520 2403336192 4810866688 9630121984 19277021184 38587596800 77242302464 154618822656
309506080768 619549032448 1240171806720 2482491097088 4969277161472 9947144257536 4977867096064
2491081031680 1246614257664 623843999744 312190435328 156229435392 78181826560 39124467712
19579011072 9797894144 4903141376 2453667840 1227882496 614465536 307494912 153878528 77004800
38535168 19283968 9650176 4829184 2416640 1209344 605184 302848 151552 75840 37952 18992 9504 4756
2380 1191 596 1193 2388 4780 9568 19152 38336 76736 153600 307456 615424 1231872 2465792 4935680
9879552 19775488 39583744 79233024 158597120 317456384 635437056 1271922688 2545942528 5096079360
10200547328 20417871872 40869298176 81805705216 163745628160 327759691776 656056254464
1313186250752 2628519985152 5261334937600 10531259809792 21079699488768 10548439678976
5278514806784 2641404887040 1321776185344 661424963584 330980917248 165624676352 82879447040
41473277952 20753416192 10385096704 5196742656 2600468480 1301282816 651165696 325844992
163053568 81592320 40828928 20430848 10223616 5115904 2560000 1281024 641024 320768 160512 80320
40192 20112 10064 5036 2520 1261 631 1263 2528 5060 10128 20272 40576 81216 162560 325376 651264
1303552 2609152 5222400 10452992 20922368 41877504 83820544 167772160 335806464 672137216
1345323008 2692743168 5389680640 10787749888 21592276992 43218108416 86503325696 173140869120
346550173696 693637218304 1388348178432 2778843840512 5561982648320 11132555231232
22282290331648 44598940401664 22316650070016 11166914969600 5587752452096 2796023709696
1399085596672 700079669248 350308270080 175288352768 87711285248 43889197056 21961375744

10989076480 5498732544 2751463424 1376780288 688914432 344719360 172490752 86310912 43188224
21610496 10813440 5410816 2707456 1354752 677888 339200 169728 84928 42496 21264 10640 5324 2664
1333 667 1335 2672 5348 10704 21424 42880 85824 171776 343808 688128 1377280 2756608 5517312
11042816 22102016 44236800 88539136 177209344 354680832 709885952 1420820480 2843738112
5691670528 11391729664 22800236544 45634027520 91335163904 182804545536 365877526528
732291923968 1465657589760 2933462663168 5871220293632 11751030521856 23519240912896
47072841564160 94214402605056 47141561040896 23587960389632 11802570129408 5905580032000
2954937499648 1478542491648 739808116736 370172493824 185220464640 92677341184 46372225024
23202889728 11609833472 5809111040 2906652672 1454374912 727711744 364118016 182190080 91160576
45613056 22822912 11419648 5713920 2859008 1430528 715776 358144 179200 89664 44864 22448 11232
5620 2812 1407 704 1409 2820 5644 11296 22608 45248 90560 181248 362752 726016 1453056 2908160
5820416 11649024 23314432 46661632 93388800 186908672 374079488 748683264 1498415104 2998927360
6002049024 12012486656 24041750528 48117055488 96301219840 192736657408 385741750272
772020371456 1545114484736 3092376453120 6189047873536 12386685681664 24790551232512
49615462203392 99299643883520 198736726720512 99437082836992 49752901156864 24893630447616
12455405158400 6231997546496 3118146256896 1560146870272 780610306048 390573588480 195421011968
97777614848 48922361856 24477958144 12247367680 6127878144 3066036224 1534066688 767557632
384040960 192151552 96141312 48103424 24068096 12042240 6025216 3014656 1508352 754688 377600
188928 94528 47296 23664 11840 5924 2964 1483 742 1485 2972 5948 11904 23824 47680 95424 190976
382208 764928 1530880 3063808 6131712 12271616 24559616 49152000 98369536 196870144 394002432
788529152 1578106880 3158310912 6320816128 12650020864 25316818944 50667192320 101401493504
202937204736 406142844928 812822560768 1626718863360 3255585210368 6515465388032 13039520710656
26096221290496 52226802319360 104522324115456 209182087184384 418639052275712 209456965091328
104797202022400 52432960749568 26233660243968 13125420056576 6567004995584 3285649981440
1643898732544 822486237184 411511554048 205889994752 103012106240 51539607552 25786580992
12901679104 6455033856 3229614080 1615855616 808452096 404488192 202375168 101253120 50659328
25346048 12681216 6344704 3174400 1588224 794624 397568 198912 99520 49792 24912 12464 6236 3120
1561 781 1563 3128 6260 12528 25072 50176 100416 200960 402176 804864 1610752 3223552 6451200
12910592 25837568 51707904 103481344 207093760 414449664 829423616 1659895808 3321888768
6647971840 13304332288 26625441792 53284438016 106635984896 213406187520 427080810496
854698491904 1710470725632 3423088934912 6850472837120 13709535608832 27436251086848
54906861912064 109882443300864 219902325555200 440079529017344 880708813848576 440629284831232
220452081369088 110294760161280 55181739819008 27608049778688 13812614823936 6910602379264
3457448673280 1729798078464 865435910144 432986390528 216627412992 108380815360 54223962112
27128758272 13572767744 6790578176 3397386240 1699741696 850395136 425459712 212860928 106496000
53280768 26656768 13336576 6672384 3338240 1670144 835584 418048 209152 104640 52352 26192 13104
6556 3280 1641 821 1643 3288 6580 13168 26352 52736 105536 211200 422656 845824 1692672 3387392
6778880 13565952 27148288 54329344 108724224 217579520 435421184 871366656 1743781888 3489660928
6983516160 13975420928 27967619072 55968792576 112004694016 224143605760 448555646976
897648164864 1796370071552 3594887626752 7194070220800 14396730376192 28810640621568
57655640981504 115380001439744 230897441832960 462069761572864 924689278959616 1850478069547008
925788790587392 463169273200640 231722075553792 115929757253632 57999238365184 29016799051776
14516989460480 7262789697536 3633542332416 1817844908032 909459324928 454998097920 227633266688
113883742208 56975425536 28504489984 14260633600 7134511104 3569352704 1785724928 893386752
446955520 223608832 111869952 55967744 28000256 14008320 7008256 3506176 1754112 877568 439040
219648 109888 54976 27504 13760 6884 3444 1723 862 1725 3452 6908 13824 27664 55360 110784 221696
443648 887808 1776640 3555328 7114752 14237696 28491776 57016320 114098176 228327424 456916992
914358272 1829765120 3661627392 7327449088 14663286784 29343350784 58720256000 117507620864
235149459456 470567354368 941671579648 1884416901120 3770981285888 7546257539072 15101105012736
30219389894656 60473139527680 121014998532096 242167436017664 484609749942272 969769255698432
1940638023024640 3883475069304832 1942837046280192 971968278953984 486259017383936
243266947645440 121702193299456 60885456388096 30459908063232 15238543966208 7623566950400

3813930958848 1908039221248 954556481536 477546676224 238907555840 119520886784 59793997824
29913776128 14965276672 7486832640 3745513472 1873805312 937426944 468975616 234618880 117374976
58720256 29376512 14696448 7352320 3678208 1840128 920576 460544 230400 115264 57664 28848 14432
7220 3612 1807 904 1809 3620 7244 14496 29008 58048 116160 232448 465152 930816 1862656 3727360
7458816 14925824 29868032 59768832 119603200 239337472 478937088 958398464 1917845504 3837788160
7679770624 15367929856 30752636928 61538828288 123144765440 246423748608 493115932672
986768736256 1974611214336 3951369912320 7907034791936 15822659518464 31662498906112
63359357550592 126787434577920 253712308109312 507699494125568 1015948744065024
2032996999757824 4068193022771200 8140784092053504 4072591069282304 2037395046268928
1019247278948352 509898517381120 255086697644032 127612068298752 63840393887744 31937376813056
15977278341120 7992934137856 3998614552576 2000381018112 1000727379968 500632125440 250450280448
125292249088 62679678976 31356616704 15686696960 7847542784 3925868544 1963982848 982515712
491520000 245891072 123011072 61538304 30785536 15400960 7704576 3854336 1928192 964608 482560
241408 120768 60416 30224 15120 7564 3784 1893 947 1895 3792 7588 15184 30384 60800 121664 243456
487168 974848 1950720 3903488 7811072 15630336 31277056 62586880 125239296 250609664 501481472
1003487232 2008023040 4018143232 8040480768 16089350144 32195477504 64424509440 128916127744
257966473216 516201381888 1032939634688 2066953011200 4136053506048 8276401979392 16561393893376
33139967655936 66314295050240 132697309577216 265532058107904 531338994122752 1063227744059392
2127554999746560 4257309022748672 8519016092008448 17046828277039104 8527812185030656
4266105115770880 2134152069513216 1067625790570496 534087773192192 267181325549568
133659382251520 66864050864128 33449205301248 16733192585216 8370891259904 4187593113600
2094870298624 1047972020224 524254445568 262261440512 131197829120 65632468992 32833011712
16424894464 8216641536 4110417920 2056257536 1028653056 514588672 257425408 128778240 64421888
32227328 16121856 8065024 4034560 2018304 1009664 505088 252672 126400 63232 31632 15824 7916 3960
1981 991 1983 3968 7940 15888 31792 63616 127296 254720 509696 1019904 2040832 4083712 8171520
16351232 32718848 65470464 131006464 262144000 524550144 1049624576 2100297728 4202692608
8409579520 16827547648 33671872512 67377299456 134821707776 269777633280 539823702016
1080184274944 2161442291712 4325032067072 8654359101440 17317308137472 34651796144128
69337952026624 138744623529984 277626686013440 555528249933824 1111606255681536
2224312022990848 4450823069237248 8906044184985600 17820884462993408 35659361112031232
17838476649037824 8923636371030016 4464017208770560 2233108116013056 1117103813820416
558826784817152 279550831362048 139844135157760 69956427317248 34995393527808 17506286698496
8757438316544 4380866641920 2191507062784 1096290402304 548413636608 274341036032 137237626880
68652367872 34342961152 17179869184 8594128896 4299161600 2150629376 1075838976 538181632
269221888 134676480 67371008 33701888 16859136 8433664 4218880 2110464 1055744 528128 264192
132160 66112 33072 16544 8276 4140 2071 1036 2073 4148 8300 16608 33232 66496 133056 266240 532736
1065984 2132992 4268032 8540160 17088512 34193408 68419584 136904704 273940480 548143104
1096810496 2194669568 4391436288 8787066880 17582522368 35181821952 70397198336 140861505536
281857228800 563982893056 1128502657024 2258079055872 4518305595392 9040906158080 18090402250752
36197984370688 72430328479744 144929376436224 289996191825920 580267261558784 1161084278931456
2323268069490688 4648735162236928 9301868370984960 18612532834992128 37242657856028672
74520500084146176 37277842228117504 18647717207080960 9328256650051584 4666327348281344
2334263185768448 1167681348698112 584115552256000 292195215081472 146166327017472
73117523247104 36575941492736 18296560680960 9152575307776 4578435137536 2290291310592
1145682526208 573109698560 286689067008 143411642368 71739375616 35886465024 17951621120
8980004864 4492099584 2247098368 1124073472 562298880 281280512 140705792 70385664 35209216
17612800 8810496 4407296 2204672 1102848 551680 275968 138048 69056 34544 17280 8644 4324 2163 1082
2165 4332 8668 17344 34704 69440 138944 278016 556288 1113088 2227200 4456448 8916992 17842176
35700736 71434240 142934016 285999104 572260352 1145044992 2291138560 4584374272 9172942848
18354274304 36725325824 73484206080 147035521024 294205259776 588678955008 1177894780928
2356863303680 4715874091008 9436043149312 18880676233216 37778532335616 75591424409600
151251568295936 302640575545344 605556028997632 1211661813809152 2424423139246080

4851045301747712 9706488650006528 19421773393035264 38861138972114944 77757462316318720
155585293376815104 77827831060496384 38931507716292608 19474549951168512 9741673022095360
4873035534303232 2437617278779392 1219358395203584 609954075508736 305114476707840
152625957830656 76347338653696 38190849196032 19104014532608 9556302233600 4780298600448
2391223042048 1196148391936 598342631424 299305533440 149719875584 74893492224 37463523328
18740150272 9374269440 4689231872 2345664512 1173356544 586940416 293601280 146866176 73465856
36749312 18382848 9195520 4599808 2300928 1150976 575744 288000 144064 72064 36048 18032 9020 4512
2257 1129 2259 4520 9044 18096 36208 72448 144960 290048 580352 1161216 2323456 4648960 9302016
18612224 37240832 74514432 149094400 298319872 596901888 1194328064 2389704704 4781506560
9567207424 19142803456 38302384128 76638322688 153343754240 306821726208 613911887872
1228360646656 2457795035136 4917737553920 9839770075136 19688130084864 39393440038912
78821239816192 157711199109120 315559837171712 631394552250368 1263338860314624
2527777232257024 5057753487769600 10119905022050304 20248606137122816 40514804460290048
81064793292668928 162199955329515520 324540648147386368 162340692817870848 81205530781024256
40620357576556544 20318974881300480 10163885487161344 5084141766836224 2543170395045888
1272134953336832 636342354575360 318308616241152 159223027597312 79645873537024 39840116637696
19928648253440 9968619094016 4986457030656 2494302257152 1247687999488 624112435200 312190435328
156162326528 78114717696 39074136064 19545456640 9776922624 4890558464 2446327808 1223688192
612106240 306184192 153157632 76611584 38322176 19169280 9588736 4796416 2399232 1200128 600320
300288 150208 75136 37584 18800 9404 4704 2353 1177 2355 4712 9428 18864 37744 75520 151104 302336
604928 1210368 2421760 4845568 9695232 19398656 38813696 77660160 155385856 310902784 622067712
1244659712 2490368000 4982833152 9969860608 19948109824 39912996864 79859548160 159786205184
319706628096 639681691648 1279900254208 2560874250240 5123895984128 10252086935552
20512763805696 41042707480576 82119774699520 164308268875776 328753976705024 657782831316992
1316115418447872 2633330348523520 5268859720302592 10542117487116288 21093031067254784
42203654320553984 84442493013196800 168955354770571264 338051447029497856 676384369035706368
338332922006208512 169236829747281920 84653599245729792 42344391808909312 21180991997476864
10594894045249536 5299646045880320 2650922534567936 1326011023097856 663280389455872
331777633681408 165957536317440 83013127897088 41523743817728 20770461843456 10389525889024
5196910428160 2599528955904 1300301348864 650419109888 325343772672 162738995200 81403052032
40718303232 20367540224 10187964416 5096079360 2549088256 1275068416 637796352 319029248
159580160 79822848 39927808 19972096 9990144 4997120 2499584 1250304 625408 312832 156480 78272
39152 19584 9796 4900 2451 1226 2453 4908 9820 19648 39312 78656 157376 314880 630016 1260544
2522112 5046272 10096640 20201472 40419328 80871424 161808384 323747840 647757824 1296039936
2593128448 5188354048 10380902400 20770193408 41557164032 83147882496 166362873856 332859965440
665988366336 1332513603584 2666100948992 5334349381632 10672993730560 21354577395712
42726334660608 85487029059584 171042777595904 342222994145280 684720866197504 1369991488208896
2741082488045568 5484363999346688 10973126045204480 21955048183431168 43927688552906752
87890561477902336 175851491699982336 351843720888320000 703968916753350656 1408500783460122624
704531866706771968 352406670841741312 176273704165048320 88172036454612992 44103610413350912
22060601299697664 11034698696359936 5519548371435520 2760873697345536 1380986604486656
690768180150272 345521529028608 172829483991040 86449101733888 43241730736128 21629455302656
10819022618624 5411658792960 2706903138304 1353988440064 677262655488 338765545472 169449881600
84758495232 42396024832 21206401024 10607394816 5305794560 2653945856 1327497216 664010752
332136448 166133760 83099648 41566208 20791296 10399744 5201920 2601984 1301504 651008 325632
162880 81472 40752 20384 10196 5100 2551 1276 2553 5108 10220 20448 40912 81856 163776 327680
655616 1311744 2624512 5251072 10506240 21020672 42057728 84148224 168361984 336855040 673972224
1348468736 2697986048 5398069248 10800332800 21609054208 43234885632 86503325696 173073760256
346281738240 692831911936 1386200694784 2773475131392 5549097746432 11102490460160
22213570854912 44444321579008 88923002896384 177914725269504 355966889492480 712208656891904
1424967069597696 2851033650823168 5704266324901888 11412930696314880 22834657485651968
45686907157348352 91408998686785536 182888366117748736 365917469723852800 732116414424416256

1464795778802253824 2930717457511350272 1465921678709096448 733242314331258880
366761894653984768 183451316071170048 91760842407673856 45898013389881344 22957802787962880
11483299440492544 5743848743501824 2873023883378688 1437061697503232 718805726658560
359540302282752 179838870618112 89953795047424 44994077392896 22505628631040 11257109282816
5630702125056 2816424804352 1408749273088 704643072000 352455753728 176294985728 88181047296
44107300864 22062039040 11035213824 5519704064 2760900608 1380974592 690749440 345505792
172818432 86441984 43237376 21626880 10817536 5410816 2706432 1353728 677120 338688 169408 84736
42384 21200 10604 5304 2653 1327 2655 5312 10628 21264 42544 85120 170304 340736 681728 1363968
2728960 5459968 10924032 21856256 43728896 87490560 175046656 350224384 700710912 1401946112
2804940800 5611978752 11228151808 22464692224 44946161664 89925877760 179918864384 359971946496
720212328448 1440961527808 2882996797440 5768141078528 11540577124352 23089744183296
46196668235776 92427696209920 184924111896576 369985662746624 740246203400192 1481042162614272
2963183836856320 5928566696968192 11861531440447488 23731858973917184 47481310133878784
94997804639846400 190065978023870464 380272693536096256 760826862048903168 1522216674051227648
3045559248009297920 6093370295832281088 3047811047822983168 1524468473864912896
762515711909167104 381398593442938880 190769665465647104 95420017104912384 47727600738500608
23872596462272512 11940696277647360 5972547162079232 2987373092667392 1494236302147584
747393028980736 373833953443840 186985696198656 93527207837696 46780783788032 23398981828608
11703785881600 5854040424448 2928093954048 1464583847936 732560359424 366414397440 183274307584
91670708224 45852131328 22934454272 11471421440 5737807872 2869952512 1435500544 718012416
359137280 179634176 89849856 44941312 22478848 11243520 5623808 2812928 1406976 703744 352000
176064 88064 44048 22032 11020 5512 2757 1379 2759 5520 11044 22096 44208 88448 176960 354048
708352 1417216 2835456 5672960 11350016 22708224 45432832 90898432 181862400 363855872 727973888
1456472064 2913992704 5830082560 11664359424 23337107456 46690992128 93415538688 186898186240
373930590208 748129615872 1496796102656 2994665947136 5991479377920 11987253723136
23983097380864 47983374630912 96001109000192 192070937477120 384279313907712 768833505722368
1538216767258624 3077533046145024 6157265115545600 12318928277602304 24646652648226816
49310897482498048 98656979337084928 197384327418347520 394909392325050368 790100259626811392
1580763469207044096 3162652838320930816 6327557476455546880 12659618552538464256
6332061076082917376 3167156437948301312 1584141168927571968 792352059440496640
396316767208603648 198228752348479488 99149560546328576 49592372459208704 24804982322626560
12406889207824384 6205643627167744 3103921325211648 1552510418419712 776530087116800
388402482511872 194269960732672 97169340104704 48601849921536 24309514895360 12159052414976
6081673691136 3041910587392 1521492164608 761014517760 380641476608 190387847168 95227478016
47630516224 23823646720 11916017664 5960105984 2981101568 1491075072 745799680 373030912
186580992 93323264 46678016 23347200 11677696 5840896 2921472 1461248 730880 365568 182848 91456
45744 22880 11444 5724 2863 1432 2865 5732 11468 22944 45904 91840 183744 367616 735488 1471488
2944000 5890048 11784192 23576576 47169536 94371840 188809216 377749504 755761152 1512046592
3025141760 6052380672 12108955648 24226299904 48469377024 96972308480 194011725824 388157669376
776583774208 1553704419328 3108482580480 6219112644608 12442520256512 24893630447616
49804440764416 99643241267200 199355202011136 398847842975744 797970563858432 1596490883530752
3194081278689280 6390361580634112 12785121207779328 25579038508580864 51175669203206144
102386522778501120 204843414301179904 409827566090715136 819936607158140928 1640436164269703168
3281998228446248960 6566248256706183168 13137000113039736832 26283007425334214656
13146007312294477824 6575255455960924160 3288753627887304704 1644939763897073664
822751356925247488 411516415950979072 205828576719667200 102949472731922432 51492328552005632
25754960369025024 12881878231023616 6443138138767360 3222668581011456 1611884046319616
806216901066752 403245889486848 201691664220160 100880191848448 50457275793408 25237227831296
12622908882944 6313601925120 3157874704384 1579474223104 790005547008 395136991232 197635604480
98851356672 49442455552 24729616384 12369002496 6186598400 3094347776 1547698176 774111232
387186688 193658880 96862208 48447488 24231936 12120064 6062080 3032064 1516544 758528 379392
189760 94912 47472 23744 11876 5940 2971 1486 2973 5948 11900 23808 47632 95296 190656 381440

763136 1526784 3054592 6111232 12226560 24461312 48939008 97910784 195887104 391905280 784072704
1568669696 3138387968 6278873088 12561940480 25132269568 50281316352 100596187136 201259483136
402653184000 805574803456 1611686477824 3224446697472 6451040878592 12906376724480
25821343383552 51659866636288 103354093010944 206776905498624 413691249950720 827657377808384
1655864511430656 3312828534489088 6627856092233728 13260110230978560 26529016554979328
53075625296003072 106186434964094976 212443238672367616 425027214833090560 850335904642891776
1701234759239204864 3403595418385252352 6809442636584189952 13623388872795750400
27255784944846241792 54529584288201965568 27273799343355723776 13641403271305232384
6822953435466301440 3412602617639993344 1706864258773417984 853713604363419648
426997539670065152 213569138579210240 106819753661693952 53427469016891392 26722530601467904
13365663347245056 6685030696878080 3343614860066816 1672357185847296 836453470830592
418364174368768 209250806661120 104659763068928 52347061403648 26182120636416 13095355285504
6549825126400 3275986305024 1638530023424 819533447168 409900941312 205017579520 102542344192
51287949312 25652363264 12830375936 6417285120 3209691136 1605369856 802947072 401604608
200867840 100466688 50249728 25133056 12570624 6287360 3144704 1572864 786688 393472 196800 98432
49232 24624 12316 6160 3081 1541 3083 6168 12340 24688 49392 98816 197696 395520 791296 1583104
3167232 6336512 12677120 25362432 50741248 101515264 203096064 406323200 812908544 1626341376
3253731328 6509559808 13023313920 26055016448 52126810112 104287174656 208641458176 417417134080
835102703616 1670742278144 3342558298112 6687264079872 13378823127040 26766236188672
53549652246528 107133664231424 214336047939584 428809534832640 857893947572224 1716337650958336
3433774813544448 6869748650344448 13743895347200000 27496586787422208 55010765760888832
110056715893866496 220183800531910656 440508338552176640 881298152081063936 1763159254115549184
3527444408137940992 7057140616089567232 14118784831806504960 28246576862867750912
56511168124244983808 113058365045508931584 56547196921263947776 28282605659886714880
14145806429570727936 7075155014599049216 3538703407206367232 1769914653556604928
885238801755013120 442760138365861888 221450437927108608 110760403335643136 55397793853865984
27707693019955200 13858244556488704 6931321301499904 3466760162377728 1733929837002752
867239796408320 433757337157632 216947388055552 108508053766144 54271206752256 27144193310720
13576391622656 6790343294976 3396245389312 1698659565568 849598218240 424933326848 212533772288
106300440576 53166997504 26591887360 13300137984 6652166144 3327131648 1664090112 832307200
416284672 208207872 104136704 52084736 26050560 13029376 6516736 3259392 1630208 815360 407808
203968 102016 51024 25520 12764 6384 3193 1597 3195 6392 12788 25584 51184 102400 204864 409856
819968 1640448 3281920 6565888 13135872 26279936 52576256 105185280 210436096 421003264 842268672
1685061632 3371171840 6744440832 13493075968 26994540544 54005858304 108045271040 216157650944
432449519616 865167474688 1730871820288 3462817382400 6927782248448 13859859464192
27728308862976 55473797595136 110981954928640 222032629334016 444202697621504 888680273149952
1777910302113792 3556920115855360 7116039254966272 14236476556443648 28481749205909504
56981090597863424 113997365567815680 228065099879809024 456270937247973376 912823349472657408
1826209648898736128 3653545197704314880 7309342195222315008 14623187990072000512
29255383179398742016 58528780757306966016 117093590311632896000 234259238217303719936
117165647905670823936 58600838351344893952 29309426374927187968 14659216787090964480
7331860193359167488 3667055996586426368 1834090948246634496 917326949100027904
458804212038369280 229472474763362304 114771421753769984 57403303062929408 28710447624486912
14359621858754560 7182009952632832 3592104487944192 1796601999785984 898575877799936
449425377853440 224781408403456 112425063940096 56229711839232 28123445854208 14066017894400
7035156430848 3518651957248 1759862849536 880199860224 440234147840 220184182784 110125645824
55079600128 27548188672 13778288640 6891241472 3446669312 1723858944 862191616 431226880
215678976 107872256 53952512 26984448 13496320 6750208 3376128 1688576 844544 422400 211264 105664
52848 26432 13220 6612 3307 1654 3309 6620 13244 26496 53008 106048 212160 424448 849152 1698816
3398656 6799360 13602816 27213824 54444032 108920832 217907200 435945472 872153088 1744830464
3490709504 6983516160 13971226624 27950841856 55918460928 111870476288 223808061440 447750340608
895769116672 1792075104256 3585223950336 7172595384320 14349485735936 28707561406464

57432302682112 114898965102592 229866649681920 459870738317312 920016354541568 1840582464897024
3682264441421824 7366727906099200 14737853858709504 29484503810441216 58986599806926848
118008383985942528 236087136716062720 472315010920480768 944911496817672192 1890385943588765696
3781897787084374016 7566047373982433280 15136598347592237056 30282203894439215104
60582422187387912192 121200873171794788352 242473803937627504640 485091723063330865152
242617919125703360512 121344988359870644224 60690508578444804096 30354261488477143040
15181634343865942016 7593068971746656256 3797660385780170752 1899393142843506688
949978046398464000 475129760687587328 237635249087971328 118852808916074496 59443996644081664
29730794415063040 14869795254042624 7437096650276864 3719647836766208 1860373674196992
930461715005440 465368296456192 232752867704832 116410793590784 58222576664576 29119878266880
14564234100736 7284264534016 3643206008832 1822139875328 911338373120 455803404288 227968811008
114017959936 57025757184 28521267200 14264827904 7134511104 3568304128 1784676352 892600320
446431232 223281152 111673344 55853056 27934720 13971456 6987776 3494912 1747968 874240 437248
218688 109376 54704 27360 13684 6844 3423 1712 3425 6852 13708 27424 54864 109760 219584 439296
878848 1758208 3517440 7036928 14077952 28164096 56344576 112721920 225509376 451149824 902561792
1805647872 3612344320 7226785792 14457765888 28923920384 57864617984 115762790400 231592689664
463319597056 926907629568 1854352130048 3709778001920 7421703487488 14847701942272
29703993819136 59425167507456 118884694753280 237838108983296 475813656920064 951902191747072
1904354139308032 3809807790243840 7621814603743232 15248027253997568 30504850601017344
61027293388079104 122089771148247040 244249911040671744 488640559569698816 977562594116108288
1955688138185637888 3912502176278118400 7827256152369922048 15659015904367214592
31327039007989170176 62672092414487822336 125380213625994608640 250832484846027145216
501809084880130146304 1003906400136412004352 502097315256281858048 251120715222178856960
125596386408108392448 62816207602563678208 31417111000536580096 15713059099895660544
7858781349761515520 3930516574787600384 1965821237347221504 983192093650321408
491736784313516032 245938760900935680 123004564822556672 61519874597322752 30768733391683584
15388764742352896 7696581394432000 3849390208843776 1925244860235776 962897308024832
481586092965888 240861765959680 120465242718208 60249801228288 30133490548736 15071040241664
7537667604480 3769907544064 1885490642944 943013756928 471641096192 235887656960 117977382912
59005468672 29511122944 14759755776 7381975040 3692036096 1846542336 923533312 461897728
231014400 115539968 57786368 28901376 14454784 7229440 3615744 1808384 904448 452352 226240 113152
56592 28304 14156 7080 3541 1771 3543 7088 14180 28368 56752 113536 227136 454400 909056 1818624
3638272 7278592 14561280 29130752 58277888 116588544 233242624 466616320 933494784 1867513856
3736076288 7474249728 14952693760 29913776128 59844329472 119722213376 239511535616 479157288960
958583013376 1917702897664 3836479537152 7675106557952 15354508083200 30717606100992
61452392071168 122939143880704 245947007238144 492031453429760 984337784766464 1969225325346816
3939550162321408 7881299347898368 15766996742307840 31542789577637888 63103171341320192
126241527054729216 252553422853636096 505247583195627520 1010776641367965696
2022116232689352704 4045358365285548032 8092968530384781312 16190440660396933120
32389888520048607232 64797791438606696448 129631611674232356864 259335280942502641664
518814677073081139200 1037917584522313990144 2076411629796931403776 1038494045274617413632
519391137825384562688 259767626506730209280 129919842050384068608 64977935423701516288
32497974911105499136 16253491055180120064 8128997327403745280 4065624563608715264
2033375231757778944 1016969090855600128 508625282916155392 254383010202255360
127222689473216512 63630936922652672 31824264554348544 15916530323685376 7960464185098240
3981331604176896 1991215557902336 995882656858112 498078767382528 249108103168000
124588411322368 62311385530368 31164282699776 15586436317184 7795365642240 3898756562944
1949915152384 975226011648 487747223552 243940720640 122003914752 61018734592 30517755904
15263072256 7633633280 3817865216 1909456896 954990592 477626368 238878720 119472128 59752448
29884416 14946304 7475200 3738624 1869824 935168 467712 233920 116992 58512 29264 14636 7320 3661
1831 3663 7328 14660 29328 58672 117376 234816 469760 939776 1880064 3761152 7524352 15052800
30113792 60243968 120520704 241106944 482344960 964952064 1930428416 3861905408 7725907968

15456010240 30920409088 61857595392 123748745216 247564599296 495263416320 990795268096
1982127407104 3965328556032 7932804595712 15869904158720 31748398252032 63513976373248
127062312484864 254193344446464 508524127846400 1017323133599744 2035196023013376
4071491557654528 8145182138564608 16294762323640320 32598320740302848 65214233666650112
130463651705389056 260997672154955776 522136081798266880 1044553638573244416
2089670227099910144 4180466354106662912 8363184508027011072 16730872615681392640
33470752430617526272 66959519259744534528 133955067316508033024 267982192227053993984
536108499642183843840 1072505229660519399424 2145586920073342222336 4292326761651291291648
2146739841577949069312 1073658151165126246400 536973190770638979072 268558652979357417472
134315355286697672704 67175692041858318336 33596853220183900160 16802930209719320576
8403716904673345536 4202984352243515392 2102055126075179008 1051309038014300160
525795256495505408 262967996991930368 131519182868054016 65777183620071424 32897387903057920
16453091998040064 8228745022275584 4115472022765568 2058285767196672 1029417761505280
514846319706112 257491879329792 128780299403264 64407329570816 32212254720000 16110422327296
8057358647296 4029753065472 2015413403648 1007975137280 504121786368 252128002048 126097555456
63065554944 31541166080 15774777344 7889485824 3945791488 1973420032 986972160 493617152
246874112 123469824 61751296 30883840 15446016 7725056 3863552 1932288 966400 483328 241728 120896
60464 30240 15124 7564 3783 1892 3785 7572 15148 30304 60624 121280 242624 485376 971008 1942528
3886080 7774208 15552512 31113216 62242816 124518400 249102336 498335744 996933632 1994391552
3989831680 7981760512 15967715328 31943819264 63904415744 127842385920 255751880704 511637979136
1023544393728 2047625658368 4096325058560 8194797600768 16393890168832 32796370272256
65609920413696 131254200565760 262577120608256 525291680169984 1050858238246912
2102266232307712 4205631976243200 8413462975741952 16831323997995008 33671444089012224
67360480364068864 134756145100226560 269582658944630784 539306055377616896 1078893585731944448
2158350121417310208 4317826142741463040 8637904085296611328 17280311770220593152
34569630739695927296 69157275877901336576 138350580552821637120 276773218699681202176
553690552587438260224 1107669335551028232192 2215915131854359887872 4432983185213326622720
886827221343586693939 2 4435289028222540316672 2218220974863573581824 1109398717807938502656
554843474092045107200 277493794640060481536 138782926117049204736 69409477457034084352
34713745927771783168 17361376563513262080 8682940081570316288 4342595940692000768
2171860920299421696 1086211935126421504 543246705051566080 271693721269960704
135882045007069184 67958614689579008 33988103437811712 16998449765416960 8501423905964032
4251811464609792 2126455488118784 1063502621966336 531888749936640 266013094445056
133040906960896 66537633349632 33277406609408 16642998272000 8323646619648 4162897051648
2081985396736 1041261133824 520764784640 260449501184 130258305024 65145929728 32581353472
16294871040 8149532672 4075814912 2038431744 1019478016 509870080 255000576 127533056 63782912
31899648 15953920 7979008 3990528 1995776 998144 499200 249664 124864 62448 31232 15620 7812 3907
1954 3909 7820 15644 31296 62608 125248 250560 501248 1002752 2006016 4013056 8028160 16060416
32129024 64274432 128581632 257228800 514588672 1029439488 2059403264 4119855104 8241807360
16487809024 32984006656 65984790528 132003135488 264073379840 528280977408 1056830390272
2114197651456 4229469044736 8461085573120 16926466113536 33861522161664 67740224192512
135514808123392 271098335723520 542334110400512 1084943098707968 2170435953229824
4341971418087424 8686141859430400 17376681765371904 34762159623766016 69541911433576448
139119007239241728 278308383222661120 556757503933677568 1113796482844065792
2228155915641552896 4457437731189948416 8917127262193582080 17838758124014534656
35686523447283810304 71391061293077102592 142818151383173169152 285708360360384266240
571560835908844388352 1143409902193840488448 2287396265139984400384 4575945451784575647744
9154196746578364989440 18313005179175157366784 9158808432596792377344 4580557137803003035648
2290855029653804941312 1145715745203054182400 573001987789602947072 286573051488839401472
143322554541438664704 71679291669228814336 35848653033869148160 17928830116561944576
8966666858094657536 4484459328954171392 2242792614430507008 1121677782191964160
560979628584337408 280560183036346368 140315275890262016 70175230131175424 35096411158609920

17552603625816064 8778500836163584 4390349929709568 2195724720668672 1098137238241280
549206058074112 274671748513792 137370233995264 68702296866816 34359738368000 17184164151296
8594229559296 4298188521472 2149631131648 1075084001280 537676218368 268905218048 134486163456
67259858944 33638318080 16823353344 8413773824 4207935488 2104492032 1052508160 526385152
263258112 131661824 65847296 32931840 16470016 8237056 4119552 2060288 1030400 515328 257728
128896 64464 32240 16124 8064 4033 2017 4035 8072 16148 32304 64624 129280 258624 517376 1035008
2070528 4142080 8286208 16576512 33161216 66338816 132710400 265486336 531103744 1062469632
2125463552 4251975680 8506048512 17016291328 34040971264 68098719744 136230993920 272529096704
545192411136 1090653257728 2181843386368 4364760514560 8731668512768 17467631992832
34943853920256 69904887709696 139844135157760 279756989792256 559651418537984 1119577714982912
2239705185779712 4480509883187200 8963218789629952 17930835625771008 35870467344564224
71758526875172864 143552238122434560 287174844989046784 574490427466448896 1149262329909608448
2299087609772638208 4599301119452119040 9200854038717923328 18406211677063217152
36821430553381175296 73660875505271832576 147357779807562629120 294787617209163186176
589719349606402228224 1179726929588956168192 2360030319930215759872 4721213561365038366720
9444732965739290427392 18894077617497008242688 37797378607030871261184 18903300989533863018496
9453956337776145203200 4728131090392679448576 2364642005948643147776 1182609233350473285632
591448731863312498688 295796423525694177280 147934240559866052608 73985134678442508288
37001574538475995136 18505290868865368064 9254897234246369280 4628574517030027264
2314850208468434944 1157706579210928128 578994027093819392 289567382291087360
144818875517632512 72427029944860672 36222311065452544 18115553579237376 9059975812874240
4531087418064896 2266093464846336 1133321610330112 566798244118528 283467841536000
141768280506368 70901320122368 35459249995776 17733919965184 8869107466240 4435627474944
2218350608384 1109443739648 554856087552 277495152640 138781130752 69407342592 34712059904
17360224256 8682209280 4342153216 2171600896 1086062592 543162368 271646720 135856128 67944448
33980416 16994304 8499200 4250624 2125824 1063168 531712 265920 132992 66512 33264 16636 8320 4161
2081 4163 8328 16660 33328 66672 133376 266816 533760 1067776 2136064 4273152 8548352 17100800
34209792 68435968 136904704 273874944 547880960 1096024064 2192572416 4386193408 8774483968
17553162240 35114713088 70246203392 140525961216 281119031296 562372280320 1125012996096
2250562863104 4502199468032 9006546419712 18017387806720 36043365548032 72103910965248
144242181668864 288553082814464 577243604582400 1154762087071744 2310073929957376
4621247371542528 9244693766340608 18493785579192320 36996367251406848 74010326688858112
148055837749805056 296182044243787776 592504825975930880 1185291126928572416
2371145203810566144 4743416307527974912 9489084414869635072 18982672429366640640
37974352057988022272 75966718514485526528 151969465825990017024 304010989246017961984
608166093680111779840 1216620417736375271424 2433817296225053966336 4868787513954714779648
9739880870918643253248 19484373427855713894400 38977970227748282564608 77974387199570274680832
38996416971821992116224 19502820171929423446016 9753715928973925416960 4878010885991569555456
2439581903748088201216 1220079182250195812352 610183706313173762048 305163910750624808960
152617984172331368448 76327006484675166208 38172510441592324096 19090758820423532544
9547631210025451520 4774941504919568384 2388033702413205504 1194298326183313408
597289900580012032 298715319034183680 149392843889180672 74714014130634752 37365803158339584
18687299625680896 9345848836096000 4674023929675776 2337561720651776 1169055738232832
584665308069888 292401373511680 146235046494208 73134703116288 36575941492736 18292265713664
9148280340480 4575213912064 2288143826944 1144340348928 572304392192 286219304960 143143206912
71588380672 35802578944 17905483776 8954839040 4478468096 2239758336 1120141312 560201728
280166400 140115968 70074368 35045376 17526784 8765440 4383744 2192384 1096448 548352 274240
137152 68592 34304 17156 8580 4291 2146 4293 8588 17180 34368 68752 137536 275136 550400 1101056
2202624 4406272 8814592 17633280 35274752 70565888 141164544 282394624 564920320 1130102784
2260729856 4522508288 9047113728 18098421760 36205232128 72427241472 144888037376 289843183616
579820584960 1159909605376 2320356081664 4641785905152 9285719293952 18575733555200
37160057044992 74337293959168 148708947656704 297486614790144 595110668533760 1190496214974464

2381542185762816 4764183883153408 9530566789562368 19065531625635840 38139859344293888
76297310874632192 152629806121353216 305329980986884096 610800699462123520 1221882873900957696
2444328697755336704 4889783295417516032 9781818390648717312 19568140380924805120
39145287961104351232 78308590320718184448 156653209438455332864 313378476470948593664
626901068129973043200 1254090366636097798144 2508757194024499019776 5018667309553604886528
10039640462116423467008 20083892610251274321920 40177008592539403419648 80372463929152516390912
160781821346452451885056 80409357417299935494144 40213902080686822522880
20111562726361838649344 10058087206190133018624 5030196524599673356288 2515674723052140101632
1258125591902221762560 629206911139186737152 314675513163631296512 157373785378834612224
78704907087926788096 39361460743218135040 19685233971236438016 9844868785431904256
4923560292622794752 2462343096264818688 1231453023109120000 615867249042915328
308003993265635328 154037181004906496 77036182688497664 38526887437271040 19267841765146624
9636119905828864 4819159464542208 2410129488084992 1205339621949440 602807249928192
301472344440832 150770531958784 75402445848576 37709812858880 18859201396736 9431748182016
4716947832832 2359010787328 1179773829120 590021132288 295077675008 147572391936 73802973184
36909875200 18459131904 9231663104 4616880128 2308964352 1154744320 577503232 288817152
144441344 72237056 36126720 18067456 9035776 4518912 2259968 1130240 565248 282688 141376 70704
35360 17684 8844 4423 2212 4425 8852 17708 35424 70864 141760 283584 567296 1134848 2270208 4541440
9084928 18173952 36356096 72728576 145489920 291045376 582221824 1164705792 2329935872 4660920320
9323937792 18652069888 37312528384 74641833984 149317222400 298701553664 597537325056
1195343085568 2391223042048 4783519825920 9569187135488 19142669238272 38293928411136
76605036691456 153244433121280 306557585719296 613252610392064 1226780098691072
2454109953196032 4909319418019840 9820837859295232 19646073765101568 39300943623225344
78619479432495104 157274143237079040 314618655218335744 629378047925026816 1259037570826764288
2518638091606949888 5038402083120742400 10079055966055170048 20162615531737710592
40334238262730162176 80686490923969806336 161409010644958576640 322890078883955081216
645924272955986018304 1292136776288123748352 2584850013328550920192 5170852948161708687360
10344011739332631068672 20692635164683689525248 41394493701404233826304 82807434146882177204224
165651761781911773511680 331377310540118385229824 165725548758206611718144
82881221123177015410688 41449833933625362481152 20729528652831108628480 10367070169424768008192
5184688006216990851072 2592920463860798849024 1296748462306551136256 648518346341351424000
324331230764713639936 162201644179375783936 81118836488197373952 40568425443353427968
20288716321304084480 10146609960465727488 5074430880139706368 2537778390023274496
1269170669988347904 634726072482529280 317433404985442304 158751886864809984 79393535618449408
39705563902246912 19857179997634560 9930789022072832 4966494022664192 2483796767145984
1242173261479936 621224069693440 310680754323456 155374736900096 77704548319232 38860864094208
19434727014400 9719510990848 4860829237248 2430951489536 1215744180224 608006307840 304070262784
152068685824 76051120128 38033948672 19021168640 9512681472 4757389312 2379218944 1189871616
595066880 297598976 148832256 74432512 37224448 18616320 9310208 4656128 2328576 1164544 582400
291264 145664 72848 36432 18220 9112 4557 2279 4559 9120 18244 36496 73008 146048 292160 584448
1169152 2338816 4678656 9359360 18722816 37453824 74924032 149880832 299827200 599785472
1199833088 2400190464 4801429504 9604956160 19214106624 38436601856 76889980928 153813516288
307694141440 615522500608 1231313436672 2463163744256 4927401230336 9856949944320 19718194855936
39444979646464 78907139162112 157848638062592 315765995601920 631669430157312 1263613738221568
2527777232257024 5056653976141824 10115506975539200 20235411997589504 40479620088201216
80976832362446848 161988849096982528 324048066938142720 648236871364640768 1296755217705992192
2594073385365405696 5189272670637654016 10380797141088993280 20766097881805357056
41541202962865455104 83100420324240392192 166236869445499748352 332545796485037424640
665235708158150705152 1330759646692453122048 2662095754137209667584 5325344429779026182144
10652994702567266058240 21310601091152959504384 42630425554342773784576 85279297852759257120768
170595489193665933344768 341264765363626704896000 682677104679843086204928
341412339316216381308928 170743063146255609757696 85389978317201514430464

42704212530637611991040 21356717951337233383424 10680664818677830385664 5341485330843522039808
2671319126174064443392 1335947793463183933440 668118011919667822592 334131063553871839232
167101560573954883584 83568794685486923776 41793404541998202880 20901205870626471936
10452854735126921216 5227553267470303232 2614339583688572928 1307451266820997120
653866370898853888 327003554193604608 163536961468891136 81786072920489984 40901832553267200
20455314323144704 10229856184827904 5116027604041728 2558563557834752 1279556656824320
639915767365632 320026603159552 160047661318144 80041010528256 40029095198720 20018842566656
10011568766976 5006858125312 2503965933568 1252251402240 626259918848 313197068288 156632088576
78332821504 39174799360 19591593984 9797894144 4899995648 2450522112 1225523200 612892672
306511872 153288704 76660736 38338560 19173376 9588736 4795392 2398208 1199360 599808 299968
150016 75024 37520 18764 9384 4693 2347 4695 9392 18788 37584 75184 150400 300864 601856 1203968
2408448 4817920 9637888 19279872 38567936 77152256 154337280 308740096 617611264 1235484672
2471493632 4944035840 9890168832 19784531968 39577452544 79171682304 158376919040 316820946944
633776111616 1267820658688 2536178188288 5073430118400 10149007720448 20302310408192
40613210750976 81243601371136 162521562480640 325111844438016 650361127829504 1300997133565952
2602544022945792 5206187557519360 10414574138294272 20833546323099648 41675888739221504
83369369664487424 166773923701063680 333618216146305024 667377169780965376 1335035814538641408
2670634579030704128 5342395057968250880 10687041915750187008 21378587431127744512
42766182061510230016 85550378521529942016 171136785840078848000 342345629274195623936
684835373736467103744 1369958977849085919232 2740494416450475261952 5482141754405557370880
10966589351820328435712 21937790389659084259328 43884804151355023294464 87788055046783756140544
175613003581714931384320 351299794139724700975104 702747162232039078363136
1405789472369257509552128 703042310137218431188992 351594942044904053800960
175834364510599446003712 87935628999373432553472 43977037871723571052544
21993130621880212914176 10998871153949320151040 5500588498479266922496 2750870709991936884736
1375723585372120154112 688005907874135932928 344075011531105894400 172073534562571911168
86054781679795437568 43036398039152459776 21522702619203600384 10763603109415485440
5382927454614585344 2692026677260713984 1346294813607067648 673288144291889152
336714440890122240 168392404817149952 84213794594619392 42115693390331904 21062244741677056
10533321394094080 5267760208674816 2634429860151296 1317489807982592 658882342944768
329509890949120 164789305212928 82411832475648 41214506172416 20611548053504 10307921510400
5155034497024 2578054119424 1289295495168 644781965312 322458091520 161262600192 80648077312
40332427264 20170407936 10087301120 5044699136 2522873856 1261699072 630980608 315555840
157810688 78921728 39469056 19738624 9871360 4936704 2468864 1234688 617472 308800 154432 77232
38624 19316 9660 4831 2416 4833 9668 19340 38688 77392 154816 309696 619520 1239296 2479104 4959232
9920512 19845120 39698432 79413248 158859264 317784064 635699200 1271660544 2543845376 5088739328
10179575808 20363345920 40735080448 81486938112 163007430656 326081970176 652298158080
1304864751616 2610266374144 5221606490112 10445360463872 20895015895040 41798621724672
83614423318528 167263206375424 334595132227584 669327703408640 1338930284724224
2678410325262336 5357920162152448 10718039347560448 21440476741632000 42889749576286208
85797091338616832 171629367049322496 343329102842822656 686798943174000640 1373879361324711936
2748321672602845184 5497769245112532992 10997790290038751232 22000084179704872960
44009175558664486912 88036365515838455808 176108759828695875584 352289577251429679104
704723269690935214080 1409734769758022139904 2820046000268347703296 5641244922041302253568
11284795687091818201088 22574203060202063790080 45157629492440982355968 90333705728955674263552
180704304946058767630336 361482396868412373467136 723112367689414423347200
1446519883284008199520256 2893630062378375104692224 1447110179094366905171968
723702663499773128998912 361925118726181402705920 180999452851238120456192
90518173169692769779712 45268309956883239665664 22638766664460047220736 11321689175239237304320
5661997509124225499136 2831575215314416173056 1416075838033359798272 708182034204755755008
354163074696415805440 177117566145226866688 88576797471122915328 44297405934816198656
22153206567035469824 11078855083331420160 5540553441572552704 2770839670739697664

1385701310346559488 692991392661635072 346566065074995200 173318216909586432 86676700640837632
43347146413441024 21677971253231616 10841184649871360 5421691836563456 2711395674095616
1355972714954752 678123796430848 339130617692160 169599668584448 84817014161408 42417097015296
21212843474944 10608569221120 5305358352384 2653216047104 1326876459008 663572447232
331853332480 165960220672 82996887552 41506832384 20757610496 10380902400 5191499776 2596274176
1298399232 649330688 324730880 162398208 81215488 40615936 20312064 10158080 5080064 2540544
1270528 635392 317760 158912 79472 39744 19876 9940 4971 2486 4973 9948 19900 39808 79632 159296
318656 637440 1275136 2550784 5102592 10207232 20418560 40845312 81707008 163446784 326959104
654049280 1308360704 2617245696 5235539968 10473177088 20950548480 41909485568 83835748352
167705051136 335477211136 671088640000 1342445715456 2685428301824 5371930345472 10746008174592
21496311316480 43001212567552 86019605004288 172073569746944 344215858970624 688569156894720
1377413191696384 2755376139206656 5511851790041088 11025902603337728 22056203253186560
44121202599395328 88259997384835072 176555179141758976 353180727027695616 706502191543746560
1413285858064203776 2827134666081828864 5655395232070500352 11313042263954685952
22630588127536742400 45270183454328225792 90558381307165933568 181152791411350831104
362377640416739590144 724899396021555036160 1450087022419261784064 2900750505590826991616
5802653932686260830208 11607613708381735354368 23219839102781898096640 46448901577600650969088
92916249899275011489792 185869393286697442082816 371812573549689722372096
743772721051969121157120 1487840590009117595140096 2976271475828593895931904
5953723543277905203167232 2977452067449311307235328 1489021181629835006443520
744658164767507179634688 372402869360048428023808 186238328168171633115136
93137610828159526109184 46578028786116617830400 23293626079076736303104 11649118882547581845504
5825712362778397769728 2913432642141502308352 1457004551446902865920 728646390911527288832
364395253049801572352 182233655321919750144 91134842059469357056 45576428228989419520
22792717714122080256 11398610656874725376 5700431228344205312 2850778564125523968
1425670757039472640 712976116008091648 356558426748223488 178314397746200576 89174791059144704
44596191622594560 22302493857808384 11153445952159744 5577822487707648 2789460999667712
1395005377740800 697640127823872 348888783388672 174478751432704 87256555585536 43636867727360
21822728830976 10913511899136 5457829691392 2729451716608 1364994293760 682631364608
341382791168 170724950016 85379252224 42698014720 21353201664 10678697984 5340397568 2670723072
1335623680 667942912 334036992 167051264 83542016 41779200 20893696 10448896 5225472 2613248
1306880 653568 326848 163456 81744 40880 20444 10224 5113 2557 5115 10232 20468 40944 81904 163840
327744 655616 1311488 2623488 5248000 10498048 21000192 42008576 84033536 168099840 336265216
672661504 1345585152 2691694592 5384437760 10770972672 21546139648 43100667904 86218113024
172469780480 345006669824 690147557376 1380563550208 2761663971328 5524401684480 11050950852608
22106196672512 44220983279616 88459146428416 176952652595200 353974024667136 708085488287744
1416445854482432 2833441464778752 5667982441185280 11338163905626112 22680725857763328
45370247808548864 90758087803142144 181551359978373120 363173088700923904 726486914890203136
1453255304757116928 2907073559467655168 5815273018842152960 11632797837497991168
23270099274623352832 46549205748501446656 93116425895512375296 186268880588043714560
372609818770125357056 745363752728326569984 1491015735832804851712 2982607932417913126912
5966368786340433100800 11935043415690079895552 23874698517398587179008 47758620406834029133824
95535687557741767819264 191108268603630954741760 382290324183556747689984
764728222319703171792896 1529751592544585696411648 3060093480899530098475008
6121367553419777608253440 12245096290080990039113728 6123728736661212430860288
3062454664140964921081856 1531522479975661813366784 765908813940420583096320
383028193946505129754624 191550990461399983980544 95793941974773701541888
47906194359423705546752 23957708865730280161280 11981160275874353774592 5991733059441783734272
2996442990473195290624 1498509725612749357056 749398977994450534400 374771546591263195136
187421802092650561536 93728915444834762752 46873464921672122368 23441236060463431680
11722869830045401088 5862560814929543168 2931843357418192896 1466203153685807104
733242314331258880 366691525909807104 183380947326992384 91708065849540608 45862829017792512

22935812555407360 11470105300959232 5736152162107392 2868625836867584 1434587796340736
717431337123840 358784388038656 179426553757696 89730456748032 44873818308608 22441204121600
11222749544448 5612448514048 2806761127936 1403648999424 701958717440 351046467584 175556788224
87795171328 43905974272 21957181440 10980687872 5491392512 2746220544 1373372416 686817280
343474176 171769856 85901312 42958848 21483520 10743808 5372928 2686976 1343744 672000 336064
168064 84048 42032 21020 10512 5257 2629 5259 10520 21044 42096 84208 168448 336960 674048 1348352
2697216 5395456 10792960 21590016 43188224 86392832 172818432 345702400 691535872 1383333888
2767192064 5535432704 11072962560 22150119424 44308627456 88634032128 177301618688 354670346240
709474910208 1419218255872 2838973382656 5679020507136 11360188497920 22724671963136
45457933860864 90933047590912 181900454920192 363869629317120 727876697587712 1456028273082368
2912606301978624 5826312115585024 11654823254425600 23314044555362304 46636885203746816
93291362593538048 186617909559164928 373306187862507520 746753113213370368 1493787701403451392
2988138352760324096 5977402605427490816 11957057010668666880 23918617620964704256
47846242441184149504 95710499280877780992 191457027358774525952 382986112311586979840
766116339811249815552 1532520909998651342848 3065618280749606109184 6132389483003819065344
12267084809016851824640 24538781304052131037184 49086785980141116850176 98192018704355943251968
196420930896859305607168 392915648770013449420800 785978871492616575254528
1572252890890412503334912 3145096077591183712321536 6291372746803084835946496
12585106676847604494499840 25174935720178078634213376 12589829043330474139713536
6296095113285954481160192 3148637852453335946231808 1574614074131847325941760
787454611018513339383808 393801092485551507898368 196937439730923173052416
98487166609535296077824 49252806676804502814720 24631015024420678795264 12317813355219553091584
6160059599114383392768 3080606260309495119872 1540591360530899271680 770439795453525491712
385291955320800673792 192682006457419300864 96359017627219132416 48188516012864307200
24098761606059524096 12051632602843447296 6026942201328566272 3014034050617704448
1507298500285562880 753789987631136768 376965362559746048 188517865651961856 94276525012025344
47147058599034880 23577927346028544 11791162696269824 5896680859762688 2948890185695232
1474719970754560 737497424330752 368817431642112 184443075559424 92238717648896 46127948759040
23068269346816 11536282157056 5769214820352 2885144281088 1442840576000 721554505728
360844361728 180455735296 90244644864 45130711040 22569549824 11286872064 5644484608 2822766592
1411645440 705953792 353042432 176553984 88293376 44154880 22081536 11042816 5522432 2761728
1381120 690688 345408 172736 86384 43200 21604 10804 5403 2702 5405 10812 21628 43264 86544 173120
346304 692736 1385728 2771968 5544960 11091968 22188032 44384256 88784896 177602560 355270656
710672384 1421606912 2843738112 5688524800 11379146752 22762487808 45533364224 91083505664
182200565760 364468240384 729070698496 1458409832448 2917356535808 5835786813440 11673721110528
23351737188352 46712064311296 93441308491776 186916976721920 373902672920576 747942784794624
1496160447496192 2992870650806272 5986840813240320 11975880649736192 23956159345983488
47921114784989184 95859821756022784 191754827884134400 383580024512446464 767300786513248256
1534883048003207168 3070329045959835648 6141783991826513920 12285819783466713088
24576143166560796672 49161293532376334336 98340601463262150656 196717231723543265280
393506521041124458496 787157157270324772864 1574602544916801257472 3149781550585905938432
6300716022676418723840 12603737888362051141632 25212087462742529671168 50433398297521914118144
100885243339117537787904 201807380166382494679040 403688547309059827564544
807524668570709331542016 1615344485046598015909888 3231279265903554737471488
6463739123427826886246400 12929839430097088595099648 25864401226677046835412992
51738247186319832961253376 25873845959642786125840384 12939284163062827885527040
6470822673152131354066944 3236001632386424382685184 1618295964098391544168448
809295556001785448497152 404721564977187562455040 202397675976741200330752
101217284732444309716992 50617865738259009634304 25313544555147932205056
12659078120583179796480 6330691981796196745216 3165922451650401796096 1583249456201352609792
791768843288752160768 395956479238414008320 198014268416225968128 99025148606622466048
49521581502565974016 24765294350910357504 12384898975268864000 6193575387541274624

3097350643724058624 1548956796838739968 774619135907725312 387379936698040320
193725152721108992 96880168546598912 48448880366321664 24228838229671936 12116618138091520
6059408580673536 3030254046150656 1515401900982272 757838389444608 378987914199040
189528316837888 94781338288128 47399259078656 23703924506624 11854109736960 5928128610304
2964601176064 1482569023488 741418729472 370776473600 185421791232 92727672832 46372225024
23190306816 11597250560 5799673856 2900361216 1450442752 725352448 362741760 181403648 90718208
45367296 22687744 11345920 5673984 2837504 1419008 709632 354880 177472 88752 44384 22196 11100
5551 2776 5553 11108 22220 44448 88912 177856 355776 711680 1423616 2847744 5696512 11395072
22794240 45596672 91209728 182452224 364969984 730071040 1460404224 2921332736 5843714048
11689525248 23383244800 46774878208 93566533632 187166621696 374400352256 748934922240
1498138279936 2996813430784 5994700603392 11991548690432 23987392348160 47983374630912
95983929131008 192002218000384 384073155477504 768283749908480 1536842377723904
3074234511261696 6149568534151168 12301336091557888 24607070229626880 49222936552275968
98463465290596352 196962114953281536 393994598650740736 788129934789836800 1576541344556384256
3153645639066189824 6308417178039222272 12619086155892129792 25242675911411630080
50494359022078001152 101006732442665484288 202049493682349932544 404171044958737793024
808486205105551441920 1617260640587254595584 3235097741926812614656 6471348405358232076288
12945002653725677846528 25894616993469783080960 51798457358976420937728
103615361462026551427072 207267616412200521957376 414609019800695882121216
829365613553981440655360 1659026375013142234136576 3318643045836643173924864
6638466683294003759153152 13279294549829442340913152 26563311466141754327040000
53136067665249247944507392 106291024796429974469869568 53154957131180726525362176
26582200932073232907894784 13293461649278051276554240 6647911416259743049580544
3324546003940230230441984 1662568149875294468046848 831431648890236910436352
415789611421413293424640 207931699198854065815552 103984296343500742459392
52001371543787226005504 26005297457912040390656 13004954571965233889280 6503630207487223791616
3252391564495915319296 1626484012624109371392 813386121500130541568 406765118344103198720
203418587969070563328 101727308383044763648 50872661390777122816 25440834295015931904
12722668947321651200 6362460373567668224 3181793136737255424 1591178043345338368
795729759161024512 397935248324689920 199002808534433792 99518996453261312 49768294319652864
24888545206337536 12446471626424320 6224335324839936 3112717418233856 1556633587023872
778454232465408 389295835709440 194682277593088 97358318665728 48687749267456 24348169601024
12176232284160 6089189883904 3045131812864 1522834341888 761551388672 380842803200 190454956032
95244255232 47630516224 23819452416 11911823360 5956960256 2979004416 1489764352 745013248
372572160 186318848 93175808 46596096 23302144 11653120 5827584 2914304 1457408 728832 364480
182272 91152 45584 22796 11400 5701 2851 5703 11408 22820 45648 91312 182656 365376 730880 1462016
2924544 5850112 11702272 23408640 46825472 93667328 187367424 374800384 749731840 1499725824
2999975936 6001000448 12004098048 24012390400 48033169408 96083116032 192199786496 384466681856
769067581440 1538403598336 3077344067584 6155761876992 12313671237632 24631637442560
49271864819712 98560909508608 197156178755584 394381076987904 788899592929280 1578074063765504
3156697883344896 6314495278317568 12631189579890688 25266777206292480 50542350505607168
101102293197258752 202239770766606336 404549910277390336 809240558043136000 1618762591062982656
3238088132079386624 6477302164065615872 12956856127944916992 25918215855517204480
51845438910289149952 103708892219087781888 207453813235194527744 414979684064426983424
830103483316929822720 1660495197010011357184 3321566854772326137856 6644286631049259122688
13290879105107731939328 26586369896233891266560 53181963164504637308928
106382373073082984169472 212801639634313387442176 425677066244921613090816
851501706442432902594560 1703298560790045158014976 3407187417390449021681664
6815555426401615454666752 13633472036044665731940352 27271666438572201109094400
54552777610110141508616192 109124444686151761598087168 218286668304166480357883904
109162223618014718759796736 54590556541973098670325760 27300000637469418980376576
13652361501976144312795136 6827361342608789567700992 3414270967114753489502208

1707430631462556097576960 853862889683867725201408 427005231818228700807168
213539509397261769506816 106788201442704594305024 53403324093389151928320
26706273732713003352064 13355442709365715369984 6678874276187464531968 3340013598846035689472
1670295029799169556480 835291630087660634112 417717872637868244992 208894965115953086464
104465496956486025216 52241755677497753600 26125381438376247296 13064942519001808896
6533597159407747072 3267361529657294848 1633962239805358080 817121857391034368
408631297439694848 204350833091936256 102193008732012544 51105300459028480 25557048276025344
12780723161268224 6391461092261888 3196280301944832 1598415028879360 799344953393152
399741196173312 199904957825024 99969658781696 49993419325440 25001004630016 12502649798656
6252398641152 3126736191488 1563636531200 781952483328 391043350528 195555229696 97794392064
48905584640 24456986624 12230590464 6116343808 3058696192 1529610240 764936192 382533632
191299584 95666176 47841280 23924736 11964416 5983232 2992128 1496320 748288 374208 187136 93584
46800 23404 11704 5853 2927 5855 11712 23428 46864 93744 187520 375104 750336 1500928 3002368
6005760 12013568 24031232 48070656 96157696 192348160 384761856 769654784 1539571712 3079667712
6160384000 12322865152 24649924608 49308237824 98633252864 197300060160 394667229184
789468676096 1579205787648 3158948446208 6318970634240 12640088752128 25284472471552
50577534877696 101172249624576 202378858987520 404826437451776 809790313857024 1619855505620992
3240260767055872 6481621045739520 12965441114734592 25935280275980288 51879356644982784
103776305476009984 207587795324108800 415245959392395264 830632656273145856 1661546787523002368
3323656524999426048 6648438949905694720 13299129699625074688 26602762998877519872
53214533197009780736 106447080792529043456 212930190382077050880 425932438358192029696
852008991904459915264 1704306214185071542272 3409188889122446508032 6819530699749499863040
13641367242508213420032 27287346171034854227968 54583915714106563231744
109186278172286836015104 218409449832721091133440 436892686641737020473344
873932947236063717359616 1748161042377306787545088 3496912380564972280741888
6995005352750661972787200 13992371888742758768181248 27989466143968387181576192
55988377020902513653579776 111995643507736505888014336 224029065947335968937738240
448133689758397852198895616 224104623811061883261157376 112071201371462420211433472
56045045418696949396144128 28027245075831344343285760 14015983721157106994249728
7009172452199270908428288 3505176521909994159865856 1752883408860176432758784
876589278382677892792320 438368426167633784602624 219221106571964311404544
109629000030055865253888 54823723387064787402752 27416473379550821089280
13710542532784624238592 6856424187896918966272 3428788554700762906624 1714682507726533165056
857485369051342438400 428814742119709147136 214443399856873537536 107239714326946250752
53628864362727866368 26818935780991303680 13411719690309337088 6706985745061511168
3354055822484176896 1677309386218799104 838795430597754880 419468084043055104
209769226393616384 104902205382852608 52459898784448512 26234347438735360 13119372742623232
6560785882939392 3280942697283584 1640746226548736 820510552227840 410323995590656
205196357533696 102615358636032 51316269252608 25662429593600 12833362280448 6417754882048
3209414311936 1604975591424 802622013440 401378115584 200722612224 100378083328 50197430272
25102909440 12553551872 6277824512 3139436544 1569980416 785121280 392626176 196345856 98189312
49102848 24555520 12279808 6140928 3070976 1535744 768000 384064 192064 96048 48032 24020 12012
6007 3004 6009 12020 24044 48096 96208 192448 384960 770048 1540352 3081216 6163456 12328960
24662016 49332224 98680832 197394432 394854400 789839872 1579941888 3160408064 6321864704
12645826560 25295847424 50600083456 101216944128 202467442688 405001994240 810138206208
1620544847872 3241626566656 6484326875136 12970801233920 25945897435136 51900384804864
103817949478912 207670258696192 415409236869120 830955912691712 1662186703290368
3324923162394624 6650945836417024 13304090696089600 26612579438690304 53233954970402816
106485502126850048 213006188625788928 426082745995755520 852306229479866368 1704893933936443392
3410350817826308096 6821827535559458816 13645906870932602880 27296317341492576256
54601641882239893504 109221298162989268992 218478625122997501952 437029307840032931840
874202730868141719552 1748693692112435150848 3497963844977173725184 6997080611458954297344

13996467065927122288640 27997545817872671965184 56004315007782198706176
112027076759638106963968 224091047007423633031168 448255880991142104268800
896659335934873884950528 1793613819774927122726912 3587817935360212951105536
7176816462341143313514496 14355994107923721449635840 28716710582330312544485376
57442865897626364379398144 114904621261184207339651072 229847021454231371841011712
459769600772188658005442560 919690317271829144657723392 459920716499640486652280832
229998137181683200487849984 115017958056773078824779776 57518423761352278702817280
28763934247159008996622336 14384328306820939320918016 7193344745031187071762432
3597262668325952241532928 1798926482068155473592320 899610814986667413209088
449879194469628544811008 224976490722961691508736 112506692105554555305984
56262569424814132428800 28135896398425493602304 14070254042221960495104 7036279942615587094528
3518716432060096970752 1759646446406200197120 879967338391175954432 440055726789625905152
220063892191831916544 110049960494425440256 55033987446467461120 27521497322861101056
13763000461244235776 6882626130528960512 3441876015217901568 1721219482585661440
860750478781186048 430445608134770688 215257988439474176 107646586405781504 53832089295912960
26920442694467584 13462420370489344 6732309696872448 3366704604250112 1683627180032000
841951028969472 421044233961472 210556476719104 105295418228736 52656299048960 26332444491776
13168369729536 6585258606592 3293166174208 1646851522560 823559979008 411847098368 205957103616
102995329024 51506053120 25757220864 12880707584 6441402368 3221225472 1610874880 805568512
402849792 201457664 100745216 50380800 25194496 12599296 6300672 3150848 1575680 787968 394048
197056 98544 49280 24644 12324 6163 3082 6165 12332 24668 49344 98704 197440 394944 790016 1580288
3161088 6323200 12648448 25300992 50610176 101236736 202506240 405078016 810287104 1620836352
3242196992 6485442560 12972982272 25950158848 51908706304 103834189824 207701934080 415470977024
831076171776 1662420779008 3325378428928 6651830599680 13305808683008 26615912333312
53240414601216 106498009071616 213030377881600 426129475239936 852396389433344 1705067656773632
3410685069361152 6822469650350080 13647138323955712 27298674694422528 54606145481867264
109229883149778944 218494950671646720 437060270087471104 874261277663297536 1748804030303305728
3498171010560032768 6997467921026908160 13997187641867501568 27998878883362373632
56006764965979488256 112031544330468458496 224099117457955880960 448270292509949689856
896684700207975235584 1793657630792102182912 3587891722336507789312 7176936366177622425600
14356178575364458545152 28716968836747344478208 57443161045531543732224
114904768835136797016064 229846431158421013135360 459766649293136864477184
919680872538863405367296 1839656892982906163560448 3679904081776171032772608
7360988755173059476848640 14724338693587553776304128 29453399753657977197821952
58916244240281693686071296 117851377946494865952997376 235740534824852689067704320
471556627513431292458827776 943264370754314413564493824 1886830972963532484422664192
943566602209218070858170368 471858858968334949752504320 235967208416030432037961728
118002493673946694599835648 59010691569939086590345216 29510068151452412940386304
14757395258967641292800000 7379878221104538057703424 3690529406362627734503424
1845559851086493220077568 922927499495836286451712 461537536724212981432320
230805661850253909819392 115421277669200664461312 57719862206637187006464
28864542789337020891136 14434577237677724139520 7218441540343468916736 3609797230924037881856
1805186845838170652672 902737538107161182208 451440826647618519040 225756442120828223488
112896235458923593728 56457124928716537856 28233066063985639424 14118784831806504960
7060518315810095104 3530822107858468864 1765692528905945088 882987001941327872
441563869714841600 220817119229509632 110426151800799232 55221871993421824 27615334043222016
13809866044866560 6906032534061056 3453566022844416 1727057889329152 863666383618048
431901911285760 215985315381248 108009837559808 54013508714496 27011049324544 13507672145920
6754909814784 3377991778304 1689264324608 844766380032 422450298880 211258703872 105646129152
52831453184 26419920896 13212057600 6607077376 3304062976 1652293632 826277888 413204480
206635008 103333888 51675136 25841664 12922880 6462464 3231744 1616128 808192 404160 202112 101072
50544 25276 12640 6321 3161 6323 12648 25300 50608 101232 202496 405056 810240 1620736 3241984

6484992 12972032 25948160 51904512 103825408 207683584 415432704 830996480 1662255104 3325034496
6651117568 13304332288 26612858880 53234106368 106484989952 213003534336 426074177536
852282572800 1704833581056 3410204033024 6821481807872 13645111099392 27294517166080
54597624266752 109212428402688 218459216543744 436987152564224 874111744081920 1748498366070784
3497546487955456 6996192487538688 13994583998332928 27993566043176960 55995928179376128
112009448544796672 224054081461682176 448178531667542016 896497800823439360 1793277076623589376
3587117103200600064 7175360106308042752 14352972012429770752 28710447624486912000
57429902448228564992 114877819294966611968 229791667386952187904 459655392367942303744
919454899923960463360 1839198030224072638464 3678972521200448700416 7359097963905504247808
14720501770820222189568 29445615227658871767040 58900453827354598309888
117819354398782906171392 235675602285713231446016 471424991547721301098496
942997557048032278609920 1886290262001243910045696 3773170819812846525743104
7547522231246410462789632 15097405645734255748186112 30199533657951381141585920
60408512048868501573599232 120835913563668481728053248 241709606059199920617816064
483494769982125755559051264 967140655691703339764940800 1934583542838310336823558144
3869771548586427988234469376 1935188005748117651410911232 967745118601510654352293888
483948117164481241499566080 242011837514103577911492608 121024808222983267536601088
605218488844457373058727936 30265646788711556174577664 15135184577597212909895680
7568772880419323866251264 3784976736020020638777344 1892783515915189672214528
946539331910184512520192 473343452931387094466560 236708619953840966336512
118372756720994192719872 59195601732533951135744 29602412552285402955776
14803512119151915171840 7402908981080564432896 3702030951292585639936 1851303706022444531712
925795968199298121728 462970041693686988800 231521049643862458368 115778539220440711168
57898276809475096576 28953642004364918784 14479072801996144640 7240662300904914944
3620894100405878784 1810728525179650048 905505000078180352 452822868783267840
226446618763722752 113240901567905792 56629246876975104 28319021484998656 14161709765754880
7081954394505216 3541526953066496 1771038354440192 885656616173568 442897027563520
221482873520128 110758616629248 55387898249216 27698244091904 13851269529600 6926708506624
3463891124224 1732213997568 866241216512 433187717120 216627412992 108330483712 54173630464
27091009536 13547601920 6774849536 3387949056 1694236672 847249408 423690240 211877888 105955328
52985856 26497024 13250560 6626304 3313664 1657088 828672 414400 207232 103632 51824 25916 12960
6481 3241 6483 12968 25940 51888 103792 207616 415296 830720 1661696 3323904 6648832 13299712
26603520 53215232 106446848 212926464 425918464 851968000 1704198144 3408920576 6818889728
13639876608 27283947520 54576283648 109169344512 218372243456 436811595776 873757409280
1747783254016 3496103378944 6993280499712 13988708483072 27981711933440 55972013801472
111961207472128 223956774682624 447982268841984 896101976637440 1792478831181824
3585507418177536 7172114347982848 14346427719221248 28697253484953600 57403303062929408
114824198311903232 229683580995895296 459437530735968256 919015798960291840 1838313072897294336
3677189095748009984 7355504091402862592 14713259982619410432 29431023564866191360
58871054328987123712 117760123056483729408 235556274909986422784 471184607414010773504
942513330016097402880 1885314890408346517504 3771206241568996458496 7543565404642599763968
15089436652294413221888 30183484990607253831680 60376193353251362439168
120770833450576434429952 241578560389300287963136 483230907754895414132736
966609389462380504678400 1933513926829940362182656 3867618149470239430017024
7736416890561196271337472 15475194964363827365281792 30955112295210524375777280
61919669323386788041981952 123858228112705054664818688 247754235157273066491346944
495584028178272047306113024 991319172083995923259064320 1982940575622895503811805184
3966485614155598322210963456 7934180154130811273596633088 3967694539975212951385669632
1984149501442510132986511360 992225866448706895140093952 496188491088079361893466112
248132024475902638108442624 124084901703882797635076096 62051895584907138107965440
31030670158936438699196416 15517696262709654172205056 7760028722975544497405952
3880604657298130954354688 1940597476554244830003200 970446312229712091414528

485296943091150883913728 242685365033722861060096 121361129260935140081664
60689788002504424816640 30349505687270639796224 15177058686644533592064 7589682264826873643008
3795417593165740244992 1897997026959021834240 949142628667586772992 474643371927831314432
237357714760934621184 118696871778976792576 59357443088743137280 29683225143998939136
14843864371813154816 7423058085813420032 3712091992860131328 1856327471406776320
928304473191743488 464222605340049408 232146487042113536 116090835707101184 58054213946572800
29031505019797504 14517951533154304 7260075278204928 3630587394916352 1815568575365120
907921726636032 454029582794752 227049151135744 113541755437056 56779467653120 28394028793856
14199161880576 7100654682112 3550864211968 1775700541440 887984488448 444059353088 222063230976
111048392704 55532584960 27770486784 13887340544 6944718848 3472883712 1736704000 868483072
434307072 217186304 108609536 54312960 27160576 13582336 6792192 3396608 1698560 849408 424768
212416 106224 53120 26564 13284 6643 3322 6645 13292 26588 53184 106384 212800 425664 851456
1703168 3406848 6814720 13631488 27267072 54542336 109101056 218234880 436535296 873201664
1746665472 3493855232 6988759040 13979615232 27963424768 55935238144 111887253504 223808061440
447683231744 895500681216 1791269797888 3583076466688 7167226675200 14336600834048
28677496635392 57363583205376 114744346279936 229523052298240 459114824073216 918367087099904
1837009052106752 3674567860027392 7350235231682560 14702669486620672 29409737019752448
58828270132527104 117674132451098624 235383449274286080 470837267292749824 941815272073854976
1883912019124420608 3768386988202262528 7537899876311367680 15078051552436420608
30160606704500211712 60330220608255164416 120678455615019810816 241392940027058585600
482857937648155099136 965859990484386054144 1932008211344923820032 3864592883442151063552
7730338688388908974080 15462983219787031642112 30930578125592490672128 61870379623221836120064
123759205990517381791744 247555305469182182686720 495184397914659203579904
990516369781908083572736 1981327887468995519971328 3963246070748349745594368
7927672733117416902492160 15857706649476268627591168 31720135665435406900396032
63449716063836553091219456 126918321593604584763293696 253874422119072126688296960
507824402101870167700013056 1015799919931192164046864384 2031902071317287985387405312
4064408605544383285362163712 8130026136908381199899033600 16262470125455991658147479552
8132443988547610458248445952 4066826457183612543711576064 2033715460046709929149464576
1017008845750806793221570560 508579980739129310934204416 254327769301427612628811776
127182774116645284895260672 63600831791288381738057728 31805138262127060514242560
15904930314304965079728128 7953645748773199951167488 3977413170196958681235456
1989001733003658693443584 994648440454419023134720 497398007203504349773824
248735897089899593990144 124386395289023506546688 62202421016548608049152
31105822194292731412480 15555216940155579400192 7778761391582396547072 3889957156543501697024
1945266808647902560256 972777519512027136000 486460817350051495936 243266437472044711936
121651233134531837952 60834623766520659968 30421815482887700480 15213159541257535488
7607705670535610368 3804415785221226496 1902489367587323904 951385421282017280
475763079385186304 237916724064681984 118975954218385408 59496773202214912 29752784647618560
14878591347064832 7440395185160192 3720747348393984 1860648552103936 930461715005440
465299576979456 232684148228096 116359253983232 58188216926208 29098403430400 14551349198848
7276748341248 3638911041536 1819723956224 909996195840 455065206784 227566157824 113799856128
56908316672 28458352640 14231273472 7116685312 3558866944 1779695616 889978880 445054976
222560256 111296512 55656448 27832320 13918208 6960128 3480576 1740544 870400 435264 217664 108848
54432 27220 13612 6807 3404 6809 13620 27244 54496 109008 218048 436160 872448 1745152 3490816
6982656 13967360 27938816 55885824 111788032 223608832 447283200 894697472 1789657088 3579838464
7160725504 14323548160 28651290624 57310969856 114638716928 229310988288 458689085440
917512388608 1835293212672 3671123296256 7343320334336 14688788152320 29381871271936
58772332478464 117561844826112 235158049390592 470384818257920 940907075469312 1882089028845568
3764727813505024 7530555138637824 15063309300531200 30131016647573504 60270829388169216
120559250962382848 241153686296854528 482377741337886720 964896220164128768 1930073915304968192
3860710780563357696 7722547461033558016 15447346721880801280 30899197043388973056

61807401286032687104 123632816970574856192 247301662738168676352 494675383070375280640
989494881328826417152 1979277993033804546048 3959132446819912515584 7919417815144431878144
15841141473298077450240 31686894632614582288384 63383012637266019352576
126784472018605748256768 253605837525358915616768 507285462027012669440000
1014718498006615015292928 2029732143918409383411712 4060054583647177472475136
8121289758915072356253696 16244940701071579535114240 32494603768626028715442176
64998652270217796721311744 130016194006367072023478272 260070166944597101208666112
520215891752920116740751360 1040582899233292062128340992 2081468029921487781550358528
4163540522752782877688070144 8328289971325180384550846464 16658997794289590027451105280
33322831291857638571601035264 16663833497568048544149929984 8333125674603638901249671168
4167167300211626765212188672 2083885881560717039899770880 1042094056507810348596723712
521122586117631088621780992 260599071990678501472600064 130318425461270729317154816
65168657463601103949004800 32589051098283421619716096 16296886732383145632464896
8149623957812290227535872 4075402274716503819419648 2037996285263431262535680
1019145716584305307680768 509646645268447492046848 254860216122371165126656
127448554805259292114944 63733500774666500833280 31871362073351677804544
15937986879685052596224 7970146361347133145088 3985649641425869996032 1993113051089086709760
996700640732619210752 498422377960347533312 249247217777192730624 124641623287105847296
62329818842807664640 31169413021031202816 15586958310329286656 7794605055071485952
3897865477489164288 1949214213721292800 974747844349001728 487444290918678528
243757329831428096 121896257101758464 60956924643901440 30482860368461824 15243629207486464
7622914115371008 3812006813499392 1906278284656640 953276581281792 476707010117632
238387864797184 119211112267776 59614146068480 29811368001536 14907831484416 7454989484032
3728031612928 1864284241920 932276338688 466205278208 233136193536 116584873984 58300825600
29154607104 14579400704 7290748928 3645898752 1823211520 911736832 455933952 227999744 114016256
57016320 28512256 14258176 7130112 3565568 1783040 891648 445888 222976 111504 55760 27884 13944
6973 3487 6975 13952 27908 55824 111664 223360 446784 893696 1787648 3575808 7152640 14307328
28618752 57245696 114507776 229048320 458162176 916455424 1833172992 3666870272 7334789120
14671675392 29347545088 58703478784 117423734784 234881024000 469829156864 939792531456
1879853498368 3760243867648 7521561477120 15045270437888 30094835843072 60198261620736
120413703110656 240861765959680 481792251396096 963721941745664 1927718761398272
3855987278610432 7713074068848640 15428347160952832 30861092368416768 61730980829855744
123479553845755904 246994292063600640 494058952871378944 988258643231113216 1976798761438937088
3954160472831295488 7909446845569433600 15821145490952552448 31646794581532475392
63302596362319691776 126623207123148865536 253282443043316695040 506636943680671318016
1013418002549418491904 2027124235474988695552 4054824931702280814592 8110802784909168476160
16223911412827550646272 32452434511673528680448 64914092395383912136704
129846631534841533825024 259730156557830486753280 519534100091955811713024
1039215774136501299838976 2078726696178181952503808 4158043688166722610659328
8317267967954162632622080 16636897119149760087851008 33278516604782389820915712
66566477942530518932258816 133151845350992516445372416 266341469633847990052454400
532758497131421894428327936 1065668109990295617503494144 2131638451435494892300664832
4263881365780797099188682752 8528971657381208827552071680 17060361166401646913453555712
34125558036081752343605936128 68260787478720421720609521664 34135229442638669377003585536
17070032572958563946851205120 8536225212298896602600308736 4268717069059255615887507456
2134660765984531465237430272 1067481498719717561265553408 533816307223584694956195840
266945932543655304639807488 133491855737759130900758528 66755372601845304740806656
33382408667405522015617024 16693565516944195830415360 8347963350092815326511104
4174571970856766368907264 2087581133333562537279488 1043938140619370945052672
522042857285980310732800 261058322131137574469632 130547607809642496786432
65283027276858103169024 32646125324447478972416 16325368505232953180160 8163837174121083437056
4082495047812845142016 2041535754282574282752 1020911992329362997248 510528053758719426560

255300055676378677248 127668042236698820608 63843028317604151296 31926017758429446144
15965260679028408320 7983756239421046784 3992441069663944704 1996502009808683008
998391742392696832 499266239940526080 249668304342351872 124851744357220352 62434668271632384
31221732182327296 15613065114419200 7807632068837376 3904365790232576 1952457773023232
976366325465088 488251882209280 244160300843008 122097330290688 61057255079936 30532922507264
15268608737280 7635378110464 3818225926144 1909381398528 954824916992 477479567360 238773338112
119403446272 59710111744 29859250176 14931722240 7466909696 3733979136 1867251712 933756928
466944000 233504768 116768768 58392576 29200384 14602240 7302144 3651584 1826048 913152 456640
228352 114192 57104 28556 14280 7141 3571 7143 14288 28580 57168 114352 228736 457536 915200
1830656 3661824 7324672 14651392 29306880 58621952 117260288 234553344 469172224 938475520
1877213184 3754950656 7510949888 15023996928 30052188160 60112764928 120242307072 240518168576
481103446016 962341109760 1924950654976 3850438180864 7701950103552 15406047690752
30816390348800 61641370632192 123299921133568 246634202005504 493337123487744 986811685928960
1973898249764864 3948346255343616 7897792022315008 15797783067885568 31599964182282240
63208724457586688 126435041101217792 252905266574524416 505880901893226496 1011902541274808320
2024086557526327296 4048736065006075904 8098598029918994432 16199447859651674112
32403399318930718720 64815805837116178432 129649626072741838848 259335280942502641664
518742619479043211264 1037629354146162278400 2075546938668476268544 4151670338089255960576
8304493597683118768128 16611293038375451230208 33227197762769329848320 66463618897575514472448
132945684539224738496512 265928262566596896096256 531930312109488630398976
1064008198171566937210880 2128311544248313227247616 4257213384306985160146944
8515607360234687731597312 17033575903710810285801472 34071874173904490216816640
68153193080774719724060672 136325275627480918028976128 272688330186824793219661824
545452218237375500762742784 1091055552202202830172323840 2182413335859309317638324224
4365431134628425949864001536 8732071195076466528902709248 17466560241792162316154830848
34937956186862783149008486400 69885583780282483331414622208 139790510373678800729624543232
69904926593396317398209921024 34957298999976617215803785216 17481067351627537866251304960
8741742601633383562300358656 4371475763726499095737532416 2186040113318153205162442752
1093171172386528431228059648 546661144056990129937448960 273368350960358022130434048
136703064946110489646071808 68360977206020984113463296 34185210969493361701945344
17094966667988115673579520 8548663925614775248093184 4274922258617746329698304
2137756277214052517675008 1069025712559615935250432 534586643256102805831680
267330215116198822019072 133683554302173120561152 66851000523123415056384
33430111947580134916096 16717361816799281152000 8359833829904247422976 4180493375704427134976
2090534918228365279232 1045411574302258495488 522777844745167175680 261424951169602551808
130730489983310757888 65374252190910119936 32691629695082430464 16348066647354900480
8175159223584292864 4088142561745567744 2044352755849494528 1022317115413102592
511228926450728960 255649647597453312 127842415984771072 63930004085407744 31969400089214976
15986899067863040 7994549045559296 3997824278593536 1999187017203712 999730947555328
499934193254400 250001456365568 125017908051968 62517543960576 31263066947584 15633680957440
7817914220544 3909493981184 1955015426048 977641930752 488888074240 244477591552 122255572992
61136175104 30572281856 15288238080 7645167616 3823108096 1911816192 956039168 478085120
239075328 119554048 59785216 29896704 14950400 7476224 3738624 1869568 934912 467520 233792 116912
58464 29236 14620 7311 3656 7313 14628 29260 58528 117072 234176 468416 936960 1874176 3748864
7498752 14999552 30003200 60014592 120045568 240123904 480313344 960757760 1921777664 3844079616
7689207808 15380512768 30765219840 61538828288 123094433792 246222422016 492511952896
985158123520 1970584682496 3941706235904 7884486213632 15771119910912 31546534789120
63101659512832 126220498894848 252475357528064 505019434532864 1010176308019200
2020627493945344 4041804743704576 8084708999036928 16171617021329408 32347632089169920
64704060271362048 129425712728768512 258886609829625856 517843588403429376 1035827914295214080
2071937303567138816 4144437557087698944 8290001014082240512 16582253827978166272
33169011255583703040 66347029710422147072 132712073819353776128 265460176435726516224

530992410465490960384 1062128936119057776640 2124546102614267265024 4249668665980837953536
8500490253466282754048 17003286349941779202048 34011184385901985792000 68031592143840826359808
136081631031755362271232 272200155551658143645696 544474098079611125497856
1089095770111811927408640 2178486688128803207643136 4357563672067965120937984
8716307935756647653179392 17434977054754730128965632 34874676475992329903144960
69758797684950399096717312 139536484835832276774289408 279110748603527510710288384
558297055070780935743995904 1116745225869013700134830080 2233792683192931057563336704
4468189829295669429714026496 8937588584410953488602759168 17877595020461136235554930688
35760025744200730987808686080 71529722894958379009015021568 143078788603030592084825341952
286196262832288852303241281536 143117474229258260218415939584 71568408521186047142605619200
35789039963871482088001634304 17896937833574970302350229504 8949677842607099780349820928
4475443384213357204762263552 2238023923561582259674808320 1119163077508242958484242432
559657096617847393565540352 279866327240786653944479744 139952053086324805553094656
69985471276128142066974720 34997458004546940678701056 17501090185514905161957376
8751725684378169992282112 4376453137999443701792768 2188521716904901203722240
1094408432405040278274048 547278003178814977343488 273675895077554907774976
136856394282851163439104 68437420513462436495360 34223321942749645635584
17113966814384036511744 8558136328696625102848 4279644625100615974912 2140110542926459699200
1070199386651305705472 535171750919690780672 267621904256864354304 133828966526941659136
66923490462725570560 33466248830990155776 16735376215308763136 8368814007561224192
4184969953734033408 2092766451843727360 1046523963410219008 523332350449287168
261701359596732416 130868271984410624 65442932085227520 32725864089124864 16365131067817984
8183665045536768 4092382278582272 2046466017198080 1023370447552512 511753943252992
255911331364864 127972845551616 63995012710400 32001801322496 16003048144896 8002597814272
4001835778048 2001186324480 1000727379968 500430798848 250248953856 125141254144 62579015680
31293702144 15648948224 7825522688 3913285632 1956904960 978583552 489357312 244711424 122372096
61194240 30601216 15302656 7652352 3826688 1913600 956928 478528 239296 119664 59840 29924 14964
7483 3742 7485 14972 29948 59904 119824 239680 479424 958976 1918208 3836928 7674880 15351808
30707712 61423616 122863616 245760000 491585536 983302144 1966866432 3934257152 7869562880
15741222912 31486640128 62981668864 125980114944 251993784320 504054677504 1008243572736
2016755580928 4034048032768 8069169807360 16140487098368 32285269164032 64579128262656
129175436394496 258385232527360 516839184531456 1033815808016384 2067906493939712
4136362743693312 8273824999014400 16549849021284352 33104096089079808 66216988271181824
132451568728408064 264938321828904960 529947012401987584 1060034762292330496
2120350999561371648 4241264949076164608 8483655798059171840 16969563395932028928
33943630391491428352 67896267982237597696 135810550362984677376 271657129522988318720
543386316640014565376 1086916748468104986624 2174121727312361684992 4348819915377026793472
8698792752258660433920 17399891347526534561792 34804394381071496511488 69618012134179847798784
139254471012433405149184 278545835513014229401600 557165458002323297009664
1114478489957236270432256 2229252127819651893690368 4459094551449662493032448
8919369694520042397368320 17841100572281519617343488 35686923511045908879900672
71383291755057557050228736 142785472976046592681312256 285608724883956142524334080
571293007631638199372087296 1142737130990728227391012864 2285776493436360112075702272
4572157449782527538738757632 9145523825384669706652221440 18293465502408568671653855232
36591766708095595860006535168 73193204822748108753410719744 146405752458610051573616738304
292850190543447771280824074240 585777752339350878828829343744 292927561795903107548005269504
146483123711065387840797933568 73251233262089610953796616192 36630452334323263993597132800
18317644018800861255147978752 9160030935220045256748695552 4580619930519829942961700864
2290612196714818628774526976 1145457214084861143034101760 572804164906156485840470016
286439861384941200081944576 143238820158402078621827072 71628854812166778601340928
35819149772566258945884160 17911936069524564295548928 8957148626382999559077888
4479164609001858485190656 2239877452406108595421184 1120086300155643974123520

560116937054116825268224 280095362015205831737344 140066127751676625420288
70042287247875167485952 35025755309956011130880 17515183497987219259392 8758744670498216476672
4379948796001411661824 2190262628376857542656 1095275429376504627200 547709772282290241536
273890914938164084736 136963471867591524352 68490743133050503168 34249875166152622080
17127189382889996288 8564720591351840768 4282923245629341696 2141743097791381504
1071012286384046080 535576511936200704 267823440340189184 133929312356139008 66973452271091712
33491124182056960 16747761114284032 8374980068769792 4188039790198784 2094294773006336
1047284825456640 523711132205056 261889925840896 130962142789632 65489661329408 32749125632000
16376710299648 8189428891648 4095251316736 2047894093824 1024081264640 512107741184 256087425024
128060489728 64038633472 32023511040 16013852672 8007974912 4004511744 2002518016 1001390080
500760576 250413056 125222912 62619648 31313920 15659008 7830528 3915776 1958144 979200 489664
244864 122448 61232 30620 15312 7657 3829 7659 15320 30644 61296 122608 245248 490560 981248
1962752 3926016 7853056 15708160 31420416 62849024 125714432 251461632 502988800 1006108672
2012479488 4025483264 8052015104 16106127360 32216449024 64441286656 128899350528 257832255488
515731619840 1031597457408 2063463350272 4127463571456 8256000884736 16514149253120
33032593473536 66073776881664 132164733632512 264363827003392 528796373483520 1057730185920512
2115735249747968 4232020255309824 8465140022247424 16932479067750400 33869356182011904
67747508457046016 135512609100136448 271060402572361728 542191173888901120 1084523085266157568
2169327645509025792 4339218240971472896 8679562381849788416 17361376563513262080
34727256726653894656 69463520652562530304 138945055703634542592 277926140204288049152
555924338002614026240 1111992791193303908352 2224273812762759528448 4449124086277822480384
8899401094060251807744 17801108031129717309440 35606827748277862006784 71222878868592578789376
142464204481258867130368 284965302450665153363968 570004391877625144934400
1140156357707839966281728 2280607863320859285389312 4561806022452077276430336
9124792636524871964164096 18251946456291178750935040 36508615279065227147083776
73026675291096193584594944 146072240048123865750044672 292182259028110688661798912
584440075919947291647016960 1169031267567346411940872192 2338364766589596481175420928
4677333996089000276938194944 9355876917997615183051096064 18714171687634459624451604480
37433179078547377765602033664 74876029563651672564601716736 149771401940417179195998732288
299581489507062026525588062208 599240350266579389318357319680 1198635443038069451171077029888
599395092771490061852719710208 299736232011972699059950452736 149887458819100183596770525184
74953400816107008831782912000 37481536111331962932590280704 18743185907305210724644552704
9372801879472219991496982528 4687005402645917310335844352 2343804932777862312461598720
1172053582116382984877637632 586102348921917406762237952 293088953392821660542828544
146563366162342308852269056 73291127814136893716561920 36650286273551316503494656
18327504320017093074354176 9164932751629263948480512 4583056671624990679891968
2291823483717674692771840 1146059315811427022798848 573103444882008349605888
286588615929151593906176 143312754708649506504704 71665600726361608028160
35837412049199231401984 17921011867608829394944 8961658855309021544448 4481405888406814195712
2240991174579558809600 1120639702477855260672 560391908832965558272 280231983213501743104
140134006005260353536 70076010201884917760 35042508700569829376 17523506150098599936
8762878974956142592 4382002437431492608 2191282693692456960 1095782084334583808
547961410911469568 274015889827823616 137025537099956224 68521564643000320 34265180368011264
17134789207261184 8568494115258368 4284796813443072 2142673284628480 1071474081267712
535805760110592 267937239793664 133985799766016 67001489817600 33505039876096 16754667421696
8378407452672 4189740597248 2095138734080 1047703584768 523918901248 261993005056 131013279744
65515028480 32761708544 16382951424 8192524288 4096786432 2048655360 1024458752 512294912
256180224 128106496 64061440 32034816 16019456 8010752 4005888 2003200 1001728 500928 250496
125264 62640 31324 15664 7833 3917 7835 15672 31348 62704 125424 250880 501824 1003776 2007808
4016128 8033280 16068608 32141312 64290816 128598016 257228800 514523136 1029177344 2058616832
4117757952 8236564480 16475226112 32954646528 65917681664 131852140544 263737835520 527542779904
1055219777536 2110707990528 4221952851968 8444979445760 16892106375168 33788507717632

67585605369856 135188390608896 270411140956160 540891001389056 1081919441731584
2164113761370112 4328777278554112 8658654068736000 17319507160727552 34643412367966208
69295620828954624 138608833843953664 277252852059996160 554576072864169984 1109292883216695296
2218867241410101248 4438297432773623808 8877720765454090240 17757693330721865728
35519890261071101952 71048787721396944896 142115589841303371776 284267208479625707520
568606474553289342976 1137357064294654541824 2275002358965460795392 4550581178683225014272
9102315278871056875520 18206936400751327444992 36418484487521082277888 72846192347079019331584
145710831438231748214784 291458556364610915532800 582990899705516669272064
1166129373363623014957056 2332553894632425382739968 4665698085075209471131648
9332576761771136353566720 18667514706783707529740288 37339751780050284704694272
74688948293066308699815936 149396786052064095980486656 298831351035991149122682880
597738259935708212568784896 1195627635598868253784408064 2391557502652640164862492672
4783719468215087644312338432 9568647862249789917799383040 19139713576138809093948178432
38284262855556076704595181568 76578197117669070442588012544 153175737048451974951971323904
306390159723131618037533245440 612857690698718572342247686144 1225870123902347817218857762816
2452049732814516979506440306688 1226179608912169162287582543872 613167175708539917410972467200
306622273480497626839076831232 153330479553362647486333714432 76674911183238240776564506624
38342291294897578904981078016 19173563499088018710839951360 9587990675363623984594681856
4794599800591619306884694016 2397602131750713310736023552 1198952181602808484014850048
599551648665130156330844160 299813603264428035327131648 149925691098145496244420608
74972290282038487412637696 37490867507502113351532544 18747794936992491498373120
9375078060116963160489984 4688129325868840285896704 2344359810839599495774208
1172327479372389424300032 586237526662489550356480 293155656819392194281472
146596275153769806692352 73307360948921758121984 36658292160479306448896
18331451923248866918400 9166878883129040306176 4584015902316823576576 2292296181534563500032
1146292205955357605888 573218160571716730880 286645109082877329408 143340568939948146688
71679291669228814336 35844149434241777664 17924326516934574080 8963289158374129664
4482207529140486144 2241385239546953728 1120833357261832192 560487047375093760
280278708059635712 140156946215862272 70087269200953344 35048032646987776 17526215346749440
8764207185002496 4382653348315136 2191601552064512 1095938214985728 548037826969600
274053273223168 137043816480768 68530498174976 34269544054784 17136919511040 8569533497344
4285303619584 2142920245248 1071594340352 535864279040 267965693952 133999624192 67008200704
33508294656 16756244480 8379170816 4190109696 2095316992 1047789568 523960320 262012928
131022848 65519616 32763904 16384000 8193024 4097024 2048768 1024512 512320 256192 128112 64064
32036 16020 8011 4006 8013 16028 32060 64128 128272 256576 513216 1026560 2053376 4107264 8215552
16433152 32870400 65748992 131514368 263061504 526188544 1052508160 2105278464 4211081216
8423211008 16848519168 33701232640 67410853888 134838484992 269710524416 539488157696
1079110533120 2158489501696 4317515874304 8636105490432 17274358464512 34553011896320
69114613727232 138246407323648 276527174385664 553123068248064 1106383575449600
2213042028806144 4426633813426176 8854367138480128 17710933300215808 35426264646942720
70861325386907648 141740242959859712 283515670291808256 567101709327794176 1134344156143943680
2268969787264598016 4538502524482617344 9078130948872077312 18158513697557839872
36321530994743050240 72652069188740841472 145322152775991164928 290680334349001293824
581432726292040515584 1163009567772156887040 2326307365920465485824 4653191192593234395136
9307535306691075637248 18617376456391364968448 37239364598801157324800 74487952569639169425408
148994351883352048402432 298025597254851515908096 596124981485997870022656
1192397536924585416458240 2385090221754350185742336 4770770739319059077136384
9542722070258835565576192 19087805323759105953759232 38180333014001081552732160
76370110760967902395891712 152759110987867283372638208 305556000907597523906985984
611187559678920962137391104 1222526235085293752921620480 2445354701625491163136917504
4891313866160789640861188096 9783836658141193910897082368 19570091167921617080143577088
39145018039121692676985978880 78299707484800302387369607168 156618757782714438841534513152

313276201191656545816659623936 626629773635768427900500443136 1253414289776447528335363276800
2507138064562716401739451334656 5014895099145075493616352231424
2507757034582359091876900896768 1254033259796090218472812838912 627094001150500445503587614720
313585686201477890885384404992 156812185913852779509487501312 78415764363483306788141400064
39212717885020111910769524736 19608776794149285213734174720 9805597322894257236041793536
4903403124356935932608249856 2452003793633371623597801472 1226153012544137640445739008
613152064135794734546288640 306613810999760324434853888 153325794965811640798281728
76672342215871559689568256 38340893474418649489997824 19172807920450759567605760
9587584551846097195106304 4794382571733407303204864 2397486433771883004428288
1198890790838531178627072 599519182395560427520000 299796484685927632863232
149916689087037525983232 74967567915555617767424 37488395643796236271616
18746503664907331829760 9374404753958272761856 4687778837731439804416 2344177649241871613952
1172232939809011662848 586188527498543759360 293130292546290843648 146583160671654903808
73300587535082192896 36654797367168466944 18329650483397918720 9165951141605801984
4583538520756322304 2292050735354871808 1146166105165791232 573153421327073280
286611895035625472 143323539703857152 71670565944950784 35839681018986496 17922039532748800
8962119278002176 4481609394814976 2241079575314432 1120677226610688 560407332782080
280238026129408 140136192933888 70076686401536 35042638168064 17523466567680 8762807025664
4381940383744 2191238627328 1095753531392 547943874560 274005491712 137019523072 68518150144
34263269376 17133731840 8567914496 4284481536 2142502912 1071382528 535756800 267911168
133971968 66994176 33501184 16752640 8377344 4189184 2094848 1047552 523840 261952 130992 65504
32756 16380 8191 4096 8193 16388 32780 65568 131152 262336 524736 1049600 2099456 4199424 8399872
16801792 33607680 67223552 134463488 268959744 537985024 1076101120 2152464384 4305453056
8611954688 17226006528 34456207360 68920803328 137858383872 275750322176 551567753216
1103269724160 2206807883776 4414152638464 8829379018752 17660905521152 35326106009600
70660801953792 141338783776768 282711927291904 565492574060544 1131122587074560
2262520052056064 4525589859926016 9052279231479808 18106757486215168 36217913018941440
72444622130905088 144906836447854592 289848857267798016 579768083279773696 1159676904047902720
2319635283072516096 4639833516098453504 9280792932103749632 18563837664021184512
37132178927669739520 74273365054594220032 148564744507697922048 297165517812414808064
594403093218867544064 1188950301625810944000 2378188833627773599744 4756954128007850622976
9515061177520308092928 19032428198049829879808 38069468082118087147520 76148159536273029070848
152314765816619767693312 304666425121386954489856 609406637219068747186176
1218960848390727170785280 2438216844686633694396416 4877023985183626094444544
9755228561987969600192512 19512818307217374022991872 39030358980917617691197440
78070162694800974672822272 156159214855533427926499328 312356208642929813014708224
624787975149585540352835584 1249727066026622909352509440 2499756363508149475998695424
5000117189926106266584743936 10001443305671827162344194048 20005304462982883583037800448
40015444629244225682774425600 80040560665045368398946500608 160100464143204570864688300032
320239613912636809862967197696 640556599077728955993115590656 1281267940660368584520593571840
2562845366330558514109911924736 5126309702680759718357273411584
10253857345400804816989445947392 5127547642720045098632172535808
2564083306369843894384811048960 1282196395689832619726767915008 641175569097371646130565152768
320626470174913491198873174016 160332577900570579666231885824 80175960356842206866513592320
40092815881699561949955620864 20048825792489010233327222784 10025621822064119745838317568
5013415373941867187506511872 2507009918425837251046932480 1253656074940370454170304512
626903595333911141408571392 313489576598818527865995264 156763677765340742513852416
78391283615636110547353600 39200364174300924918890496 19602543270391897282052096
9802452226816666052329472 4901816409218691731816448 2451203352514525218734080
1225749250209852285779968 612948412081220981096448 306511099528757909651456
153273996508452664377344 76646221626263186964480 38327722499150020870144
19166167092584224129024 9584236467796718911488 4792694694650662879232 2396635577701483151360

1198461904038817431552 599303009613446643712 299687533603742285824 149861781200380624896
749398977994450534440 374744524993498972 16 1873947804948863385 6 937086492465115955 2
468599541227900108 8 234327918111621120 0 117178032804646092 8 585960532767408128
293015450755792896 146525317563940864 73271454874992640 36640125484007424 18322261765259264
9162230394257408 4581664952942592 2291107354378240 1145691116142592 572914277548032
286491498512384 143262929125376 71640054497280 35824322215936 17914308591616 8958228037632
4479650889728 2240093880320 1120181157888 560157687808 280112398336 140072976384 70044876800
35026632704 17515413504 8758755328 4379901952 2190213120 1095237632 547684352 273874944
136953856 68485120 34246656 17125376 8563712 4282368 2141440 1070848 535488 267776 133904 66960
33484 16744 8373 4187 8375 16752 33508 67024 134064 268160 536384 1072896 2146048 4292608 8586240
17174528 34353152 68714496 137445376 274923520 549912576 1099956224 2200174592 4400873472
8802795520 17607688192 35219570688 70447529984 140911837184 281857228800 563781566464
1127697350656 2255663136768 4511863144448 9024800030720 18051747545088 36107790057472
72224170049536 144465519968256 288965399674880 577999518826496 1156136476606464
2312547831119872 4625645418053632 9252390347735040 18506979718725632 37018357483962368
74045511060946944 148108614307938304 296252412987965440 592575194720108544 1185291126928572416
2370863728833855488 4742290407621132288 9485706715149107200 18973665230111899648
37951834059851169792 75912675318957080576 151843365036423643136 303722758869866250240
607517575333770428416 1215179265855616712704 2430646762087385137152 4861869984927073697792
9724892891358754242560 19452091625726722179072 38908794937471871746048 77826813246980598267904
155672073238034906087424 311381039964217231278080 622835866904729300762624
1245819307762048277938176 2491933763429275908702208 4984457822668910523056128
9970096236958538457415680 19942553657158511737438208 39889829680799893120090112
79789104094565525530607616 159597097655062529642070016 319231974241988016445849600
638539506347701947215118336 1277230128422855723077074944 2554762488300615103447826432
5110129439511037521483005952 10221467804841689672140718080 20445353461322608602630848512
40895542625923675721960521728 81800756658404268477318692864 163620856129922371021432684544
327280397886072410176455966720 654638167024600156620093128704 1309431076554110985774548647936
2619171638118043316617822076928 5238962246255729323373093715968
10479162432550744027021086556160 20960800745180058814591971360768
10481638312629314787570884804608 5241438126334300083922891964416
2621028548176971387030170763264 1310669016593396366049447772160 655411879549153519291905081344
327744625400804427779543138304 163891655513516047956566867968 81955499163314941011681083392
40982585284935929022539366400 20493710494107193769619095552 10248064172873211513984253952
5124636549346413071579480064 2562620506128110193083416576 1281461368791506925188546560
640806242259479376917692416 320440900061602645620555776 160239339496732801391132672
80129114481332139985993728 40069279607148939638210560 20037000986815904641712128
10019681085028669732159488 5010430838324693571731456 2505510567067526138691584
1252902857486352745758720 626525215719471211085824 313299501347883024646144
156668197418015221874688 78343322081044465713152 39176272726540660244480
19590442206279543816192 9796374024644378755072 4898763473074492801024 2449669966913398112256
1224979098644774912000 612561606916425383936 306316832255231655936 153176430526125309952
76597222462317395968 38303114830786068480 19153809215206719488 9578030507510202368
4789578203708522496 2395070576830971904 1197676025903841280 598908381696098304
299489375220137984 149762279796113408 74889935991078912 37449366042050560 18726882044280832
9364540533768192 4682820022697984 2341684889255936 1170979883581440 585558661267456
292813690372096 146424025055232 73220602462208 36614596198400 18309445582848 9155796533248
4578435137536 2289486004224 1144877219840 572505718784 286286413824 143159984128 71588380672
35798384640 17901289472 8951693312 4476370944 2238447616 1119354880 559742976 279904256
139968512 69992448 35000320 17502208 8752128 4376576 2188544 1094400 547264 273664 136848 68432
34220 17112 8557 4279 8559 17120 34244 68496 137008 274048 548160 1096448 2193152 4386816 8774656
17551360 35106816 70221824 140460032 280952832 561971200 1124073472 2248409088 4497342464

8995733504 17993564160 35991322624 71991033856 143998844928 288031244288 576129597440
1152393412608 2305055260672 4610647392256 9222368526336 18446884536320 36898064039936
73804718014464 147626615898112 295287591534592 590643902545920 1181425244045312
2363125365997568 4726800487809024 9454700487245824 18911599997747200 37827598042005504
75663992177033216 151345576540110848 302726337452310528 605523043648798720 1211186824785952768
2422655124548616192 4845873199050653696 9692872298008150016 19387996395829985280
38780496391287341056 77569999981829423104 155158014362168328192 310352057521355620352
620776172636749168640 1241696460461574193152 2483681151299300098048 4967938763350903619584
9937030448206414086144 19876366739422041866240 39757345164862511120384 79523913701761877016576
159066274147597463584768 318169441783342346272768 636412670542979530752000
1272972915038548737916928 2546240977982276828659712 5093072251774912362971136
10187325095170542137245696 20377011373582519097098240 40758745113647907839410176
81526934960261554969247744 163072759386454588519350272 326183297704772134200410112
652442153273270182724239360 1305035422273992194095316992 2610373076002888045484310528
5221350614915583405555974144 10443910155650781440286654464 20890238162940792138922721280
41785312029160042794544267264 83580295464877002622486183936 167179933742867839311767666688
334398553111963346757125931008 668874477476382029781433057280 1337903697457674732097228505088
2676116879925170809263181791232 5352852729869984308663813144576
10706943399779253997602525413376 21416362679637078755754849075200
42837677119431299032609294647296 21421314439794220276854445572096
10711895159936395518702121910272 5356566549987840449488510517248
2678592760003741569812980039680 1339451122506781457440852410368 669802932505846064987607400448
334940151879150700627394297856 167489418752689184380492447744 83754380782901509223643873280
41882026094729213128520761344 20943430899003835822609793024 10472924375321532540479602688
5237066650570573584827154432 2618835556740190449707253760 1309568894097547053500465152
654860004912499441073651712 327467781388112677698535424 163752780159987817430122496
81885834812959648005488640 40947639772962693647958016 20476181069722781646585856
10239271126482108234596352 5120225859051412822949888 2560408077430885764300800
1280351612668032558563328 640249593310311117488128 320161690143302977847296
160099291815725198475264 80058869279899454013440 40034046325968154394624
20019329005993290891264 10010817424501252292608 5005985173002929569792 2503280816877616496640
1251784523626884104192 625964319407479980032 313018188500758953984 156527108648888958976
78272561523699220480 39140784361476980736 19572643980552175616 9787447890182930432
4894286895044886528 2447424922499153920 1223853198737932288 611996968113143808
306033668428660736 153034426400374784 76526009293209600 38267402693115904 19135900369813504
9569049696534528 4785074604081152 2392812179947520 1196543528927232 598340483940352
299204601708544 149619480723456 74818330296320 37413460115456 18708877541376 9355512512512
4678293127168 2339414999040 1169841717248 584987967488 292527538176 146280546304 73148661760
36578525184 18291359744 9146728448 4573888512 2287206400 1143734272 571932672 285999104
143015936 71516160 35762176 17883136 8942592 4471808 2236160 1118208 559168 279616 139824 69920
34964 17484 8743 4372 8745 17492 34988 69984 139984 280000 560064 1120256 2240768 4482048 8965120
17932288 35868672 71745536 143507456 287047680 574160896 1148452864 2297167872 4594860032
9190768640 18383634432 36771463168 73551314944 147119407104 294272368640 588611846144
1177357910016 2354984255488 4710505381888 9422084505600 18846316494848 37696927956992
75402445848576 150822071566336 301678502871040 603425725218816 1206988889391104
2414252656689152 4829055069192192 9659209650012160 19320618323279872 38645634693070848
77300065479163904 154617723144372224 309270630660833280 618611630065844224 1237363997620043776
2475009470216798208 4950581890387017728 9902289680680878080 19806831161175441408
39618165921978253312 79245339043211247616 158508692484931977216 317053413766882918400
634178885127803764736 1268501885443683385344 2537292001263518482432 5075160463279340388352
10151473848063287623680 20305253539135788941312 40615118764290005270528 81239460900616865316864
162497368545307440185344 325031630578762299473920 650137048133819437154304

1300421670220228550721536 2601138488345636454268928 5202867272501631614189568
1040691513662398063968256O 20816191456489396101971968 41637105279461661849157632
83283655291889062988742656 166586200049709604558340096 333210179031282166278389760
666495915926290246880198656 1333142947580032322407235584 2666588126614968302108147712
5333780716139743918803648512 10668770358099102466782003200 21339958567837434191913418752
42684752838953326900525662208 85379177084463570834448973824 170777696982040975735693246464
341594079590309619604977090560 683265530433074575477135376384 1366685803371059823488633143296
2733681091751940992045991067648 5467981153523524674229431697408
10937200247086334728733762519040 21876876374251240218017323286528
43758704508659621957134243069952 87527312537633526956467679133696
43768608028973904999333436063744 21886779894565523260216516280320
10944627887322047010383157264384 5472932913680666195329028194304
2736775941850154442733238878208 1368542713429987893900981829632 684348727967449283217672110080
342213049609952309742426652672 171125867618089988938008625152 85572605215601911502401961984
42791138311079414267899805696 21397987007178936392299315200 10700202429409082825324363776
5350705677614348727249534976 2675655070262078020918444032 1337978650858490839106060288
669064883292971333876449280 334570220578348624099934208 167303999755105790630821888
83661444610518634605838336 41835444671742186948132864 20920083519112528296673280
10461222351176981559640064 5231201471398849485471744 2615895883604604095561728
1308095515754891724193792 654121544853740700303360 327097665915017769254912
163567279701582594179072 81792863222828151865344 40901043297432503320576
20452827491725465354240 10227566667367339524096 5114359794435973185536 2557468127594138304512
1278878178985145008128 639511147086610432000 319791602340324179968 159913815568671571968
79965914983590526976 39987461091422633984 19995982345525002240 9999117072669343744
5000121486288093184 2500342218120757248 1250311846548733952 625226292018544640
312648330381361152 156341757376724992 78179674781384704 39094235437203456 19549316741857280
9775757882556416 4888428697092096 2444489226452992 1222382052179968 611259745566720
305664232521728 152849296130048 76433237999616 38220913967104 19112604467200 9557375975424
4779224858624 2389880864768 1195074650112 597604433920 298835771392 149434662912 74725720064
37367054336 18685624320 9343860736 4672454656 2336489472 1168375808 584253440 292159488
146096128 73056256 36532224 18268160 9135104 4568064 2284288 1142272 571200 285632 142832 71424
35716 17860 8931 4466 8933 17868 35740 71488 142992 286016 572096 1144320 2288896 4578304 9157632
18317312 36638720 73285632 146587648 293208064 586481664 1173094400 2346450944 4693426176
9387900928 18777899008 37559992320 75128373248 150273523712 300580601856 601228312576
1202590842880 2405450121216 4811437113344 9623947968512 19250043420672 38504381808640
77017353551872 154051886972928 308138133684224 616344986845184 1232827412643840
2465929703194624 4932409162203136 9865917836034048 19734034695323648 39472467437158400
78953730967339008 157925054120722432 315885292613533696 631840953971245056 1263822645430845440
2527926765838401536 5056416481630224384 10113958863167291392 20230169526148268032
40464842651923906560 80938692503102554112 161895399404714590208 323826827606448144384
647725712806934216704 1295595540801944289280 2591479311980040290304 5183535084712384004096
10368223090929374855168 20738752024867963404288 41482115735754354196480 82973454843545563168768
165965356431164835889152 331967606350477090881536 664008999677249019969536
1328165573307087716352000 2656626294519354785529856 5313842884849068276711424
10628866361318853964726272 21260093905879142752059392 42524910178241155149332480
85059265089448049589092352 170137419644827577759039488 340312618221518112679788544
680700794306762139682996224 1361552704340976108012830720 2723407640136855873319337984
5447419743183519061226029056 10896048412186652751626764288 21794514676012534761602940928
43593865055303528039904706560 87197401517163973113207062528 174414145847441780293209423872
348866977321111228720009445376 697811325894677793707200086016 1395777394294266259948762562560
2791864273598353864966249906176 5584347517216350420069949374464
11169932974471986220414797873152 22342341829022543201379393994752

44689635418202227923858384486400 89389174356718738889915961966592
178798155754066043864230309920768 89408981397347304974314347954176
44709442458830794008256770473984 22357197109493967764678183485440
11179836494786269262613990866944 5590537217412777321444444995584
2795578093716210005790947278848 1397943789363015675429836029952 699049265933963173982099210240
349563318593209255124640202752 174801002109718461629115400192 87410172461416147847955349504
43709921933986532440676499456 21857378818632495478687662080 10929898335135862368518537216
5465553630477738498846621696 2733079046693772906716987392 1366690639074338282005331968
683420877400895055326085120 341748217632310484824752128 170892998282086720993230848
85455943874009099787042816 42732694303487419538735104 21368708334985144591974400
10685534759113289707290624 5343357675367003559297024 2671973985588681132474368
1336134566746930242650112 668141070349759959531520 334107428663027398868992
167072161075587408986112 83545303909830559268864 41777263640933707022336
20890937663476067205120 10446621753242640449536 5223887337373623648256 2612231899062963535872
1306260064719557623808 653202089953816739840 326637073773927333888 163336551285473148928
81677282841991315456 40843145020623028224 20423824310125199360 10213038054969442304
5107081977438142464 2553822463695781888 1277051969336246272 638596353412300800
319333361078239232 159684272725164032 79850932455604224 39929864274313216 19967131160412160
9984665091833856 4992882301730816 2496716028772352 1248495453339648 624316446146560
312192582811648 156113471275008 78065325572096 39036957753344 19520626360320 9761386921984
4881230331904 2440883601408 1220576018432 610355118080 305211113472 152622333952 76319555584
38163972096 19084083200 9543090176 4772069376 2386296832 1193279488 596705280 298385408
149209088 74612736 37310464 18657280 9329664 4665344 2332928 1166592 583360 291712 145872 72944
36476 18240 9121 4561 9123 18248 36500 73008 146032 292096 584256 1168640 2337536 4675584 9352192
18706432 37416960 74842112 149700608 299433984 598933504 1197998080 2396258304 4793040896
9587130368 19176357888 38356910080 76722208768 153461194752 306955943936 613978996736
1228092211200 2456452857856 4913442586624 9827958915072 19658065313792 39320425594880
78649441124352 157316062117888 314666483974144 629401687425024 1258940813803520
2518156505513984 5036862766841856 10074825045311488 20151849113878528 40308096274268160
80624988641558528 161267569469161472 322570323310411776 645211015365001216 1290562768218357760
2581407011413426176 5163376972780273664 10327879845467389952 20658011490748465152
41320526581124300800 82650060361503342592 165318135121516167168 330672299040051298304
661416655674140524544 1322977426536356904960 2646243083448865521664 5293062627650034466816
10587278176804675780608 21176862196618565255168 42358336079255557898240 84725895530547970572288
169470237805169650696192 338977369098486720495616 678028525173268279197696
1356204624299126234808320 2712704396503431822442496 5425999088817222350536704
10853178769255162112376832 21708718721751759047360512 43422159809986387739934720
86853764352938514770296832 173726418171808508121448448 347490615275479973404606464
695056788414685861132632064 1390264692556823550912102400 2780831616568550759117881344
5562267696046908832823115776 11125744317913432294820937728 22253906487466093847991287808
44512648678210646212681400320 89034968762978209458760450048 178089280339070252984316198912
356217246304368174102222995456 712511863861191684471627186176 1425178470227294041477616762880
2850666425464409428023958306816 5701951820948461546185366175744
11405141581936208472645631475712 22812759043950987705841061199872
45630469848059116932781718896640 91270843216432516907762630787072
182561493473493599899923647561728 365162601028244331968644067098624
182601107554750732068720419536896 91310457297689649076559402762240
45660180409001966059379297878016 22832566084579553790239447187456
11417520982329062275394622717952 5709379461184173827834760921088
2854999215601908258986105241600 1427654350305864802027415011328 713904546405387737280888700928
356990958828921536774034948096 178514822227574602453812772864 89267082520344218260304035840
44638376963450567646850842624 22321606333364513081774833664 11162012092501871170062123008

5581610509160742899618414592 2791107486035275107102883840 1395704858745089382198280192
6979279872362706054225559232 3490017725499982598729891 84 1745197757409306085173493 76
8726933260343104354910208 0 4363938866819839141976473 6 2182205551734063053248921 6
1091220835029103267754803 2 545669447095587504442572 8 272864238338311687503872 0
136446876564414811393228 8 682308169798368895172608 341190978387331866689536
170613935937739642896384 853161913409066762240 00 426627073564717654999 04
213336595212450964439 04 106679826821271550689 28 533456780181588095795 2 26675721312840921907 20
133393018083012195123 2 667037148000909890355 2 333554602801568415744 166795315799293689856
834066650989015859 20 417078361490781634 56 208561698743527669 76 104292108370832261 12
521516836849503436 8 260786565922422784 0 130407356710046924 8 652107152294412288
326088760519294976 163061972445691904 81539782315868160 40774289204445184 20389343625478144
10195771324366848 5098435417997312 2549492586905600 1274883732406272 637510585679872
318789652578304 159412006158336 79714593013760 39861591474176 19932943220736 9967545352192
4984309547008 2492423208960 1246345822208 623240019968 311653564416 155843559424 77930168320
38969278464 19486736384 9744416768 4872732672 2436628480 1218445312 609288192 304676864
152354816 76185600 38096896 19050496 9526272 4763648 2382080 1191168 595648 297856 148944 74480
37244 18624 9313 4657 9315 18632 37268 74544 149104 298240 596544 1193216 2386688 4773888 9548800
19099648 38203392 76414976 152846336 305725440 611516416 1223163904 2446589952 4893704192
9788456960 19579011072 39162216448 78332821504 156682420224 313398394880 626863898624
1253862014976 2507992465408 5016521801728 10034117345280 20070382174208 40145059315712
80298708566016 160614597001216 321263553740800 642595826958336 1285329092870144
2570933063647232 5142415883108352 10285931277844480 20574061578944512 41152521204400128
82313838501822464 164645269189689344 329325722751467520 658721814247112704 1317584365982580736
2635450206941872128 5271463363837165568 10544052627581173760 21090357054976032768
42185217709579436032 84379442618413613056 168776899635336708096 337589828067692380160
675251713729422688256 1350647542646921232384 2701583315669994176512 5403743092092291776512
10808639105689190400000 21619584054387594493952 43243779794793616375808 86496782961624087527424
173012012667321884606464 346060918822791188316160 692195624621877214838784
1384538823196344106090496 2769372794297867565006848 5539335884406093835665408
11079852360432905082634240 22162065904107244987875328 44328854174697359620964352
88667153082360458532356096 177353195630652395645566976 354744170193167748452843520
709563898250061411229106176 1419278912227574651105050624 2838860055910052959503777792
5678324574729913233594908672 11357858075279441096364523520 22718134002198111451078459392
45441103707674681418855743488 90891878821906279871109136384 181803100456926393809013571584
363644886540080455751617740800 727367144332616247770416676864 1454889031170143168075195744256
2910087547350107681219116269568 5820794064719858052575682101248
11642826069479001485426263326720 23288128019036573731402324901888
46581207798230288983904246300672 93172319116774861010007685595136
186364445274178288104413757177856 372768504629613708377624286330880
745616237421741681092842116612096 372847732792127972715217830281216
186443673436692552442007301128192 932317402386605592632028435578 88
466208218794874211527010182758 40 233128868198222813369003073863 68
1165768134995042604872505281740 8 582945964499485571449997597081 6
291503930750724920231871276646 4 145767439625853527369371877376 0 728914569381722973114040582144
364495970317089154690610888704 182267327971658411412100743168 911433353923861227394480209 92
455765033994715198864228352 00 227906695513749892015608299 52 113965437015071092299551211 52
569887631366336192956491366 4 284974038828658462207613337 6 142502130987074413968490496 0
712586212799097984165871616 356330885331411949244645376 178184332131637453203177472
89101610798784465892016128 44555527765875102591221760 22280125066178986118217728
11141243124710210470412288 5571211858165463940857856 2785901076987911323254784
1393098112446545338040320 696622843199567507226624 348348315087931172716544
174192604288039295909888 87105525516056502730752 43557374444046678753280

21780993065032553070592 10891649454020883382272 5446401187762745114624 2723488824257524269056
1361888527316837990400 681016321252456923136 340544189423247425536 170290109110133194752
85154061754321338368 42581534476788039680 21293019038207705088 10647635419010695168
5324380659458768896 2662471804706095104 1331376639841402880 665758688664879104
332914528704528384 166474856538308608 83246224362176512 41627510227599360 20815954137055232
10409076580155392 5205088045891584 2602818900852736 1301546889379840 650842164166656
325455441821696 162744900780032 81381040324608 40694815129600 20349555048448 10175851266048
5088462503936 2544499687424 1272384061440 636259139584 318163124224 159098339328 79557558272
39782973440 19893583872 9947840512 4974444544 2487484416 1243873280 622002176 311033856
155533312 77774848 38891520 19447808 9724928 4862976 2431744 1216000 608064 304064 152048 76032
38020 19012 9507 4754 9509 19020 38044 76096 152208 304448 608960 1218048 2436352 4873216 9747456
19496960 38998016 78004224 156024832 312082432 624230400 1248591872 2497445888 4995416064
9991880704 19985858560 39975911424 79960211456 159937200128 319907954688 639883018240
1279900254208 2560068943872 5120674758656 10242423259136 20486994001920 40978282971136
81965155876864 163947491622912 327929342984192 655927405445120 1311992249843712
2624259377594368 5249068511002624 10499236533633024 21000672090521600 42005742227554304
84020280548130816 168058153282306048 336151490936700928 672373350617579520 1344887438723514368
2690056352423739392 5380675654800900096 10762477209508642816 21527206218830970880
43058916037289312256 86126839273833365504 172271692946176212992 344579414689371389952
689230886972780707840 1378605889133637271552 2757500008643426254848 5515576478039155933184
11032305877582918713344 22066917598175051120640 44138446882368529629184 88286117136773914034176
176590681017621537619968 353218255523390494343168 706510298023075826892800
1413168169998741330198528 2826631487902662013222912 5653853271615682732097536
11308887134852082875498496 22620135452945600573603840 45244993272374070792421376
90499431277713880875270144 181017752021359240331395072 362073282974581437824499712
724222123812888789972418560 1448595363353229408591675392 2897492958161362474477027328
5795590379232532263541407744 11592389684284679156257521664 23187197220208587570864455680
46379230143695633658427736064 92768131693948184350253121536 185555606201010202767301541888
371149898028248073668193681408 742377167308951483603568558080 1484909077122813639741499506688
2970127639255448624551723794432 5940874248530539939240897150976
11882986437100365258756693426176 23768448754279301278063185100800
47541849268715744077225966698496 95093602057745771196651126390784
190207011156120108477700638769152 380453636393497349124198049513472
760986500949508962585989642977280 1522131458224046453847166373855232
761144957274537491261176730877952 380612092718525877799385137414144
190325853399891504984090954694656 95172830220260035534244670341120
47591366870287159288221931667456 23798159315222150404660764082176
11900317597650360582605281165312 5950777768844822981440090144768
2975698369432232835788769853440 1488003927221027090428747317248 744079334862968881481554853888
372078353057712108874368024576 186058519341969888503979311104 93038931077541861285387304960
46524301242049389159392477184 23264568472663923838045650944 11633493162151576548197531648
5817351043985595588686118912 2908977753447701451636736000 1454639992451302554465206272
727395554089377191556022272 363735555976551552939720704 181886667454207255050715136
90952778460069366815784960 45481111596517553053106176 22742916981500211349159936
11372639082370823085883392 5686909836995770248593408 2843750066403064477122560
1422022607154121914974208 711085090553355795693568 355579438764825316950016
177808166126486368026624 88913306435280038789120 44461264903658446782464
22232938294838437085184 11117622068923825389568 5559387495214216118272 2779981977983259770880
1390135104179705741312 695139609683890798592 347605833638964363264 173820931217991663616
86919472808250572800 43464240003752656896 21734371801690013696 10868311800751849472
5434718850329346048 2717640900141383680 1358961187559047168 679550962523701248
339810665633939456 169922925003014144 84970258594529280 42489527343775744 21246962695143424

10624580859199488 5312840185413632 2656694970613760 1328484924260352 664311181606912
332189950541824 166112155140096 83064667504640 41536628719616 20770461843456 10386304663552
5193689202688 2597113036800 1298690736128 649412476928 324739792896 162386673664 81201725440
40605057024 20304625664 10153361408 5077204992 2538864640 1269563392 634847232 317456384
158744576 79380480 39694336 19849216 9925632 4963328 2481920 1241088 620608 310336 155184 77600
38804 19404 9703 4852 9705 19412 38828 77664 155344 310720 621504 1243136 2486528 4973568 9948160
19898368 39800832 79609856 159236096 318504960 637075456 1274281984 2548826112 5098176512
10197401600 20396900352 40797995008 81604378624 163225534464 326484623360 653036355584
1306206928896 2612682293248 5225901457408 10452876656640 20907900796928 41820096561152
83648783056896 167314745982976 334663851704320 669396422885376 1338930284724224
2678135447355392 5356820650524672 10714740812677120 21431680648609792 42867759343730688
85744314780483584 171506221747011584 343047627866112000 686165624476401664 1372471986441158656
2745225447859027968 5491013845671477248 10983153591249797120 21968558982313279488
43941621564253929472 87892250327762599936 175802515054034681856 351641058905088327680
703354175404214583296 1406852465996505022464 2813993162369161756672 5628562785490626936832
11258278492485860720640 22518862827980935135232 45042337341980297658368 90093898055997450092544
180206242856068609736704 360449379200284638576640 720972545376864115359744
1442092664706317907132416 2884480477317815167090688 5769551250445989039833088
11540283092512695490969600 23082927368266825804546048 46170577103016521254305792
92350598938998781799038976 184720087343929042178932736 369477953619721041519575040
739031465103167997362569216 1478214045933787823371976704 2956730323322479304037629952
5914065109554765922662612992 11829339144929146474499932160 23661096141497522207349276672
47327027986273502931397378048 94663727379103922896192405504 189346797571321679859180109824
378732280768871027851950817280 757541932790197391971082829824 1515238608085305456476528050176
3030786701180432258021780881408 6062192372380507206181011324928
12125622684800299792636921774080 24253721249679170345823641796608
48512394259515482212746880090112 97034692039345247467692953174016
194089191119319061019784292335616 388217996319895254208365356646400
776515220802304772754324257243136 1553188897929638074183835602386944
3106694708509333205718045380575232 1553505810579695131534209778188288
776832133452361830104698433044480 388455680807438047221145988497408
194247647444347589694971380236288 97133727242488077889684883111936
48571815381401180465942038052864 24288383570779160993520817274880
12145429725428865877035307761664 6073333832734075628655103442944
3036976401376859159396276502528 1518642943193340252232500641792 759398842849125462383431516160
379738107050790399325306355712 189888396338509033729448476672 94953869575811433898121887744
47481770491184175465759768576 23743303097231316991229296640 11872860474435273124789354496
5937034700127443876982030336 2968819581518625595784691712 1484560906486764626539184128
742356011107108227593011200 371215784485417070958215168 185626781708640014059962368
92822835587285746320408576 46416140160125742805417984 23210431263304306225315840
11606396223272870523961344 5803788407446793967632384 2902189351628576336642048
1451242249766877844733952 725694911859733760573440 362884349418014299389952
181460621453080859246592 90739534098577284399104 45374378735307069587456
22689495210662748487680 11345900526835981090816 5673526724170293968896 2837051592461298696192
1418669911418725203968 709407013303400529920 354739535448719228928 177387782122869096448
88702898260689289216 44355952729972015104 22180228164799692800 11091239982306689024
5546182941106765824 2773372945530093568 1386827210253402112 693483973870878720
346777171307528192 173406177839808512 86711885012926464 43360340552974336 21682369299742720
10842284161499136 5421691836563456 2711120796188672 1355697837047808 677917638000640
338993178738688 169513769238528 84765474553856 42387032244224 21195663605760 10598905544704
5299989643264 2650263257088 1325265846272 662700032000 331383570432 165708562432 82862669824
41435529216 20719861760 10360979456 5181014016 2590769152 1295515648 647823360 323944448

161988608 81002496 40505344 20254720 10128384 5064704 2532608 1266432 633280 316672 158352 79184
39596 19800 9901 4951 9903 19808 39620 79248 158512 317056 634176 1268480 2537216 5074944 10150912
20303872 40611840 81231872 162480128 324993024 650051584 1300234240 2600730624 5201985536
10405019648 20812136448 41628467200 83265323008 166547423232 333128400896 666323910656
1332782039040 2665832513536 5332201897984 10665477537792 21333102559232 42670500085760
85349590106112 170716360081408 341467079901184 683002879279104 1366143197511680
2732561272930304 5465672301674496 10932444114976768 21867087253209088 43738572552929280
87485941198880768 174989474583805952 350014133539700736 700098635823579136 1400338009135513600
2800957493247737856 5602477936448897024 11206081772804636672 22414415345422958592
44833334290473287680 89675675780201316352 179369365958912114688 358774760714843193344
717621579023724314624 1435387273235524485120 2871062776847200681984 5742702014446704787456
11486556950398016421888 22975419743805246537728 45955451173628920463360 91920125719294695702528
183858698182663100956672 367754289853473621016576 735582366683242080239616
1471312307319073836892160 2942919762543327026610176 5886429820897012758872064
11774040233414742929047552 23550441650070920680701952 47105605666624711006617600
94220656066215161303662592 188460201598361801188179968 376958182128586559538069504
753991922120899033399558144 1508134959969249895445954560 3016572151393403448185585664
6033748765696614210958524416 12068706457212843051091755008 24139830766064915360532922368
48284497235408289237764669440 96578665877373495508926988288 193176674567860825084649275392
386392034761949318302889148416 772861440776353972872959492096 1545877624057618618280281374720
3092064733125058581629287530496 6184748436269759853396024623104
12370734812578805087066948370432 24743945505236180934683694989312
49492842770629503390466986475520 98995589061573289823133165944832
198010985163775145730664717877248 396061584408807423630126207729664
792202396980129111597845959409664 1584563250285286751870879006720000
3169443413220630561092132189241344 6339520651741375236885012730085376
3170077238520744675792880540844032 1585197075585400866571627358322688
792677765955214697623407223111680 396378497058864480980500383531008
198209055570060806574648577753088 99114431305344686329523481870336
49562167412829484685861337432064 24783559586493313103480466964480
12393017733285941932015132606464 6197127836662613656145015865344
3098873403341128173141232713728 1549591444175474759104978747392 774873093340192715819670568960
387475232296324026043425882112 193756958961275847088508239872 96888150887194840577651769344
48448911146875878805524709376 24226873425077168661111767040 12114645638358198959730589696
6057927282088906794452647936 3029265872499357054520000512 1514784051977130355906838528
757467583852291092276838400 378771570858008503300128768 189404674894935730230919168
94711782180433604405886976 47360613456699671848157184 23682667911591270746685440
11842514547416352784646144 5921847569518535097974784 2961218932664446901813248
1480757040284813127319552 740452307118701401866240 370263047047498120036352
185149970267822769569792 92584208505948239560704 46296715938992547168256
23150663812505487278080 11576484827757350486016 5788818874630978666496 2894697667691641044992
1447492949033896378368 723818532110986117120 361945294852512022528 180990661824765493248
90504338111637487616 45256672655446114304 22630588127536742400 11316419963675213824
5658772931791028224 2829667940872224768 1414974707924467712 707557722706411520
353814045725294592 176924615048691712 88471103617368064 44239949855195136 22122173950853120
11062186487054336 5531642999341056 2766096377577472 1383185627742208 691661533347840
345865126412288 172949743075328 86483461472256 43246025703424 21625160335360 10813653909504
5407363825664 2703950348288 1352109391872 676121804800 338094456832 169064005632 84540391424
42274390016 21139292160 10570694656 5285871616 2643197952 1321730048 660930560 330498048
165265408 82640896 41324544 20664320 10333184 5167104 2583808 1292032 646080 323072 161552 80784
40396 20200 10101 5051 10103 20208 40420 80848 161712 323456 646976 1294080 2588416 5177344
10355712 20713472 41431040 82870272 165756928 331546624 663158784 1326448640 2653159424

5306843136 10614734848 21231566848 42467328000 84943044608 169902866432 339839287296
679745683456 1359625584640 2719519604736 5439576080384 10880225902592 21762599288832
43529493544960 87067577024512 174152333918208 348339027574784 696746774626304 1393630988206080
2787536854319104 5575623464452096 11152346440531968 22306891904319488 44618181855150080
89245159803322368 178507911792689152 357051007957467136 714172384659111936 1428485506806579200
2857252488589869056 5715067927133159424 11431261754173161472 22864775308160008192
45734054215947386880 91477115631149514752 182972245660808511488 365980520118635986944
732033097831309901824 1464210310850695659520 2928708852077543030784 5857994164907389485056
11717141251319385817088 23436588345647985328128 46877788377314398044160 93764800126665650864128
187548046997405011279872 375132987482957441662976 750339761942209721532416
1500827097837009119477760 3001949343579197591781376 6004488982968753889214464
12010158557558225189732352 24022678298357885202071552 48050078963198640049356800
96109602659363019389140992 192238094784657517359136768 384513968501177991879983104
769103494866081898083385344 1538358105459615624813608960 3077018442374134906920894464
6154641347658077128429142016 12310491621135768886032990208 24623401093910767030415392768
49251637891099992577529610240 98512947188756902188456869888 197045237190627638443709038592
394129160007482945021008674816 788335691267421226309198544896 1576826125039753125152759480320
3153961735089327595374243741696 6308542440198297880885937045504
12618322820435881142046773215232 25239121520950333044643344678912
50483194802057807610386285854720 100976293124429898262971764703232
201972393289488362610341915394048 403984400660233857389480602763264
808048029482981979116554749476864 1616254515290992486908296586854400
3232825943232042031166967349510144 6466285711764198177034683050622976
12933839074128624583470862804451328 6467553362364426406436179753828352
3234093593832270260568464052715520 1617205253241163658959419114258432
808681854783096093817303101079552 404380541472805179077448322514944
20221007777031155623122547245056 101114942408829860853760466616320
50562422964572071947979829805056 25283687362364606734539713150976
12643081621221588747544755699712 6322159780630437063909827411968
3161389375325039877023638487040 1580849430167430611046181634048 790502086336170641790272012288
395289728794312989028726603776 197664207210270328581158600704 98841775011692081323976949760
49425723209124499178687299584 24715279456201478847693062144 12358848653920354053021237248
6180028789869984341097971712 3090316626389895827842662400 1545309428922399742568169472
772730272324925785607503872 386402915094325849965461504 193220347013094403563585536
96619618239512941072220160 48314531486239340181323776 24159626926361104913268736
12080994054801269867937792 6041087323210993639620608 3020838809510676172636160
1510566978707927762731008 755357276330258719571968 377715531653276778889216
188876212570712098996224 94447329657392904273920 47228276514714879524864
23616444100366653456384 11809374971687933575168 5905263946596270211072 2952920203674286817280
1476604217025219264512 738374166106647560192 369223111850342744064 184629570323680854016
92323792361095168000 46166399780174954496 23085451689901162496 11543851744857423872
5772488822382133248 2886525886167777280 1443403680572243968 721772209030299648
360921288887238656 180478236629663744 90247914407854080 45128355250438144 22566376648474624
11284287835865088 5642693673746432 2821621714780160 1410948296343552 705542867648512
352805793562624 176420076650496 88218628259840 44113609097216 22058952032256 11030549757952
5515811749888 2758174310400 1379221372928 689677795328 344872452096 172453003264 86234890240
43121639424 21562916864 10782507008 5391777792 2696151040 1348206592 674168832 337117184
168574976 84295680 42151936 21078016 10540032 5270528 2635520 1317888 659008 329536 164784 82400
41204 20604 10303 5152 10305 20612 41228 82464 164944 329920 659904 1319936 2640128 5280768
10562560 21127168 42258432 84525056 169066496 338165760 676397056 1352925184 2706112512
5412749312 10826547200 21655191552 43314577408 86637543424 173291864064 346617282560
693301673984 1386737565696 2773743566848 5548024004608 11097121751040 22196390985728

44397076938752 88802743812096 177622667493376 355279694725120 710628108926976 1421393656807424
2843062191521792 5686674138857472 11374447789342720 22751094601940992 45506587250393088
91021970593808384 182061533373661184 364158251119411200 728386870983000064 1456914479454355456
2914110433885421568 5828783817724264448 11658693535355371520 23319638870524428288
46643781340676227072 93296569880607195136 186611154159723872256 373258337116466708480
746588731826971344896 1493321578842018545664 2986931388060188803072 5974439236872681029632
11950031395249968906240 23902368633509151506432 47809348953036730400768 95627921278110315577344
191274289300294340706304 382585472088736100515840 765244731153767039238144
1530637036260123754889216 3061569220425426862604288 6123728736661212430860288
12248638064943142273024000 24499637313127719368654848 49003996992738308382523392
98017438718442356055474176 196053766902816190691803136 392145312737495338545315840
784366183338716591414050816 1568883482404885011474939904 3138069196264673680243556352
6276742855439154675074465792 12554694636697923979323637760 25111807125035077216996687872
50228449953348612950692200448 100466571313254142934782050304 200952485439622119936359399424
401943656505471908006309396480 803964684263399152279799988224 1608084111031708977093962366976
3216477707073239299256649515008 6433574384166121288650748592128
12868386708371527957576396308480 25739249296821626675702590865408
51483450353800394872504778227712 102976804227915072787208749449216
205973415496458711658815884886016 411986445074174555486428541747200
824052118310863375310450627444736 1648262692946755279296088342790144
3296842298543567615942550861381632 6594318422387249346585850074365952
13189904495374726922573196851937280 26382344291949910303949387110285312
131924397965751833813761902583480326596853723587705805388843480776704
3298743774443909960044795916189696 1649530343546983508697585045995520
824844399936006018686386066948096412461814049260141511989805449216
206250714065258636840393288712192103135260552943601462395837349888
5157258203662894225229751517184025788766898393041886698555834368
12895621389235806323624177041408 6448429664637545851949538082816
32245243173285942710434938224641612416901169207808056109301760806285821837059240295235846144
403181596544757288281208520704201610141085492478207399559168100814741949303156137097428992
504122066779300365852475392002520852119060424755097318195212605469521121738404661297152
630333922347067651691800166431519718431902419157526773761576137037322572786523176960
788144076525012307585007616394109817194369110954213376197073798063116034057961472
985463437645237563194081284927789424874474780491776024641308307613808725065728
1232183474542762177383628861615076685241695925698563081048982167264149110784
1540672065036221750968320770409819494405713690624385241803235350275948544
1926393483617488475258889632889755291127853875248169060462474066657280
240868360742462470225921204457095862773035827260228619400661686026243011719200409236013056
1506003715392693862400753073915290384859136376572986442211393536188304507619615178752
94161261009062330368470851341041585356802354481885189295308811773535325853319168
588733061288008089629439467814167511041472114128196730880736127432842543104
368098900793360384184067042582724608920423173843845124602555673870336023014977392607232
11508588207931392575484385977958428776968077967361438985842851840719561640902656
35981518018969617992476996403289970974916608449897824256002249703869644811249593090048
5625333415936 2812935143424 1406601789440 703368003584 351717556224 175875555328 87946166272
43977277440 21990735872 10996416512 5498732544 2749628416 1374945280 687538176 343801856
171917312 85966848 42987520 21495808 10748928 5374976 2687744 1344000 672064 336064 168048 84032
42020 21012 10507 5254 10509 21020 42044 84096 168208 336448 672960 1346048 2692352 5385216
10771456 21544960 43094016 86196224 172408832 344850432 689766400 1379663872 2759589888
5519704064 11040456704 22083010560 44170215424 88348819456 176714416128 353462386688
706991882240 1414117982208 2828504399872 5657545670656 11316165083136 22634477649920
45273250267136 90555090468864 181127360806912 362289081352192 724646882181120 1449431203315712

2899137284538368 5798824324890624 11598748161409024 23199695346073600 46403788738658304
92816373570338816 185650339326722048 371335863025532928 742742094795243520 1485624927078842368
2971531329134395392 5943625608222212096 11888377116351266816 23779006032516218880
47562515664659808256 95134038528574357504 190286091455658196992 380608211708335357952
761288481010708643840 1522721077209493143552 3045730384795137998848 6092037230342579421184
12185227382189765689344 24372760607388745072640 48750132900795917533184 97509489173628689842176
195037425091331089235968 390111743670809597575168 780297274317914033356800
1560742122588417743126528 3121779393082014839078912 6244149081974388383809536
12489478755569494178922496 24981318694380423180451840 49967359755243716006117376
99944164243453171302662144 199907217952837821186179072 399852214837538599534067712
799779987538803113391554560 1599711090805058055429947392 3199724413065019768153571328
6400053289039846850894495744 12801315503899308330963697664 25605048859437845920276807680
51214933422154150357252440064 102439538250865217747902529536 204898419314844269562600357888
409835524255916207258791313408 819748419764287750784763822080 1639651582033486174103890034688
3279612649076793693276504850432 6559844268173230076690459262976
13120926476385745533655817650176 26244328832850061827861433548800
52493609425857265176822463594496 104997122372028813395844120182784
210014051784686192876086626353152 420067717650629517920970024681472
840214663463773300179533593313280 1680587783252575129034254274527232
3361492479155207315418882724855808 6723618783610528745538513801314304
13448505217821285720478524305833984 26899545736843027899760042018078720
53804162076086968717126070848978944 26904616339243940817366028830900224
13453575820222198638084511118655488 6727421735411213433743003910930432
3364027780355663774221876131266560 1682172346502860415786125153533952
841165401413944472230656120717312 420622314788229368284124832333824
210330964434743250226460802154496 105175385737685908155429594071040
52592644629000095598814393532416 26298798194578618559956995014656
13150637037328594660253396631552 6575937488683940020264147877888
3288278229351791355200798720000 1644293857180806350134761750528 822224299842858511334562070528
411150835547656923800871632896 205594760586942295967231115264 102807051700028065017013207040
51408361553292491025205428224 25706598628285474770952126464 12854508239962352014650769408
6427858582890983321912737792 3214231522900395318250045440 1607266877177649487771860992
803708996452550658209349632 401892277158138286266384384 200965028045000621714046976
100491958755466050147450880 50250701744215894718939136 25127712055349382182076416
12565036619295408502341632 6283108605458062956822528 3141849450634210831237120
1571072299269695092031488 785609936611142384222208 392841861793718611214336
196439377640933015158784 98228912192503362355200 49119067782270108565504
24561839734144267976704 12282072788576740835328 6141612855040673841152 3071094657896488632320
1535691444136320172032 767917779662198013952 383994918628117970944 192015473712568467456
96016744055538974720 48012875627396857856 24008689613512114176 12005470706662899712
6003298303284871168 3001930626619146240 1501106050797928448 750623394143141888
375346881443659776 187691032907874304 93854312546959360 46931554319990784 23467976183250944
11735087603253248 5868093557440512 2934321656627200 1467298267267072 733717853110272
366893286293504 183463823015936 91740501442560 45874545688576 22939420327936 11470783905792
5735928823808 2868232847360 1434250641408 717192429568 358629769216 179331661824 89674219520
44841304064 22422749184 11212423168 5606735872 2803630080 1401946112 701038592 350552064
175292416 87654400 43831296 21917696 10959872 5480448 2740480 1370368 685248 342656 171344 85680
42844 21424 10713 5357 10715 21432 42868 85744 171504 343040 686144 1372416 2745088 5490688
10982400 21966848 43937792 87883776 175783936 351600640 703266816 1406664704 2813591552
5627707392 11256463360 22515023872 45034242048 90076872704 180170522624 360374599680
720816308224 1441766834176 2883802103808 5768141078528 11537355898880 23076859281408
46158013530112 92324616994816 184666413858816 369367187456000 738803094388736 1477743627730944

2955762133368832 5912074022551552 11825247556730880 23652694136717312 47309786319945728
94628368732913664 189274329651871744 378583843675832320 757238056095842304 1514616849680039936
3029515174336790528 6059593298627002368 12120312497160847360 24242876794135379968
48490257187898130432 96989521575051001856 193997057548611485696 388030143894241935360
776132345382521798656 1552408805953119453184 3105105842282390618112 6210788145317084659712
12422729212138776166400 24847764267286766026752 49700140220591959441408 99409503813220773658624
198837454370515256868864 397711802229177932840960 795497391434650703888384
1591142356821891084189696 3182579861548961521205248 6365750018908281748062208
12732680629437280907427840 25467722442115996637462528 50940167250714862920138752
101889779234395465130704896 203798447934722408842264576 407634674801307774846238720
815344907466341464015896576 1630840930660134756678631424 3261984092775173170650939392
6524572648460153655889231872 13050354222739921940953169920 26103126297119073140255752192
52211088297516604797210329088 104431848001590126627818307584 208883038816294087322431913984
417804763258815842778454425600 835686897770087021824090046464 1671528538045084716182542483456
3343366561099990777433809747968 6687352092219624245005069058048
13375942124478533870285037240320 26754360129035638501119872729088
53513672018228418523339341955072 107037247556771120088877876903936
214094302154170806262154139795456 428228218389598744693105051566080
856535664941711753723803647082496 1713229786208452036122794382065664
3426776485066961129595962939932672 6854186795434036373892674231468032
13709641241468300977186845166141440 27421817784137058413176683738693632
54848706170675029743959354290208768 109707553546151885323130682206060544
54858847375476855579171327915851776 27431958988938884248388657364336640
13717247145069670353595825385373696 6859257397834949291498661044289536
3429945611567531703099704697946112 1715131262108794380225039436873728
857644859216911454450113262387200 428862043689712859393853403168768
214450828885484995781325087571968 107235317963056780932861736779776
53622610741685531987530464886784 26813781250921336754315030691840
13408128565499953757432414470144 6704683252769619568853656797184
3352651111394631129495553179648 1676480298202226237282138980352 838317520353568454908250685440
419197445803011895587715940352 209618065714619781860653268992 104818704263866807963724283904
52414187835211862498560966656 26209511769245160507629895680 13105964810442194882989654016
6553586868130904756082180096 3277095665520356035334766592 1638698948487629846314221568
819425032107540837480529920 409750294985633375901974528 204894036958748166531842048
102456463212339822556348416 51232953972652780923387904 25618838169567825284300800
12810599676404630053453824 6405890134012673732378624 3203240214911516219015168
1601767681408347785920512 800957627680468731166720 400515707328381784686592
200276300408264601894912 100147373576169155723264 50078298474103005249536
25041455080060716318720 12521880461534965006336 6261516691519785926656 3131046576136044675072
1565667403256098193408 782905759222087024640 391488908408062476288 195762468602540720128
97890241500525101056 48949624349889921024 24477063974758645760 12239657887286165504
6120391893596504064 3060477421774962688 1530379448375836672 765260092932096000
382665230838136832 191350207605112832 95683899895578624 47846347994300416 23925373020405760
11963786021830656 5982442766729216 2991496261271552 1495885569589248 748011504271360
374040111874048 187037235806208 93527207837696 46767898886144 23386096926720 11694122205184
5847597973504 2924067422208 1462167928832 731151073280 365609091072 182821322752 91419049984
45713719296 22858956800 11430526976 5715787776 2858156032 1429209088 714670080 357367808
178700288 89358336 44683264 22343680 11172864 5586944 2793728 1396992 698560 349312 174672 87344
43676 21840 10921 5461 10923 21848 43700 87408 174832 349696 699456 1399040 2798336 5597184
11195392 22392832 44789760 89587712 179191808 358416384 716898304 1433927680 2868117504
5736759296 11474567168 22951231488 45906657280 91821703168 183660183552 367353921536
734774951936 1469684121600 2939636678656 5879810228224 11760694198272 23523535880192

47051366727680 94111323389952 188239826649088 376514013036544 753096745549824 1506330930053120
3012936738013184 6026423231840256 12053945975308288 24110090973872128 48224579994255360
96457956081532928 192933504349110272 385902193070309376 771874754884796416 1543890247257948160
3088061969492606976 6176686888938635264 12354499677784113152 24711251155381911552
49427005910391193600 98863019020037128192 197744052438583738368 395552413367418644070 4
791120324942410809344 1582384765072897474560 3165057760521946660864 6330691981796196745216
12662536885097000337408 25327379613203214368768 50659370912424856125440
101327965196886567026688 202674377137846843604992 405385647763841106313216
810845082503977050832896 1621837738960543778078720 3243970625826266908983296
6488531547462892523618304 12978243686546502458540032 25958848556334439739686912
51922419479151749124587520 103854283691269237539602432 207727456848469953660059648
415492692628802864481828864 831060943121331643287076864 1662273001970115115220992000
3324848235395133887735660544 6650300933700075090058674176 13301810793219764809292054528
26606039438078758876933521408 53216914579435976270565867520 106443500565428869574529384448
212906343943971573215854067712 425851373514170814565298733056 851780118280796965397778661376
1703714979066504603329919713280 3407739443142830551728564207616
6816097856305303793594577977344 13633433652649892967464055078912
27269343185378356695477908406272 54543638130913854912055413309440
109097179782141992866310019612672 218214166604912551817018425212928
436467947291082235802833622401024 873015122744678735943260788752384
1746188701814386000561708665405440 3492694316278829058473791506612224
6986022457857772231648331364827136 13973312566315772692698159432859648
27949160433832001844199312272130048 55903391470064916606004611357081600
111816924144931659047221196339806208 223654130699466969764866339930898432
1118372065545353107176451435910922 24 5592367387966856827642855860836761 6
2796437224103474059701727271059456 0 1398345377111759852791013305850265 6
6992360710858913378655814880854016 3496497268079513746678281616228352
1748407090364785402014327896014848 874282773344906965344757491957760
437181000753710614841175517954048 218610307417483873504986144964608
109315057229056219794692265476096 54662480374685251418445729234944
27333716067421196469772662865920 13668095973749883615161230557184
6834666956894584497718064840704 3417642963457113593927757201408
1708976224233467469498240991232 854565483369189071016301690880 427321427310822203641741443072
213680056468524935887666020352 106849699640819384977230659584 53429685523688151005314154496
26717260613483304761006489600 13359839232561267009677950976 6680524079190440819426328576
3340564271050124067006840832 1670433251252513862150258688 835292183489982845398548480
417683870676854379860983808 208860824804358668511346688 104439857135145073546100736
52224650934055406418264064 26114686650269138031738880 13058523916755286427172864
6529852254188001919238144 3265221274999180312444928 1632758211452179832635392
816452892702384754524160 408263339839339796365312 204150116663743607734272
102084281703908658642944 51046752537972756709376 25525682111995592048640
12763993977502402871296 6382573449503504859136 3191574955127904141312 1595931592752027926528
798037853970051891200 399054955782044909568 199545492289531936768 99781753344020709376
49895380271637725184 24949941935632547840 12476096867723116544 6238611383814979584
3119587166884200448 1559934320930455552 780037529209405440 390053948976791552
195044566674440192 97531079430242304 48769937761632256 24387167904071680 12194683463663616
6097891487645696 3049220621729792 1524747749818368 762442594385920 381255656931328
190645008334848 95331094102016 47669842018304 23837068492800 11919607988224 5960340865024
2980438867968 1490353651712 745243934720 372655521792 186344538112 93180657664 46594523136
23299358720 11650727936 5825888256 2913206272 1456734208 728432640 364249088 182140928 91078656
45543424 22773760 11387904 5694464 2847488 1423872 712000 356032 178032 89024 44516 22260 11131
5566 11133 22268 44540 89088 178192 356416 712896 1425920 2852096 5704704 11410432 22822912

45649920 91308032 182632448 365297664 730660864 1461452800 2923167744 5846859776 11694768128
23391633408 46787461120 93583310848 187183398912 374400352256 748867813376 1497869844480
2996008124416 5992553119744 11986179981312 23974507446272 47953309859840 95915209654272
191847599177728 383729558093824 767527835664384 1535193110282240 3070661098471424
6141871952756736 12284843417141248 24571885857538048 49148169761587200 98305135616196608
196627863418437632 393290911208964096 786652191162105856 1573445119812567040
3147171714601844736 6294906379157110784 12590938658221064192 25184129116255813632
50372761832138997760 100754530863532736512 201527076125574955008 403090181048168873984
806252419690375675904 1612648954568827207680 3225586139513806127104 6451748739779915677696
12904650401064438202368 25811606645138090098688 51627824976294607585280
103264873324626069946368 206548193393325849444352 413133280274799117991936
826340347525893074190336 1652828269004375824793600 3305951685913931002413056
6612493667638220710477824 13226167926897158832259072 26454697037035752487124992
52914116440554374619463680 105837677614074488529354752 211694244694080455639564288
423426268320023868440838144 846928094503773651205095424 1694007304734999131057029120
3388316840924901919407734784 6777238144759611153402822656 13555685215338836935980351488
27113788282316903130310115328 54232412267912264777319055360 108474495942381446588035760128
216968334697876727242866819072 433975355021981122619324235776 868028081296417581505829666816
1736210905097745835546021724160 3472731295205313016160768229376
6946081560430268722458986020864 13893401060899822825192871165952
27789278001878216410935540580352 55583507763913574342970677657600
111176919048141431728140548308992 222373645136911429540679482605568
444786904355079991250155737186304 889653036872674246837905018322944
1779464530070377022350997124546560 3559245972790811102052368424894464
7119125770881736318805485201391616 14239519192363700867012467105988608
28481573685927858192827927618387968 56968217974256629303261842049597440
113946577153315084441735657724837888 227913436716233820553895262700961792
455867438251674944448638419904495616 227954001535441123894743157203533824
113987141972522387782583552227409920 56998641588662106808897762926526464
28501856095531509863251874869673984 14252195698365983161027434138042368
7126731674483105695214465420623872 3563682749891609904957606886113280
1781999831270833481153990530957312 891079143797931004914588809428992
445579185980222634626091176689664 222809400030739883397443974332416
111414603535684224740921180160000 55712253527999253891560186576896
27858602644078197706329891536896 13930539262078384233439844892672
6965888601058834806857372008448 3483253785539238748497410785280
1741781635274530046783067783168 870968188889720359658715086848 435522780071087847962948141056
217780732848657758048269369344 108900037830885796057532334080 54454854618721356545464991744
27229845160999907531081908224 13616131506319568394715660288 6808670216069591511945183232
3404637339489699413266268160 1702469785472301535279972352 851310450599876681963405312
425693004231801298143412224 212865391581832127652560896 106442140523881803116707840
53225792628423771203567616 26615257497453320424390656 13308809340347377623498752
6654994965984047517401088 3327792630897203111526400 1664043889401191232176128
832095731676890454294528 416084759326592646250496 208060826407370032676864
104039636575721871114240 52024429973879362945024 26014520829948895166464
13008413336479054430208 6504783128991830638592 3252679794872067031040 1626484012624109371392
813314063906092613632 406693060750065270784 203364544773542117376 101691279586025799680
50850143392640270336 25427323496133820416 12714787647973752832 6357956773940297728
3179259861946859520 1589770668461785088 794955702975070208 397513035859623936
198774110115856384 99395851150950400 49702323621986304 24853360834248704 12427779928752128
6214439720189952 3107494738001920 1553884807954432 777011123453952 388539921465344
194287140601856 97152160235520 48580375085056 24292335026176 12147241254912 6074157498368

3037347184640 1518807810048 759471013888 379769061376 189901307904 94959042560 47483715584
23743954944 11873026048 5937037312 2968780800 1484521472 742326272 371195904 185614336 92815360
46411776 23207936 11604992 5803008 2901760 1451008 725568 362816 181424 90720 45364 22684 11343
5672 11345 22692 45388 90784 181584 363200 726464 1453056 2906368 5813248 11627520 23257088
46518272 93044736 186105856 372244480 744554496 1489240064 2978742272 5958008832 11917066240
23836229632 47676653568 95361695744 190740168704 381513891840 763094892544 1526324002816
3052916441088 6106369753088 12213813248000 24429773979648 48863842926592 97736275787776
195489731444736 391013822627840 782096364732416 1564330168418304 3128935214743552
6258420185300992 12517939882229760 25038078787715072 50080555621941248 100169907336904704
200357406859853824 400749998091796480 801570364927770624 1603281467343896576
3206844409664503808 6414251769282428928 12829629438471700480 25661510676757086208
51327524953141542912 102664057105537826816 205346128609585135616 410728286016189235200
821528629626416398336 1643201374440908652544 3286690979257969016832 6573958419268241457152
13149069760041089761280 26300445363091393216512 52605502412201213820928
105220228196439282417664 210458903136952274386944 420954699762051967877120
841983186500398773960704 1684113946953387224334336 3368523041811953801494528
6737636379434266308640768 13476453350489250028584960 26955267884219934879776768
53915258134922739404767232 107839961002811218099961856 215698811471553914780778496
431435401874970786723266560 862946361613667487769952256 1726043838954786804186742784
3452389909364477265667162112 6905384281638761845921677312 13811977489097138321018060800
27626372829833505900385533952 55257581362945470317469892608 110524834132447857668337434624
221069011078009549403470168064 442176707782246766940530933760 884430786816948870148243062784
1769016316138808412830848516096 3538342117287438170730421813248
7077303204594519031598293188608 14155844349228323443471485501440
28314164578535217647492769251328 56633280917227576816085134999552
113276465354769436674369462992896 226572737750167439433137311973376
453185089581592011035071395921920 906449407325698286407736335794176
1813057270976425101490659759489024 3626431454602907260331693694779392
7253496734505928635364135741161472 14508261119612085500129768185528320
29019057540424627459062529777467392 58043185683250167835731046367756288
116096512571302161506674066361155584 232213307552207974683772079973597184
464467179923623252708392054449766400 929015489485661112098479897904676864
464548309562037859390087843454910464 232294437190622581365467868978741248
116157359800113116517945908115013632 58083750502457471176578940870328320
29044410552429192047092463841574912 14523472926814824252947728623992832
7262370288707526241174612663599104 3631502057003820177937680507600896
1815909484826938617644027341701120 908033970575983573159607214800896
454056599369248918748600379375616 227048106725253025458698575675392
113533956882940795771548480831488 56771930201627539406873836912640
28388440980892340463986716704768 14195458430485455612268257476608
7098348185262370496271578300416 3549483577641006593204513931264
1774896531325413969136619356160 887525636915162320835490873344 443801504083808828551336034304
221920094855018248342463315968 110969718834066041204629307392 55489695120311479119013478400
27747265411794968817856151552 13874841631717099038102781952 6938025278768356833638744064
3469314870839082074113048576 1734808551146992865703362560 867479833437222347175100416
433777695650474130749259776 216907737291168543955484672 108463313378550011268169728
54236379055757875279298560 27120550711120372462256128 13561455947180903642431488
6781318269400810526867456 3390954282605584616259584 1695624715255381984542720
847886144603985830477824 423979965790140334342144 212008429639143876722688
106013438191608793137152 53011330781822823956480 26507971233920625672192
13255138538464919683072 6628145729984763265024 3314361095368533344256 1657324662872342528000
828734389030209191936 414403223312123559936 207219626054571261952 103618820226540371968

51813913712897556480 25909208656262463488 12955730228038074368 6478428063972458496
3239495506962939904 1619888490969825280 810014614229090304 405042491486633984
202538837929361408 101278215057702912 50643505575362560 25323951810936832 12663075417096192
6332087464361984 3166318610087936 1583296743997440 791717091475456 395892905476096
197963632607232 98990406238208 49499498086400 24751896526848 12377022005248 6189047873536
3094792372224 1547530403840 773832310784 386949709824 193491632128 96754204672 48381296640
24192745472 12097421312 6049234944 3024879616 1512570880 756350976 378208256 189120512 94568448
47288320 23646208 11824128 5912576 2956544 1478400 739264 369664 184848 92432 46220 23112 11557
5779 11559 23120 46244 92496 185008 370048 740160 1480448 2961152 5922816 11846656 23695360
47394816 94797824 189612032 379256832 758579200 1517289472 3034841088 6070206464 12141461504
24285020160 48574234624 97156857856 194330492928 388694540288 777456189440 1555046596608
3110361628672 6221260128256 12443593998336 24889335480320 49782965927936 99574521790464
199166223450112 398366806638592 796802332753920 1593742104461312 3187759086829568
6376067929473024 12753235370573824 25508669764403200 51021737575317504 102052271243657216
204122134673358848 408279453718806528 816629276181790720 1633399289851936768
3267080054680584192 6534723059314589696 13070572018536022016 26143395836885729280
52291295273398829056 104591597746052399104 209201209890614280192 418438448578247524352
836948954750532976640 1674042024689141809152 3348372279754435330048 6697321020261174083584
13395794962026955014144 26793895767063123722240 53592403220144674832384
107194029812326204440576 214406506368726118432768 428849906225599655968768
857773599427494150144000 1715694772807577976700928 3431684693520335306227712
6863959682851029318107136 13729099957322776047517696 27460561097886986917642240
54925844562256843480498176 109861133857479426251423744 219741157180890331083702272
439520093293643619329114112 879115744451013152981647360 1758382604629478134610132992
3517067440713859926513942528 7034739344337527167615238144 14070687614494668964405182464
28143793080628567187159777280 56292421864535592891018379264 112594515135628102815434407936
225208373084370039697664114688 450455431794967747528918827008 900988234842390831325018849280
1802131212189692335184400089088 3604571909389206015437524959232
7209762788798054721012499480576 14420763517635394822299898085376
28844002915349360405149594419200 57692957590855862331398785335296
115395818702026007704996763664384 230811444444680581494391913316352
461662502970618295157580598607872 923404234103750854652754741166080
1846966924532530237980696570232832 3694250761715117533311767316267008
7389135348730349181324282984136704 14779538348060926592050062671478784
29561611997322309642903118749368320 59128294597045532203412224311558144
118266730398892890242036422248759296 236553743207389432154496791748804608
473148051233986167649841478000181248 946377232106386941981378745005506560
1892916723489603097326149068021301248 946539491383216155344770323015794688
473310310510815381013233056010469376 236675437665011342177040475256520704
118347860037307496923732211253903360 59179000621054661379472092439773184
29592035611727787148539039626297344 14797285456464121803671016516354048
7399276553532175016536256609779712 3699955189416144565618502480691200
1850136051033100811484438328246272 925147253679064670079812708073472
462613240920789467208703126011904 231326427501023299688749948993536
115673117270825932886574167490560 57841510395570107964386680242176
28923231077863624742743138369536 14462853478971097751646468308992
7232045709505191565960683716608 3616332339762417128049066639360
1808320912386119236558895710208 904237827445514954546629050368 452157599348985145406905122816
226098142487606406770247860224 113058742650360120418521579520 56534207028458518725959614464
28269521365868488621329219584 14135969608753858939839315968 7068589267286736784507011072
3534596865098272049547182080 1767449548276587853420429312 883800332002019841033633792
441937944932872877678526464 220987861932367917420118016 110503375699149698000486400

55256410216057718645456896 27630566291270294145335296 13816463737255864483971072
6908822164438290947637248 3454706230124324826644480 1727500689014752089735168
863824131483670883074048 431948959229982860640256 215992926359065139871744
108005686551569424711680 54007454961803139743744 27006033323910783565824
13504169583459998629888 6752661252482302738432 3376618856617303080960 1688453543496727396352
844298829342401626112 422185443468219777024 211110736132619370496 105564375265564426240
52786691232409583616 26395597416018477056 13198924607916081152 6600025253911461888
3300294101932441600 1650287788454576128 825214262971465728 412642315857821696
206338750114955264 103178171150499840 51593483621761024 25798940834136064 12900569928695808
6450834720161792 3225692237987840 1612983557947392 806560498450432 403314608963584
201674484350976 100845832110080 50427211022336 25215752994816 12608950239232 6305011990528
3152774430720 1576521433088 788327825408 394197467136 197115510784 98566144000 49287266304
24645730304 12323913728 6162481152 3081502720 1540882432 770506752 385286144 192659456 96337920
48173056 24088576 12045312 6023168 3011840 1506048 753088 376576 188304 94160 47084 23544 11773
5887 11775 23552 47108 94224 188464 376960 753984 1508096 3016448 6033408 12067840 24137728
48279552 96567296 193150976 386334720 772734976 1545601024 3091464192 6183452672 12367953920
24738004992 49480204288 98968797184 197954371584 395942297600 791951704064 1584037625856
3168343687168 6337224245248 12675522232320 25353191948288 50710678863872 101429947662336
202877075193856 405788510126080 811645739728896 1623428918411264 3247132714729472
6494815185272832 12990729882173440 25983658787602432 51971715621715968 103952227336454144
207922046858952704 415879278089994240 831828924924166144 1663798587336687616
3327878649650085888 6656320249253593088 13313766398414028800 26629784596641742848
53264072792910856192 106537152785076453376 213092319968662388736 426220668734343741440
852513395062725410816 1705170905313526677504 3410630041003205066752 6821836542758713556992
13644826007022033960960 27291957857053281615872 54588527400124990619648
109186278172286836015104 218391003088647381581824 436818899665442182266880
873711586307179202740224 1747570746566948081893376 3495436641039075516612608
6991463577888509738876928 13984107747397736889057280 27970576678036908600721408
55945875722556686846656512 111901196178079112983740416 223821281822089704548335616
447680342576042366258380800 895436243015810646840180736 1791023601759073122327199744
3582349434973049901948076032 7165303332855907118483505152 14331815591531428866141716480
28666049034702086990632845312 57336933772682632497964515328 114683538951922182029326680064
229386420716958198125448658944 458811527060144064384487915520 917700425372743465036157026304
1835555593250397602606676443136 3671420671510616550282077667328
7343460313040875790701604896768 14688158566121036961678108917760
29378793012320644683906016083968 58762537784798430888911628664832
117534979089911144820022450323456 235089765220450855724443286634496
470219144522158843617683345244160 940517517206831951572960234438656
1881193490738692431821107556777984 3762703894127441920992589289357312
7526041613554997956685926930317312 15053350877710224142773350563840000
30109237056620904744349694534090752 60223544715642722406305375881003008
120457230636087270647822725387649024 240934743681778192966069398026584064
481910052182763689272986690555740160 963901234003941985227669170116624384
1927964727284713183818729918243536896 3856253973123084794364242992507650048
1928289245838371610545513074264113152 964225752557600411954452326137200640
482153441098007509318074057571172352 241097002958607406329460976036872192
120558642684105528999942461644079104 60284391944453677417577217634861056
30144731273427295167591602223841280 15073633287313875813197297815126016
7537450468957052021299397259165696 3769042147128583068000072805384192
1884679529889320062675223490592768 942418993107174295675205289246720
471249110634844280006399416598528 235644362358050706087598094286848
117832084699339636085998240137216 58920994109826959564098716565504

29462972934992050542599156531200 14732724407535310651574477389824
7366981173787298015924688257024 3683800071903470353031068909568
1842054778456645849049896845312 921104760480778260792129617920 460591065866616798529655406592
230314875746422233331623002112 115167109279768033699209150464 57588390343162475366303399936
28796613023220466941501112320 14399515437429848099925262336 7200362181624731364549984256
3600483322267269339568668672 1800392776861086498431172608 900271946294269163539005440
450173752078997538931212288 225105765505430248046460928 112562327485680863313657856
56285886109323301302042624 28145304237903085473628160 14073832710572260148117504
7037506651096488779710464 3519048473453423742681088 1759671810679301547753472
879909692315945612083200 439991739646120225144832 220014316567133822124032
110016381655603765837824 55012802513820310306816 27508707099919368847360
13755506471464291270656 6878329696484449058816 3439453078618376241152 1719870654497263976448
860007384842669916160 430039721218353922048 215037875007686443008 107527944703097962496
53768475951176351744 26886489775401861120 13444370787607773184 6722748343757307904
3361655646855364608 1680968560916037632 840554649202196480 420312508973187072
210173846672637952 105095719429341184 52552257761181696 26278327903846400 13140263463550976
6570681487589376 3285615621701632 1642945249804288 821541344378880 410805031927808
205419695833088 102718437851136 51363513892864 25683904430080 12843025956864 6422049849344
3211293360128 1605780897792 802957557760 401512333312 200772943872 100394860544 50201624576
25102909440 12552503296 6276775936 3138650112 1569456128 784793600 392429568 196231168 98123776
49065984 24535040 12268544 6134784 3067648 1533952 767040 383552 191792 95904 47956 23980 11991
5996 11993 23988 47980 95968 191952 383936 767936 1536000 3072256 6145024 12291072 24584192
49172480 98353152 196722688 393478144 787021824 1574174720 3148611584 6297747456 12596543488
25195184128 50394562560 100797513728 201611804672 403257163776 806581436416 1613297090560
3226862616576 6454262104064 12909597949952 25821343383552 51646981734400 103302553403392
206622286675968 413278933090304 826626585657344 1653390610268160 3307056098443264
6614661952700416 13230423417028608 26463045857312768 52930489761136640 105869775615295488
211757143416635392 423549471205359616 847169311154896896 1694479359798149120
3389240194573008896 6779043339099439104 13559212578105720832 27120676956025126912
54245857511677624320 108500722222609989632 217019458843729461248 434074946484477886464
868221950562993700864 1736588016314063257600 3473464263004278226944 6947504986760859877376
13896162895026326601728 27794631633061866897408 55593874952142161182720
111196973276321177141248 222412393296716063834112 444861680081579546771456
889797147139453931749376 1779741868231497539911680 3559778884368174432649216
7120148064546707570950144 14241476720714132553203712 28485314624669699929014272
56975351615822269503242240 113960147964610278296911872 227939185395152035174678528
455916149722167027511066624 911907857308059969345552384 1823966830343571767337943040
3648235892142047191969562624 7297076247193901698526478336 14595361420207418026227662848
29193140692054065310804738048 58391117087386589138308300800 116791905581330095310014251008
233603153975774024686823800832 467244993577775717507238199296 934567358408006771281657593856
1869289459320924215097677578240 3738888403651669775264079937536
7478395777322982240665609437184 14958029494685249861606117998592
29918534869449070483762034245632 59842021499055282488623664988160
119693946518424848019446522970112 239407700077478262123291431927808
478855014236213656415379635830784 957789256634941577168352815611904
1915736969594911683011892719124480 3831790851839880423374159614050304
7664215528979874961449067579703296 15329698708559978152299631862611968
30661932718320412763402257131634688 61328936039041738444410501076090880
122668013282885302724032975777824768 245356308975374257118489898806935552
490753182769955817577827692116443136 981587495178326241837351173238030336
1963337249633481697038093924486348800 3926999017820621820802971004993273856
7854647072748560495059508322027700224 3927648054927938674256537317034426368

1963986286740798550491660236527501312 982074273008813881927525907268894720
491077701323614244304610848137019392 245559133071410773822729371319795712
122789707740507212746576659285540864 61399924472654519290894316455591936
30702497537527716104250151634206720 15352516419364086281526572520308736
7676892034982157255464034611757056 3838762930141135685082391481679872
1919539921395596371216382828740608 959849188860312449945784958320640
479964208511413357141689251135488 240001911296335244655243011555328
120010859168481905369820698771456 60010381344398094206009945882624
30007666552277617863554771189760 15005071216178094312052284719104
7503154578108689846163591921664 3751886774064166268150520741888
1876098129536993806609622761472 938126436020952239571992576000 469101903636703787919586885632
234570294631465728026588741632 117294818722289781046692020224 58652245064423349040044834816
29328540383850903778371829760 14665479117745066518360621056 7333344021782340573767663616
3666974242346073944177508352 1833638236900488800735592448 916894676313970314691215360
458485117088848114507317248 229261448010355535834513408 114640168738143507207684096
57324806735554623249055744 28664764551018746447134720 14333562867130090634870784
7167371729375404023087104 3583981012592881364369408 1792138080249030358597632
896142827100810017505280 448108307038552427855872 224072600263349923479552
112045523503711816515584 56027373437874335645696 28015992561946381516800
14009149202477797605376 7005151061991202226176 3502863761371752824832 1751575995873952268288
875860055531014062080 437966056562525995008 219001042679772479488 109509528539140980736
54759267869197860864 27381885734412615680 13692068767113150464 6846597333509996544
3423580141731708928 1711930808354209792 856035772921282560 428053070832730112
214044127602409472 107030859894226944 53519827993624576 26762113020067840 13382156021661696
6691627766644736 3346088761229312 1673181819568128 836659629260800 418364174368768
209199267053568 104608223461376 52308406697984 26156350832640 13079249158144 6540161449984
3270349160448 1635308797952 817721507840 408894308352 204463931392 102240354304 51124371456
25564282880 12783190016 6392119296 3196321792 1598291968 799211520 399638528 199835648 99926016
49967104 24985600 12493824 6247424 3123968 1562112 781120 390592 195312 97664 48836 24420 12211
6106 12213 24428 48860 97728 195472 390976 782016 1564160 3128576 6257664 12516352 25034752
50073600 100155392 200327168 400687104 801439744 1603010560 3206283264 6413090816 12827230208
25656557568 51317309440 102643007488 205302792192 410639138816 821345386496 1642824990720
3285918416896 6572373704704 13145821151232 26293789786112 52591874539520 105192339013632
210401857896448 420838075531264 841744870539264 1683627180032000 3367529237970944
6735608231755776 13472315975139328 26946830973534208 53898059993579520 107804916080181248
215627424346406912 431290033064902656 862650434873982976 1725441607236321280
3451164689449353216 6902892328852127744 13806910557611098112 27616072915035881472
55236649429699133440 110482306058653007872 220982626515815497728 442001281828649959424
884074621251337846784 1768293357690751549440 3536874945757654810624 7074326352267613044736
14149805626039832936448 28301917095088879566848 56608445876196186521600
113226115124429227819008 226470676992932165189632 452978247474011749482496
906030281924318337171456 1812208137801226350755840 3624711423507632054337536
7250013142825622814326784 14501206877271963039956992 29004774937785360902520832
58014272242053591450255360 116037989217072922190938112 232094867900077322962731008
464227514732017603087171584 928530587327761120497762304 1857212290382974069642362880
3714726812220851796578402304 7430058087351510907744157696 14861325100522636444663021568
29725068052684502147675455488 59454971808647462812049735680 118919615023851842657497120768
237858572860817519381789540352 475755831347862706897169678336 951589033948180750061520551936
1903332810401272172657403494400 3806975105812365690383531769856
7614569181644374070904513101824 15230376303328033522083925327872
30463228486734637804717648904192 60931408733626417130534894305280
121872720987567117303268981604352 243765249015762800690936349196288

487570112112782733550669470367744 975219452388079731438932484685824
1950597361101187991553052057272320 3901511634852433040456478290345984
7803657095004980195613704932294656 15608581840610188620628906567794688
31219698982420833700060806542000128 62444468567242580317727599896821760
124899078339286986470667173419286528 249818439088177624611758294089859072
499677442995562552564364482682290176 999436015629539711810424754369724416
1999034290535908636984241086749736960 3998393099625475700695265329520050176
7997435236358268254844096971081252864 15996168546931170216595326566244810752
7998733310572901961751229595163557888 3999691173840109407602397953602355200
2000007846196883917164590554811465728 1000085052736856565263991066410876928
500083091187635585972843427708010496 250061828003421444656845661105291264
125041055206512548163634804178288640 62525598205657186999423388901965824
31265334404029049958514687857393664 15633934852614753208658840631902208
7817601251607490719030168667553792 3909117538453802416865458509578240
1954717225551929737107916342689792 977437840938479132891551715295232
488758534550496698614572629622784 244399074315876915391684700798976
122209440678252740738041543393280 61109672099283511890120368193536
30557311929720326705609982345216 15279893904899448733079890296832
7640565922469367056677394710528 3820592446244504873407422136320
1910450965627163109238073458688 955302854066036890886217924608 477690112659246113576699559936
238864399142736890855145078784 119441870977925362460970188800 59725771192241139747183919104
29865303447759799131941371904 14933860649699514195145392128 7467534787759564412160049152
3734069625334685863373701120 1867185928394794760333688832 933668522061123294490263552
466872039962424604406841344 233454909447143780784275456 116736899456537629682565120
58373172094751684486496256 29188947230617277065854976 14595654206929355944230912
7298417399275036677767168 3649503847542697691709440 1824899497723938522267648
912523535838264099340288 456298661407279468773376 228167777447713443938304
114093112095893576744960 57051167733965215760384 28527889709991821574144
14265097776500517634048 7133125349002562240512 3566850904877432832000 1783569567626792271872
891856841407434063872 445964449500735995904 223000239148877479936 111509126773693480960
55759066986474110976 27881785293050740736 13942018546432212992 6971572223169527808
3486067586561474560 1743174530769092608 871657634128723968 435864001436450816
217949592904269824 108983592545157120 54496194319089664 27250296182800384 13626247603027968
6813673557327872 3407111656570880 1703693267238912 851915353096192 425992036286464
213013198012416 106515188940800 53261889437696 26633092202496 13317619843072 6659346792448
3329941831680 1665105133568 832619675648 416343392256 208188473344 104102625280 52055506944
26029850624 13015973888 6508511232 3254517760 1627389952 813760512 406913024 203472896 101744640
50876416 25440256 12721152 6361088 3180800 1590528 795328 397696 198864 99440 49724 24864 12433
6217 12435 24872 49748 99504 199024 398080 796224 1592576 3185408 6371328 12743680 25489408
50982912 101974016 203964416 407961600 815988736 1632108544 3264479232 6529482752 13060014080
26122125312 52248444928 104505278464 209027334144 418088222720 836243554304 1672621326336
3345511088128 6691559047168 13384191836160 26770531155968 53545357279232 107099304493056
214215788855296 428465937448960 857000594374656 1714138627702784 3428552133312512
6857654022438912 13716407556505600 27435014136266752 54874426319044608 109757648731111424
219532889648267264 439100963668623360 878272296081424384 1756685329651204096
3513652134279118848 7027867218511659008 14056860336930160640 28115972473674006528
56236448546975383552 112481904293205508096 224981822984920498176 449999674766859960320
900071407127757848576 1800286929443591553024 3600862089263334817792 7202300639278973059072
14405754200062552965120 28813814243134319624192 57632240172287066636288
115273703716610988048384 230565854177295685648384 461168601842738790400000
922410990661772419006464 1844969555276134514425856 3690234258457448381677568
7381058812725255469006848 14763298217071228349317120 29528957617383891521241088

59062637601250652687695872 118134719935467044665819136 236288329336865567912493056

218064079552512 436162518843392 872393757163520 1744924953280512 3490124784467968
6980799324749824 13962698161127424 27927595345510400 55859588737531904 111727973568086016
223473539322216448 446982263016521728 894034894777221120 1788210527042797568
3576702529062305792 7153968008078032896 14309061916062908416 28620375631939502080
57245254863506374656 114499516926267490304 229017048251044462592 458070125299107889152
916212308192253706240 1832568731572583268352 3665425693521318248448 7331427847794939920384
14664008617094486687744 29330323077198187069440 58665257840414801526784
117339739052866457829376 234697924849806625210368 469432743187760669523968
938939273351816177254400 1878026120656222030921728 3756347389217623414669312
7513285074245605534990336 15027750740111928481284096 30057862663465291785175040
60120447693413453215563776 120250340119792645721554944 240519569705516770023964672
481076918342896497209638912 962229394549518908742696960 1924609904826489646132232192
3849522041107882949558140928 7699648545125573213703634944 15400506016070761056581976064
30803429883780751371513364480 61611695470839961259725553664 123233062348236839552848756736
246485467509587513172492812288 493009620645402694478576222208 986096612543260725224333639680
1972347967591432122983029669888 3945005420192685591034784120832
7890629810405013872207017803776 15782497560849313124688934731776
31567471001777197009927667712000 63139893763711535540954931920896
126289691047737354124109056835584 252599189136103274332616499658752
505237992353463680834029771292672 1010555212869441626005653086535680
2021268882063911780686493260972032 4042854676777880618723360697745408
8086343178855875352147469747093504 16173954008311978933696436197392384
32350443317824414326195865801195520 64705957238049741569997718415212544
129422055680901308975207410456068096 258864393771406269620838768163422208
517769352362019842582525430829416448 1035619834362454291846746650663976960
2071401928001737797056884879338242048 4143128374557134020840552914697060352
8286905786221584895134672141435273216 16575109646657803497176476906952851456
33152815441744874408167219062070312960 66310823180348283643962968620469846016
33158007738603409235795749558399533056 16580301943516338324805007403282071552
8290800008865486015856070013682188288 4145724522986401434654818162861670400
2073024520770029930690800659441123328 1036593390023429572027096118725705728
518337259830922089354395953865424896 259188912325064696347621924183998464
129604597367334174009022935717642240 64807369286067999922117454671642624
32406219944234456419861720742232064 16204377622717456439332357074321408
8102822636658842334366926888763392 4051728230979478224533837620183040
2026022571814767640942105897992192 1013090514069898084808646492946432
506584871116206174573120018448384 253312242598731653370958395211776
126666024819680109727678390599680 63337964169997196384938791796736
31671457965077168953019194146816 15836966922577869856784496197632
7919102431308577618529697660928 3959860700664110154333573611520
1980085092836965749701149196288 990119917670938211117755793408 495098644461696773692468494336
247568665043962220913029545984 123794003928538027489912422400 61901837667547472261655035904
30953336685412965389176930304 15477877268526097323763171328 7739543097172855976468938752
3870073780041331645528145920 1935188005748117651410911232 967669560737784740028874752
483872559300755327176146944 241955169116309142168928256 120987029291120310374891520
60498237012043024832659456 30251479689262947238936576 15126920436252191030771712
7564050513936454221037568 3782320404873406463344640 1891307776389292908085248
945727675170941292249088 472900731073618065227776 236468812280882742165504
118243629512478225858560 59126426442257540317184 29565519064137983852544
14783912453573598773248 7392532687539102810112 3696554574145703116800 1848421402260927414272
924282758724501635072 462177408159269781504 231106718478144372736 115562366438326927360
57785686818790834176 28895095209209102336 14448673504511393792 7224899702209118208

3612731326081269760 1806506400528990208 903323569008672768 451696968876425216
225866076624257024 112941834405150720 56475315249086464 28239856647798784 14121027835527168
7061063673577472 3530806714695680 1765540796301312 882839117627392 441453918552064
220744139145216 110380659507200 55194624720896 27599459844096 13800803663872 6900938702848
3450737786880 1725503111168 862818664448 431442886656 215738220544 107877498880 53942943744
26973569024 13487833088 6744440832 3372482560 1686372352 843251712 421658624 210845696 105431040
52719616 26361856 13181952 6591488 3296000 1648128 824128 412096 206064 103040 51524 25764 12883
6442 12885 25772 51548 103104 206224 412480 825024 1650176 3300608 6601728 13204480 26411008
52826112 105660416 211337216 422707200 845479936 1691090944 3382444032 6765412352 13531873280
27065843712 54135881728 108280152064 216577081344 433187717120 866442543104 1733019303936
3466307043328 6933150957568 13867375656960 27736898797568 55478092562432 110964775059456
221946729988096 443927819714560 887924358905856 1775986156765184 3552247191437312
7105044138688512 14211187789004800 28424574601265152 56853547249041408 113715890591105024
227449373368254464 454933931108597760 909938230961373184 1820017199411101696
3640315873798914048 7281194697551249408 14563515295009341440 29129282389832368128
58263068379292106752 116535143957838954496 233088302314187390976 466212633425393745920
932497324444825419776 1865138764077726695424 3730565758531605102592 7461707977815513628672
14924568877135634104320 29851443597280481902592 59707498880579391193088
119424221133195637161984 238866889010464983875584 477770671509077386854400
955615129994449611915264 1911377833941488900243456 3823050815788157153312768
7646691927386673012277248 15294564446394063435857920 30591490076029561694322688
61187702518541993033859072 122384849770049725358145536 244788589006030929297145856
489614956943924815756001280 979305471751575545835421696 1958762059230602920317681664
3917826349916109497929039872 7836257162742026310445432832 15673723251303667250065571840
31349864354246563758480556032 62704564411771586033659936768 125418800230100089100717522944
250856943273314012268230344704 501752572172855692670051287040 1003582515598166721607283769344
2007319773701244115748929929216 4014949032412309576566584639488
8030517034844261843270618841088 16062272009727809066816136806400
32127019899534188894182071861248 64258991559225519309463740219392
128527886638765321661126673432576 257075580318159209406651732852736
514190774717575550982100237680640 1028460777597665366301794019311616
2057080011520359261278775126523904 4114476935690775579907924428849152
8229587696681665274516597209300992 16460443043963558778434691121807360
32923421389127574015672375650025472 65851913380656060948950738112872448
131713967966113947733113449851387904 263448218341831547136650846954061824
526937001502870397614149588410695680 1053955132644155401909994965826535424
2108072524565140017183381509663358976 4216469567683938461093546175347294208
8433588172475193775640658662735740928 16868474419165021258188449949553786880
33739544986759309930191165147272183808 67484282270377154688010860790873587712
134978949134471379031278782574405615616 67494666864094224343267921783532027904
33749929580476379585448226139930624000 16876262864452823499631245694047617024
8438780469333728603269189159064961024 4219714753220522728361377735553056768
2110019635887090577544080445786816512 1055090947581959895453736011898552320
527586038610187251067715900451848192 263813301714697277204281897477210112
131916792062150464437352922364248064 65963466633476145136282447994945536
32984268617938529026944217403883520 16493401959569492742873605405147136
8247334805084860486137551054176256 4123984315192487300419149702889472
2062150613921272178884761939345408 1031154535123150353779974513623040
515616881642832309058784028786688 257828247862044720613790400380928
128924027451336643349094393184256 64466965485825463195646793089024
32235958622991302358373194792960 16119217251534936559461496520704
8060227595787110969868197822464 4030423282903376830002823692288

20153663839565990875357742366672 10077605632307548800350683136000 5039189672416051081511247544432
2519788264339163881423576760032 1259990846235151111045764874224 630043780150360140689870684416
315046068591572362928429465560 157535123553982327755961794456 78773606406089237023854428816
39389825517593655084863979952 19696423916071345828900372248 98489675366729320576843777760
4924861557655095600459284448 2462619673486862586038190008 1231404284073088685923368896
615749365701373039413821444 307898294683100867932979200 153960953257757608079523884
76986379586982391096279044 38496141272542989076398088 19249546375797391302328832
9625511057661644033228800 4813124463712296207646672 2406746699296885199339552
1203465583368811147427844 6017789085445898475929296 3009125127023870607336000
1504677855662395988377776 7523965739064283365376 376227109990829339444432 188127966514222255530888
940711890165149204480 470391973879593566208 235214001338306265088 117616007868407873536
58812507533831307264 29408505566729338880 14705378683271512064 7353252291589177344
3676907620771299328 1838594547874004992 919367642681180160 459719005712678912
229877095042383872 114947343614214144 57478069853618176 28741233950064640 14371716486660096
7186407999143936 3593478877478912 1796876877692928 898507158323200 449287938899968
224661149319168 112339164594176 56173877264384 28089086115840 14045616799744 7023345270784
3511941070848 1756104753152 878119485440 439093297152 219563425792 109790101504 54899245056
27451719680 13726908416 6863978496 3432251392 1716256768 858193920 429129728 214581248 107298816
53653504 26828800 13415424 6708224 3354368 1677312 838720 419392 209712 104864 52436 26220 13111
6556 13113 26228 52460 104928 209872 419776 839616 1679360 3358976 6718464 13437952 26877952
53760000 107528192 215072768 430178304 860422144 1720975360 3442212864 6884950016 13770948608
27543994368 55092183040 110192754688 220402286592 440838127616 881743364096 1763620945920
3527510327296 7055557525504 14112188792832 28226525069312 56457345105920 112923280146432
225863740162048 451761840062464 903592399601664 1807322238156800 3614919354220544
7230388464254976 14461876440137728 28925951903531008 57856301853573120 115721399800168448
231460391786381312 462955967944851456 925982304633880576 1852105346756116480
3704492168488943616 7409547286931308544 14820220473769459712 29642692747352604672
59289889094332579840 118588785387919900672 237195585174349283328 474427199145717530624
948926455885472989184 1897997026959021834240 3796282284294195380224 7593141029340694183936
15187434980185995214848 30377175803381204123648 60758963292780835635200
121527149957598526046208 243072746659270761644032 486182386806688942391296
972438560589672722989056 1945024695131935122391040 3890344538169049597607936
7781279372148457900867584 15563739335917633213038592 31129839855076701248684032
62264402076636272142581760 124538248886238283575590912 249095387238408045732036608
498228553408679048625782784 996532664681084011574984704 1993216445089619851796807680
3986735121634143360887291904 7974074706178094036361936896 15949358338175802701898579968
31901134527990834662146572288 63807104759260127840991969280 127623880925077172715381587968
255267104663268179497558474752 510572894952764027128707547136 1021223161157983390524596289536
2042601064820877453583554969600 4085511614651576252235834720256
8171642199322795194609119002624 16344522338684875769493137129472
32691520557448322299536072507392 65387992875053786120171741511680
130785889270421855282542676017152 261591585581472276649483738021888
523222785244201685467764248018944 1046524798650917635273122039988224
2093208053626863799221431167877120 4186733019903784655793236511555584
8374099865107683426287221374713856 16749467380815595081975939452633088
33501470062831646622754872311676928 67008010728064206163115731436175360
134026162660930238161443436497993728 268072607731464127993310820247273472
536185780282135559327469534997118976 1072452690202685725336634858999382016
2145067639682200664036661296009052160 4290459797918059754800105748038680576
8581568632943436363053777808118513664 17164435340101506433014688240319332352
34331466828632280279843641728803274752 68668125954123095387315813953935769600
137346636501963260429888688900529979392 274714042191360660170291499786376839168

137367405689397399740402810885846859776 68688895141557234697829935939252649984
34347043719207884762729233217790935040 17174819933818576088271749232977772544
8588059004016604897589440928530038784 4294354020561960875521503620285595648
2147339269557809651124143388153085952 1073750764417319432243767483081687040
536915947027867019462731636043415552 268478255923537161401789765272993792
134249269166570406536106856262139904 67129705185686116185659414943891456
33567387894043514551632700878356480 16784961597621985505217847142383616
8393114624111106867309671922794496 4196874224705610491005210137198592
2098595568677833774177792156499968 1049377012501431151426489622200320
524728120331972707882041583075328 262383867206614920025419177525248
131201837123621743054908781756416 65605870321968013048553987375104
32805411041062577284826791936000 16403943460570574022688295092224
8202590700304929701481597108224 4101604835162286195809523335168
2050957160086053770439124058112 1025555951295482221486743224320 512816661273968778876962209792
256427673450098223505276403712 128223508131606028786035851264 64116589769081472909716750336
32060712736179965713207787520 16031565293909597485778599936 8016387109864606057476653056
4008495786387206686032003072 2004399008921055171662839808 1002275062324253500154839040
501175310093989707239129088 250606544512926332200419328 125312716989428905390637056
62661080861197322340532224 31332901613840095992872960 15667631398540765407739904
7834405995080741409521664 3917498145445550057586688 1958896646675364705206272
979522110313977190809600 489797948645136014508032 244917421066641716805632
122467933905357713178624 61238578638697283977216 30621595162357855682560
15311950502683534688256 7656551712094070767616 3828564086423187095552 1914426158399669403648
957285136793872629760 478678597193955278848 239357312995487121408 119687663696998301696
59848335448126521344 29926419523876945920 14964335661845315584 7482730780876079104
3741646865414750208 1870964170195730432 935552453842042880 467811411293110272
233923297832599552 116970445009321984 58489620551172096 29247009298841600 14624604161048576
7312851836338176 3656700796076032 1828487836991488 914312637972480 457190678724608
228612519231488 114314849550336 57161719742464 28583007354880 14292577419264 7146825580544
3573681225728 1786974830592 893554524160 446810816512 223422185472 111719481344 55863934976
27934064640 13968080896 6984564736 3492544512 1746403328 873267200 436666368 218349568 109182976
54595584 27299840 13650944 6825984 3413248 1706752 853440 426752 213392 106704 53356 26680 13341
6671 13343 26688 53380 106768 213552 427136 854336 1708800 3417856 6836224 13673472 27348992
54702080 109412352 218841088 437714944 875495424 1751121920 3502505984 7005536256 14012121088
28026339328 56056872960 112122134528 224261046272 448555646976 897178402816 1794491023360
3589250482176 7179037835264 14359149412352 28720446308352 57445187584000 114898965102592
229815110074368 459664579887104 919397879250944 1838933197455360 3678141272817664
7356832301449216 14714764114526208 29431727252307968 58867852551127040 117744501195276288
235506594576596992 471048373525282816 942167115794743296 1884474969077841920
3769231413132394496 7539025776218210304 15079177452343263232 30160606704500211712
60325717008627793920 120660441216510328832 241338896831530139648 482713822460079243264
965499702514196414464 1931143520216468684800 3862575270809089081344 7725727002370481586176
15452606926245570019328 30907519695500353732608 61819651077019134853120
123648525526075124482048 247315497796223958515712 494667889080595336134656
989409565137485510475776 1978966704227560697364480 3958228556360300747554816
7917047408530960200761344 15835275408682637812826112 31672912000606710448259072
63350546367696290541731840 126710537468358320373891072 253439964402648119328636928
506917707737159195818983424 1013910973338044305961385984 2027973062403540440569610240
4056248356261984538432897024 8113101175433776391453147136 16227411276687167412081000448
32457240405013564082511413248 64919316513305586681721651200 129848304433168090396840951808
259715951679450014860477202432 519470588985127697854545002496 1039018549222710731976271200256
2078191840950332136486904791040 4156693166910485618042534363136

8314005303840613926222518288384 16629248547720513232719935700992
33260972975519597225989669650432 66526897711196335973078935797760
133063698942706954988357064589312 266147204926042476061112515166208
532334023933342084291021802307584 1064747276029198432919637148565504
2129653008383425394514461385031680 4259622929416907846379296945864704
8519879684133929807459342243332096 17041027018868087844320181189869568
34084589338936632147443355786149888 68174249280274177212492698385121280
136358639765350180260197370395885568 272737561940304012190818688043057152
545515688699815327722485270588686336 1091112507038045262126666330182516736
2182387273352919737616724238375321600 4365099065259497901960231632771219456
8730847167626312657374029577583591424 17462992409467259021655191779249487872
34928580967363785457124648806663585792 69862354231586105741877828109656391680
139735093056889281139012717211971223552 279490955301212701588539556409259327488
559023448977293681798107356789152415744 279532493676080980209567800379893088256
139776631431757559760040961182604984320 69893508012737314707649011087631712256
34949350154797924767638770791980466176 17475973151613596090726518020072538112
8738635612914114898816825322077421568 4369642325010715876135195817059287040
2184983421782187151430989486539931648 1092572840529508182397190532275109888
546326985083961394539443160640126976 273183774951584348940145527571349504
136602028680594000305284737411317760 68306084942697913070248355518480384
34155577772549412993927171165650944 17079056536874934726365082286030848
8540162093737581477883289494618112 4270397959518847796292018923110400
2135357436084452426821196549455872 1067757946204740477748191818678272
533918587183627371042892681314304 266979100632442251605844726644736
133499453836535408845121556316160 66754678678424845943660374654976
33379815219290993732379985575936 16691145549684782246464891912192
8346191744862033813369895518208 4173405357440838251753672540160
2086857421225329798411198660608 1043506081865120235472780525568 521791726558787785869980860416
260915206092507727001785729024 130467274452810780534290513920 65238472929683848783844081664
32621654316481153650271453184 16312036084060191454310432768 8156622504939903041742569472
4078613483924855178164961280 2039457857689879417729318912 1019804486708665623188078592
509940022286195768755748864 254988900609029362958729216 127503895037480420769792000
63756669885223080030109696 31880696125852974837661696 15941528654547204830134272
7971354623083961120718848 3985972459447159913185280 1993133803676169633005568
996640688814379654709248 498357237895337246457856 249197065691742332780544
124607756217908021166080 62308489794972437970944 31156550740495432679424
15579428291752323186688 7790290606628465016832 3895433533690384220160 1947860882033267965952
974002498610671910912 487037278102354919424 243536653449686941696 121777333924098211840
60893170561676476416 30448837080651923456 15225544440232804352 7613335170069823488
3806949060011622400 1903615267494166528 951878002491260928 475974185617719296
238004684994904064 119011138590474240 59509967341748224 29757182694129664 14879690858692608
7440395185160192 3720472470487040 1860373674196992 930255556575232 465162138025984
232598248882176 116307714375680 58158152155136 29081223561216 14541685522432 7271379632128
3635958251520 1818113343488 909123780608 454595444736 227314499584 113665638400 56837013504
28420603904 14211350528 7106199552 3553361920 1776812032 888471552 444268544 222150656 111083520
55545856 27774976 13888512 6944768 3472640 1736448 868288 434176 217104 108560 54284 27144 13573
6787 13575 27152 54308 108624 217264 434560 869184 1738496 3477248 6955008 13911040 27824128
55652352 111312896 222642176 445317120 890699776 1781530624 3563323392 7127171072 14255390720
28512878592 57029951488 114068291584 228153360384 456340275200 912747659264 1825629536256
3651527507968 7303591886848 14608257515520 29218662514688 58441619996672 116891829927936
233800839725056 467636039188480 935340797853696 1870819034660864 3741912947228672
7484375650271232 14969850812170240 29941900647596032 59888199341703168 119785194776428544

239587981738901504 479211147849891840 958492664443961344 1917126066376278016
3834533607729266688 7669630165411954688 15340386230730752000 30683024261275189248
61370552122177748992 122750111443610238976 245518237285729959936 491072503368478883840
982217064330995695616 1964578243850067247104 3929444718076286205952 7859465896904875835392
15720084715314358517760 31442475273637930729472 62889562233294288846848
125788347838625432469504 251595142421324574490624 503227178330796568084480
1006528143637887974375424 2013203861228365625163776 4026702870361910603153408
8053996036534179911958528 16109172664689077235220480 32220706512619589293047808
64446135391722048231309312 128901715516409835753046016 257822320498751150086946816
515682419929365257335603200 1031440397722456428994625536 2063031911172364686636089344
4126366053799633030565855232 8253336570509073375719063552 16507882066837761380612833280
33018181985314752019575078912 66041199673907962555848982528 132092070754372842145095614464
264203484321859518356986527744 528445654269946704847563653120 1056968679792348745962308501504
2114092102089608164458979393536 4228493689189037673986683568128
8457606348397718038110816698368 16916450636834721456496532520960
33835377153748013673542863290368 67675706067653168868185323077632
135361315655620620778569839149056 270742438351869807641538064285696
541524490784996747451872900546560 1083128209732507759241339345043456
2166414875790044047157865777987584 4333146664230145151666105731776512
8666927153760404418032959815155712 17335121958121037065467416333516800
34672779217442530589737826073444352 69350629037285974097081638959710208
138711399279373774029375251545063424 277443080968351199729174450341412864
554926726755909702799196795185397760 1109934583150234012280089379375939584
2220031425577297237923570336762167296 4440387369708252902573923829544910848
8881423776523822658601413971130974208 17764145627262279024109960566344253440
35530887402953825462034186380853116928 71066967102766185751696903258035453952
142144318799249441158650867508729348096 284309406785933021627815857002775576576
568660351946734321876659957976184913920 1137403780643205200995376403893637349376
568743428696470879118716445917452435456 284392483535669578869872344944043098112
142206626361551859090193233464679989248 71108505477634464372725147228669214720
35556848887246499600176838862499217408 17779722517837883506995552055331913728
8890510296026258606951342339707109376 4445579666566787730202454325874130944
2222952092560223078464618740947353600 1111557175918526145914005159478820864
555819152778470376297850474241982464 277929858798838839819349184372277248
138975070604221245744886565811781632 69492605904511535790049269718712320
34748838253456224353827628265766912 17375686777328340406315310836088832
8688477213964284317858403769647104 43444555519632199216279576060624896
2172436216141128136814975118213120 1086297336233078332745081103056896
543188282197796298541337323503616 271613948139526715355067047739392
135816877590077640719732716863488 67913390555195961880965954928640
33959171157676551701032775712768 16980823518877561230791286980608
8491030729458423305533093052416 4245824849739032997835271307264
2123067167374427171451998044160 1061610954939668921993180217344 530844163096062129130180706304
265441424361144898631885651968 132730383587129366349340475392 66370027496843141691369062400
33187431600060800104033943552 16594924725850014681191677952 8298066825834814655183192064
4149335644372310984885272576 2074818937913607321089474560 1037485026820529574868156416
518780292342127744595787776 259409035636995350878748672 129713962551463414729801728
64861703642214577010114560 32433213004348723327664128 16217787093795079075135488
8109483842707898243219456 4055037069259128474435584 2027666108582153913630720
1013906841267371795021824 506990314121833316614144 253513603804990367858688
126766025274532038705152 63387624323284446740480 31696118004651437064192
15849211923830325379072 7925182422667466113024 3962879441709884768256 1981583836043018240000

990863975615547047936 495468016604792487936 247752022700905725952 123885018549707603968
61947012874481172480 30975758237054271488 15489005018433978368 7745065459170410496
3872814204561915904 1936547839769313280 968344288628834304 484207328686505984
242121256529297408 121069424357670912 60539110225346560 30271754135928832 15136976579592192
7569038045609984 3784793900711936 1892534389309440 946335914131456 473202316804096
236618338271232 118317759070208 59163174502400 29583734734848 14792941109248 7397007425536
3698772148224 1849520291840 924827254784 462447181824 231240368128 115628572672 57818480640
28911337472 14456717312 7228882944 3614703616 1807482880 903806976 451936256 225984512 113000448
56504320 28254208 14128128 7064576 3532544 1766400 883264 441664 220848 110432 55220 27612 13807
6904 13809 27620 55244 110496 221008 442048 884160 1768448 3537152 7074816 14150656 28303360
56610816 113229824 226476032 452984832 906035200 1812201472 3624665088 7249854464 14500757504
29003612160 58011418624 116031225856 232079228928 464192012288 928451133440 1857036484608
3714341404672 7429219680256 14859513102336 29721173688320 59446642343936 118901874622464
237820929114112 475676217966592 951421155409920 1902979749773312 3806234377453568
7613018510721024 15227136533069824 30456472089395200 60917342225301504 121843480543625216
243704553273294848 487444290918678528 974958950581534720 1950058638651424768
3900398752279560192 7801360454512541696 15603846808931926016 31209945417677537280
62424394434982445056 124857796069219631104 249733606536948744192 499503241870916452352
999078541335870832640 1998301197859817521152 3996890626095786754048 7994357712943876931584
15989868347392360710144 31982042537793935114240 63968696761606297616384
127946616895249450008576 255911680534572609568768 511860254557292638240768
1023794296090880114688000 2047736166134349905788928 4095767480173879164403712
8192125256158117034459136 16385431103936951480221696 32773223391115337783050240
65551169148713545211314176 131111783030392829713055744 262242455526717138006966272
524522689985297233175642112 1049120937834320380674703360 2098392991396092589996244992
4197088214247088837286166528 8394780891403984989159686144 167907770708627584607494078464
33583959268894398473337569280 67172754241067255463373963264 134355179888691427960145575936
268729702590496689987086450688 537498090807221048107763499008 1075073552866897432482708193280
2150301848238705537499778777088 4300913181487232420068282335232
8602445332994107530274014232576 17206128606027500440822927589376
34414733092133571642195653427200 68834417944424284805490903351296
137678739409162852653180999696384 275377285858954271390760385380352
550794185799165674950317542735872 1101667599760845614238228629422080
2203493655846719757151644346744832 4407304224343496571653662869291008
8815242273987107258008074090184704 17631752198574442745417644883574784
35266039698349341949638283173560320 70537149999099596816882553159942144
141084441203001019468977079945527296 282189164815605690608378107142340608
564418894450418684557604108787253248 1128918918539251975796904006579650560
2258000096355333164957199591169589248 4516324711264324756641182338359754752
9033298459635966366735930988760662016 18067894993486566440378994601603629056
36138386135402400294572254451371868160 72281964567663335416773039399072956416
144574313729043740488803139790804353024 289169396645521620288120401566925586432
578380331665911519197269047104484933632 1156843740081559595636594582150237388800
2313853633662592305757302140183009820672 1157009893581032710120707558032772431872
578546485165384633681382022987019976704 289294011770126456151205133478826868736
144657390478780297730859627732071874560 72333887536248683693058344362365157376
36169539916553609260343437429347188736 18086068032491438337078851338755899392
9043683053353036021992991981419102208 4522166045230176437723279146730127360
2261245281891917432225031151375351808 1130703770584373322794211364692819968
565392450111393964737953576848982016 282716507465300634039400735675777024
141368394937452142854912341463531520 70689268071126984345062157544587264
35347169336763948631334072178704384 17674852318982202545068532792557568

8838059984791215387235014747881472 4419346905045664750967881549742080
2209831908847860904159127862771712 1104995182586444716417157475336192
552537205374479490377375509643264 276288409727868311273086140809216
138154108384248438678742263398400 69082005952281360860470728196096
34543478856219251190785162346496 17272977368148910975667480297472
8637107654094098177971189710848 4318863312056870434054319636480
2159586398533345889561522208768 1079870570519128281047942299648 539973970885791808657561747456
270006328256009738395576172544 135012835534561786231185735680 67511253470559351632291692544
33758044586918905074495258624 16880231219279067166422335488 8440720072549340897798520832
4220662267729574106192936960 2110482249592238881743306752 1055316682659845355195072512
527696120261785634759245824 263866949596824295960477696 131942919531377887270666240
65976182132171813280546816 32990452249327341462880256 16496406716284388142743552
8248793653952552777023488 4124691974881455741337600 2062493561393317547081728
1031320567672953611747328 515697177324624224976896 257867035406385822040064
128942741075229765795840 64475982223633310285824 32240296954825868836864
16121301398917541265408 8061227160211074056192 4030901810481688739840 2015595020428920225792
1007869567808498040832 503970812701267984384 252003420749143474176 126010717573826478080
63009862386540609536 31507182993083990016 15754717396448837632 7877921648177840128
3939242299065630720 1969761887021170688 984951312254763008 492510840499470336
246273012435779584 123145302310912000 61577049201967104 30790723624239104 15396461323747328
7698780417687552 3849665086750720 1924969982328832 962553710641152 481311215058944
240672787398656 120344983633920 60176786784256 30090540875776 15046344179712 7523708960768
3762122915840 1881195675648 940664946688 470366027776 235199791104 117608284160 58808336384
29406265344 14704181248 7352614912 3676569600 1838415872 919273472 459669504 229851136 114933760
57470976 28737536 14369792 7185408 3592960 1796608 898368 449216 224624 112320 56164 28084 14043
7022 14045 28092 56188 112384 224784 449600 899264 1798656 3597568 7195648 14392320 28786688
57577472 115163136 230342656 460718080 921501696 1843134464 3686531072 7373586432 14748221440
29498540032 59001274368 118010937344 236038651904 472110858240 944288825344 1888711868416
3777692172288 7555921215488 15112916172800 30227979829248 60460254625792 120929099186176
241875378241536 483785116221440 967638951919616 1935415342792704 3871105563492352
7742760882798592 15486621277224960 30975441577705472 61955281201922048 123919358496866304
247856309179777024 495747802731642880 991565974207463424 1983272685903282176
3966826846783275008 7934216643519971328 15869559186946785280 31741370173707255808
63487243947041882112 126983495093338505216 253985004585186492416 508006037967391948800
1016084133528821825536 2032312382245719506944 4064912994867590725632 8130402450487484874752
16261957822479576596480 32526221487968366886912 65057054661955161161728
130123332695947177099264 260265112135968063750144 520567117760083546603520
1041208022496461931413504 2082563618945513539239936 4165422385796206431305728
8331435067402771568263168 16664050726426260547829760 33330462636093955918266368
66665647638670781481746432 133340740010307302253920256 266700369486546083088695296
533438517904955123339100160 1066952593673636161001619456 2134056303074724150650077184
4268414837604351958593830912 8537434138118511231775014912 17076077202056637092724736000
34154572255752503443798884352 68313980214783465404296593408 136637631836123847841990836224
273294606485361529750776971264 546627898596950727635144540160 1093333168446356791537470275584
2186821079397624255609302941696 4373951643805069856287330664448
8748522257629782402712110891008 17498282455298850185699120906240
34999040790676271131948040060928 70003033341509683784995676618752
140015970203333650612190546231296 280051747447295867308779478450176
560143108975848866786355728875520 1120365446114211997910305001701376
2240889348553452524495797091303424 4482095609756962106341968358408192
8964825044814038327384685068419072 17930917740228304884170866840043520
35864370781657066227144727086497792 71733812165715045371895440985817088

14347776553623191657900285559727718 4 28697581348206748482842965844584038 4
57399219178334227299770721139425280 0 114806551320509915267711021179364966 4
22962932856870275187176120015975874 56 45929110899277134641620071592157511 68
91864712169627437817775806304726548 48 18374240508140121270462293885027614 720
36751077164709509954738853018219839 488 73507346626277554737106236532768899 072
14702507784627217912946953405819623 8336 29407092487997849756945319010170935 7056
58818338813482527375993462417405247 4880 11764498530193871047619257362893724 71296
23530658595382473240079644484612799 85664 47064640260754408769841548486876300 57472
23533981665371935529761904002263500 71808 11767821600183333373015168805444255 7440
58843261838403494548610408799785503 9488 29423707837945161205356616598424440 0128
14712892378344287568204014398478064 0256 73569654188579972668648602488719540 224
36787423242719253748138566492524380 160 18395009695574260580976415870344495 104
91981538848944471439417742472134000 64 45994014610008819986976702796272762 88
22998629897772702127122267178239262 72 11500126245270497130378091479171072 00
57504687708273215985975246846112563 2 28754372095096973160030018148184883 2
14378200168028669163536206436656742 4 71896071442544258735287018996105216
35950571022472585826446502904463360 17976553161836521142624748155437056
89889104062183746860131224293212 16 44947721157592444003569353904619 52
22475445142046507288536547831316 48 11238514852648396287644209355161 60
56196535671367694655100723973324 8 28100248539746703935990200585420 8
14051114621904780272215019592089 6 70260524869681042882174694457344
35132738314919092201637145477120 17567607097498831481093471862784
8784422518769058430684185493504 4392520744394350560410817527808
2196415114702085952739771154432 1098284928603498312637066772480 5491811499279768244521239838 72
2746099177771022462928572907 52 1373146302951080401798262947 84 6866215085083247860661197209 6
3433349327705546856165539840 0 1716795556434734891000240537 6 858458224508348176958855577 6
4292593353996644542087954432 2146447792725774099690815488 1073299454226612964168826880
536687506045169439246123008 268362642488516198203916288 134190765977223838392385536
671001053550947888414064 64 335524138607888292433100 80 167773875220151320329584 64
83892840568179247221309 44 41949371763141417138913 28 20976161621096605333585 92
10488818680311251048857 60 52447782750370997154611 2 26225736049592869532467 2
1311137903620001202438144 6557356349601902860697 6 3278908759101872799744 0
16395696717013970845696 8198424819259288846336 4099500640005796134912 2049894435190973923328
1025019275189524889600 512545666391781408768 256290847594400186368 128154430996454834176
64081719097854787584 32043111348741079040 16022681574277382144 8011903737092112384
4006233343522766848 2003257409249738752 1001699073369047040 500884721056612352
250459952714350592 125238772450197504 62623784271609856 31314091159060480 15658145091158016
7829622301392896 3915086028603392 1957680453255168 978908946104320 489488832790528
244761596264448 122389388066816 61198989000704 30601641984000 15301894733824 7651484237824
3826010554368 1913139494912 956636856320 478351982592 239192768512 119604772864 59806580736
29905387520 14953742336 7477395456 3738959872 1869611008 934871040 467468288 233750528 116883456
58445824 29224960 14613504 7307264 3653888 1827072 913600 456832 228432 114224 57116 28560 14281
7141 14283 28568 57140 114288 228592 457216 914496 1829120 3658496 7317504 14636032 29274112
58552320 117112832 234242048 468516864 937099264 1874329600 3748921344 7498366976 14997782528
29997662208 59999518720 120007426048 240031629312 480096813056 960260734976 1920655687680
3841579810816 7683696492544 15368466726912 30739080937472 61482456842240 122973503619072
245964187107328 491962733953024 983994187382784 1968125813719040 3936526505345024
7873602766503936 15748305044635648 31498809112526848 63002016271564800 126012828636151808
252043249458348032 504121683288784896 1008313735321747456 2016768208131850240
4033817891240411136 8068198732434243584 16137523364775329792 32277298529364344832
64559100658356060160 129127208515966861312 258272431430443204608 516580891657905373184
1033233840909848674304 2066611797007773204480 4133511824391698120704 8267600109535699664896

16536353140576006176768 33075012124161226047488 66154635934340879482880
132318495240718613741568 264655437225510937034752 529347767939169293172736
1058769322854633424551936 2117686219661856525516800 4235667587228892403859456
8471925470268143513370624 16945031532157004438044672 33892424247555443698696192
67789570861593757042606080 135558858645615325337563955 2 271196062378237985332133888
542429903688338927825977344 1084935365240403769975373824 2170021846208259368597585920
4340345923871422394488848384 8681296310652652103565049856 17363801547124918836304805888
34730020945889066930959024128 69464877595056592378616872960 138939426596670101790631395328
277898196006454037648058089472 555835077639135743429706776576 1111747526530726823126594748416
2223649795566364318787551887360 4447609076142549982643828555776
8895837122304742655425106673664 17792912184648770691125112471552
35588300249376112142800023191552 71181552258909365806699642880000
142373008038133014655598478753792 284765823116894595395595343495168
569571260315046322959987458965504 1139221748792606910257568461881344
2278601953910242349190324011663360 4557520820470541755731022199128064
9115675466241197626162792749858816 18232618583082623481727082202923008
36467772467365703422257157812256768 72940615537132319762120302437335040
145891372279066465359452578500313088 291803026967736582389329104251912192
583646618754680468119506103006396416 1167374367147775542920707995017936896
2334910993572380299204807568046161920 4670146505698419025136398292112900096
9340942048504154903726362896266952704 18683182171222943514359858416616210432
37368960490875154442533982081397030912 74743113278608843712696494659123281920
149496611150934757080650050310905004032 299013991489303653471814222607126888448
598069521353475585564656689184887537664 1196222119456687728371369866311042596864
2392610392412848571226852708504620236800 4785553091824643371421931368774310559744
9571770797647179200780314641078761291776 4786217705822535829358383272304450732032
2393275006410741029163304612034760409088 1196720579955107071823708793958647726080
5984018283524218145328826409499576238O8 2992216833636450465769554424602956922 88
1496212262755395929437347822280628633 6 748158054346283312994959216077323632 64
374104988657434330635622260520307916 80 187065475070863502386882456500977008 64
93539227906504919727976891370900029 44 4677285913878904413125627724565577728
233880521621628141992620544029307699 2 116948373774655531664479850915168256 0
584782433692484961663247149078413312 292411499255846132502047521790492672
146215890832724892086235734520889344 731130160187633589607238540732661 76
36559043310582135939164920443043840 182807893058912961989839569247272 96
9141028478245762214192726813966336 457083115177293816444673758278451 2
2285574032211497610898555879292928 114286624426826306978687148359680 0
571472736215388667062232513773568 285756175148322899615514642874368
142887991094475732849956514430976 714489473073950079460778537123 84
357269495337760747335887251046 40 178647127069273227470692616765 44
89329753234833040636720804003 84 44667971467514733769047649812 48
22335533158806473609867448811 52 11168540291927790167605536358 40 5584657002226171765138674155 52
27925219292442242232372900659 2 13963576786876812819526215270 4 6982271963766252261432990105 6
34913777670470490565514362880 17458097761054859911931887616 87296533434372372705532968 96
43651289031735222925703249 92 21827155673142129749320007 68 10914333415208324017894195 20
54575444969227915805641932 8 27289611431207105760906444 8 13645750188900126809495961 6
68233473310983503692693504 3411909783873318666895360 0 1706072951098731074578022 4
85309550513040140785418 24 4265772673557186392096768 2133033910731182872461312
10665907423418862744371 20 5333322646590905563217 92 2666845790736189877125 12
1333515129088463486320 64 666803681404416017039 36 333424899132300145459 20
1667239787811961411993 6 833677539981211048345 6 416867593028220695347 2 208448208032917933260 8
104231309775862759424 0 52119257767633276108 8 260614303236675862528 130316158817592672256

65162583008423706624 32583543304025538560 16292897551919611904 8147011725913227264
4073787337933324288 2037034406455017472 1018587571971686400 509328970357932032
254682077365010432 127349834775527424 63679315434274816 31841856740392960 15922027881824256
7961563696726016 3981056726269952 1990665802088448 995401620520960 497735169998848
248884764868608 124450972368896 62229781151744 31117038059520 15559592771584 7780333256704
3890435063808 1945351749632 972742983680 486405046272 243219300352 121618038784 60813213696
30408704000 15205400576 7603224576 3801874432 1901068288 950599680 475332608 237682688 118849536
59428864 29716480 14859264 7430144 3715328 1857792 928960 464512 232272 116144 58076 29040 14521
7261 14523 29048 58100 116208 232432 464896 929856 1859840 3719936 7440384 14881792 29765632
59535360 119078912 238174208 476381184 952827904 1905786880 3811835904 7624196096 15249440768
30500978688 61006151680 122020691968 244058161152 488149876736 976366862336 1952867942400
3906004320256 7812545511424 15626164764672 31254477012992 62513248993280 125035087921152
250087355711488 500209071161344 1000486861799424 2001111162552320 4002497203011584
8005544161837056 16012187835301888 32026574693859328 64057547434229760 128123890961481728
256265374109007872 512565932590104576 1025202233924386816 2050545205337128960
4101371885650968576 8203306721255358464 16407739342417559552 32817730484648804352
65639964568924979200 131288936337104699392 262595887072718880768 525227802942456725504
1050527663478951378944 2101199442145978613760 4202687114668108939264 8405950690088521302016
16813054301681649451008 33628414446372512595968 67261440578763452579840
134532104529563759935488 269082655803201229422592 538202205094549877948416
1076478197165394594103296 2153103968283378864619520 4306503084471937082064896
8613596464754232869781504 17228373521129183150866432 34459108225499801124339712
68922938817482471893893120 137855322367930683078213632 275729534201792844737282048
551496847335448646636273664 1103069252534623207595966464 2206289620796698243838771200
4412881473048300144971218944 8826367409006407604529790976 17653943743832429838234288128
35310305339304088934817988608 70625446381886636386334801920 141260564170330189806067253248
282540471153774213678929805312 565119627933776095491450208256 1130316627120007527250081611776
2260787996744925727034525614080 4521885478499672799137776009216
9044389927018988288413001580544 18090017794077261957100902285312
36182511468233094674751602819072 72369974696623330870602802135040
144749852913560944783404797263872 289519512867750455651207980515328
579078639816758043471212733005824 1158236507796030351280019009961984
2316631471917089231235225107824640 4633579856484235519820824391450624
9267793538268585154342397134503936 18536854727137398538086290972213248
37076244755475253534975575350837248 74157560113351419987557137514496000
148325261431504665810326248654635008 296670805272612983291076444560556032
593382175364433269923000783623684096 1186845480367281146527697356252512256
2373853220011391506418786290515312640 4748030958576441439564355737051201536
9496710954260199732582277786143555584 18994719982735033172071688196369416192
37992036113899333757957641640903442432 75989264524657202343543813778136104960
151988913643031474342344688548930650112 303998596473497087995203499083178180608
608038731321862454611435242136990121984 1216160539393461466464926972215247765504
2432487232286396047413966920313030574080 4865306771571738323796159792391131234304
9731278157141369105528771488312402640896 19463885542278523126930446783685085626368
9732607385137154021401675295372682985472 4866635999567523239669063599451411578880
2433484153283234734318644775608240832512 1216825153391353924401378875745387937792
608454115070545240821717681843327729664 304247826722706759721372962906980745216
152134297955070449515943542446148812800 76072341274393759585600301719403626496
38038766785626147206614416107866423296 19020681467027707310214340678015516672
9510989770621170508560736651048910848 4755819403864243681007151481545031680
2378071961208951053866967318782803968 1189117110242890133615179448396546048
594599119940652370148437618700845056 297319842379929836744642756601708544

148670062394766744207533351926497280 743401017997842850213726627760701 44
371725862010925989694893247944458 24 185875607511465277141461591004282 88
929441420087337797177382790181683 2 464752401308674604323728812670976 0
232392046286840155029383115125555 2 116203945959671503948450911957811 2
581059343879614651911051331764224 290549478980435892039924051869696
145284643010532229062161218928640 726472732654232560521802059612 16
363261125127901987866399012290 56 181642941964343847735948497387 52
908276606823683507693487443148 8 454169251912823888353616199680 0
227100100206903011430244338892 8 113557787228697039341840288972 8 567827621769712864842792042496
28393315369797026648819132006 4 141976248255542050277493309440 709929598310494836554454794 24
354988977671639710860721520 64 177506578094016001722107822 08 887593336761060740069274419 2
443826891526020735764004864 0 221928557335755550746686259 2 110971834454250366805685043 2
554896951203114791190134784 277467365067488874175922176 138743127266710176378388480
693762859998379578344079 36 346905041831604137398108 16 173464326832009242812088 32
867380663741082084625612 8 433719846661058977595392 0 216874680725788456438988 8
108444719060523712040140 8 542260488790765979303936 271148691139456699203584
135583568941765204377600 677963961569010295767 04 339005039214597284823 04
169514048822344710881 28 847627890186953896755 2 423842768131092119552 0 211935795584353645363 2
105975103551580615475 2 529911546554922041344 264973787675970502656 132495901037239992320
662524541182473666 56 331284788589373685 76 165653653293755269 12 8283245614641184768
414190428229730304 0 207109287863700684 8 103561680806268108 8 517843588403429376
258939386387759104 129478489286901760 647436426899619 84 323740203682365 44 161881096957460 48
809460460368691 2 404757717975040 0 202392602882867 2 101203173389107 2 506050226683904
253042293211136 126529736540160 632691632373 76 316367291023 36 158194382929 92 791025601740 8
395539644416 0 197783243980 8 988983328768 494525218816 247279386624 123648081920 61828235264
30916214784 15459155968 7730102272 3865313280 1932787712 966459392 483262464 241647616 120832000
60420096 30212096 15107072 7554048 3777280 1888768 944448 472256 236144 118080 59044 29524 14763
7382 14765 29532 59068 118144 236304 472640 945344 1890816 3781888 7564288 15129600 30261248
60526592 121061376 242139136 484311040 968687616 1937506304 3875274752 7751073792 15503196160
31008489472 62021173248 124050735104 248118247424 496270049280 992607207424 1985348632576
3970965700608 7942468272128 15886010286080 31774168055808 63552631078912 127113852092416
254244884054016 508524127846400 1017116975169536 2034371389292544 4069017656492032
8138585068797952 16278269649223680 32558738321702912 65121874689916928 130252545472856064
260522683131756544 521080550635601920 1042231470015381504 2084603677519118336
4169488830014947328 8339540609983315968 16680207119873474560 33362666039560634368
66729835678748639232 133468678556752019456 266955371512013520896 533946771821046005760
1067965601236129939456 2136075317660335734784 4272438865696823181312 8545454192145949786112
17092061305796506419200 34186428454602226532352 68377468595222880452608
136764160562482615681024 273546767869038940913664 547130429226225300930560
1094334645428745440067584 2188816864810080556548096 4377928877525340465922048
8756448050861039637495808 17514076693342796686295040 35030514569927028195196928
70065751506336926035607552 140140947745639591361642496 280300784957210661304139776
560639348846284279769989120 1121354255556294473863397376 2242859626840040776373633024
4486021485134985210040942592 8972647433179777734669238272 17946503792179170098513182720
35895425435997569455375777792 71795686575273597427450380288 143601044557104111888298409984
287221431927322057843392118784 574481549480871783820374835200 1149040470214198903907930865664
2298235682933308480350224121856 4596780850876438305769173024768
9194180671772519301675795611648 18389599283584323983626490347520
36781674447247218727802778943488 73568300654651578976705154383872
147146504829617440995609501761536 294312816699863448075617389510656
588665247480984028320031550996480 1177409723124482320977656645943296
2354977902573993170630500379787264 4710272717798043398611374935375872

9421179260896200911923498222354432 18843626172392630053248493147914240
37689787645985716565299979702239232 75384645894372346048205946217299968
150779432993546517931623866060242944 301579148396696687533671679371771904
603198861612600678408191253246115840 1206478852863615963498078295497375744
2413119965004061140359548169005039616 4826564448561780707445879494030655488
9653777934230878268345325300102463488 19308853942676390243597783224287232000
38620304033782047901009831696739074048 77245800364422630629648193889807368192
154501985322562330914553448772273176576 309024739832558801139621019529863233536
618091018039985880900270283030360227840 1236265112829708319042597054001987977216
2472696379158889752569307083886510997504 4945725065316725734106840119538092081152
9892114744631343926150132142606324334592 19785558717258472768173168092272929013760
39573775890508515368092143798666418716672 19788217173250042599918975706393489702912
9894773200622913757895939756726885023744 4947718907310403107916195830128512598016
2474025607154674668442210890946791342080 1237095880327073891463161933414663192576
618589478538405224352609210677965357056 309315508456636751486818727324299558912
154668138822035445398666424654808219648 77339261707876257526961742823733329920
38672227002367396177295136660031275008 19337411575398331795554700954097942528
9669354824806482751230916789090123776 4835001930956899802342241550565638144
2417663224755279114534512353293107200 1208912742016054163948951965651697664
604496935827234385315323877328420864 302268750323220844328085885915496448
151144516366412247999254916583391232 75577328785607036917233445104517120
37791199694003974917419715958669312 18896867497602215688111354682540032
9449067574101221958756425692872704 4724850699700668036728587022237696
2362583806175362547039480599019520 1181371131250195537857333843460096
590725179706354901097463693705216 295382396893806016633130232840192
147701101967217291358764309413888 73855502743765787200481751203840
36930227251961464360790673850368 18466351566020017560670236049408
9233794753029651470472567586816 4617206861524647080305008574464
2308758173267234212686866677760 1154456457886072442610614534144 577266914569263889438897864704
288652800097745778786244231168 144336071455429806426519764992 72172871430993361729958707200
36088853567135910123328765952 18045635709387569690839089152 9023422317603592160006897664
4512013390256699737297125376 2256157810855801697295400960 1128154463291626762971119616
564115010577676338647269376 282076394754769647904489472 141047642110350563242672128
70528543421658151266549760 35266632894070510455881728 17634497038655972639244288
8817838815138345025273856 4409214555474351865462784 2204754851689765609144320
1102451212821177642778624 551262499898736240492544 275649696693441829797888
137834071718757769674752 68921647545397312225280 34463129615707869806592
17232717729358541750272 8616935325431574298624 4308755893091938861056 2154522061734045286400
1077333088461060571136 538702573027549249536 269369300912284106752 134693657655396794368
67351332427325767680 33677918013476569088 16840084906645127168 8420605403275984896
4210584176614703104 2105432825795706880 1052786781642031104 526428575193104384
263231879782596608 131624735984320512 65816766038671360 32910582042591232 16456390532923392
8228745022275584 4114647389044736 2057461133475840 1028799286214656 514434002845696
257234181292032 128625680580608 64317135257600 32160715112448 16081431298048 8041252519936
4020894695424 2010581565440 1005357891584 502712500224 251373027328 125694902272 62851645440
31427919872 15715008512 7858028544 3929276416 1964769280 982450176 491257856 245645312 122830848
61419520 30711808 15356928 7678976 3839744 1920000 960064 480064 240048 120032 60020 30012 15007
7504 15009 30020 60044 120096 240208 480448 960960 1922048 3844352 7689216 15379456 30760960
61526016 123060224 246136832 492306432 984678400 1969487872 3939237888 7879000064 15759048704
31520194560 63044583424 126097555456 252211888128 504457330688 1008981770240 2018097758208
4036463951872 8073464774656 16148003291136 32298154065920 64600603099136 129209796132864
258436772134912 516907904008192 1033884527493120 2067906493939712 4136087865786368

8272725487386624 16546550486401024 33095299996057600 66194998038626304 132398792170274816
264815176526594048 529665537425276928 1059401443594731520 2118943624677818368
4238168724332347392 8476900398618116096 16954926697143074816 33912105194099834880
67828713987827040256 135666435174908821504 271350884748327124992 542737798293673213952
1085547654181384355840 2171239423550844567552 4342767077477840846848 8686110615707985117184
17373374152920577081344 34749054148850367856640 69502719983719163101184
139014663339475180978176 278047773423024071507968 556132440334195562119168
1112338667644685962444800 2224824909241961601302528 4449944966389102555430912
8900480228588563816513536 17802141048797845044330496 35606643280837124911267840
71218008928157119467749376 142445462589279978225926144 284909814644491435032707072
569857408220845827227123712 1139790374305417568777666560 2279731864338286966202171392
4559765960131477589698019328 9120136383172762493983391744 18241481692165139617141489664
36485381235969508492632391680 72975598175217475501963608064 145960867756991868037324865536
291941078327097570141445029888 583920842280422808416480657408 1167919055813300953100142510080
2335992854131512578734647410688 4672295193272846502538019602432
9345209356565335695213488766976 18691656653169956770701876658176
37385789186418484301953551564800 74776530132994110125006699626496
149562963786302503292212592246784 299145734613233572668823570481152
598331083307724277506443912937472 1196741394777962819350481369825280
2393641245880954167376149827551232 4787599404411965392102673830903808
9575832634124044898906096013410304 19152932918848318027213688730025984
38308401138897092513230370866462720 76621872880195097944066728545746944
153253886965192021723345430717136896 306528056339987695117114808685559808
613096677499182693575077511873691648 1226274484636779993831850812752527360
2452711228550389201027093203515342848 4905746975654436828780969563051261952
9812142988416190511015505438143676416 19625584051047014728938143500369657856
39253764250523296871690552248903925760 78512720797905128571009634994137071616
157035826189527326797276330980932583424 314092421566488792905066783947182047232
628226381507845864431161811864997855232 1256535839765428286104380111671263232000
2513237833030329686692873199225061507072 5026807973059605602353972350215193100288
10054280560117103662644396603960526372864 20109890348229992241161697014981333090304
40222439152451554314069201644083226869760 80450195216886248291630018516407575117824
40227756064434693977560816872324348248064 20115207260213131904653312243222454468608
100582682441044584102631080251413674065920 50294664290511754340997799643357537894400
251489936802506083153400295805041193779200 125753276076226697300905796696647349043200
628807918756001765125557227453870505984 314424728565435021873292735712252133376
1572227488764345805919034288487845068800 786165667350758251235802449207214735360
393108795159671799756043877085253468160 196567378321982236947093264783449784320
982901795320642870080822955121364172800 491483349515687277713089793162739712000
2457579006855265601928840543823986688 1228870633066047407646116060917137408
614475881352231007163905924961140736 307258223085719155252376909731856384
153639252747661403461400428491571200 76824696976231614648306201058607104
38414883789316263782956093935714304 19208709545258360120879543671062528
9604988597929294175140520187133952 4802811211614704144920634269368320
2401564062132380601135504222584832 1200861259228704564905345655242752
600470243695609414621469599596544 300254928888433273395133185785856
150137367964530919739765785886720 75073635742422601390982489440256
37539293751289871456041042968576 18770884815684221108295420608512
9386061377861753244285159866368 4693340173940697967211304714240
2346824829475259656140014747648 1173489785990085164337188569088 586783578621270250302184882176
293411132123748959217887739904 146715237468431396642341519360 73362454437494156837869584384
36683645070386307677284204544 18343031461012768467816808448 9172120193416191548495757312

4586362328162999431541555200 2293332279808951544417615872 1146741697768201686532227072
573408627815963800427823104 286723203373913378794766336 143371046419922428687810560
71690245576444083989118976 35847483971463476817166336 17924922577352455819886592
8963051584486586615595008 4481820940148472660623360 2241058044026826006724608
1120602808989707841568768 560338297983001339887616 280187595735574379495424
140103021239824044523520 70056122305930449649664 35030366995974438518784
17516336419491826106368 8758744670498216476672 4379660565625259950080 2189974398000705830912
1095059256594390843392 547565657094214385664 273800842945616674816 136909428672063078400
68459217935658909696 34231860767643140096 17117056283728412672 8559091091817627648
4279827020885524480 2140054247931117568 1070097492709736448 535083930726957056
267559557549522944 133788574867783680 66898685480402944 33451541763457024 16726870393356288
8363984952492032 4182267354152960 2091271116029952 1045704277491712 522886498484224
261460429111296 130738804490240 65373697212416 32688996089856 16345571786752 8173322764288
4086929817600 2043599126528 1021866672128 510966890496 255500222464 127758499840 63883444224
31943819264 15972958208 7987003392 3993763840 1997012992 998572032 499318784 249675776 124846080
62427136 31215616 15608832 7804928 3902720 1951488 975808 487936 243984 122000 61004 30504 15253
7627 15255 30512 61028 122064 244144 488320 976704 1953536 3907328 7815168 15631360 31264768
62533632 125075456 250167296 500367360 1000800256 2001731584 4003725312 8007974912 16016998400
32036093952 64076382208 128161153024 256339083264 512711720960 1025490550784 2051115319296
4102499074048 8205535019008 16412143779840 32826435043328 65657165053952 131322920042496
262663019954176 525360399646720 1050789518770176 2101716476493824 4203707830894592
8407965417603072 16817030346833920 33636259716923392 67276917480357888 134562631053737984
269142854293520384 538320892959129600 1076712154662436864 2153565046813229056
4307411568603168768 8615386087159758848 17231898074226360320 34466047948266405888
68936599496160182272 137882206191575105536 275782426781659693056 551600882360338350080
1103273822314714628096 2206691759817505112064 4413671750011161935872 8827919960774627295232
17656992843053861437440 35316291529116936568832 70637194744252300525568
141283612860541455826944 282585672465156621205504 565208238418460661514240
1130490263813216161234944 2261128101579021998882816 4522551351063223350591488
9045692997936805406834688 18092566587494328224972800 36187494358230091272552448
72379711082943052190318592 144768866898851843671064576 289556623263635165922983936
579151025459133289007677440 1158377608781992492338774016 2316906333291436813324386304
4634114898037777283942449152 9268834258985361882472251392 18538877443790338394119208960
37080172739219906046587830272 74165181181718270609874485248 148340033769993458253146619904
296699410353100750573088538624 593437506332429169279767674880 1186952383917313674826716545024
2374059510339538022187795480576 4748428505688897389444315742208
9497475981397437469026081046528 18996189902834160318327061217280
37994855685746891397203920683008 75994663131650924315507437862912
151999229783616131673214068719616 304018266607860829430826523426816
608076147296978791030449818828800 1216231522756471846398493181607936
2432621501837972221472173451116544 4865559916326001500294721078034432
9731753657952117115290190507671552 19464774966504462459981877718548480
38932085234209381378766748843507712 77869241070819675675139484499836928
155748623346441177185490942625316864 311517529102486006041405832501919744
623075623024179315423659559506411520 1246232375686773237529014908017967104
2492627010650375688421421394046222336 4985578539854409803569625944113020928
9971806116816136460592818200267194368 19944910307846906628092769024616693760
39892416764123080669999803297397997568 79790025825104696167628137091125215232
159590436243926461990513335174908870656 319201641675287063291540792335134621696
638444821725442405204109828640903004160 1276972720200621367650276145223073529856
2554111593900715849784665266328682102784 5108555494800377928537556484422434291712
10217775603598648315011564872375008755712 20436880435193081545896033551810297856000

40876419326377732923537874717741156401152 81758155564738605510567364663723434180608
163526944953443490348117959783929111117824 81768789388704884837550595120205676937216
40887053150344012250521105174223399157760 20444855803167791041133456394171979923456
10223092515581787978503180100616130134016 5111878564789840218219816002073135153152
2556105435894393223594020976919102619648 1278135794696933169039066976400818831360
639109435723334863140561732171043176448 319575487049101570880794988070838468608
159798128118267855095654555028077674496 79904256355992462375455808010368057344
39954724326425498601542169253348638720 19978660237427383007678217250756624384
9989979155821008357292674937419464704 4995314096464162605373120624730308608
2497819307508910516049951890375442432 1248990783392869864706671734192865280
624535956515642235694183761599004672 312288260667424769517515828050788352
156154271538514210593969887651037184 78082206371658018214590930638340096
39043638487029465566098458725580800 19523086894114961012450726065995776
9762177272357594620926111384600576 4881405548828854367813429868101632
2440861230739455712581902021951488 1220509843532242120628544554926080
610294535847378192483069049438208 305167074964317662325932910706688
152593441002473114205165648347136 76301672261393698623682420670464
38153312010775420072391008583680 19077893945426995416470403416064
9539565942733140398372651270144 4770092456376391544255050416128
2385200970693106444661887598592 1192677856599008558598124994560 596377613925731947432653094912
298208149775979807783121846272 149113746294546820924958572544 74561708850551868979178110976
37283272276915163747938467840 18642845064277196503143940096 9322026995048405566159323136
4661315728979106440373338112 2330808980217005048833507328 1165480047972228438740172800
582777802917977176531795968 291407790924920066846752768 145713340195425772713803776
72861392464195756002115584 36433057415339312823664640 18217709299290373823135744
9109444945455545617219584 4555017620632952161435648 2277656384269065757130752
1138901979110827716771840 569487883043561277489152 284762388265854348296192
142390417504964028923904 71199820438500441849856 35602216062259434618880
17802260952634324156416 8901706937069465501696 4451141698910884462592 2225714964643518087168
1112929539915796971520 556500798754917449728 278268413775968206848 139143214087238844416
69576110643246792704 34790307121437081600 17396279460625383424 8698702680266113024
4349632815109767168 2174957145043238912 1087548941265797120 543809655004987392
271922419688538112 135970005937291264 67989401015156736 33996899530833920 16999549277044736
8500324394336256 4250437075075072 2125355976491008 1062746707722240 531407713599488
265721036668928 132869108269056 66438849101824 33221572034560 16611859759104 8306466750464
4153501810688 2076885123072 1038509670400 519288389632 259660972032 129838874624 64923631616
32463912960 16233005056 8117026816 4058775552 2029518848 1014824960 507445248 253739008
126877696 63442944 31723520 15862784 7931904 3966208 1983232 991680 495872 247952 123984 61996
31000 15501 7751 15503 31008 62020 124048 248112 496256 992576 1985280 3970816 7942144 15885312
31772672 63549440 127107072 254230528 508493824 1017053184 2034237440 4068737024 8137998336
16277045248 32556187648 65116569600 130241527808 260499832832 521033220096 1042133549056
2084401315840 4169071067136 8338679005184 16678431752192 33359010988032 66722316943360
133453223821312 266923627511808 533881614761984 1067831949000704 2135801336954880
4271877551816704 8544304859447296 17089709230522368 34181617484300288 68367633015111680
136744062123245568 273505716432535552 547046617237159936 1094163603218497536
2188467943925350400 4377217362827411456 8754997675608244224 17511121251123331072
35024494302060347392 70053492203748065280 140115991606750871552 280249997612011225088
560536024021041414144 1121144105636120756224 2242432326460317368320 4485152883296786448384
8970882227345876320256 17942917376196359487488 35888140595401932668928 71780892876822292725760
143571009125681440227328 287160464995436590006272 574357823479020599115776
1148789433934336036438016 2297726441821261749288960 4595748031547702851403776
9192086358905764408459264 18385353309432246228221952 36773067802105927279050752

73550857970694724203315200 147111160674355187697057792 294241210814641853974970368
588520200561146665111650304 1177115958986019244546719744 2354383033699490317740277760
4709068298853884292774232064 9418741060617575900135817216 18838691047054766429446340608
37679799945748762117242093568 75364435594775982751183011840 150738542596108882535763673088
301496428005331599138322644992 603031541636890866410235887616 1206140454526237069087652970496
2412435651557384810709668331520 4825180788124590966488061444096
9650980546268824623113572450304 19303199032576934626502044024832
38608873945232440013553886298112 77222699650622021548207369093120
154455302821558326138613931180032 308930412683745218361626248347648
617900439448747568892049268670464 1235880107060009402121692081291264
2471918670445047332918571250483200 4944154253540151723187516676767744
9888942332380417561075781705138176 19779152315361063351553060113481728
39560839931922583161909113633374208 79126750466246079241424214079569920
158263642137293984318060401784782848 316547566684191620306544750820851712
633135698187590543953937396144275456 1266352526013595694589570581293694976
2532867311304020602542532740597678080 5066059141161699631811848637215932416
10132767319430716117077263586473017344 20266832713076065941061659797028339712
40536261574581399295937584842221289472 81077715446021333419503700180771799040
162165815485759736494264461354202038272 324352400158953612299043044693720956928
648746338692775503219114333358075674624 1297575754135287563680285154657418870784
2595317661770048241844683285197372784640 5190967630539042712657592522159815655424
10382599875075977883251636947849771483136 20766528978147740682376177702759823310848
41535716412287051196498163019640207310848 83076749736557242056487941267521536000000
166164133297080763439959112991525314756608 332349534242094085533884686896015115026432
166185400945013322093925573904489800269824 830980173844898007104544021804860215132l6
415516671482364701869730087043635714457600 2077716280211402000935940815924206067456
10389246015054902462616155983151173206016 51949553145263974602763039433406566891l52
25976438107626718446222649475528633876648 12989049821310724795531889617176992l5360
6494940294404045183976227248294833684448 32476778390763639850932548440005856460l8
1623942765475352689099198031926877224962 81202330570626169282588432092673081344
40603761433742352055108481294501150720 20303178791085809734461373271332880384
10152238432650221720684252947707592704 50764437348787692870689096298743726088
25383841267162138568978463929474744432 12692731929965215351306189854788812880
63467716131746807090615738724201267 31735886306833768712350264087229235
15868957273897066939696329406178918 79349856971886247616087633843716096
39677463787143580266846810328268800 19839999544172018362824901867339776
99206335973861232961131992852725766 49606337113431187054069738184376326
24804753119965878813786739971194888 12403168841608082050269305425100880
62019805616166123468226204323020888 31011883512145918342552940760268888
15506932108104387475496389679513686 77539612300679078898581544894464
38772282030418110209840570695680 19387378955248340485195184472064
96943084476438129327350417981448 48474637088317278114362456801288
24238865969207745782524852305928 12120206697128426253934238105608 60604902048264898083030250291288
30304385305443832448194655027288 15153159793377607927437092454488 75770634670166498153884286976
37887735186722478335291555840 18945076519180853796820484096 9473142722500234212997595136
47368735927050207637924741128 23685879120799622105430753288 118436951390370701959495680088
5922225358837164669591879688 29613015740778971206044876888 148074523436860595320651776
7404198408491316730553958488 37023353225698018475376640 18512857204469726648991744
9257018898045222030147584 46288045969277903678996488 23145498724164848603627528
11573487231845372683878408 57871125508041605329715288 28937407428428173620019288
14469626051417772287590488 72352741943107288825856 36178676814562858106880
180904913287860359004168 90458221251453213736968 45231992929488123985928 22617437616624820551688

1130943938425278955520 565507998009658441728 282772013403338702848 141395013900924092416
70702010550089416704 35353257074858393600 17677754437336039424 8839440168621441024
4420001559287431168 2210141517132070912 1105141127310213120 552605748027195392
276320466199642112 138169029192843264 69088912642932736 34546655344721920 17274427183988736
8637763347808256 4319156551811072 2159715714859008 1079926576906240 539997648191488
270016003964928 135016591917056 67512590925824 33758442946560 16880295215104 8440684478464
4220610674688 2110439555072 1055286886400 527676997632 263855276032 131936026624 65972207616
32988200960 16495149056 8248098816 4124311552 2062286848 1031208960 515637248 257835008
128925696 64466944 32235520 16118784 8059904 4030208 2015232 1007680 503872 251952 125984 62996
31500 15751 7876 15753 31508 63020 126048 252112 504256 1008576 2017280 4034816 8070144 16141312
32284672 64573440 129155072 258326528 516685824 1033437184 2067005440 4134273024 8269070336
16539189248 33080475648 66165145600 132338679808 264694136832 529421828096 1058910765056
2117955747840 4236179931136 8472896733184 16946867208192 33895881900032 67796058767360
135600707469312 271218594807808 542471549353984 1085011818184704 2170161075322880
4340597028552704 8681743812919296 17364587137466368 34731373298188288 69467144642887680
138943085378797568 277903762943639552 555842710259367936 1111755789262913536
2223652316014182400 4447586107005075456 8895735163963572224 17792596227833987072
35587444255481659392 71179392110590689280 142367791420436119552 284753597239381721088
569543223275782406144 1139158504145602740224 2278461123479281336320 4557210477334714384384
9114997415421732192256 18231147752348071231488 36464601347705356156928 72933814381429139701760
145876852134895134179328 291772151013863977910272 583581195515875374923776
1167236178008045588054016 2334619929968680852520960 4669535007842541057867776
9339660311495440821387264 18680501214611599054077952 37363363612464632930762752
74731449591412135506739200 149472343915790010303905792 298963577297511499188666368
597964933526885955539042304 1196005424917497825401503744 2392161965562447479449845760
4784626162579798616193368064 9569856788069404546974089216 19140922501958423723122884608
38284262855556076704595181568 76573361414390611925889187840 153156394235338140885176025088
306332131283790115837147348992 612702948193807899807885295616 1225483267640071135882951786496
2451121277785052944300265963520 4902552040579927233669256708096
9805723051179497157475962978304 19612684042398279695226825080832
39227843964875130151003448410112 78460639689907401823106493317120
156931182900129086688412179628032 313882172840886739461222745243648
627803959763030611091242262462464 1255687147688575486520078068875264
2511532751702179501715343225651200 5023382416054416060781060627103744
10047398657408946236262869605810176 20096064965418120701927235914825728
40194665232036697862657465236062208 80394401066474308642920917284945920
160798943337750443121053808195534848 321618169085104537912531563642355712
643276902989416379165911021787283456 1286634935617247365013517832579710976
2573432130511323943390427243169710080 5147188779576306313507637642359996416
10295026596259929480468841596761145344 20591351266734492667844815817604595712
41185298681898252749503896883373801472 82375789660655040326636324263076823040
164761963915027150308529709518812086272 329544697017488439927573541022941052928
659130932409845158476175326016515866624 1318344941569426874194407139974299254784
2636856036638326862872927255831133552640 5274044380275599954714080463427337191424
10548753374549092367364612830384814555136 21098835977093969650602129467829909454848
42200330410179509132950066549780379598848 84405977732342157929391748327801880576000
168822589288650595185766727112086003908608 337666446225233749025499915137136493330432
675375427746332615358932752100201957687296 337708981521098866333432836963065464356864
168865124584515712493699648938014974935040 84437879204240995910341439697248608845824
42221598058112067786916527462744865112064 21112128257051818809331167538432712900608
10556728742523801862602035672746496622592 5278696678260847160269243788138318397440
2639514492629896694618734869951694241792 1319840323064684904551423922917114642432

659961699907210730896740205429191081984 330001619141039504758884224699912421376
165011194164236822034699173342614650880 825107893789769458449781171676365455536
412579908379177403363033238319828828816 206302934931735038750587945400737464432
103157957836940687909829635820780257288 515822241040069282221826494705958912 0
257927346447717562447252405154008268 8 128971786187700241891795781477518540 8
644899495757708512799826801890164736 322470030288457908070337348196368384
161245156349030779870380647723827200 80627648776916302852796310674735104
40316359689658607885201148743778304 20159447495429532172002071075094528
10080357573014880200701783889149952 5040495699157497157701266120376320
2520406305903777107525820148088832 1260282381114402818100503617994752
630180804638458541219048580972544 315110209359857836693922676473856
157565008200243201389160531230720 78787455860278742215679862112256
39396203810217941868389729304576 19699339845148256314469763776512
9850288892593770847372331450368 4925453931306706768754890506240
2462881708158264056911807643648 1231518225331587364723085017088 615797798292021350495133106176
307918241959124509314361851904 153968792386119171690578575360 76989231896338044361988112384
38497033799808251439343468544 19249725825723740348846440448 9625467375771677489010573312
4813035919340742401798963200 2406669075397823029546319872 1203410095562637429096579072
601742826713181671709999104 300890302822522314435854336 150454596144226896508354560
75232020438596317899390976 37618371402539593772302336 18810366292890514297454592
9405773442255615854379008 4703181869032987280015360 2351738508469083316420608
1175943041210836496416768 588008414093565667311616 294022653790856543207424
147020550267465126379520 73514886819750990577664 36759749252884708982784
18381027547946961338368 9191090234725784092672 4595833347739043758080 2298060789057597734912
1149102452122836795392 574587254858437361664 287311641827728162816 143664828113118822400
71836917656186781696 35920710627907076096 17961481213860380672 8981303556883611648
4490933253418516480 2245607364197613568 1122874050842984448 561472209793581056
280753697082834944 140385644634439680 70197220363730944 35100809205121024 17551504114188288
8776301812908032 4388425784360960 2194350331133952 1097243885043712 548656302260224
274345330999296 137181255434240 68594922684416 34299608825856 17150878154752 8575975948288
4288256409600 2144262422528 1072198320128 536132714496 268083134464 134049955840 67029172224
33516683264 16759390208 8380219392 4190371840 2095316992 1047724032 523894784 261963776
130990080 65499136 32751616 16376832 8188928 4094720 2047488 1023808 511936 255984 128000 64004
32004 16003 8002 16005 32012 64028 128064 256144 512320 1024704 2049536 4099328 8199168 16399360
32800768 65605632 131219456 262455296 524943360 1049952256 2100035584 4200333312 8401190912
16803430400 33608957952 67222110208 134452609024 268921995264 537877544960 1075822198784
2151778615296 4303825666048 8608188203008 17217450147840 34437047779328 68878390525952
137765370986496 275547921842176 551130203422720 1102329126322176 2204795691597824
4409866261102592 8820282278019072 17641664067665920 35285527158587392 70575452363685888
141159700820393984 282336993826832384 564709172025753600 1129488712795684864
2259118163079725056 4518517801136160768 9037598552225742848 18076323004358328320
36154897808530341888 72314299216688054272 144637605632630849536 289293225663771181056
578622480124561326080 1157317017843160580096 2314778150874397016064 4629844532124945743872
9260265525002194911232 18521683971508996669440 37045673786027207032832 74095959258072841453568
148201141888182537682944 296420730520438784917504 592878354529024988938240
1185830496034344816082944 2371808566021279308578816 4743912279947737969983488
9488414855705834645618688 18978010303032386702540800 37958381789306208227688448
75921485945095286100590592 151852416623156311491608576 303723722712244101564071936
607485224356351160289853440 1215046006576428234903126016 2430243128880308298453090304
4860788489215520254199857152 9722181441340847822987067392 19445571808501310275148840960
38893561468641849808647094272 77791958640562158133993013248 155593588687681233301383675904
311206520188476300669562650624 622451726003180269472715898880 1244980823258815875212612993024

2490116389022542422959588376576 4980542263054906190987901534208
9961703496129455072113252630528 19924644932298195524501404385280
39851765744674961809552607019008 79708483249507065140204810534912
159426870019328413322608814063616 318873547079285392729616014114816
637786708239827917628028800204800 1275652644642170099593651144359936
2551463745609368727862489376620544 5103244403868794513075352929042432
10207122633037703140851454209687552 20415512916675634511104405122580480
40833561134551725481011803651571712 81672192871504363879629594115964928
163354526947810553594471161857572864 326729336305224758859366270966431744
653499237429656821059580436435435520 1307079604497728248800856661876015104
2614321468272285710965104901762318336 5228967455098229848656992959545212928
10458583947303776550767552231131578368 20918465968822186808442237086345461760
41839528086073641030698739420855533568 83684248469005816889026009338040287232
167378881531728703433309079668739014656 334778532250891546177132281322794909696
669598602876651370975292806616223580160 1339280282503039299192642101173714681856
2678726718505551712869397178229964406784 5357785744010049654707020308224998899712
10716236102017991767350492519980137971712 21433801432031768450573888847020556288000
42870261320055106732893585308161673265152 85745839552093353129278785844564467908608
171502312928152985585540802145611178573824 343025893504238529825048065204186842660864
686094322304342176958029052234302656348160 1372273715200414588531923948120463254749184
686179392896072411573894895886160598401024 343110964095968764440913908856044784713728
171566115871950661547440184884504635113472 85788374847958470437211707670493438935040
42896845879970805050351661449367280156672 21449752167981187441048734531743920422912
10725540697988486178460819169402100383744 5363102655993189318198635536466120278016
2681717481496067773583430744115595182080 1340941817497770444033771859999065112576
670512447123753500637914173970166317056 335276992749310889629471208970400038912
167648880968372514469992665477858459648 83829632781044792062624863235258449920
41917412538951663445126696865793835008 20960004343690465429470481056979222528
10480651208952549568188806840530763776 5240650123029933210821186576285958144
2620487320791795818773984866153267200 1310324790034312516068688222081777664
655202959836363561375192005543460864 327621762327785432358019950023016448
163821022368694542014221948637151232 81915581786748183924716961131397120
40960326194574548421161473972109312 20481430747887502439982233689260032
10241349199243865334691865196232704 5120991512271989724696306773917696
2560654212461023391023340474859520 1280406334393025959849263781380096
640242781277770112093428662665216 320141197679513622131112717320192
160080502360071094107755551653888 80045202940192688574977372323840
40025077350174915048038484410368 20013776615126742904294141329408
10007507277583014142284520226816 5004063123801328416210984894464
2502186304405574880639854837760 1251170523455242776587108614144 625623947353849056427144904704
312831316490038362280367751168 156425329651576098173581524992 78217500529066507603489587200
39111168116172483060094205952 19556792983905856159221809152 9779000954862735394198257664
4889802708886271354392805376 2445052470170587505843240960 1222601792949019667245039616
611338675406372790784229376 305688227169117873972969472 152853558317524676276912128
76431501525245207783669760 38218111945864038714441728 19110236564552736768524288
9555708578086727089913856 4778149436948542897782784 2389222292426861125304320
1194684933189725400858624 597379360083010119532544 298708126785578769317888
149363286764826239434752 74686255068431547105280 37345433377224987246592
18673869610117100470272 9337511265810853658624 4669043863281578541056 2334666046828865126400
1167405081008470491136 583738569301254209536 291887299049136586752 145952656723823034368
72980831961538887680 36492667780583129088 18247459790198407168 9124292845052624896
4562427897503023104 2281354686239866880 1140747711864111104 570409040304144384

285222112338116608 142619852262080512 71314324177551360 35659361112031232 17830780067643392
8915939789635584 4458244772724736 2229259825315840 1114698632134656 557383675805696
278709017772032 139363098820608 69685844377600 34845069672448 17423608578048 8712341159936
4356439015424 2178353725440 1089243971584 544655540224 272344547328 136180662272 68094525440
34049359872 17025728512 8513388544 4256956416 2128609280 1064370176 532217856 266125312
133070848 66539520 33271808 16636928 8318976 4159744 2080000 1040064 520064 260048 130032 65020
32512 16257 8129 16259 32520 65044 130096 260208 520448 1040960 2082048 4164352 8329216 16659456
33320960 66646016 133300224 266616832 533266432 1066598400 2133327872 4266917888 8534360064
17069768704 34141634560 68287463424 136583315456 273183408128 546400370688 1092867850240
2185869918208 4372008271872 8744553414656 17490180571136 34982508625920 69969312219136
139947214372864 279911608614912 559857576968192 1119783873413120 2239705185779712
4479685249466368 8959920254746624 17920940021121024 35844079065497600 71692556177506304
143393908448034816 286805409082114048 573646002536316928 1147362373816811520
2294865485121978368 4590012445220667392 9180587840394756096 18362301580696354816
36726854961206394880 73458213522040160256 146925434243335061504 293868882885179604992
587773794567378173952 1175619646728794275840 2351383408645664407552 4703055047667480526848
9406686556087264477184 18814526033679135801344 37631357910367485296640 75267327506753397981184
150543878385543650738176 301106203515161011027968 602249300518469441159168
1204572388013233720524800 2409292349979057117462528 4818879847863293587750912
9638349991536945881153536 19277880574694609173610496 38558122332630653169827840
77120967031744175984869376 154251378796454091260166144 308521647058839661101187072
617081073049542279364083712 1234237703962810473051586560 2468626523653072774750011392
4937555278761049206793699328 9875715020431905728174751744 19752638966683426085524209664
39507695785006081429397831680 79020227273290621375494488064 158050125953138159784386625536
316119594719390153635568549888 632277875065007975404727697408 1264633121382471287076636590080
2529420985269853246687635570688 5059151455549527838443995922432
10118921881118698367025441406976 20239081702276682114325781938176
40480639284631934989201362124800 80966230329421011499502320746496
161942364179156306041203834486784 323904535398941178166806054961152
647848684879139488502408881897472 1295776597920793241342411307745280
2591711652166615011360009703391232 5183740216983287080070393582583808
10368114259266688274841535516770304 20737496169133604779084567736745984
41477527639467666016972128879902720 82960125881336244951550244572626944
165930392967474315738312462770896896 331881068344552283147048872793079808
663802701508311869634945640088731648 1327686532655038345951587069182607360
2655535324586905905266565716375502848 5311395167727470237259914588771581952
10623439372562257327973395489584316416 21248176819339148362853923603250937856
42498949787107564139522112454666485760 85003091871073663106672755405662191616
170016568335864395868602571803982823424 340053905859162931047719265593282527232
680149350093194140716466775157198815232 1360381776936124838674990038255665152000
2720929707371722791834093052393865347072 5442191721742391812636412056552800780288
10885048057482676083209276016635741732864 21771425342961137082291455840331763810304
43545509141913843996328719294784088309760 87096335195810827656149053817809297997824
174203304215587934639281338092100838752256 348427876079108427932529137097166163017728
696898287454081973172991196020261297061888 1393881645499894180961848235692380536176640
2787933432183248831155428158688476956459008 1394051786683354650193579922996096420282368
697068428637542442404722883323977181167616 348555481966703779856327902574953076097024
174288374807318169255147181743958780805120 87149504315642224291065206100220511780864
43577410613812681977278410664230816579584 21790034534902125904512109139175688634368
10895681881448955410192506473117984489472 5448173247723423934064479188324062330880
2724252777361185081516352570044566208512 1362209465430329098000232772963550625792
681146271090032827621144630452409073664 340593904732450553121086437211521417216

170307336959942346215800279598419148800 85158860776829707935528670295538794496
42582026536844121381578600395934007296 21292311342636694397696432822049308672
10646804708425664052301782723065806848 5323726872766490452877674517553479680
2662025695660074439802228836787027968 1331093977468451826582810207398658048
665587553553433216632252998201901056 332814059186320259986550446352236544
166417170797961955828487196801761280 83213656001381890831849585213702144
41609363301891401874727786013261824 20805949301545929166765389709836288
10403608476073078698083443206520832 5202121150686596406392095779061760
2601219031668326731871234977431552 1300688743996677630273211032666112
650383986079595947305402288308224 325211800080426539737099530141696
162615803560527552910748958064640 81312853540420917976474075529216
40658902650289029748786836013056 20330689265183800254668317130752
10165963602611542817471608127488 5083291286315592753804528844800
2541800385662707049436626812928 1270977564083808860985494601728 635527467668132098626337898496
317783076647179883379964248064 158901209730146858723379773440 79455440568351887878388711424
39730138135815173197543768064 19866277993727201227946590208 9933743459773407928560648192
4967173961341607621574000640 2483738096398255639433838592 1241944606062853734040338432
621010081963289824181878784 310523930447576390671794176 155271409956753934626324480
77640427344859836958375936 38822574855671353301794816 19412468019456394062200832
9706824305538555736752128 4853707300674457221201920 2427001224289818287013888
1213574399121203981713408 606824093048749409959936 303430493268448414531584
151724470006261062041600 75866846689148958408704 37935729187583692898304
18969017515296453296128 9485085218400530071552 4742830839576416747520 2371559534976284229632
1185851825082180042752 592961941338108985344 296498985067563974656 148258499733036728320
74133753466145734656 37069128532886552576 18535690166350118912 9268408033128480768
4634485491540951040 2317383483258830848 1158762110373593088 579416239558885376
289725711965487104 144871652075765760 72440224084393984 36222311065452544 18112255044354048
9056677277990912 4528613516902400 2264444197404672 1132290818179072 566179768827904
283107064283136 141562122076160 70785356005376 35394825486336 17698486484992 8849780113408
4425158492160 2212713463808 1106423840768 553245474816 276639514624 138328145920 69168267264
34586230784 17294163968 8647606272 4324065280 2162163712 1081147392 540606464 270319616
135168000 67588096 33796096 16899072 8450048 4225280 2112768 1056448 528256 264144 132080 66044
33024 16513 8257 16515 33032 66068 132144 264304 528640 1057344 2114816 4229888 8460288 16921600
33845248 67694592 135397376 270811136 541655040 1083375616 2166882304 4334026752 8668577792
17338204160 34678505472 69361205248 138730799104 277478375424 554990305280 1110047719424
2220229656576 4440727718608 8881992368128 17765058478080 35532264439808 71068823846912
142146237628416 284309655126016 568653669990400 1137376059457536 2274889557868544
4550053993644032 9100657743101952 18202414997831680 36407029018918912 72818456084348928
145645708261720064 291309008709484544 582653201791057920 1165376772326293504
2330894282140942336 4662070039258595328 9324703028470611968 18650531956848066560
37303315713509818368 74611135026647007232 149231277252548755456 298480568903606992896
596997166604232949760 1194066390802503827456 2388276896793083510784 4776842023962318733312
9554260508676940890112 19109673938858488627200 38221653720726190948352 76447919127470809284608
152905061626978473345024 305828569998030656241664 611694033484208731586560
1223461853944712301379584 2447071281842014279172096 4894437711589207911170048
9789465718988774527991808 19580112029598266467287040 39162585242437967757180928
78329892851358805159575552 156669230435683349609578496 313357350337298177800011776
626752479606459312761733120 1253580517076644539846885376 2507312149880740908340609024
5014926531216385473974894592 10030457525342578262537142272 20062123976504771154248990720
40126665804648771566847393792 80258167312576001650393612288 160526006031708920334184873984
321071354876531674735165046784 642181395379291017603920691200 1284440162011037371475022577664
2569035066526985415484407545856 5138379618063792176037539872768

10277378206147227042212529307648 20555994352333739464699957739520
41114464584746049689949713727488 82233880929649240900999023951872
164477665379612764844197240897536 328975137799854095772792867782656
657989889680965323714382507540480 1316059007524444911766358559031296
2632276471373918352207904205963264 5264869855397893761766182587727872
10530373536095901638233113527058432 21062014722792031505867723757322240
42126564746784519470538440921055232 84258200095969951858682868654931968
168526541396741729552577710935506944 337073365203087110775579369122299904
674187295225381524892006632747171840 1348455720089177656465709054499487744
2697073699455184526294809687009263616 5394471917464027479316402530039103488
10789592872035371812086371372119359488 21580483818285377331079875368321024000
43163563785000022075974015984806658048 86332319866858578979576562465942536192
172675024327434227614410185924543512576 345370817842302594539334493834403905536
690783174059473467699697231639441571840 1381649424868683492641450951220150665216
2763465003236840099767014878322836373504 5527262313472626428502255708410742833152
11055189240943145314940963320351625838592 22111707709882075545754830447763532021760
44226073875755720923255468509647624732672 88457464663494581510002552247536370843648
176925563150955442346988334951554984443904 353872393949843443347943130816074454401024
707787323195552004003819183458077879828480 1415659716982834242623504210568013701709824
2831489575149128954478740108439743287525376 5663319432665178847420943591486918343262208
2831829857516049892942203483047175055736832 1415999999349755181086967585175445469921280
708042534970742707851416714413651705987072 354042535133303912579674818119790338506752
177031901390618235616820639516377412009984 88521267607292257471901934986429827383296
44263292259637698567696775107335474380800 22132975357814634199721291360728017534976
11067152292905209557797097583894148939776 5533908453451551007866774743712144556032
2767120380225248618417500347738607321088 1383643266862360866450806661810571182080
691863171806048711846431574875919351808 345952355090458495233729909423276556288
172986562138946317272122015704296718336 86498473366331693463689538348477579264
43251832831595114145659034422403399680 21627214490012190779736649835284004864
10814256282113412243321891229683154944 5407452659610364548387728770862153728
2703888589082011487557255963441364992 1352025424179420350460323770725826560
676053276908917478571009779865485312 338046920864062390955928837184028672
169033601636833021313176392217657344 84521871420817423574194182921650176
42263471011609168245900084867235840 21133003156404812352351539136823296
10567135403502520290876517920014336 5283884614401317202788633135808512
2642100763525687130069503655804928 1321129609925357829372345371852800
660604419043936046854969457901568 330322016562596589511883114938368
165170911801612577798140750462976 82590407660963430420169971728384
41297679710560285970634784112640 20650077795319428365592291180544
10325657867679356872933595152384 5163138418849499781535522357248
2581723951929660563302123569152 1290939347217285617918242979840 645508359234870477092712087552
322773522430549072613151342592 161396432621831453339973320704 80703052014194185186685485056
40353943858736321851692154880 20178180855187775555020783616 10089694890503695092097744896
5045149676706751203342548992 2522725954080827430318112768 1261438534904139629482475520
630757046383932771902947328 315397412657897864532328448 157708151061914671556591616
78858797897440205423509504 39431760131961537534361600 19717060657601486178484224
9859120624611101794893824 4929855460210730250272768 2465075304057954801549312
1232611439005272238981120 616342612990783538593792 308189753239465478848512
154104099991769594200064 77056661681903224487936 38530636683960825937920
19266471263485019815936 9633812092494813331456 4817194276623558377472 2408741253499855044608
1204442684343965450240 602257370969001689088 301146699883010326528 150582357140759904256
75295682170007322624 37650092884817346560 18826172342315515904 9413649121111179264

4707106035532300288 2353693755254505472 1176917246371430400 588493807557804032
294264495964946432 147141044075495424 73574920084258816 36789659065384960 18395929044320256
9198514277974016 4599532016893952 2299903447400448 1150020443176960 575044581326848
287539470532608 143778325200896 71893457567744 35948876267520 17975511875584 8988292808704
4494414839808 2247341637632 1123737927680 561902518272 280968036352 140492406784 70250397696
35127296000 17564696576 8782872576 4391698432 2195980288 1098055680 549060608 274546688
137281536 68644864 34324480 17163264 8582144 4291328 2145792 1072960 536512 268272 134144 67076
33540 16771 8386 16773 33548 67100 134208 268432 536896 1073856 2147840 4295936 8592384 17185792
34373632 68751360 137510912 275038208 550109184 1100283904 2200698880 4401659904 8803844096
17608736768 35219570688 70443335680 140895059968 281806897152 563647348736 1127361806336
2254857830400 4509984096256 9020505063424 18042083868672 36086315220992 72176925409280
144362440753152 288742061375488 577518482489344 1155105684455424 2310348807864320
4620972493635584 9242494743085056 18486088997797888 36974377018851328 73953152084213760
147915100261449728 295847792708943872 591730769789976576 1183531908324130816
2367204554136616960 4734690583249944576 9469944116453310464 18941014132813463552
37884280065440612352 75773063730508595200 151555134660271931392 303128283719053344768
606292596235125653504 1212657250064289234944 2425458615316654325760 4851205461009460363264
9702987382771224150016 19407127687047055147008 38816561217103323987968 77637734120225075363840
155284691612487005503488 310587829969047720558592 621212553426242860220416
1242498893828780558647296 2485145361610150793707520 4970585871125480940240896
9941762038061320586133504 19884704667743358583570432 39771770518728151989747712
79548263403939173624709120 159105971540844086539845632 318230832547619651660546048
636499444027102260482801664 1273074445917930435289022464 2546300007563312699224883200
5092902246581529055743442944 10186408956072865426074238976 20374026837965345481323184128
40750471527569920220995780608 81505778758418298958690385920 163021228923393514950778421248
326061800659900863968352141312 652162286946029396070294880256 1304401945144514128407770955776
2608958632793938929349904302080 5218226750597699203768533385216
10437072471215041097674516332544 20875382882469367575623931789312
41753241645017305911797661827072 83511435050191753344694920151040
167032773620697789731589033295872 334085354282024145547576452579328
668210322645305423263949677133824 1336499873453125110865492898217984
2673158203231278750406172884336640 5346633319112614558162719944474624
10693900463525343231026188240551936 21389068577650914691453873184309248
42780672456502285841710739775029248 85566415515405484601027466362880000
171142972235612795037266906351403008 342306226880829241744957759954092032
684653018580865786830763414410756096 1369387166800146180343222617826656256
2738936592877121574049836813663600640 5478197704307901574826456783347777536
10957044445723120003106479878736707584 21915386965660873713120092381555720192
43833370079751014840054450011276050432 87671932456360564507737430518881320960
175354249506438198670731922030421082112 350729268200310536651977966046159044608
701500074775489351924984176062951849984 1403083226300715261092024840067171221504
2806332606100903636668162656016877486080 5612997519200753502304551263798825058304
11226659652399399462545554431127790288896 22454648532794583840964012669315860922368
44911955521580737513673832952752282533888 89829227955144614690839281133745686446080
179669089734255508708661792723973615648768 359359447116443576071290046360911716810752
718761429528752269450513014547752404647936 1437607929649234773516891872747362751348736
28753860004819300162655515432798441386803200 5751112283330780970994494240204314541817856
11502905131395403818915915229623492620058624 5751792848064622847921420989419178078240768
2876066565215771893192442182013304923226112 1438118353199616181212086934658510403665920
719101711895673207913976389155184172859392 359572123595769162610954655490556571942912
179796695621850860632460558201760528728064 89903664722908569979721894329121385742336
44954490817445854821606754778681253560320 22478574636718712326676281196400907124736

11239951932357248621274592501730593734656 5620308273177570539605522202630366953472

2810320290088258384286874077197718519808 1405243221793865749385493526540126781440
702663149271801153313775007240697151488 351352343823334715967401625605665456128
175686556505384427638957873795491168256 87848470549550748647107467394074804224
43926831423204641737367998945202012160 21964713785816954575591132096683311104
10983005930015794141249132360382808064 5491827483561555497351349336211980288
2746076001057606962039066246116278272 1373119130167218087701228912063283200
686600129902816347191462350534213632 343320347361011825266155122518392832
171670314885307738468289534884839424 85840228045054782151750754255241216
42922649323727847534678370534031360 21462592312464151996740681970221056
10731929981532190113071089336713216 5366281903416152113885918844157952
2683299408033104585618146509979648 1341728932179066557146666798940160
670904080170790410742130171445248 335471847126023771455463471710208
167745827083326168769930928848896 83877865301820225906065060921344
41941408530988683713582328709120 20971942205533627237066063478784
10486590072786456308670481301504 5243604521403049499403965431808
2621957003206435422236345106432 1311055872855673047385353748480 6555566622054064191826267471872
327802653840145929979929034752 163910998326629882023362166784 81960334866593399528379908096
40982585284935929022539366400 20492501568287579140444389376 10246855247053596884809547776
5123729854981702099698450432 2562016043218302878496063488 1281083579472877353571450880
640579568668301633947435008 320308673800082295554572288 160163781633006887067713536
80086613182986313179070464 4004566774734591412142080 20024014478988013117374464
10012597535304365264338944 5006593915557361984995328 2503444531731270668910592
1251796052841930172661760 625934919909112505434112 312985906698629962268672
156502176721351835910144 78255700046694345342976 39130155866356386365440
19566230854682800029696 9783691888093703438336 4892134174423003430912 2446211202399577571328
1223177658793826713600 611624858193932320768 305830443495475642368 152924228946992562176
76466618073123651584 38235560836375511040 19118906318094598144 9560016109000720384
4780289529477070848 2390285502226890752 1195213119857623040 597641744300900352
298838464336494592 149428028261269504 74718412177145856 37361405111828480 18681802067542016
9341450789584896 4671000272699392 2335637575303168 1167887507128320 583978113302528
292006236520448 146011708194816 73010149064704 36507222016000 18254684749824 9127879245824
4564208058368 2282238246912 1141186232320 570626670592 285330112512 142673444864 71340916736
35672555520 17837326336 8919187456 4459855872 2230059008 1115095040 557580288 278806528
139411456 69709824 34856960 17429504 8715264 4357888 2179072 1089600 544832 272432 136224 68116
34060 17031 8516 17033 34068 68140 136288 272592 545216 1090496 2181120 4362496 8725504 17452032
34906112 69816320 139640832 279298048 558628864 1117323264 2234777600 4469817344 8940158976
17881366528 35764830208 71533854720 143076098048 286168973312 572371501056 1144810110976
2289754439680 4579777314816 9160091500544 18321256742912 36644660969472 73293616906240
146595823747072 293208827363328 586452014465024 1172972748406784 2346082935767040
4692440749441024 9385431254695936 18771962021019648 37546123065294848 75096644177100800
150202084447223808 300421761080492032 600878706533072896 1201827781810323456
2403796301109002240 4807874077194715136 9616311104342851584 19233748108592545792
38469748016998776832 76943999633624924160 153897006466504589312 307812027331518660608
615660083460056285184 1231392224514150498304 2462928564216376852480 4926145358808905416704
9852867178370114256896 19706887278244835360768 39416080399498884415488 78836772485016196218880
157682768342069247213568 315383983428212203978752 630804860344571827060736
1261683507665438492327936 2523514589283466661068800 5047324326472112674963456
10095238948754584055578624 20191658489129885522460672 40385678161501205867528192
80776078689485281380270080 161561602111936302050967552 323142093689804082682789888
646321966311471122527289344 1292719490486668159377997824 2585590096700788147402833920
5171482424856479952099344384 10343569312622767218786041856 20688347551065149066746789888

41379112953769527391842992128 82763061610817513300384808960 165535794628191943634167267328
331090932069497721335129833472 662220549765223110803850264576 1324518470782901557874881724416
2649191684070713788284125839360 5298692853151248921636976459776
10598004676322140533411402481664 21197247292683566447097704087552
42396970465445703654745206423552 84798892691048548830590009344000
169607688902411380703379211681792 339235184845451327491156809351168
678509983772159787151110390677504 1357099195706833838639814325305344
2714356847738696205954815738511360 5429030608127449469260005652824064
10858695041555013053220759657250816 21718657733710254335843016017707008
43439850768620965130489025441824768 86884772139642843178584037696471040
173779685484087512192380049018585088 347579653377778676055184045288456192
695199871574764655451215985079484416 1390480872787943917584127759164112896
2781124004852717048531647096338513920 5562572528259092523790077348697604096
11125794093625501901033721009436360704 22252886261465637508974574642955026432
44508368671360542431763414534074662912 89021929639579619691155359564478545920
178054243872876309037567780121615532032 356129256933186757385649682228547944448
712300052241241793392327608427729649664 1424683181232220144026711704796726820864
2849532515963913402537536385475988684800 5699397338926773034043298722717047455744
11399459291851438526023049348964235083776 22800247811698661967919002504988750512128
45603154079388893767583812624098061713408 91211625070760927198659240476437244805120
182433883965488133724301711409356732366848 364889035578908826102569883731677950246912
729820606453682769513072689289284871520256 1459726283499095773642011222230427685093376
2919622708181652016515754131764571254292480 5839585698730224971494971638136574276796416
11679851962194291819916870025488012090015744 23361065053856267393687593549405751252877312
11681213091661975573770723523917739162861568 5840946828197908725348825136566301349642240
2920643555282414831906144255586866558926848 1460406848232937650568937971445291221516288
730245959412333942592401907548574581784576 365144247354099529950167414687251776405504
182582757501016044302066937800108130959360 91296695662491161814525084128295186857984
45651006287237150739008349678268154118144 22826832371614360285377078646194357403648
11414080799805072600624991226627318874112 5707372706901482529280721565078729523200
2853852506950214379124473758421899804672 1427009330224843746804293367152217423872
713546203487290152023174927546742472704 356793870931079215322101585758688116736
178407320059256677316307853872002498560 89208852326486873485782457432330469376
44607022311672704156705493964329844736 22304809230050985785259879606247227392
11153053652132809746083506115164766208 5576851344620063299768536213602959360
2788587931586860863247659684811767808 1394375095431845038305525631411027968
697228112535129822493610710208086016 348634338677168562917229302355329024
174327310543386107293826624803307520 87168725874093966564519299214475264
43586898238247439741062643013648384 21794716769723948099932818210029568
10897992210162088164667157456617472 5449313017731101139683952904110080
2724814965190579098517163539955712 1362486710757803813596175313928192
681282969460159038966884428939264 340661291770708085567840600457216
170340549405668325826119493222400 85175226462991304434159343108096
42590089111574222977629469802496 21296282495826396869089634025472
10648760217932841124682266574848 5324689593976241907409858068480
2662499539493031626239291424768 1331327140998971149386826907648 665702256125713242827004051456
332870470875970455480297324544 166444906844542144773546311680 83227289125549530903471980544
41616062414413994710085402624 20809240133026611984217407488 10405224529423113306696056832
5202914496166460310641704960 2601608363810681983967690752 1300879739769066906307264512
650477648816396410315341824 325257713874129683738525696 162638301670030581159690240
81323873201498160225058816 40664297783990514935136256 20333329483615974878871552
10167255037618346145087488 5083922666714352425369600 2542108907309765889097728

1271128240631177782755328 635601013803736310480896 317818953645941864792064
158918700195007787171840 79463961783522320973824 39734286734770374180864
19868296288889793937408 9934724605197200392192 4967650532974751907840 2483969381675451809792
1242056748431763832832 621064403012900880384 310550215904959922176 155284115151734702080
77646561175494721536 38825532387561046016 19413892093687365632 9707508996797104128
4854035973375262720 2427158724175986688 1213649730832171008 606860049788174336
303447617080131584 151732604633088000 75870700363055104 37937549204783104 18969874114019328
9485486812823552 4743018284318720 2371646581112832 1185892010033152 592980364754944
296507362246656 148262271057920 74135430496256 37069862731776 18536005107712 9268539424768
4634538147840 2317403291648 1158768754688 579417931776 289725743104 144871260160 72439824384
36222009344 18112053248 9056550912 4528537600 2264399872 1132265472 566165504 283099136
141557760 70782976 35393536 17697792 8849408 4424960 2212608 1106368 553216 276624 138320 69164
34584 17293 8647 17295 34592 69188 138384 276784 553600 1107264 2214656 4429568 8859648 17720320
35442688 70889472 141787136 283590656 567214080 1134493696 2269118464 4538499072 9077522432
18156093440 36314284032 72632762368 145273913344 290564603904 581162762240 1162392633344
2324919484416 4650107404288 9300751679488 18602577100800 37207301685248 74418898337792
148846386610176 297709953089536 595454265917440 1190977251311616 2382091941576704
4764458761060352 9529467277934592 19060034067496960 38122267158249472 76248932363010048
152506660819042304 305030913824129024 610097012020346880 1220264392784871424
2440669523058098176 4881620521092907008 9763803992139235328 19528733884185313280
39059719568184311808 78123942735995994112 156256892671246729216 312531799741002940416
625099628279024844800 1250271314152087617536 2500686743492251090944 5001661717360653893632
10003899895473611210752 20008952712451829268480 40020211267912872230912 80045034221844171849728
160099291815725198475264 320217030375524106502144 640470954239195632107520
1281015695454686102421504 2562178964861961881255936 5124653077629103115337728
10249896451068564936327168 20500973493757847283957760 41004308170757129390522368
82013338707997128426258432 164036122148959996142944256 328091133763851470866743296
656220046459565898895196160 1312515650782857712113811456 2625182417293167252874461184
5250667066041238163042598912 10501938594992283640672550912 21005086115804181910519808000
42012590083247593079389028352 84030015869773644675476881408 168069703146104206384351412224
336158749105322246835498123264 672356183836872161804586844160 1344789738926199659876354883584
2689734220357309992287072157696 5379777925724441329642869096448
10760174821468525349423187755008 21521587582976336079121274634240
43045651046031242918792347516928 86096253852219627358684291530752
172202411224753537759567776055296 344424629490135641603533938098176
688888873061528415375864648171520 1377856974285571095089322840293376
2755872404896170718853832768487424 5512061722442398495058039712776192
11024757270184911104816827777155072 22050782190970050439035152257515520
44104099683140557336873297921441792 88213269968682027591352582655705088
176436681142165881017917138937053184 352893644693935413706258225125392384
705827854207078130753364344753356800 1411736838052570868188424478511857664
2823635935381970949740240535034003456 5647596389317600326207264226088583168
11295841815742517505868094764218318848 22592981705699668718643322152518942720
45188559559828604851100909553202495488 90382311416515744529830349602734211072
180775007426748558714917760198126862336 361570784040931256740349642381570605056
723183106456730792101727528733774970880 1446449289663198141445511545408817463296
2893064732825869397375136066700169969664 5786461772650685023718498085165410025472
11573588159299262505373448073860960223232 23148505546594309926619799954782200791040
46299669549180189684985407523684962271232 92604656010343519033462430275611045920768
185219945844653317393908091007704334598144 370461159337239193441782642928373154709504
740964853970343504191498207682675280445440 1482014778532417242998862259017208502943744
2964199698248294955229456205338132889993216 5928739678863510848922375785283697548197888

11858159922460863574771678319782258632818688 23717680974389410903397210137994244338483200
47438084207714189314502127272847942822658048 23720403233324778411104917134853698484174848
11860882181396231082479385316641712778510336 5930781373065036479703156032928288157466624
2965560827715978709083309703767859962839040 1482865484449719589157520695535787923472384
741475277520724911886693269593822932762624 370758906408295014597313095709875951894528
185390087028113786625639778311420218703872 92700360426040032976311504383951230730240
46352838669011586319901559806096176054272 23177748562501578075823683710108368371712
11589538895248681495848293758584324358144 5795101754623286976892372831057232265216
2897717030811116602930299391411151175680 1448941592155294858707206183646843109376
724512334452515707974631335794055315456 362276936413691993297829789882344538112
181148852800563066304171955933830709248 90579618697140067979714508463244574720
45292405496999301403671519479786897408 22647500822714284408742892363975753728
11324399448464459057825012494029029376 5662524242785887955639289403035090944
2831424380669773191183036279527833600 1415793319973301202273213928769060864
707937224805857904477454858887102464 353988894812532603909151376694837248
177004588611068127789787661973061632 88507364907934976812499817799352320
44256217755167944865052902306086912 22129376528184200661927947856248832
11065322089392214445664722279727104 5532977957346164280182735315664896
2766647434998110668766554745733120 1383402945661569598720870916816896
691741086912041931529232230383616 345890350496649531849014501179392
172955078768639048966706443583488 86482491144476666004452818288640
43243721452316903762776207392768 21623098666197737261663002820608
10812168303118511320968950972416 5406393636569077005553200267264
2703351560789449175310962524160 1351753151647179923922662457344 675915261449817630094921826304
337976973538022649114256211968 168998158175568241590525755392 84503914791062579311961702400
42254375247170518914330263552 21128396549404874086339837952 10564802737612244357757272064
5282703600261025836172312576 2641502915857964746732994560 1320827015792708287689916416
660451286828217101006667776 330244532880040029084188672 165131711172985753832521728
82570577952975746561474560 41287650159729308103344128 20645005671485371462975488
10323093131553044437139456 5161841713681701571395584 2581068430793440462110720
1290608002373015069261824 645340894674654953734144 322688894081401186418688
161353670412737447985152 80681446892387151380480 40343029289202789384192
20172667566106001539072 10086910243805304193024 5043743352278803808256 2522015791327477760000
1261079953257776807936 630576005425907367936 315306017111463165952 157662015754986323968
78835511477120532480 39420007538373951488 19711129669093818368 9856127784500330496
4928345367226875904 2464313421101793280 1232227079295074304 616148724019625984
308091954195857408 154054773190950912 77031784641986560 38518091344248832 19260145183752192
9630622347689984 4815586051751936 2407930464829440 1204033951891456 602051335684096
301042847711232 150530013790208 75269301862400 37636798414848 18819472949248 9410273345536
4705405108224 2352836771840 1176485494784 588276301824 294154928128 147085852672 73547120640
36775657472 18388877312 9194962944 4597743616 2299002880 1149566976 574816256 287424512
143720448 71864320 35934208 17968128 8984576 4492544 2246400 1123264 561664 280848 140432 70220
35112 17557 8779 17559 35120 70244 140496 281008 562048 1124160 2248448 4497152 8994816 17990656
35983360 71970816 143949824 287916032 575864832 1151795200 2303721472 4607705088 9215934464
18432917504 36867932160 73740058624 147488505856 294993788928 590021132288 1180109373440
2360352964608 4720974364672 9442485600256 18886044942336 37774237368320 75552769703936
151114129342464 302245438554112 604525236846592 1209119193169920 2418375825293312
4837026528493568 9674602812801024 19350305137229824 38702809297715200 77410016641941504
154828829376905216 309675250939854848 619385686251798528 1238841741247774720
2477824219983904768 4955929914944520192 9912422779842461696 19825971459591766016
39654194718997217280 79312893037621805056 158634793274498351104 317287600947506184192
634611230692031332352 1269294518978100592640 2538733153144277041152 5077754536664705794048

10156085534081715011584 20313323989668036870144 40628953822345287434240 81262519330709002256384
162534262033454859288576 325086970810983428128768 650210835110114275360768
1300495457196523388928000 2601138488345636454268928 5202572124596452261363712
10405734545003263228379136 20812649681627243868061696 41627660546495922558730240
83260043459474714762674176 166529531651915168815775744 333077952769761816212406272
666193684471386589586522112 1332462926806499093496463360 2665076969340450015639764992
5330456170135803688573206528 106661516803181414691733766144 21324242532182444012642238464
42650902916004117283633889280 85306641535286693083966603264 170622954477130303201330855936
341265251767374440469457010688 682569189160976549072504619008 1365215749574408434412190433280
2730586241653727541358743257088 5461481968317276427786211295232
10923582906654195545709872152576 21848403753347676471694643429376
43699283386773923703939085107200 87403518533704988928977766711296
174816940587724260900154726416384 349653688216077087884707838820352
699346990513411307938212449615872 1398773209189336880214018443182080
2797704874703702289103223974264832 5595726662057461635556822124331008
11192087149415037385814392600264704 22385441949430303001030281903734784
44773419200061062460863557213880320 89551909002523037839333101240582144
179113959209847901513878176106807296 358248200829299454698180299464900608
716536966477806212737208493432373248 1433155062594027032156112775869890560
2866472384464883277675617129750069248 5733269287483424982078017415520714752
11467187612074166817609601143082582016 22935673298362967342126334910247469056
45873942745155202098066935068659548160 91753077787168939023762400633648316416
183516540168054947702781862259955073024 367053849523544034716077846505227026432
734149237421956348053183936981087813632 1468381551593649253348424361903443148800
2936929256686771621180961699689421340672 5874190820372489471330149351143912767488
11749046254742871400596750605817965707264 23499421737481527717066405018696211759104
47001501930954625265878617651512984207360 94008320773892390195248850531267089793024
188027275371751059717480931519016422342656 376075818391434678088928323950997330198528
752194172078734473485789569727923631423488 1504473414749199181587444983107705204899840
3009116970681858832406621653519126293905408 6018574223730638603276706681645684356022272
12037829012195119083480340112506232248467456 24077019153857921920814533723442191569780736
48156760566651211349336774443743837285253120 96318965651173157714088962881206582861889536
48162205084521946364752188437462745576636416 24082463671728656936229947717161099861164032
12041912400598170345041900607795413467004928 6021296482666006110984413678505138501713920
3010818382516463524723938526556285134962688 1505494261849961996977835106930000509534208
752789666220846115796850475290929225793536 376416100758355616552391698558429098409984
188218684203144087603179079735696791961600 94114659013555183465081155096089517359104
47059987962769161564286385162165319368704 23531323209380365698016096388142940028928
11766326218688075306944500097601610186752 5883495416342983882440476000565875179520
2941913861670965055704350976165472632832 1471040007585219085094231976024003837952
735561542167477821168144231982635679744 367801540271173049894586237976634720256
183911154729303594602550179980975800320 91960769661510332128903620486817120256
45982980979184433478266075491573170176 22992788563806850446040170369868890112
11497043319010742076473651496975597568 5748846178059029464963608904508375040
2874585348306343945845196030264475648 1437373803791586579604293804137381888
718727466715000593142994796571262976 359384015767103948241921345536917504
17970214908835379956172646394101760 89856145146577812895692310009872384
44930607874489362906649148411346944 22466571587844909682726070908878848
11233919619222568956063783806042112 5617276722261341535382266078822400
2808796817455699296366320127311872 1404477636890363912520753607606272
702278432526439088429173575778304 351159023303848110298985173876736
175589415172238338191691779932160 87799659346276310616945486462976

43902305553216726069022541479936 21952390716647648414786169864192
10976814328343466897530534494208 5488716649181554793833992028160
2744513067095688069451358404608 1372333904800299370992860397568 686205638026377353630020796416
343122161826302510881805697024 171570752319708172474300497920 85790211863132544753849073664
42897523783205501635273949184 21449970817422365446811680768 10725589871620990037993193472
5363097167265398676290273280 2681699699360151166791974912 1340925407543801497719406592
670500482703763706021412864 335269130817813331591561216 167644010141872405086208000
83826727437419072188317696 41915724901950970916765696 20959043042596202869686272
10480111817108460140494848 5240351056459409423073280 2620323102182294387949568
1310235338067442032181248 655154562521868435193856 327595728005007927148544
163807087374540818350080 81908155373288836562944 40956383529653631975424
20479344686331422834688 10240248803918014840832 5120412632335159132160 2560350431355655421952
1280247273271865638912 640159665432951783424 320097847114985373696 160057930756747427840
80033468978001084416 40018986288814227456 20010619044313956352 10005872472110399488
5003217711031910400 2501749593004310528 1250945165246332928 625507766995255296
312771475683672064 156394533934858240 78201665013940224 39103031530225664 19552615276740608
9776857394184192 4888703574999040 2444489226452992 1222313332703232 611191026089984
305612692914176 152814936391680 76411763163136 38208029065216 19105088274432 9553081008128
4776808939520 2388538687488 1194336452608 597201780736 298617667584 149317222400 74662805504
37333499904 18667798528 9334423552 4667473920 2333868032 1166999552 583532544 291782656
145899520 72953856 36478976 18240512 9120768 4560640 2280448 1140288 570176 285104 142560 71284
35644 17823 8912 17825 35652 71308 142624 285264 570560 1141184 2282496 4565248 9131008 18263040
36528128 73060352 146128896 292274176 584581120 1169227776 2338586624 4677435392 9355395072
18711838720 37425774592 74855743488 149719875584 299456528384 598946611200 1197960331264
2396054880256 4792378195968 9585293262848 19171660267520 38345468018688 76695231004672
153399051943936 306815283757056 613664927252480 1227398573981696 2454934586916864
4910144051740672 9820837859295232 19642775230218240 39287749483692032 78579897013895168
157168590120812544 314354772427669504 628744729227427840 1257559827199033344
2515260391886422016 5030802258749554688 10062167467452530688 20125460834811904000
40253173469437493248 80510850538502356992 161030708276259454976 322079430951028391936
644194890699075747840 1288461838992189423616 2577067793172454703104 5154423816721061117952
10309424094194425659392 20620001109893458165760 41242308062796130025472 82489227811610687438848
164987678995258229653504 329993804734590168858624 660024502957327756820480
1320122792890950351847424 2640393159734490380107776 5281081467374160113041408
10562753230558678931734528 21126687052738075274772480 42255735288717585372151808
84516192943918040389517312 169041830620801820069462016 338102550707535118719778816
676242880346933194601267200 1352561318557592303525953536 2705273752842636435698745344
5410849737140176528691167232 10822303937190160371969687552 21645816800199935373114081280
43294051452039100004577574912 86592938607356658525853974528 173195548621270234085105598464
346410440055654302237006495744 692859565737536272607603589120 1385796502727527881482388373504
2771747747959966435499139137536 5543804980929754216067003056128
11088228931879151122271455674368 22177695803797587624817810472960
44357867487673746010185419194368 88720686735504633541470434885632
177451276991323550125140062765056 354922361023275666334678511517696
709884336127808464838153795010560 1419847900418131194013901133971456
2839854257161290916702989355843584 5680025426972638890756352887488512
11360684679245391896213454126579712 22722637009091012021828404956364800
45447809319382480502459803319140352 90900689241165873922525593451102208
181811519687133573680263160527847424 363643321783870799030950268306980864
727327208386948901402748431116533760 1454735546412312409487192651238211584
2909633352101454032337776880486711296 5819591222756566491402336916993998848
11639831482620449836258240146029150208 23280961039455533379423612916140605440

46564518227340334172661491080445820928 93134228751539203172951512657220861952
186278842096795476001160086307100164096 372578453381025091312834294599517208576
745198445136918461246696833169668177920 1490479967023573479735450154280603877376
2981126087546620073955013284443742797824 5962584482092186376878252520652555681792
11925833578182265211692956944835251535872 23852996384360315339258817696730783416320
47708651224712200510263443007582127521792 95422619361407540684018501243405376421888
190855872546781360695020232943292995600384 381733012741495280044006926799550476713984
763508560778855677395946775425029924454400 1527102192149441589407759394501917790961664
3054374525482343648047250476307551466029056 6109089333331608234557964327222534700269568
12218859231397058346042855403659932936962048 24439079592261800445939564305749592946769920
48880881443458968399586835608358640039231488 97767207404788671814589085210436188369846272
195545303845318813660008998408310193322459136 97778096440530141845419913197874004952612864
48891770479200438430417663595796456621998080 24447246369067902969062685296327955383844864
12224303749267793361458269397378841228345344 6112492157000817619192598073296852382384128
3056416219683869278828030723952142075297792 1528293180433664874029881205627928979701760
764189125512697554322873524639893460877312 382115830404281335815403223232911215951872
191068549026106947234684842072937850732544 95539591425036613280834036264710046744576
47772454168509876472162825746475584061440 23887556312250723151954316680298072375296
11944442770123254033913610243679176359936 5972553692060573245925031073604658266112
2986442999529759737446628512684864176128 1493304576514616425965370744283699609600
746693826632176491603713616112483565568 373367682503522385112370930041558663168
186694225845478262211442526013437771776 93352305219597665933349793503048105984
46678748758228100380489161999688663040 23340672453328683897151713623926636544
11670985263771658802029423124004470784 5835817150439487827741494718022811648
2918070834496573127234138937021693952 1459116546886701170298765257515991040
729598838262557888490230523260567552 364819701540882595915539208881569792
182419991975243123792981578066427904 91215066590022474814096775846035456
45610068596211693865851381329428480 22806301948706075162327187367919616
11403784799653151695864342035562496 5702209312476632905282545193582592
2851263112563344981316459684691968 1425710784444186754995823386296320
712895006303350509666708465123328 356467310192303820917752618549248
178243558616466193501075502268416 89126731068390238271637347631104
44565841414273689896368472064000 22284158647176130328459135156224
11142698293607707854367017140224 5571658631813675272252233351168
2785984058411748308660479066112 1393069400458329490597420728320 696573385855392413432300961792
348306035740810040782945779712 174162689276961937424870539264 87086180341759427229134094336
43545508022518942872916459520 21773962937079086065632935936 10887585931449350347403821056
5444095197179578830995587072 2722198714317241244144631808 1361174915022346536395735040
680625236443036225359577088 340331507687449591260643328 170175198576690534920749056
85092321654828137105588224 42548522010655503375400960 21275441596948469099003904
10638311094284593255153664 5319450695047475980402688 2659872921476327666614272
1330010247714458671513600 665042017345376754860032 332539455416762086981632
166278951080417898266624 83144087226227376521216 41574349456122901954560
20788327649566057824256 10394740285535332335616 5197658373143817879552 2598973301759984795648
1299558708474030325760 649815383034034126848 324925705915526545408 162471860157018013696
81240433678136377344 40622468638881873920 20312360219347779584 10156743059627311104
5078653004790366208 2539467239883538432 1269803988685946880 634937178715062272
317486181543575552 158751886864809984 79380341478916096 39692369762713600 19847284392984576
9924191952306176 4962370854060032 2481322865983488 1240730152468480 620399435972608
310216897855488 155117038862336 77562814398464 38783554682880 19392851083264 9696962412544
4848749641728 2424509038592 1212321628160 606194368512 303113961472 151565369344 75786878976
37895536640 18948816896 9474932736 4737728512 2368995328 1184563200 592314368 296173568

148094976 74051584 37027840 18514944 9257984 4629248 2314752 1157440 578752 289392 144704 72356
36180 18091 9046 18093 36188 72380 144768 289552 579136 1158336 2316800 4633856 9268224 18537472
37076992 74158080 148324352 296665088 593362944 1186791424 2373713920 4747689984 9495904256
18992857088 37987811328 75979816960 151968022528 303952822272 607939198976 1215945506816
2432025231360 4864318898176 9729174667264 19459423076352 38920993636352 77846282240000
155701154414592 311419488698368 622873337135104 1245815393746944 2491768226447360
4983811330801664 9968172417417216 19937444346462208 39877087716179968 79758573478871040
159525943050764288 319069478287572992 638174140947234816 1276418650638647296
2552978038765649920 5106237552508010496 10213038054969442304 20427202009845727232
40856655819505139712 81717815238637649920 163444637676530040832 326907289751569563648
653850608300158091264 1307773274194354110464 2615690663576784076800 5231669557529719865344
10463915575811743154176 20928984073128093155328 41860273989265400004608 83725159664549227397120
167459542701135309570048 334937532146344328691712 669911957780836076486656
1339897702537966991179776 2679942979028523658772480 5360181105962226670370816
10720952507734812046393344 21443085607090341504090112 42888532397422117830787072
85781787161327105306787840 171573019055619949904003072 343164927577171378388860928
686367634086205713939431424 1372810826036137342202281984 2745772767799726513051402240
5491847767054356683396481024 10984299997018520681380315136 21969808919856655991935336448
43942035691352541242220085248 87888907085983541001138995200 175787485578523999035675639808
351594313970161832138146578432 703227313566551332409883754496 1406531998385558001086948704256
2813218739276026674708259799040 5626746963561874694485244379136
11254112897143392079107938320384 22509463734326069538490775764992
45021403348730709837531349778432 90047758457618561196162296053760
180105420435551405434523785101312 360230647911731376953445956190208
720500909904719886075688684355584 1441081047971954036488970912661504
2882320552268936601653128913223680 5764958017187930260656632002248704
11530549859675974636014012356100096 23062367369952177501429521415405568
46127270041104811461662036237221888 92259610684610535840930059287265280
184529362574022897517072092200173568 369079007557649446704568131651633152
738198579934506196749984157805838336 1476478289507427000181664104616820736
2953118838291683213726719787243929600 5906562195137024854180222730508435456
11813773427381366561814011773058023424 23628844928977366830535156170198351872
47260286006384001074884577588561313792 94525764309626536977397685673451847680
189061913212970143610052432339562135552 378144595613374426530618986664441151488
756330729601617131682266217299516063744 1512744535952970820606588922540299649024
3025655225405414755697290820963134341120 6051642757809775740362807593691338768384
12103950129617443938662067090912817709056 24209229487230672793197037988885915762688
48421117430452915418139883591892392214528 96847551772888970499771382412025905807360
193705737369744220326525995280534054371328 387432742387420999307018451474032594255872
774908020070707115921969824773994159538176 1549901110733144466459805493199846261129216
3099972362649749402151342673703408406364160 6200285007666419742766148722014248580939776
12401250580066681362459224193243360698302464 24803862289601046478772301884916448469450752
49610446838137460465252310766692351084593152 99226338194145655945920035527103610460569600
198463565424032781922670899041645037503905792 396948908919548503907003454058165708173344768
198485343495515721984332555016520670669438976 992481162656285960075816915019792436261027840
4962678039174966551149855274784907595874304 2481475132534251650960312987235426505221734
12408056227405100131728491685391996062531584 6204368396069471004327709217303429799477248
31023543392181959713955862959554307838443520 15512622402008282203136589916295733339750400
77567365539627922746476241764071563801395200 38785809534607217238634766973332230452019200
19393968149700236552015706532314339501670400 96975157660484322423570147889812818886656000
48490237286233731043530881559026970132480000 24246447871112650437638344586573765410816000
12123888549554217676755624196817022877696000 60622765817760550673460380501735815249920000

30313044443875006481571320009693258055568 1515735298943486881320622488425930424320
757909187846611719281339488183598972928 378975363110739998951183866077116366848
189498066149087069130848994031216623616 94754225371402069393053027511937531904
47379708834130302110340779004133376000 23691152491279784762077522126148993024
11846225282747209234492327375115649024 5923437159927263043972946843578400768
2961880839240460735349864999799488512 1481021549258644974356628288904888320
740551339448529790519162038955016192 370295952133868546930004966728794112
185158117271736099300214456990040064 92584129238268962567713215307841536
46294599920334937742659601060331520 23148567610767697100731297233371136
11574917630683962665066396968288256 5787775727992038389883572659945472
2894046320321047723616973417873408 1447102388323038126146080252887040
723590808242776195241836898418688 361815211162016663705316835196928
180917509101322614894857610592256 90463706310818448968528401793024
45234329035487795244813999144960 22618402457783183002681898696704
11309820198911234191478398910464 5655219584465438440807924236288
2827764534737629892938324508672 1413959638621270282736343449600 707018504936862809501762322432
353528595281545238817676460032 176773969047329536442235879424 88391820226943226737816764416
44198327965110842627257794560 22100372908375035942803603456 11050790917097325285989154816
5525697690003566300288253952 2762999960729234978790965248 1381575538228343403718901760
690825548046034659021160448 345431663488948808091435008 172725276477440143336144896
86367360605202941313286144 43186041485842905479249920 21594201334542170150928384
10797690963081443781115904 5399140629445901243383808 2699717888675540298104832
1349932731314064987258880 675003259145179912732672 337520076316663665917952
168769261530368687734784 84389242451202771255296 42196927068610599321600
21099616455809906507776 10550384688657256677376 5275480574704780050432 2637884402540465881088
1319014258864270868480 659543158229154398208 329789593513086681088 164903803955798081536
82456405577526411264 41230454588576890880 20616353194195288064 10308739547051065344
5154651248502243328 2577466361739476992 1288803549613916160 644436959179046912
322236071775567872 161126831980806144 80567814036914176 40286106041712640 20144152532484096
10072626022055936 5036587888934912 2518431383420928 1259284411187200 629676565331968
314855462535168 157436321202176 78722455568384 39363375267840 19682761375744 9841917558784
4921227214848 2460747825152 1230441021440 615254065152 307643809792 153830293504 76919341056
38461767680 19231932416 9616490496 4808507392 2404384768 1202257920 601161728 300597248
150306816 75157504 37580800 18791424 9396224 4698368 2349312 1174720 587392 293712 146864 73436
36720 18361 9181 18363 36728 73460 146928 293872 587776 1175616 2351360 4702976 9406464 18813952
37629952 75264000 150536192 301088768 602210304 1204486144 2409103360 4818468864 9637462016
19275972608 38554042368 77112279040 154232946688 308482670592 616998895616 1234064900096
2468264017920 4936796471296 9874129813504 19749333368832 39500814221312 79005923409920
158020436754432 316058053378048 632150466494464 1264369652465664 2528876743884800
5058028365676544 10116606487166976 20234312485961728 40470823995179008 80946046036869120
161900888166760448 323819368519565312 647673921411219456 1295418211566616576
2590977160621588480 5182235796219887616 10365034542393196544 20731194984693235712
41464641769200156672 82933787138027683840 165876581475310108672 331771177349129699328
663578383495278362624 1327228824584594653184 2654601764357265162240 5309491759090682036224
10619559978933667495936 21240272879371941838848 42482851601753097371648 84970314889524622131200
169949853151086099038208 339918153046245907628032 679873199580639234359296
1359820186137573306925056 2719787946227736290263040 5439871040360651933351936
10880332376531662572355584 21761845344684042556014592 43526051872609519934636032
87056826111701909514485760 174123096956369558319398912 348265083378670595219652608
696567945689204147601014784 1393211449242134209525448704 2786574014211720247697735680
5573450259878344152689147904 11147504982666495619965648896 22296218891152605869106003968
44594855633944440996561420288 89194546971167340509821665280 178398765348891598053040979968

356816873510897030172877258752 713672432648021728479345115136 1427422236548498793225871425536
2854999215601908258986105241600 5710307916213637863040935264256
11421234802446918416219320090624 22843707544933122212713539305472
45689890969944815185976876859392 91384733700046771893053350215680
182779370920407826828305893425152 365578548881444219741010172837888
731196711844145571650817117650944 1462472651850805407639227779252224
2925103760026639343953642646405120 5850524432703335745257659468611584
11701682690706785605216067288825856 23404633032013799439833631280857088
46811801365228055338470255968124928 93628673332857023594546498749071360
187267487870515873024304971123785728 374555258150635397719033889498857472
749151081120478098778915673500286976 1498383291879370804239527136005718016
2996928843035570821842445850021724160 5994182204624800070411674856064024576
11989013446356916994276916024169201664 23979324966928467695460964672420708352
47961246082286202804736194593006026752 95927684461430940437100919682341273600
191865753516578950529458900357340987392 383752276220592040369431922699998855168
767546090816052359359892089370631471104 1535175258381841275961840666682530463744
3070516670263155666407794309247595970560 6141365647525257561783814570260262027264
12283395909048407581504081044050664226816 24568121046092600078881065895161608798208
49138900548176769989507939404443778285568 98283118008336679642507494037128677949440
196576869840639638611998218530739598655488 393175007329211835877962897974443682824192
786392549954288789063858717774816336674816 1572870170500307812743583279201490615402496
3145910482184076094718898245706697114910720 6292161246735073127901259866020825998032896
12585003058203988132729446481256515532488704 25171367245875660019312746460942758137823232
50345456750686687546333199918744970421338112 100696358019244110108081813831208849134059520
201403605074229690246994455649855514850885632 402828988219942320555650567274586662867304448
805701532582850521234624446498924592065675264 402872544362908200678973879224337929198370816
201447161217195570370317767599606781181952000 100729025126468520200574297793522298882359296
50367234822169627607994855893620603586871296 25184978540552497557851281445240028866281472
12593169835010090655852567471834877969563648 6296925199871966266389747110524870752993280
3148632741119443602426605242566151260602368 1574401441151452035829168464934933572354048
787243255871591135222517154293395757203456 393642895583728126265225038059662364114944
196832081615830342459595749486313424814080 98421357719898310893289489971397833785344
49213337315940725278390552599819477581824 24607997885966147555068180106970019135488
12304663556980966235470541957015149740032 6152664085489429346703496930272644956160
3076498196244187787835861441018857521152 1538332174871830451159987208450696282112
769207625810783504201021848195981901824 384624582092825891411025046083307831296
192322675640130015360769584034312355840 96166530116923542508013322513485398016
48085861206891038667820926504907309056 24044228677660153040817595876535959552
12022763375937393373862364250309132288 6011706206522355113657965281175142400
3006015362538006770192374218597859328 1503088810907417991777882898304073728
751584970272916299229789343654608896 375812767546061801285318619078590464
187916524977832726477871283164938240 93963333091317276156541628395290624
46984201846859094537073807604056064 23493368574029775497938400505233408
11747318112315001863669948604219392 5873975968807557989185348477911040
2937146440728807523267861326856192 1468652448526918025971524207378432
734365838344716145154558875664384 367202726212986638661677823819776
183611266626807602373038104903680 91810585073560942707618648948736
45907768416859042114359122722816 22955122148468806437454460485632
11478180044254045908864679804928 5739399507136844299501064683520
2869854496073332822284894732288 1435004619289121747409628561408 717540995270788541838404878336
358789840448508104985997737984 179404591630810969526396518400 89707131518683943279897083904
44855983610981200898297954304 22429200731310215078323683328 11215204828564914853749194752

5607904645737361084168273920 2804103438596132370730975232 1402127277161792099688906752
701101417512759007006162944 350569598222310982083936256 175294243844121230332395520
87651844288543484811411456 43828283327513177228312576 21915322255377306025459712
10958251423499011718381568 5479420859654685212016640 2739858003779932282421248
1370002788866260979417088 685038287921277908811776 342537590704712663957504
171278018724393186754560 85643621048215020765184 42824116367116724076544
21413211105062968885248 10707182013283787866112 5353879237018045644800 2677083733697098678272
1338613924442587267072 669342991018312597504 334689509907665780736 167353762153087631360
83681384676171186176 41842944137899278336 20922597968856481792 10461861934381662208
5231212442167541760 2615746958572126208 1307943848030240768 654007108387209216
327021146379649024 163519369282846720 81764082687934464 40884240367222784 20443219695239168
10222159603433472 5111354679623680 2555814778765312 1277976108859392 639022414168064
319528386953216 159772783411200 79890686672896 39947490820096 19974819151872 9987946446848
4994241658880 2497255047168 1248694632448 624380870656 312207212544 156111994880 78060191744
39032193024 19517145088 9759096832 4879810560 2440036352 1220083712 610074624 305053696
152535040 76271616 38137856 19069952 9535488 4768000 2384128 1192128 596096 298064 149040 74524
37264 18633 9317 18635 37272 74548 149104 298224 596480 1193024 2386176 4772608 9545728 19092480
38187008 76378112 152764416 305545216 611123200 1222311936 2444754944 4889772032 9780068352
19561185280 39124467712 78253129728 156514648064 313046073344 626125701120 1252318511104
2504771239936 5009810915328 10020158701568 20041391144960 40084929773568 80174154514432
160356898963456 320730977796096 641496315330560 1283061350137856 2566260139229184
5132795156365312 10266140068544512 20533379648716800 41068958320689152 82142314687889408
164293425468801024 328604443123646464 657244070619381760 1314558509982941184
2629257757454237696 5258796989885186048 10518156929723793408 21037439759354429440
42077131318522544128 84158766236672458752 168326539672599658496 336671093743708798976
673378216284436561920 1346828490162911051776 2693801095513897959424 5387890421403947630592
10776357303560198684672 21553867528625004216320 43110040900259222126592 86224693486536871641088
172458610345110598057984 344935667434294905667584 689908228356737230438400
1379890243689769299083264 2759928061332128274579456 5520151270569435901984768
11040892836949230509621248 22082966265519178430545920 44168293714279791683698688
88341309795042453012611072 176692064323050645315649536 353403018112032769212153856
706843815155928495586017280 1413763188175582905495453696 2827677492078617639637745664
5655657215612138936569167872 11311918894134085187725688832 22625046714087785004626083840
45252511279814799267601580032 90509858262908057051901984768 181029387932373031137201618944
362078118677859896341198536704 724194922981947460815987671040 1448467217216350257899156537344
2897089176937611188332675465216 5794487838885043721734075711488
11589594647789730133605600985088 23180427235618745647486101094400
46363330351316062055522000437248 92731612462789265632143597371392
185473128445892814306486387736576 370966063932414194697371161460736
741971741946085521563539094896640 1484022712054685307464671733743616
2968203880434399143604530555387904 5936724673518855344559435286577152
11874083172337824803819618924756992 23749433995275877837040734552719360
47501403291752212132884462511849472 95007877185905337183374911836520448
190025895576612500201961797298683904 380072073562828652074347541848653824
760184711944864607489542978199879680 1520450553528143821660781745404903424
3041063366333116856684955068820094976 6082451251219892140096693293660766208
12165551539547101133646952899362684928 24332401153308835974201038422807674880
48667398455046939362216342093779959808 97339989206952413552061214683889139712
194690363007621896759379490360436719616 389401495202677932829273102706190319616
778844528780224144279574449383014400000 1557772134310184845801205386707296321536
3115710422119842806086523749297127686144 6231753151238631841141273450359325458432
12464170916475156140218998804248791089152 24929671060946097196310901415557862522880

49862000577883764224367610445236285734912 99729318067750668112226836118713692848128
199469269959467615551436902693909628452864 398959807566867789756840266300783742418944
797962150429600696821613454427496455864320 1596009371450931628259092752506850853781504
3192188884085323725749917192317417591668736 6384718050537568389963297759242266951548928
12770116665808978656853522267699397439520768 25541594461085641067560898033828521951887360
51085911181106649642829503064516498049466368 102177266880084034301074420122751904390316032
204365422795909538632979668232941625363398656 40875262366330201732762099244075883892330496
817548803469569914778565296831269034115727360 1635184719225071589803777217562040600893587456
817635915755501675025211920730771566777860096 408839735949233777574267616340261416554463232
204430757010358358817964636157568524859998208 102220823023049914424397732072503170721382400
51113133770460324719906573033111039506382848 25557928014697846113807140014985246826037248
12779644572082764933830496756707486949441536 6390162568408303405378711752961175242932224
3195251425387612171921087563784303505571840 1597710783285536320576409625544009694838784
798897926938633277596137734597933818445824 399470231117249197452035328211931394736128
199745749382590878053000894562447940124672 99878191603278578689992062509465091440640
49941754257630859176741838868853106409472 24972206356811214504243823241486833549312
12486767792403499710058363524273556946944 6243716203200696083997407713901848559616
3122024255099821156482816832833459322880 1561095204299647135483464904357997182976
780589140524691846362760696149632352256 390315339449780062491894470060133056512
195168054318607100901204296022724968448 97589219456162085278230678507691704320
48797205876510310052929604502010462208 24399901012469788733371934875087536128
12200599543342211220139533749584920576 6100624290224764036796550030813036544
3050474404389211231761666593416806400 1525318331833020222562529085713547264
762699730735717414622112437359345664 381370147777462358981480165930958848
190695215093533005325952056591122432 95352678149167415580582015108382720
47678874375784164249094000960602112 23840704838492310353948497183506432
11920986244546269291674996943355904 5960810034923191703187872647479296
2980563473786624380269123411640320 1490360965055826454472155249770496
745220096609170359404874396860416 372629855345213745786835584417792
186324831192921155935616985202688 93167367356617719488908089098240
465861595558387430505003842797568 23294317719233000632776820523008
11647777829636143006525859823616 5824198399827892848331654692864
2912253942418857096700189736960 1456204342461883884617276063744 728140856857169610442228629504
364089771241698639287909613568 182054557027406236677352456192 91032114216981576855375052800
45518474960130017686036938752 22760446405884623472193175552 11380827665852119050683940864
5690716064380963182635646976 2845509147917933419964661760 1422830131822692624305750016
711452844843209269314584576 355745311887536113238147072 177882100676733795909500928
88945772704849767599964160 44475247535666318622588928 22238804359453876722597888
11119992475537297066950656 5560291385673827886301184 2780293266789503619563520
1390220420371046647988224 695147103673670743097344 347591998580909081100288
173805222662491395325952 86907223017264125050880 43455917351641276219392
21729111597325244956672 10865132259414925901824 5432854360083614662656 2716571295229883187200
1358357705208979521536 679214881401508724736 339625455099263844352 169821734748886663168
84915370974070702080 42459937286849036288 21231094543331360768 10616110221619101696
5308336585786261504 2654309030381486080 1327224883934920704 663647626339549184
331841405355819008 165929498770931712 82969147431976960 41486772739244032 20744485881249792
10372792696438784 5186671226126336 2593473052016640 1296805245485056 648436982480896
324235671109632 162126425489408 81067507712000 40535901339648 20269024411648 10135049076736
5067792973824 2534030704640 1267082461184 633574785024 316804169728 158410473472 79209431040
39606812672 19804454912 9902751744 4951638016 2475950080 1238040576 619053056 309542912
154779648 77393920 38699008 19350528 9675776 4838144 2419200 1209664 604864 302448 151232 75620
37812 18907 9454 18909 37820 75644 151296 302608 605248 1210560 2421248 4842752 9686016 19373056

38748160 77500416 155009024 310034432 620101632 1240268800 2480668672 4961599488 9923723264
19848495104 39699087360 79402369024 158813126656 317643030528 635319615488 1270706339840
2541546897408 5083362230272 10167261331456 20335596404736 40673340293120 81350975553536
162710541041664 325438261952512 650910883643392 1301890486763520 2603918412480512
5208111702867968 10416773161549824 20834645834727424 41671490692710400 83347379431931904
166703554956886016 333424702099816448 666884588571721728 1333839545887621120
2667819829263597568 5335921133503905792 10672405216961232896 21345936333829308416
42694124467472302080 85392752534571974656 170794512268398690304 341607038935306862592
683250106667632689152 1366572270929303306240 2733288657046682468352 5466865544469516648448
10934307549691336720384 21869768020887280287744 43741841884783774269440 87488295455585975926784
174985814283208806629376 349990075310491322810368 700017044109130064723968
1400107875194554967654400 2800363324341699611721728 5601021796588578576269312
11202633888987515858190336 22406448369595749127684096 44815257922432933077975040
89635238211348735801163776 179279921155663210892754944 358578731777257900366364672
717195242486378757894438912 1434466042836483430112296960 2869083201400418688871432192
5738468634255741035036540928 11477541731421289384660434944 22956292388662193398495576064
45915002628963616055340564480 91834840961205690627379953664 183679353328968298288157556736
367378049471050430643110412288 734794784568328529419811422208 1469666940389112395106804039680
2939488623283135462747970469888 5879286731576092270564665720832
11759192433171827231266781003776 23519622806382939842808461131776
47041721492844450446166720512000 94088394745846042413433037520896
188186693012006367869065268035584 376393193064641301822528922058752
752826000210539735813854616092672 1505731228583593735965302776135680
3011620913492216000605792640172032 6023558739634489058561959456145408
12047751304569092231824667263893504 24096770259738412693050831230992384
48196075820677281844904655868395520 96397222243755476607415298549612544
192804585692312779050042570724868096 385629453794229209770509088701022208
771299472407665722881866071904616448 1542680074453746052445427932814376960
3085522408184321318254247443639042048 6171369334922301063235278043298660352
12343387706951918979924122398638473216 24688073488118471666755377421359251456
49378743124666210747325020090883112960 98762678546190956322278570678095446016
197535741686098982299814202348849332224 395092252559632103910142526683015544832
790226043494132486441313297336664850432 1580535163738001530124683082614597222400
3161236480975476174733479141111729487872 6322805268949898578435184233988529061888
12646275151897689614806820371507198296064 25293879531791164145486544550074676936704
50590417519573898122718896714269914562560 101186151951130935908929408656780950503424
202382937726228151144842047770044143763456 404787143100388860943650556453052773040128
809616821496642839195234034732034517106688 1619318713585015913006333913115926976266240
3238807568353492295244399513535569836638208 6477955419073905528952262401678571441487872
12956591402881652934831451552572006419398656 25914543935230989623516756603573739911643136
51831810129397346754741220204006933968977920 103669064776665428524897854401732776229339136
207349018589072327080626536790903369041444864 414719815249627594222914729556682371248422912
829483186642221068569152771063116008827912192 1659053485570373897384952166025734550317957120
3318281195712611315263197579850474165960179712 16592277101422374178782454138247396156422222592
829657411214084589062446018862121074152177664 414850483678525234592884665405936170241622016
207436130875004087327273160690405901703577600 103723509955372778679051994338921859143172096
51864477236621756847233704166320383717277696 25933599747778562177470705581589918931484672
12967480438623122965662279540009823002165248 6484080501678482421294603144612343269294080
3242210392022701679879033259609887518752768 1621190266603081074555382473456801701429248
810637668597405654585624158554329821741056 405340101946635385946778540190129396383744
202680684797283972300372500551546940948480 101345659310625125813677865504014591852544
50675488111304132738584740366127856615424 25339073283647851285165273990124208652288

12670201255821818100519088898592244498432 6335432934909855279227770401061192335360
3167882620954400754097998176413131210752 1584024387226936934291055576147833126912
792053731988336745766556032044550324224 396047635181602512193792138007592042496
198034202184518325752153129996454461440 99022293389117697703705095494556450816
49513742842988116265666812995442835456 24758169495708691839740539121803722752
12379733784961662773323835872943013888 6190191411034489813388701092492083200
3095257964794074120057742124256329728 1547710112035451666710566851133308928
773895620836933136696131320069226496 386968092828070220018489607285899264
193494187618836935844456777268592640 96752164411819380839834375447117824
48378617507110146878720181129969664 24190576404155301668761587268190208
12095922027377764949081541985697792 6048277926338939531891145168650240
3024297419494498294620759672225792 1512227937909763411647973380063232
756153583036138837992783462006784 378096598558697985080790116990976
189058202799663275582594251489280 94534053159988779312396722241536
47269502460072960416748159369216 23635989170075765588648978808832
11818613555057525484461938966528 5909616262538584087299694264320
2954962873774202716184209522688 1477558808139556694359285956608 738818089696006015313233575936
369428387661116841723412086784 184723865237115337895103692800 92366768321836127464250671104
46185802012557292990474747904 23094109932098261124412080128 11547659428958937876793393152
5774131945934372595690373120 2887217088694638126492024832 1443684102211044977569431552
721879830037385445946425344 360958804484624201554067456 180488846975277840067461120
90249145854121789678944256 45126934110302329662078976 22564647646771882242342912
11282914119196299826823168 5641752207503329266237440 2821023677704254309531648
1410585625828421992972288 705329706402358415589376 352683299945252917346304
176350873344663313448960 88180048358350084112384 44092330022184255750144
22047317932596734722048 11024235427050670784512 5512405943901487104000 2756347087138819407872
1378245601163447631872 689158829378742779904 344597429087880871936 172307721743195176960
86158364471224958976 43081434035426164736 21541842917619924992 10771484408763383808
5386023679358402560 2693152577167556608 1346646657327955968 673358513036066816
336696848704077824 168357220445061120 84183008269041664 42093703157776384 21047951090515968
10524525301071872 5262537528442880 2631406203174912 1315771821064192 657920270270464
328977315004416 164497247436800 82252918685696 41128606826496 20565377155072 10283225448448
5141881159680 2571074797568 1285604507648 642835808256 321434681344 160725729280 80367058944
40185626624 20093861888 10047455232 5023989760 2512125952 1256128512 628097024 314064896
157040640 78524416 39264256 19633152 9817088 4908800 2454528 1227328 613696 306864 153440 76724
38364 19183 9592 19185 38372 76748 153504 307024 614080 1228224 2456576 4913408 9827328 19655680
39313408 78630912 157270016 314556416 629145600 1258356736 2516844544 5033951232 10068426752
20137902080 40277901312 80559996928 161128382464 322273542144 644580638720 1289228386304
2578590990336 5157450416128 10315437703168 20631949148160 41266045779968 82536386527232
165081362989056 330179905847296 660394171432960 1320857062342656 2641851563638784
5283978005184512 10568505766182912 21138111043993600 42278421111242752 84561240268996608
169131276631015424 338280145448075264 676595475268239360 1353261319280656384
2706663376049668096 5413608227076046848 10827779404105515008 21656684708117872640
43315621216049430528 86635746031726231552 173280499262707204096 346579012923923890176
693194054644866744320 1386460166883771416576 2773064448955618689024 5546417128287389089792
11093410717327081603072 22187974356158770053120 44378254555326753800192 88761120796671934988288
177531464965380724752384 355081376674835159056384 710199646837817737216000
1420473080651930312638464 2841093735256450301689856 5682482618418079956205568
11365555532646518618062848 22732291656913754647429120 45466944497068944117465088
90938611360620757880143872 181886667454207255050715136 363792224374345988682285056
727622227680554934526279680 1455320013224835783375978496 2910791142177123395398795264
5821884515809150448091267072 11644373494528108210769887232 23289955914875831050714480640

46582329681390891359778373632 93169495066060241236255571968 186348661538677399505908793344
372716665890468633078612885504 745472017407164934290816368640 1491021406066785204848813932544
2982197554638481082231990255616 5964704594286783509532705292288
11930028158593209709202860146688 23861294257225704798680619417600
47725064394529980357911037083648 95455080549217102236921670664192
190920064618748487516042534322176 381859936278125541116483454631936
763759486637508214401763681239040 1527598201437530693141120906428416
3055354859200089914957428900757504 6111026631050236887265231977316352
12222687087400587889231212306235392 24446641825401404007863921315676160
48895818952003264474530836037763072 97796708506407441866667658888347648
195603558217616709568547291402338304 391227398844837070807518530055962624
782495362508881444955884954614497280 1565071854656177496593465698234138624
3130305968589184206550322974478565376 6260936455732026839827429104977707008
12522521948571370533108424521996566528 25046341971357374773123981668075438080
50095280091144016960062228584315486208 100195752479146568747752987664960192512
200401889552010207150763036322578825216 400824548291454553612040194630474530816
801690634957777385845108633231582822400 1603464346665291328932273754404433166336
3207094846830055772348660484691401375744 6414522000659057773665546921147872837632
12829708615316008005267545745825885847552 25660746458627800926407995298712052039680
51324151373247171684561798211544664768512 102653619658477483032615211651330450915328
205317873140921245392213653759143144587264 410657013929775049438393768431250774687744
821356563155415216184720458688430520401920 1642798196902560666985306761028718982856704
3285766534988581803202345209361153849819136 6571873352344084544868153793329739467849728
13144427269422010966663234335874342472122368 26290215668311705687180322170178412017090560
525831535955558778882068351337216278179872768 105171751708988292779552116668151464651128832
210354392453718055589935061323740745885024256 420730562978919051241531778622357124935581696
841504682100803982606386869194465516202229760 1683096476487539725459420362288433565066592256
3366367177546942971412133972375872195457449984 6733082804237612983810854440349754521563430912
3366715626690670012398720467973882326105980928 1683444925631266766446006857886443695715123200
841766018958599263346326740892973114188627968 420904787550782571734825026421362190259847168
210463282811132755898243341198118911712690176 105237085923437112964537084592778364147728384
52621265220653923989976249293248636219555840 26311993739794645748841978145054045182623744
13156677434631164751347915821741886127734784 6578678999682503314137421285478374832078848
3289509641024712126300442330042903300145152 1644839891104086297766087008673309592125440
822462480847908266190976426162583767089152 411252508071886691749454673994256369057792
205636887859909625201710567453610427285504 102823760841937952264346898955046335021056
51414538876960545963919257091643728199680 25708598666476057897832532352882144444416
12854963947235921406852718079971212394496 6427814280616906932394584991750676283392
3214073293807926580681405471757873184768 1607119723653699847582759223820204113920
803601400201718202412407855880735817728 401821469288293240516718049925684789248
200921119237863689913616085955500834816 100465751915790379784436573474079637504
50235472106324457306032551985204428800 25119034127376862359923408616684519424
12560166100795748033415270620383412224 6280407568951532443434418466212282368
3140366043752595435080600811116429312 1570264151514712324221996194563358720
785172640576563465451845991784251392 392606602697885384396346943143411712
196313442553744518033385445197348864 98161791879273171934298709411495936
49083431240837042425952348112158720 24542983271018749442377670759284736
12272125460809488835889583731245056 6136379643054801475295166041423872
3068348277852429266322770108612608 1534253367088728897498978598256640
767166297625621580918286071103488 383602955853439356543541421539328
191811381447033961313969903763456 95910642483674122178084548378624
47957797121915631849592072437760 23980136500997101305070935343104

119906872205181933426729172336664 59956530952689180164051833978888
2997981290139369680736954089472 1499068016322140176635658240000 7495726937872977564514197176332
3748056897067627122925051576332 1874125162599382731796502282224 937110938332475951065239388816
4685796476826302681161138176O 2343019130995112803498039705 6 1171570011788537133207755161 6
585815229039758932333245235 2 292922726092624649031306444 8 1464689188326849159479951360
732382373095287536901685248 366210076013575247031697408 183114482739753362806276096
915619637363595510483517444 457833430514212103467827720 228928521173313225846947844
114470163544760199979991044 572380332514318935182540 8 286204923652418435232563 2
143109840523838701436928 0 715586096107340926287872 357811494797744172695552
178914970770908941123584 894620970714728979496 96 447333543787456626688 00
223678301108774381813 76 111844915161910225141 76 559253398847166296883 2 279641110942390734028 8
139827761230599159808 0 699174834950014763008 349605431873516863488 174811723136013172736
874103651676339568 64 437074343836306636 80 218548430917221744 64 109279844958145085 44
546427372288396492 8 273227759893033779 2 136620916820934656 0 683139768476762112
341587476424425472 170802534305234944 854056651991285 76 427050316228198 40 213536153230376 96
106773574173327 36 533895358657331 2 266961423224012 8 133487583559680 0 66747227753676 8
333753318637568 166885249253376 83446919593984 41725607280640 20863877382144 10432475561984
5216506216448 2608387325952 1304260771840 652163940352 326098747392 163057762304 81533075456
40768634880 20385366016 10193207296 5096865792 2548563968 1274347520 637206528 318619648
159318016 79663104 39833600 19917824 9959424 4979968 2490112 1245120 622592 311312 155664 77836
38920 19461 9731 19463 38928 77860 155728 311472 622976 1246016 2492160 4984576 9969664 19940352
39882752 79769600 159547392 319111168 638255104 1276575744 2553282560 5106827264 10214178816
20429406208 40860909568 81726013440 163460415488 326937608192 653908770816 1307884650496
2615903518720 5232075472896 10464687816704 20930449375232 41863046234112 83730387435520
167469364805632 334955909480448 669946178699264 1339961076875264 2680059592704000
5360394063314944 10721337882443776 21443775276515328 42889749576286208 85783897199083520
171576590491189248 343170773168422912 686376730708934656 1372823830162046976
2745788397812449280 5491858270601609216 10984279491156639744 21969684882220122112
43941621564253929472 87887746728135229440 175784500655525199872 351587015709559881728
703210060216138727424 1406492178026315382784 2813128471240706621440 5626545172857564954624
11253666806467433332736 22508486534439473512448 45019278911888160718848 90043169509794748825600
180095562391626352427008 360209571527326414405632 720456036542800247914496
1440985860061895334035456 2882119294076380344483840 5764533736057940041793536
11529657767926238789238784 23060496127473194989780992 46123353438187824802168832
92251429242858519249551360 184512303218682777789530112 369043495903297034159915008
738124770738457025481539584 1476325099340639965286498304 2952801314408731759219834880
5905904860272367175733346304 11812414183454541666054045696 23626037292728697961282797568
47254492437096625180915007488 94513820577471708878528839680 189037312561500334790455328768
378093967936114503647705956352 756226621498456675429002510336 1512530614249368687125186215936
3025215971003648046784734822400 6050741427017117438638194425856
12102101824053877567413838413824 24205441588147040515102575951872
48413359056372651790754950152192 96831669872902445102609496801280
193673243266119173247418186596352 387366293572866912579234759180288
774772201226990957327266290335744 1549623630616496178992126124621824
3099405717558020886659439337144320 6199128347766098830669252850089984
12398890520832311776039254051782656 24799048692264851781480004806770688
49600632685730160021763003019952128 99206335973861232961131992852725760
198422813152524291757475959331094528 396865908714652235185375865913475072
793772382248511773711599626329522176 1587625894135438154104895041664188416
3175414047547705521573181661338664960 6351152613649069469873146478697906176
12702954264405455793199859269436964864 25407206603025545293306851162956234752
50817009354480358000427967574077079552 101639211005819250828484465644483379200

203288806605355571312225992281625198592 406598382398145281934966106548567277568
813238303171158842490960457067768315904 1626559683092054242223977402076804153344
3253285519683581598932067780036143349760 6506903346366109426832361511837356785664
13014471306730111311601174927204853743616 26030271841456007539075253661469987831808
52063202138903584909896314937060536352768 104131721189790309483284245102362194083840
208274076203546898293551720661206630924288 416569420055026355241069902235377747361792
833181375405917827790072726296684465750016 1666447821403565890196011296245226873552896
3333065783990592249623754279794169631211520 6666471850348105437710971934195771030634496
13333624265430052752348870617606405597691904 26668609660327789258551594733642538268229632
53339941579590946024810896464144530682150912 106668532767705262706503720692200796965568512O
213381544389846724160905241831453755894136832 426784866851176388383472139637783144953806848
853613289845318656890267591225317556238680064 1707313691976569074027181806350137645139492864
3414801608525001668547656860499280355603251200 6829951666193730378081900216596570841855033344
13660600230674914838136973424389161945007128576 6830648564481184460055073207792591103152095232
3415498506812455750520829851695300616900313088 1707836365692159635507061549747152841112289280
853961738989045697876854086823327686887211008 427002647566005789000088699386539476609138688
213512212818744364530875177680707554887335936 106761550927242917280853002834072685735051264
53383497722556826148134208413895797013217280 26693109990746096827920957705377625579454464
13347235560106890290887405601903676326150144 6673958062420366083907166175559269931286528
33371491723936435111185314775083350849748992 1668659656788551990208523231193533366927360
834372363690141112412194537422695654490112 417207449493003114860063729624312312758272
208614358570467836757015095268638399135744 104312496197217058041999162862560320946176
52158906554600098852745389045400721162240 26080782505295834342245598329760640925696
13041055866645809629059251068410460635136 6520860240321851043497851485970300403712
3260596273660398636233038718867685244928 1630381213579935875358575847375110144000
815232145164836216300316167658188832768 407636841769852247460672205814411296768
203828805478643193385593163899864088576 101919595036180131520425112446261264384
50962393666519333174026821471295242240 25482494907474300293920543359729926144
12741896490844467000413837991906115584 6371272763975891926933702151973634048
3185798641264775176830242653997105152 1592980450270802195096817116003696640
796530789954608400889256452504420352 398285677386907852115052173503496192
199152979898255751892738060377391104 99581560551528788863975017001517056
49793315576964850890790501907169280 24897925439082653674796747656790016
12449596544841440952099122179997696 6225115185070777533399935265800192
3112716048860417295375154720800768 1556437252592722912025170904350720
778258240377618588181382224150528 389148927229437860175089498062848
194584367135033213129743942025216 97297135327673748085971567509504
48651043543915444803535582003200 24326759711997007782042690125824
12163998826018146581158794625024 6082308898018894635648122093568
3041309191514357990358423437312 1520731967009634331446392913920 760404669131044833856787054592
380221677378636250995188826112 190120510095875042530992062464 95065090751215979782194855936
47534963227247219149446840320 23768690539443224203898126336 11884949732631419416536416256
5942777097770613365561884672 2971539664612758511427780608 1485845390170105170037309440
742960474016915542180364288 371499126474389249671036928 185759007970160364125945856
92884226351563051708186624 46444474359022960676700160 23223417771132197749653504
11612299181376457580478464 5806444738593408143065088 2903369943249293747945472
1451758758600941712179200 725916272788618275192832 362976583138382847148032
181497514941228278349824 90753369156632566562816 45378990421325496975360
22690648132167355334656 11345900526835981090816 5673238493794142257152 2836763362085146984448
1418453738636611420160 709262898115324674048 354649463456171819008 177333738927340650496
88671373063297695744 44337938331462533120 22170095065638109184 11085610482772475904
5543086716362948608 2771684095669829632 1385912416579092480 692991392661635072

346513288516861952 173265440351453184 86637118222237696 43320758134374400 21661478578814976
10831289045221376 5415919400517632 2708097139212288 1354117289082880 677093004279808
338563682009088 169290430939136 84649510436864 42326902702080 21164525092864 10582799417344
5291668144128 2645968289792 1323051253760 661559181312 330796367872 165406572544 82707480576
41355837440 20678967296 10340007936 5170266112 2585264128 1292697600 646381568 323207168
161611776 80809984 40407040 20204544 10102784 5051648 2525952 1263040 631552 315792 157904 78956
39480 19741 9871 19743 39488 78980 157968 315952 631936 1263936 2528000 5056256 10113024 20227072
40456192 80916480 161841152 323698688 647430144 1294925824 2589982720 5180227584 10360979456
20723007488 41448112128 82900418560 165809225728 331635228672 663304011776 1326675132416
2653484482560 5307237400576 10615011672064 21231097085952 42464341655552 84932978278400
169874546491392 339766272851968 679566905442304 1359202530361344 2718542499676160
5437359877259264 10875269510332416 21751638532292608 43505476087840768 87015350222192640
174039496537407488 348096585260859392 696228354893807616 1392527078531792896
2785194894551941120 5570671264080592896 11141905478114607104 22284936856136056832
44572125512085798912 89148754623798968320 178306516446852677632 356631047292214837248
713298123381448638464 1426668304356935204864 2853480723901946265600 5707249678180044242944
11415075817112391909376 22831304555729390665728 45664914954467995025408 91334441594954417438720
182678106561945689653248 365374659867965088858112 730786213224077596819456
1461646213424450031845376 2923440000801489740103680 5847175149508158833033216
11694940594826676371718144 23391061781274070154739712 46784484745789575132086272
93573691858062019909386240 187156828449089779109199872 374332546364111036799254528
748702871660085030760218624 1497481301183895975843856384 2995113718095243780334551040
5990529667645391217962778624 11981663798200589750512910336 23964536522220794130200526848
47931490896080817518750466048 95867817495440093554199756800 191745306397437104141797163008
383509955607988042350389624832 767058596842203752834369847296 1534194564936862841935920889856
3068543872378636356406204170240 6137397229767094057881133121536
12275413429553830805899715805184 24552064799146946992074330734592
49106605478372464744698459717632 98218162716902071010496515932160
196446228954118425063192224858112 392912264948865416210782835703808
785864143978987964590362443382784 1571807516120490193518318430715904
3143773488566008915711823949332480 6287863889782074888774022074466304
12576361604864263892248792500535296 25153990860328756013899081704275968
50310517021857968486601156814962688 100626104646116849890808300442746880
201262350497035525616828574511136768 402544983403674702904081096273559552
805130531626556709149010087049691136 1610342192891528024979715963104526336
3220846645059885263322823504219340800 6442017808673428953372430164459257856
12884684654454174760198426640959668224 25770667383122983227303985906001641472
51543930914675233868422237060167892992 103093054126209002564473004616665006080
206196492846135074784203070225988452352 412413754879704288878920262437293785088
824869048134276856378868768845221330944 1649821173018290269999794025631710183424
3299808499536053654483701027145955409920 6599949306071053537935628006056980905984
13200563226139999533807707915644101984256 26402455680275783983488319638348484313088
52807569816543137798722446890817529315328 105620456545069415260936509009876180008960
211251546914105109848856248476234602774528 422524361476142778351678957865433691062272
845091258248150674011290837556796353150976 1690267587088031582638447518765450648354816
3380705315359523634508626724834617180815360 6761750913085968207480716824276666129842176
13524182390905778291888360397768195796107264 27049725911279240337630574293966118665060352
54102174081493848182968855584791691475812352 108209792680858431381353125163302291243008000
216430474397458332793537078314042399068782592 432882726866399605648735812602960431303098368
865809009875765091420794937155672128937263104 1731705132037461943088236498210846790536658944
3463584488646787406669766244220698646397583360 6927517426437301854326118984039407423443697664
13855731751162057790625410959274835108184457216

27712857298899023745197167900941710738963038208
13857125547736965954571756941666875630778580992 6928911223012210018272464966431447946037821440
3464629836077968529629525731014729038343176192 1732402030324916025061409489406867051833720832
866244571305423892654028056653184792247926784 433144063724194886388675684301468029289496576
216582920897838913225168670138171831227514880 108296904966790191627999749062804823905140736
54151174742330463321707581528261866098262016 27076948500632915414707644262560660121976832
13539154815050299584280748880495193597411328 6769917689892070730603837814855028566917120
3385128986129495834533650594731230167564288 1692649563656478151882691141017473025835008
846367317124104193249278492334665483943936 423204926209984655278605707080297227485184
211613096928958606966286083996630856499200 105811865376462443146634657226556549627904
52908591144222791405063136227398835503104 26455624800107180618404471920759698096128
13228477014051482767138687863909989220352 6614570814024687612537569883720064696320
3307451560511816920752897917742567391232 1653808857005645017618505446812551217152
826945966877690787430280967376909369344 413493752626279533025654605673771565056
206757260906856836168084363829544222720 103383822750286952911670712411101331456
51694507523572743869649621453715275776 25848551836001005641731943350939942912
12924924955107819674319537987511123968 6462786996107568263886552149776138240
3231555757330613345306667652898357248 1615859008303721279335029615454322688
807970068971067943008362702229733376 404005316895137623174605298366152704
202012799652370637422514622808719360 101011470428586231628863298217181184
50508270515493572273234642515001344 25255402908347014366018817960706048
12628335279473621297710157331955712 6314484552386867706205452841779200
3157400732518462381777913508790272 1578779594421745455226550298345472
789429411292129859782071921147904 394734512686693495975434346561536
197377159863661031029916366274560 98693531691987657036057779634176
49349241726072399278578688065536 24675858803075485019564243156992
12338548371557385199919571140608 6169583670788513945028510351360
3084946577899167645048617566208 1542550660202039158791489978368 771314015727247247529335586816
385676350676737457831463092224 192847846744925645949129195520 96428759075741281491263422464
48216797389509870003981123584 24109607620574549631165267968 12055408273197082130169987072
6028006368053444722378670080 3014154299754174189836173312 1507152707740813009241505792
753614132802269461782462464 376825955867066209472086016 188422422666498844026470400
94215933699732291658448896 47110328033107580651831296 23556344608174507737219072
11778762599897612574261248 5889676447853985639956480 2944985797879582496391168
1472566685916086086402048 736320236446190462304256 368178564967168940703744
184098505855621325127680 92053864613829089951744 46029238149923758669824
23015771996466486181888 11508462458985546514432 5754519459868924968960 2877403845122538340352
1438773980155307098112 719423018874672513024 359729523835845738496 179873769117177610240
89941388158216175616 44972945878921773056 22487598839367729152 11244362369637285888
5622462659795353600 2811372067386032128 1405756402437193728 702913385590685696
351474284981387264 175745938583715840 87877367338369024 43940882692440064 21971540857847808
10986320184737792 5493434970275840 2746854924091392 1373496181522432 686782450499584
343408405118976 171712792494080 85860691214336 42932493090816 21467320287232 10734197014528
5367366942720 2683817689088 1341975953408 671021531136 335527542784 167772160000 83890274304
41947234304 20974665728 10487857152 5244190720 2622226432 1311178752 655622144 327827456
163921920 81965056 40984576 20493312 10247168 5123840 2562048 1281088 640576 320304 160160 80084
40044 20023 10012 20025 40052 80108 160224 320464 640960 1281984 2564096 5128448 10257408 20515840
41033728 82071552 164151296 328318976 656670720 1313406976 2626945024 5254152192 10508828672
21018705920 42039508992 84083212288 168174813184 336366403584 672766361600 1345599832064
2691333881856 5382936199168 10766409269248 21533892280320 43069932044288 86144159055872
172296908046336 344610995961856 689256351662080 1378581422800896 2757300284555264
5514875447017472 11030300649848832 22061700811325440 44125600645906432 88255599338323968

176519994769670144 353057581725384704 706150347822858240 1412371064389894144
2824882866268143616 5650047207512997888 11300657364979417088 22602440629865676800
45207133059545038848 90418769718717448192 180846546636689637376 361711107671888756736
723458244140796477440 1446988545875630882816 2894121206939337621504 5788530644254826954752
11577637749261957332992 23156428420028521512960 46315162683066256719872 92634937052150940827648
185279097476338736431104 370576641696751182413824 741190176881649783930880
1482454140739594406068224 2965055855431778488549376 5930406858768736329924608
11861404013347831365500928 23723988618316380142305280 47450338419874195107217408
94905399206231259859648512 189820243145428259009724416 379659375756787996600303616
759356530445438950362316800 1518788618754603815048052736 3037728353236659458742943744
6075758937928222574779564032 12152122338766252464146481152 24305453603352119557467668480
48613325058343468373284749312 97231485819965395263268323328 194472643046487707559934296064
388964628906089249186663890944 777967943438406166506918379520 1556013258129267669281017954304
3112181258763446011096398299136 6224672002536713367261521379328
12449962975093069424660492320768 24901163890225424229595883765760
49804803660529419219741565779968 99614559081215979960582728056832
199239021682746242963364649107456 398497850406121052011127684202496
797035314893499236191052140380160 1594149857949512736719697824710656
3188458172224054002114582737321984 6377233257098165061579539650445312
12755100339496444237859827652493312 25511468329593116705121152008192000
51025471960386689869045297422794752 102056014523174292655696581658411008
204122170251150411146605136942465024 408264622911904473963634221136216064
816569810643016251268116336775004160 1633220750924447109217928462555152384
3266603761125723431799248503120592896 6533532040805105290325280162261762048
13067713118717527434104126636564676608 26136724311649688575115385897211658240
52276044771728644564045037042587926528 104557281840315823955718604581505073152
209124948274348717566694270155668586496 418270665736131574443902662296654053376
836582869847131427508833568563941867520 1673248816443999412259723625069151256576
3346663786387471939003560226020837556224 6693659879773890106975346403806745198592
13387984373545672671887144711143630569472 26777297975087130259647193229347541483520
53557254406165830351040194072815643656192 107119825724314800365572003373872408690688
214250285272595880058127237204227060137984 428521838193124318770220935321418605789184
857086211682113754848374792468766182604800 1714257493955957744312615428589390307262464
3428685129095375957856962544482496498630656 6857710540557672854177388463572424765472768
13716101645849187585281703676359713067368448 27433564421166058924417260851149153207582720
54869851101267485356542228699157760560857088 109745146720405705728499871392034429413097472
219501182476552881487830570771506675408961536 439024143024588703037322797517888983983456256
878091842192143286197968906985529234297978880 1756270796670218332642584437870561001258090496
3512715817912300185778462123540127067840446464 7025780084968327412543510742678264266329423872
14052257068224108907060194476552548793955909632
28105907933023125978066734935497138110505943040
56214603459196068284026161835778357266200133632
28108695526172942305959426900281219155694190592
14055044661373925234952886441336629839144157184 7027870779830689658463029716266325050220609536
3514109614487208349724808105932167590434570240 1757141919529535935109050676865586327879417856
878614515907733847677848650382544430270775296 439329036025349863900585981166147848300920832
219675407048416401981123818570511740733227008 109843148042078936005977323278974778657996800
54924296279974835510696368636346843474690048 27463509269455101509202037816603148810190848
13732435199461392631527945657516437941518336 6866557882097617254227436203365650738970624
34333449082232269096345449788986541253591040 1716809611707864782788590738145128568848384
858447341149797508702228290898493255450624 429244938222831313005080606362211113238528
214633102935381935829523533637587799375872 107321868379674107578253382047035021066240

5366359264582862362087249863763807122272 2683312555091009672630915312587931595712
1341722738945294082109102846646979815014 4 6708946001725416639513740184999969161216

3354639154362181434240983068382519623680 1677402653930827274362548022132527333376
8387428653402819158023022550368974274 56 4193922018575750972116652495037655941 12
2097064855225046182610896857445412372 48 1048584350581108439581733733685998387 20
5243181367748468939290095193246452940 8 2621720491295697840335760859031456972 8
1310925149358580605513237060719843737 6 6554950265346561454292968459619794944
3277637391950109940509875807820185600 1638899825613469576936633692915236864
8194904776259420918091647409601904 64 4097655212225746975750063177313812 48
2048929018160891746227151324913336 32 1024515215104455002289635530584883 20
5122829605642320657328476993565491 2 2561541567881183151604388167103283 2
1280834166470602987272268918711910 4 6404487745003071993711718769360896
3202402328826564525531046472581120 1601280392575796527103116780240896
8006798103691553957203551620956 16 4003597122252062639445759670353 92
2001897596329174150144871765114 88 1000998315766158490283431847526 40
5005239166838649527472139062476 8 2502743377423253301763559443660 8
1251433585713590919895524678041 6 6257477413577775944546348171264
3128893449293798644807536476160 1564524095899354658670949433344 7823007335759049974690653143 04
3911697096010663328013279559 68 1955945262070900834340616273 92 9780209880682350023372963 8400
4890346725505097937521423155 2 2445294255334510431678182195 2 1222707573958235947297826406 4
6113840101246083393782808576 3057071166350493525538242560 1528611141038972677092540416
7643433494513492957079797 76 3821905641916061264348446 72 1911047268287688025078497 28
955570857808672708991385 60 477809040736750702721761 28 238916326284582525473914 88
1194640661003948497934745 6 5973498452924921842499584 2986896800415050597662720
1493522187183820137037824 7467979870800574876221 44 3734174402841024533626 88
186717943514088081457152 933635834430624681164 80 466840975645404477521 92
2334320170377483072307 2 1167217731263971878502 4 5836376886696011104256 2918332558536081408000
1459238336862078631936 7296551972280582799 36 3648456130125386219 52 1824318137055240519 68
912204104523893964 80 4561245702600838348 8 2280735441291103436 8 1140424015640893849 6
5702401553181179904 2851341514078945280 1425741125783650304 712905747263913984
356470465818001408 178244029002022912 891264125475225 60 445654052970168 32 222838021601361 92
1114245083588198 4 557150029584793 6 278588758687744 0 139301251291545 6 69654061619609 6
348287487967232 174152333918208 87080461926400 43542378446848 21772262965248 10886668353536
5443602612224 2721935523840 1361034870784 680550989824 340292272128 170154524672 85081456640
42542825472 21272461312 10636754944 5318639616 2659450880 1329790976 664928256 332480512
166248448 83128320 41566208 20784128 10392576 5196544 2598400 1299264 649664 324848 162432 81220
40612 20307 10154 20309 40620 81244 162496 325008 650048 1300160 2600448 5201152 10402816 20806656
41615360 83234816 166477824 332972032 665976832 1332019200 2664169472 5328601088 10657726464
21316501504 42635100160 85274394624 170557177856 341131132928 682295820288 1364658749440
2729451716608 5459171868672 10918880608256 21838834958336 43679817400320 87363929767936
174736449470464 349490078810112 699014517358592 1398097754193920 2796332947341312
5592940772589568 11186431300993024 22373962113613824 44750123250483200 89504644547477504
179018085187977216 358053762561998848 716142709496086528 1432355787736350720
2864852312961056768 5729986100898824192 11460535151751069696 22922196203408982016
45846644206631649280 91697792012890669056 183404591225036079104 366827196848581640192
733690422494182244352 1467452902582402416640 2935049920352880689152 5870388071081913090048
11741352602916129603584 23483858127336866054144 46970022097682945802240 93944655881384318992384
187898535134805492760576 375815517013684695072768 751667927515516809248768
1503409642007328456704000 3006966857967246589820928 6014228863839672532467712
12029048023489703770587136 24059276638600124952477696 48120914460441684727562240
96246551287366239100338176 192502547307698217491103744 385023984081327913563062272
770085747094518784287834112 1540247052052763482899087360 3080645219832978794445012992

6161592671120861246183702528 12323789805151529806954758144 24648788536122674243084222464
49299994923884577744517857280 98604825551047614005734539264 197219322508652145044866727936
394457987830418124156528754688 788954661287063916446648107008 1577986693826583169160477409280
3156128130158077010855317209088 6312565745325975366779359199232
12625750460671593423696167960576 25252738861382472227667235045376
50507953602843515215884268339200 101020858965844171952868133175296
202051621452002626947935459344384 404123049944633819980269304676352
808285713970524772129335381327872 1616650656103563808596264306606080
3233459768532156145867715701112832 6467236449714369349085805578027008
12935106724728852812872359507656704 25871481100057933855146215718518784
51745497501316324169095424843448320 103496065605033561255796836499718144
207002272414868948346805646625079296 414024827239341548364035240501444608
828090219297890400068918375505461248 1656261568234195406819532540016066560
3312685395745220027002456658042421248 6625695310044098480731696472105418752
13252039657195513814916959256251990016 26505377388605661336741051136586285056
53013350925640590087296367521337180160 106031894148139715002221265539003580416
212074172889996499659699592070665601024 424169114967427138629913306126648082432
848379768309722555880854856223929925632 1696842613369181669003766200389127372800
3393851380237836452491645376660789788672 6788035067474619133951516705086649663488
13576734748947130725839485313703439499264 27154798725890046367551874434467159343104
54312255907771662566849556483054879375360 108629828727526464797190728194350880129024
217270291279019208921364686845184003014656 434561850205970976496695834603332491542528
869166235707807070301324591032593954111488 1738417542007344375218515025717045850275840
3477005225198149219668761738737807584657408 6954350732763219377800986852083046937526272
13909382030260280632528900453380957411475456 27820125189988245018911654405191641895796736
55642972638911857545531015807242737937285120 111291389795694450106477445608204384165953536
222593668627130370243785719203846584914673664 445209115325743680549233094382568802994880512
890461786794453241221789500714888872320827392 1781010685874838242690225625329280277303787520
3562195596321540005873744498457565619931840512 7124739641786807052734075492513141370512211968
14250176181861068187441323976222303002321485824
28501746160297044538828993934836646527237095424
57006279913743905405550679834457374099662438400
114018135013787443466886743598482910289701371904
57011855100043538061336063764025536190038933504
28507321346596677194614377864404808617613590528
142543575715857926792803619233984245701038571527 127527234936623380626767457297222415700459520
356393784204017521080667697644761627317449523217 8205603330601936564998511212331066924938035
2891071572795975562948315868011406600955756544 4455575644694707215358195899805789336434114 56
222789671270476830798740622977727283404472320 1114002801531091504147857254825825499936194 56
557028623354899427151005697381507291425013762 7852792297212655111404138367505091644096512
1392707671334016943262899593296740935847116869 63878639037005654777961341091136447447040
348210946070196329662071235784928410782924817 4113980094271188292622202257649999596748 8
870612435767221058771043933114178969010176 435327485531543088039488427470053970018304
2176743765897378233467274441915092277657601 0884250520685205133685533732399573526118 4
544239110594175955001734762761184283197442 72132847577045826659596419451194945044 48
1360730699285018379091627287608988742451268 039858034240381244263623898100137984 00
3402159055211492176697294170787541942272170 1162604355482645590703573335038492672
85062284055260960141638003063815300710442 5332189463738940018704137304393383936
2126764793255865396646091296448551321601 06343431959651804659933095318756786176
5317431212825516974378081290754300313626 5884541383422185787975390778538065 92
13294876106278426142852335850968055808 6647762571692871498152951081504604160
33240435451232649624398671187625902081 66210290220004708790162934838643 9168

831092015919230847291662568695791616 415566290369219075316255231599181824
207793286389411363493339589425233920 103901713797106594664275781525438464
51953392199753753790940884169129984 25977963750477105124871938787770368
12989615700538666677136717745487872 6495124762919390395918733048545280
3247720837784723726634553612173312 1623939647054876127654870350036992
812009437608695195996231946993664 406024525844976164082514359484416
203022166442802365083456372736000 101516034981558324062827782864896
50760493370857732791963689680896 25381484625468151776256743964672
12691361282753718578265821544448 6345990126386680634201635553280
3173149805698250989635180167168 1586652274101580831084771278848 793364822677018083675976237056
396701754151622875904783417344 198360548482368354985789358080 99185109944462636009593503744
49594972823870547263146164224 24798695337754888260747788288 12399952131787251444961247232
6200278297348529379774300160 3100290264401716518533988352 1550220690064584173590413312
775148123964155043956916224 387592951448009000559312896 193805920456970239570083840
96907682594967989430255616 48456202480725429537734656 24229281831983432180170752
12115231211802074795737088 6057910753806216750694400 3029102950855698051760128
1514625262404143864086528 757349524690219351146496 378693209089183385124864
189355827916628547338240 94682525644332701057024 47343568665175564222464
23672937254092388958208 11837045087798497902592 5918810774275400663040 2959549502325776187392
1479846808756926021632 739959433175481974784 369997730986250469376 185007872692379975680
92508439945817358336 46256471772722364416 23129361786268024832 11565243843087433728
5782903396520427520 2891592435748569088 1445866586618462208 722968477681319936
361501831026704384 180759711606374400 90384253849698304 45194325948104704 22598262485680128
11299680998653952 5650115377233920 2825195127570432 1412666283261952 706367501369344
353200930553856 176609055211520 88308822573056 44156558770176 22079353126912 11040213434368
5520375152640 2760321794048 1380228005888 690147557376 345090555904 172553666560 86281027584
43142610944 21572354048 10786701312 5393612800 2696937472 1348534272 674299904 337166336
168591360 84299776 42151936 21076992 10539008 5269760 2635008 1317568 658816 329424 164720 82364
41184 20593 10297 20595 41192 82388 164784 329584 659200 1318464 2637056 5274368 10549248 21099520
42201088 84406272 168820736 337657856 675348480 1350762496 2701656064 5403574272 10807672832
21616394240 43234885632 86473965568 172956319744 345929416704 691892387840 1383851884544
2767837986816 5535944409088 11072425689088 22145925120000 44293997723648 88592290414592
177193170763776 354403521396736 708841402531840 1417751524540416 2835640488034304
5671555853975552 11343661463764992 22688422439157760 45379043901571072 90762485849653248
181533767792328704 363085127770701824 726205439913492480 1452481248571162624
2905103234630680576 5810487944238071808 11621538838429564928 23244203576765972480
46490658953345630208 92985821506318630912 185980650211892002816 371979314822293487616
743994658441605939200 1488061374477249806336 2976266864142575468544 5952821958661302648832
11906220378074908721152 23813593677654424289280 47629493198318062272512 95263598082654551932928
190536419537345958641664 381091285818765626834944 762219465125678672773120
1524512717227652183752704 3049173008407894043918336 6098641164720967440662528
12197872625252293586976768 24396925842125304585256960 48796212867492043993120768
97597148101466957631455232 195203740935899654553337856 390426371337730787687530496
780890521607324532536770560 1561856601078374979396960256 3123864317884201787440758784
6248030867223307232175194112 12496666197356421778937741312 24994541320532458187050188800
49991500492704145632449789952 99987836688686749781598404608 199985344783930416596594458624
399990032380974667259984216064 800018750388177002653559029760 1600114872028809341574299254784
3200384486562529355682960900096 6401078458134880056434646581248
12802775886289402803006742724608 25606789712618090986288384573440
51216055305314752733126567395328 102437062370786646987352731287552
204884028261887577016904655568896 409787863564403720118207697125376
819615341210064572405212166225920 1639309910582643409148017876402176

3278778277490315346971222840705024 6557873467630687751292819857211392
13116380760561489617286388066025472 26234029171723207463974272835256320
52470593644646871386751539076923392 104946257891694655691109064966668288
209902656988191137217430103558979584 419825596385985926105284154369245184
839691757591179155551416203241062400 1679464644820772917784528195487268864
3359091548918375048932447968984825856 6718507616390408524591679093990227968
13437664269888133902636924500021608448 26876626613990901512180981624125521920
53755849376411070438176228496415653888 107516891049680675703980987489160527872
215044166693078421063219035970979495936 430109102573590981436952193927275872256
860259743522050241494932631825185505280 1720602563793837040231921751591638532096
3441371281087147194947956479065812107264 6883074869173240618864138909896694300672
13766814352344373695664729723323528773632 27534957932684532307202363253707337891840
55072574321360634446150534121535236472832 110150465554704408555792683471311594323968
220311564933375096438568597399105431404544 440644397514682751531103655711175348322304
881331330325230620370140233248279667671040 1762747731242191475356146310148417277394944
3525665603667843419944024307600550438895616 7051671489702607778351511989808532646002688
14104023544139057433629950728831928828428288 28209408217745798621113754956093584729702400
56421538694426964749935216909046623605096448 112848521906724664515285847811812155501576192
225707932849190799061402523611062127585918976 451437643769864538184466703196999888337371136
902918843682694956492256718343751043005808640 1805924799651321673231160060587004618673750016
3612023823874506866955613368973014302671765504 7224396096892740774897813233544038735992061952
14449489092072935631768799458284097733281185792
28900371980720779427483944898960235989156495360
57803531554591375182860581762704553023501238272
115612638295482383021506547454977268137378971648
231236426963564031354583862769090860455510933504
115623788668081648333077315314113592318131961856
57814681927190640494431349621840877204254228480
28908734760170228411162020793312479124721238016
14455064278372568287554183387852259823657680896 7227880588330011184763678189524140042477371392
3614114518736869112875132342561075086562951168 1807144371654366316684212795180040075943608320
903615741970149038465429709539771304302870528 451829649056557459294376510744761285316968448
225925713564020199678019083359818459241250816 112968301299880834854424955673628137912008704
56486872908875784934920184833673523101696000 28244797583905576221313945915266488623693824
14123079356686629987583899706848107848269824 7061879960710235932255413228031485692346368
3531110121538578435359438301319458730278912 1765640131361019452295584994311587307192320
882862600976374843455725418981722624622592 441452568136119980381829170403825797824512
220736917892026269517897815658395141668864 110373775857996274422440523057438692212736
55189546384989707042966069142839906795520 27596102420490638437355938378480233742336
13798715824243211676614421092770257043456 6899690219120552067275436498150198607872
3450011263059749148121831224957634347008 1725088708279611131302972100420084695040
862585892514673844272514294180676108288 431313715444771061446771269075654934528
215667242316102600378642695530485907456 107838813454909835016949878261572173824
53922002875884184922289204378950696960 26962299512156726168051734813557653504
13481798793185679937479433718819979264 6741223915146498395466500015430565888
3370774216850078411096641585725571072 1685468238063453812230016581867929600
842774683850934209455856185436536832 421407624335070756398352039969554432
210713953372337204034387993610420224 105362047288569514934799983618031616
52683558945485213926202985215426560 26343047123342835192502989310918656
13172157386971531710952243007062016 6586395606135822912826495679332352
3293356259392939985088434927566848 1646757357858984256881811007733760
823418293010749260609702275842048 411728953546003196389249523908608

205874380293315881236823954948096 102942141906815082139511573970944
51473546833486111830305585233920 25738011356782341295427691741184
12869624648410813337851295432704 6435121809215228013994372497408
3217715647112524679531548639232 1608935194808717676032955514880 804506283030586506150068355072
402272484328407087141829476352 201145913570760460604312387584 100577792488658688818855018496
50291314095968573667776921600 25146865973803901463063166976 12574037449811758046118936576
6287320956360782680353144832 3143811593907843168823410688 1571981354817647498735124480
786028456340686706529271808 393033117636274831845490688 196526003551103155213172736
98267724142034447251800064 49136223254258658448506880 24569292218750046635556864
12285236405185382023430144 6142913350497870364540928 3071604249201524858683392
15355875911577057267548160 767974849276676052877312 384005871382411735990272
192012159063242722770944 96010691217639788773376 48007651451829108080640
24004978647419160887296 12003065784461883867136 6001821122607093645312 3001054676491622678528
1500599395839849267200 750335726716943597568 375185877756981280768 187601946077745381376
93805476638500061184 46904990119063715840 23453620959438700544 11727373429672771584
5863968189813096448 2932124832394903552 1466132784941629440 733101576842903552
366568380607496192 183292986396770304 91650891244896256 45827644645703680 22914921834479616
11458010673053696 5729280214433792 2864777546170368 1432457492561920 716263106019328
358148732878848 179082956374016 89545773154304 44775034060800 22388590772224 11194832257024
5597684563968 2798976499712 1399555358720 699811233792 349922394112 174969585664 87488987136
43746590720 21874343936 10937696256 5469110272 2734686208 1367408640 683737088 341884928
170950656 85479424 42741760 21371904 10686464 5343488 2671872 1336000 668032 334032 167024 83516
41760 20881 10441 20883 41768 83540 167088 334192 668416 1336896 2673920 5348096 10696704 21394432
42790912 85585920 171180032 342376448 684785664 1369636864 2739404800 5479071744 10958667776
21918384128 43838865408 87681925120 175372238848 350761254912 701556064256 1403179237376
2806492692480 5613253820416 11227044511744 22455162765312 44912473014272 89829240995840
179667071926272 359351323721728 718737007181824 1437542733840384 2875222906634240
5750720691175424 11501991138164736 23005081787957248 46012362599170048 92029123244851200
184067042582724608 368151677351493632 736338539075076096 1472747446894329856
2945635631277015040 5891552737530740736 11783668425014902784 23568462749936648192
47139177299686981632 94282858199001333760 188574723597257408512 377167461593024299008
754370951983067561984 1508813961560173051904 3017772038308421959680 603582306992995631104
12072241074738294685696 24145635070981196218368 48293575984971606130688 96591763655961639649280
193192750683960134074368 386403948111993977700352 772844789712135374503936
1545763366400565587214336 3091674306753720850841600 6183643761412621054509056
12367877818635600814669824 24736936228891919040643072 49476233641025272903892992
98957189648533415452999680 197923824030032570196426752 395866537525996618973708288
791770853983856195109126144 1583617265831438304541671424 3167385647390328437730181120
6335073526235560532754038784 12670751515380928380095430656 25342711956581471389365567488
50687841764802172037080547328 101380519232882802590859919360 202770709872322522215117488128
405560762557758878497030275072 811160210741745425127651147776 1622397792735946186522483490816
3244950327976803045579329372160 6490210140963427436227383525376
12981039251946497562592216612864 25963316443932280505459332349952
51929108767943131771468462948352 103863169296043405064036522393600
207736242112401093170272237780992 415492291265430752424942861549568
831024196612118637018682495074304 1662127621386751538374958534098944
3324413699098531605425104156098560 6649144310847120268200582487998464
13298922446994354651101913327599616 26599112544588937531605323358404608
53200760390378331522013640123219968 106406591383157575961633267059261440
212823323971116977758478507744165888 425666930351837607187380962739617792
851374425522882517715609819981807616 1702829980684179642112915428968759296
3405822220645188497589222435947806720 6811968959844035421905228027916189696

13624586956795387697264022367873531904 27250471987805409101435177359829368832
54503540124040085616684619967823347712 109012272544938706060997770431975915520
218034929683594481777252601856610271232 436090628554623102865019325698537422848
872222795484114484351066895367708606464 1744528667717965525944190278676684734464
3489223488935404166372493533235904512000 6978779284869754561713213018236879110144
13958223183737401581362877940003898392576 27917775595470588078598659687068077129728
55838209646932745988943126988256714948608 111681736205848631641377869204754551275520
223374106235663542609738968865991345307648 446769480119259643873444398644947176128512
893581495534384405054821719115823323283456 1787248061660499044725509281883504588619776
3574666264504458558682750251070725061345280 7149672811375838055828963876748881890902016
14300026187485517988584854502712627318226944 28601413504438719731023562503854981709299712
57205549267812806969754832004569417564291072 114416543053496348954925078002857743419965440
228843975142734167940680983993153303422697472 457709728356951275943023623961182240010928128
915463012856868432009370559872115746352922624 1831013137999668624265387743643734025367977984
3662200500571200769024068735086473116060221440 7324749450286128579034723965770956362768973824
14650195798859711240042620922737932986835009536
29301785394294330644031587827867906496264142848
58606358381738477615955867620519894037716533248
117218291949776587887697119170607950165809561600
234447734272152441086965006200352224512372113408
468917769289503412797071548118977097386250207232
234470035017350971710106541918624872873878093824
117240592694975118510838654888880598527315542016
58623083940637375583312019409224380308846018560
29312935766893595955602355687004230677017133056
14657164781734252059774350834698135599805628416 7328930840010853070873761912947077930551345152
3664639644577290055930174204272544030599938048 1832406934574576788211733726035774547962101760
916247023430254274229190174967638540312117248 458145289786610077176256743458694903321591808
229083533929046508618959199716785268243562496 114547211482393989324895013852111542413164544
57276328000132362170155213922915225352273920 28639525129533864838931460459887339748982784
14320443129500774296392656979158533410914304 7160561847117308086659791864186698473668608
3580451064742114512561627619397065120940032 1790310602962787490896679653350390502522880
895197836777258862756272748501124222287872 447620186036561990032102835163526596657152
223820726842247274343034648038245541085184 111915680333106776835008939247363891920896
55960498622544958249250277237802506649600 27981578539268264040498042425961533669376
13991453883632024478185473116510907006976 6996059248814958468060962510020523589632
3498195777906952348514594230892796837888 1749180965703212731499353603387665940480
874632021226474644370705045664466731008 437336779800671461495866644817550245888
218678774494052800403190383401433563136 109344579543884935029223722197046001664
54674885920371734928426126346687610880 27338741034400501171120195797426110464
13670019554307567439013664210754207744 6835334295707442146233615261397680128
3417829407130550286480199208709128192 1708995833203689749921795393359708160
854538481421052178301745591182426112 427289523120129740821296742842499072
213654902764866696245860345046892544 106832521984834261040536159336267776
53418796293617586979071073074544640 26710665797409021718937033240477696
13355966724004624974169264971841536 6678300274652369544435006661722112
3339308593651213300892690418761728 1669733524988120914783938753331200
834906376575317589560766148640768 417472995328287360864781460307968
208746401184457963474589923147776 104378152352386123258394558070784
52191552056271632389747077283840 26097013968175101575148437766144
13049125954107193477711668445184 6524872462063418083924559003648
3262590973536619714496641892352 1631372858020765193515502141440 815725114636610264891341668352

407881900131418966512466132992 203950621472266400289630715904 101980146439411658661514182656
50992491071345058589106503680 25497454461492143923727958016 12749331693655879276451332096
6374968078282843295519342592 3187635154868873476406509568 1593893135298162652526673920
796984346580944283425046528 398511062756403620293378048 199264976111167549437116416
99637210422066644363771904 49820966394274757004492800 24911663788758095913549824
12456422190189406662426624 6228506242999882684039168 3114400695452531018432512
1557274134702560347422720 778673960839427592814592 389355427163787505958912
194686936953930607755264 97348080162983731265536 48676345924501079326720
24339325883755146510336 12170239402629876678656 6085407931691090051072 3042848081033620881408
1521496098110848368640 760784077852443148288 380410053324731056128 190214033861620269056
95111520530437505024 47558012065032437760 23780131932423061504 11890628916164952064
5945595933059186688 2972938704017948672 1486539720753152000 743305044748664832
371670114560376832 185843853373210624 92926324733116416 46465361389813760 23233780206534656
11617439859081216 5808994807447552 2904634842677248 1452386140815360 726227430146048
363130894942208 181574037405696 90791313670144 45397804318720 22699975901184 11350524821504
5675530846208 2837899640832 1419016929280 709542019072 354787786752 177402281984 88705335296
44354764800 22178430976 11089739776 5545132032 2772697088 1386414080 693239808 346636288
173326336 86667264 43335680 21668864 10834944 5417728 2708992 1354560 677312 338672 169344 84676
42340 21171 10586 21173 42348 84700 169408 338832 677696 1355456 2711040 5422336 10845184 21691392
43384832 86773760 173555712 347127808 694288384 1388642304 2777415680 5555093504 11110711296
22222471168 44447039488 88898273280 177804935168 355626647552 711286849536 1422640807936
2845415833600 5691100102656 11382737076224 22766547894272 45535243272192 91074781511680
182158152957952 364333485785088 728701331308544 1457471382093824 2915080203141120
5830435284189184 11661420324192256 23323940160012288 46650079343280128 93304556733071360
186617909559164928 373253411304374272 746542006980837376 1493154382705852416
2986449502900060160 5973180480776830976 11946923911507083264 23894973722921009152
47792199245655703552 95588902090938777600 191186811381132296192 382391637160774074368
764819303118567112704 1529710663831172153344 3059565442850420162560 6119419116076992036864
12239414692906287497216 24479982307317181841408 48962270457643577376768 97929152601305582141440
195867528574648019058688 391753503893369747668992 783543901274886914441216
1567161589526068667088896 3134470753004727010590720 6269236653914633374007296
12539063603639625453666304 25079307798899968318636032 50160976781041371459878912
100326675928565612564971520 200662796590096964420370432 401344482646125407421595648
802726744224113772004900864 1605529046311953458333220864 3211209208351358745313280000
6422720648157621147920236544 12846045759225049610427826176 25693300444269713850030358528
51389018740178656958410129408 102782873183635772433519083520 205575417773828461900435816448
411170178360770757867666931712 822379042347769183868924461056 1644835455947993704005030117376
3289825654400898080544422625280 6579960793811617506157570031616
13160540557642877702452589625344 26322319055325040785180078374912
52647113990728652330909954998272 105299179741614446182919506493440
210608263003543175408038205980672 421236333047714916900474797948928
842512280176686965969746367873024 1685103788515888196277086279696384
3370366033356804921229359647293440 6741048979363666899809093470388224
13482731784027447914318935292379136 26966731218655124058039367287963648
53935997738510704574881727982338048 107877066079422322067369442777497600
215764273363646469969950859180638208 431548829136896591610325665612562432
863138223093000486561499225727696896 1726357575824415579804694240460537856
3452877410925660372972780058931363840 6906079340404979172672343273883303936
13812807717917275198798252859807760384 27626913510049184104503638343697825792
55256423168527635622821541935560261632 110518038633913806073271614367449743360
221046461861544681801800289727557926912 442113692910523502914114701440432734208
884268924195915284449257646851499229184 1768620925141567126140571781644265979904

3537408003782607366765256539171067002880 7075148314564160962498739030107204091904
14150961243126214382933929963744548356096 28303251714248213681740763734549377056768
56609161884487997195227335083219314802688 113223640680959134053946285394679750983680
226457915185884547434875801245841744723968 452937098019701653523718063404647974961152
905916731335268424355369048635224920948736 1811918533262267083326603940922307783950336
3624007207707994635884939569148331452006400 7248354697782910210233342512904094672224256
14497389960299662297393611775023052880871424 28996141050067008348641077048475832834588672
57995004359069384204989861093811119814868992 115995453236009503425395136181341147921121280
232001795507760476881621100350120112425009152 464025369087003893824903856675115858015551488
928094294316973667773131025299982982362169344 1856275700919870957929086744994684973864714 24
3712725626411621712079110596797942060097208320 7425799701966970465144807689193894250842947584
14852296302221395012262788369583808762982957056
29705986401017698188471922721559658048560037888
59414760395185212704836537407903397142308323328
118835095976670058065458458745374956374993141760
237681342325939381442487685349886236930739273728
475384985397077293508116906418045122222984527872
950814572284551648262516884272635541168981016576
475429586887474354754399977854590418945996488704
237725943816336442688770756786431533653751234560
118868547094467854000170762322783928917252112384
59437061140383743327978073126176045503814303744
29719924366766779827935382545480063274501275648
1486065908167084399594086426393605189854769971 2 7430677989979149038957018627566036079922380800
3715513219561438039971802561582023105285455872 1857843722066650780232547904690514085304860672
928965417176291270239597264295008308983496704 464504486659628575181460288122379787657281536
232263132365555757621560972048627710411407360 116137010700648613826195900018032763497086976
58071227609259674420805657005875835894235136 29036974934097520964256682001367645019963392
14519168031782602359055267749898686046404608 7259924298258222117991097249556774791413760
3630132290312571528227280312082103279812608 1815151215748015998729505999692909581959168
907618143169873116672685921672383762006016 453830339232869116990309421749156366516224
226925803440400837822137941331060426014720 113468218632183558574560585893771334385664
56736767772083349119026100561006227881984 28369713114037459475385954087563394285568
14185521171016622195629428947311837315072 7093092892507257326782940425420988743680
3546712599753101777875583188593029414912 1773439376626287446179848082237782228992
886761226688012001710952285089524875264 443401382531440140165990264530079318016
221711075859437139738252193257698099200 110860730226577104696754627125178269696
55432961261717819762191578810753744896 27717778705073543588002922029459177472
13859538389644088647455027326770741248 6930093713375702750454296819405946880
3465209115964680588590539987713261568 1732685687620754900976965782861774848
866383408629584753829330785933459456 433211986724396028585089340218015744
216616134566999840127756643734650880 108313137885900832981484308680146944
54159104244150872949545147746484224 27080819772675664704174070576447488
13541043711637946466787783639826432 6770838768469030290744265995714560
3385577840559543674047320085757952 1692868148442286101361253586829312
846473688302400182849423565389824 423256651191828657509110168682496
211638229116228611796754277335040 105824066318271447419476735164416
52914509039214294470288165830656 26458492459646432615418982039552
13229865199842858997846940581888 6615242084931250843992195072000
3307757849705360945304599265 28 1653965263737723383532411158528 827021317495089359899796176896
413530001560658514016693387264 206774672186886174041744343040 103392171796721545537570996224
516985037500000020271349104 64 2585046080081961564274216140 8 1292583486331961513595843379 2

6463219663114711225272893440 3231760947284807441283284992 1615956031506129634965061632
808015794684927774644240384 404026786808395365902974976 202022838137163422241914880
101016141435064580766171136 50510431900773725205692416 25256396542007580014149632
12628788566814148712726528 6314689431312253709189120 3157492289608716531007488
1578819931780653103710208 789446859378473970958336 394741876433310695030784
197380161588692202291200 98694692480364528533504 49349652083191477960704
24675978963100345827328 12338565942302476337152 6169571201527389880320 3084929715951770796032
1542536915569923325952 771304486581980626944 385670257689499795456 192844136044004638720
96426571621629689856 48215537610628530176 24108894705221107712 12055010302563975168
6027786626258698240 3014034050617704448 1507087394053029888 753578881398603776
376807032885346304 188412312535695360 94210554314358784 47107476180434944 23554837601845248
11777968556736512 5889259156275200 2944767017091072 1472452228022272 736260473749504
368147416743936 184082298306560 92045444120576 46024869543936 23013508513792 11507291127808
5753913999360 2877091217408 1438612717568 719339913216 359686733824 179851755520 89930072064
44967133184 22484615168 11242831872 5621678080 2810970112 1405550592 702808064 351420416
175718400 87863296 43933696 21967872 10984448 5492480 2746368 1373248 686656 343344 171680 85844
42924 21463 10732 21465 42932 85868 171744 343504 687040 1374144 2748416 5497088 10994688 21990400
43982848 87969792 175947776 351911936 703856640 1407778816 2815688704 5631639552 11263803392
22528655360 45059407872 90123010048 180254408704 360525594624 721084743680 1442236596224
2884607410176 5769483255808 11539503382528 23080080506880 46162308497408 92328911962112
184666413858816 369350007586816 738734374912000 1477537469300736 2955212377554944
5910699633016832 11821949021847552 23644997555322880 47292194133901312 94588786314313728
189186368721649664 378390329629343744 756815843630776320 1513702056005730304
3027544849499815936 6055371173976342528 12111305297906106368 24223736495719055360
48449724791251795968 96903953182130962432 193816913563516665856 387651841525542813696
775339711848104591360 1550751481290247110656 3101647077768570077184 6203582385913291866112
12407741232578887155712 24816635386662381158400 49635576616333976010752 99275764918686379409408
198560753209409613594624 397139953162892936740864 794316799813933292584960
1588707386604161423376384 3177562347160912523165696 6355419842227004399157248
12711429980264367503966208 25424040552149452419235840 50850442287540339661078528
101705606941563548967370752 203420658616092837225168896 406860206698117153031192576
813758192328097263224094720 1627591942519920440771608576 3255335000767292710190055424
6510972232989489077673787392 13022548928888785469934927872 26046306783597185569044561920
52095031418833600396438536192 104194898540945659309575897088 208399468488448235652549443584
416818279790010305371894185984 833675245206248278877378969600 1667427861664951894021939134464
3335010465834814460578240659456 6670330416679450266225206099968
13341279803378543222587861762048 26683797546796371825450622648320
53370070973671314411451043545088 106745093707499770344001683587072
213500090935313823730202560167936 427019988911256213544803506323456
854079591903769559258403784622080 1708238411970053382854401113194496
3416635280265135294383989314289664 6833587473180327646118352804380672
13667808771660769406937453960364032 27336885193921767043276404623933440
54676305689043990545355802654277632 109357681980488894008317592121376768
218725505165779613851847157868396544 437471292741162879374118262988079104
874983150301533062089084420478730240 1750047430241480730859864629962604544
3500257119759790675083120837935497216 7000838758073239776893024831891570688
14002326553253796407239615975824293888 28005951180722226521386364575730892800
56014498509873720456586994399626395648 112034189316605975740802519295582011392
224078763226929021136862099583822462976 448178295641292181584238321152961806336
896398129657452641789504886276557373440 1792879336064641840821066260494382268416
3585924825628756796126245496871299579904 7172181958256459821220716945507669245952
14345028530510812100377885794545478664192 28691386289017409116628675396151237672960

57385431034026388065003158406423036035072 114776178980035915793497932041087193448448
229562991784038110913979094538656629653504 459147251216008780481924649990277744820224
918337037727882678271782221806484460666880 1836759146047495591159430287264826863386624
3673688433278451651550592261833369610878976 7347717148923824241564647898274170989969408
14696114862581490360056222545763205516361728 29393590854630664473966298589956138105569280
58789903968196696455640304176771730356830208 117585252454264127926696022347262369005043712
235181393944269725884222872681962554592854016 470384565960022391830107401338800742351241216
940812688063010663783538114627352751033548800 1881712488411953087813722853154208034729230336
3763599201395769696120738954107421134782726144 7527546851935266433228064403812852400213983232
150555790602157986948429301798821725061725028352
30112975000890882060804949580035490646044180480
60228737594931580449502591124855062337276608512
120463050376162793554790566179278286764929712128
240937251124924852421151900217692897710612414464
481896802995048235465445336153658443782730809344
963838207480493532177173743743862184288473579520
1927765617941781186846913630360814962022971080704
963927410461287654669739886616952777734497501184
481986005975842357958011479026749037228754731008
241004153360520444290576507372510842795130355712
120507651866559854801073637615823583487941672960
60256613526429743728429510772695872789159084032
30129700559789780028161101368739976917173665792
15065547178182344096053723675566008719883894784 7533122038234899089013448333381014490590478336
3766735243689313065000017414489512310619504640 1883454734130588292746655331144258687971885056
941770923208260026496650977521880610317008896 470907239675612953309987144735815938324037632
235464508873547946685824400355345785744785408 117737698954644708358327614171391801163776000
58871571736257721686871514082555354727579648 29437146997596544597289610539707404436635648
14719254063532114175571732019068565754740736 7359967314132978026249329384141714645581824
3680153798249949482356396379374573206896640 1840161969716704975794064033339144545501184
920123520154217605204964938495501243777024 460083027725041361256448930160715107401728
230052147686486959955207695536839796457472 115031390755226619641095462996661019607040
57518353833604879652293539112451070492672 28760506144798224742019673363285815590912
14380917686397004828946288585173047967744 7190791150197448643441370244351594070016
3595561728598197436204798098058332078080 1797863941048835275344455536970433560576
898973508899285916293256012455850541056 449507523637077097457142128213242150912
224764146412255618383828125099279515648 112387265502986344019542593045968977920
56196228899922439423585561771149099008 28099412524175853418699913509656854528
14050355299195243562803523066869579776 7025502168151280208128544689455366144
3512913343352469317427663922737971200 1756537801314649265395527750374129664
878309465476531936038611769689636864 439175015147869619689729832096104448
219597648778736635680076889673695232 109803894991769230757644431649669120
54904482797085071837625209231245312 27453509049142764148214101318828032
13727388349871496188807799011016704 6864011087585805151754273681309696
3432164000117931104552323928555520 1716161228221479816613755508228096
858120228191997040475674526089216 429079921136627086322235649032192
214549864088627826203317017509888 107279883804471054622758105251840
53642417782314098071928850874368 26822446831196334416239324561408
13411842385617809898257111842816 6706230677818726294197280702464
3353270081414273819633002741760 1676712411959592246083682566144 838394891606023791175431880704
419216788616125729654511239168 209618065714619781860653268992 104813868560588349447025459200
52409352131933403981862141952 26205884991786316620105777152 13103546958802965624640241664

6552075710856386469613797376 3276188971155645063453736960 1638170043441548446050287616
8191228006526371801868533376 4095802897922500686742814720 2047995896290907736275681280
1024045171810282564589977600 5120461977375556305210572800 2560349047849849893735628800
1280233553505960817432985600 640146291543498343999078400 320087903167008139640832000
160051330281133553641062400 800293544893815187308544 400165219190981303205888
200091832967527506378752 100050528169782180577280 50027569927900303982592
25014937885454758838272 12508045403479682842624 6254310932115993133056 3127299581246072422400
1563721848217074139136 781896952905556033536 390966490851287498752 195492252624898490368
97750629912076615680 48877566755851993088 24439909277832839168 12220517588869840896
6110540269411631104 3055410872194170880 1527775804841263104 763923086792720384
381979135582404608 190998363884224512 95503579988623360 47753989017567232 23878094020411392
11939596766019584 5970073260916736 2985174069411840 1492655754182656 746362236829696
373198298284032 186607739076608 93308164505600 46656229736448 23329188610048 11665131175936
5832834023424 2916551229440 1458342723584 729204916224 364619235328 182318006272 91163197440
45583695872 22792896512 11396972544 5698748416 2849505280 1424818176 712441856 356237312
178126848 89067520 44535808 22268928 11134976 5567744 2784000 1392064 696064 348048 174032 87020
43512 21757 10879 21759 43520 87044 174096 348208 696448 1392960 2786048 5572352 11145216 22291456
44584960 89174016 178356224 356728832 713490432 1427046400 2854223872 5708709888 11417944064
22836936704 45675970560 91356135424 182720659456 365458096128 730949746688 1461966602240
2924067422208 5848403279872 11697343430656 23395760603136 46793668689920 93591632347136
187191854628864 374400889126912 748836137992192 1497740995461120 2995619429875712
5991513737658368 11983577231130624 23968253973889024 47938706971033600 95881811988578304
191772420070178816 383562432326402048 767160049024892928 1534390466793963520
3068921671076282368 6138124817129275392 12276812584211972096 24554751068330786816
49111753936475258880 98228011472577888256 196465030144410517504 392948074687330516992
785932178171679997952 1571936413937397923840 3144016943062871703552 6288322116501895118848
12577220693756093661184 25155594309016794169344 50313494461042802032640
100631600608104031453184 201272424588244917682176 402563295920563544915968
805163485329274508935168 1610400757634843856076800 3220949089222277388566528
6442193326349734129958912 12884976948509826965569536 25771134488640371342442496
51544630160522177507491840 103093982687527224660197376 206197410108020188610822144
412413709681971855802499072 824865198295806668766707712 1649805954455339251856834560
3299763024638130332360507392 6599828280731164322014691328 13200261024372135958616735744
26401730974563886546408177664 52805879800767002351165767680 105616595304812463219030360064
211242862016181843471458369536 422505066845477521009712037888 845048819317182710153014673408
1690175009886820756573210542080 3380504762278552185680783474688
6761319009566925716430291730432 13523256989153494122998033022976
27047751918346273626270965170176 54097979716771118013091728588800
108200911193699377547283053674496 216411725907713038136765300342784
432843258856054642357928986673152 865726131793366416884654745321472
1731531491749247098106903034593280 3463221439823522724888993157087232
6926759792297102507128360489975808 13854153409894319128957469331554304
27709574470388866487316435366313984 55421684241978189433435864139038720
110848439086357291784477715090898944 221707019377516409404167403807440896
443434321164636470478758754866167808 886909207148480244298365404234907648
1773899543935375095278426597474959360 3547961347147579403920244772960206848
7096247212848817234567272701940989952 14193143462804951322588111715923132416
28387584999824536352083356055928569856 56777766148078340117980977360021749760
113560724593015215063590485216372719616 227131833779747499782438031425403879424
454284436746929138875390184836124639232 908610411868726556371808613642883039232
1817303900487189669985673715227033600000 3634773954473852454455460406336602243072
7269880215946651137879146764438274572288 14540425045891194733694745432406689316864

29082179319778174383262394671873658978304 58167017095547918598270596957867878645760
116339351103078976860032809143976878669824 232689336030124233047048848744436000096256
465399939708181024748064158401836485705728 930842414712227166804061238629601942437888
1861769900016184568223988320911061826928640 3723709941215829605679708329125839537963008
7447760164798580149822880032859110844137472 14896200894331002176572686814933085224697856
29793762918129688106999227128295897522241536 59590248095194743721706161253451249190174720
119185940708260222458827736500621406671732736 238382770452261914948486300988680629926232064
476787318976006769958634257952236893017997312 953618194094979420040591827854225052367060992
1907323500475890600327830279607952637396254720 3814821225523644721148953807014910340116774912
7629990900191016483284494109627830810882080768 15260678698669487048542161210451681883061223424
30522751193913882261030668403295404288716570624
61048289980977580849954028771374889622621388800
122102155148254794355693441472317941335619272704
244215460669108854022957650803772206851991535616
488453222083416238669056837325817062065489051648
976951045657229538584396746088179420853990064128
1953991294295253199661359635049449435154004049920
3908160994552094644307851555845080057200055943168
1954169700256841444646491920795630622046051893248
977129451618817783569529031834360607746037907456
488587026554607422407906051635452952234524934144
244304663649902976515523793676862800298015457280
1222157907011251120913547280767999562239384223744
61081741098775376784666332348783862164880359424
30542264345962596556279512156783971605034303488
15271829071268752360112929069588006063814213632 7636262984778103221043051030392013162555637760
3818305716960915131014818762995011646602084352 1909239970766389325754056005397008355963174912
954663541526160543000351314648255444312653824 477353548834563211561837313299003355321860096
238687663453023075811749484636939494243696640 119349276244382272921290156312188655413231616
59677360381126503968352785152953781852307456 29840041320030935738030246074906617998999552
14920701224749309745942049786668172535922688 7460690894741575811434488267941518036172800
3730515588554248374948975821274474902192128 1865342864868854422090353754289095393148928
932713967730292328353109798970476667600896 466378251513078722830521360398202819313664
233199759580505640742243910655583652413440 116605196702235960034613570556032947585024
58305256807109549849052592892137034481664 29153957631550559840399200253128797585408
14577643429773172378136052030094538964992 7289154021885532418036251966812339568640
3644743164442239323502238959288704827392 1822454658970856218993175967585619935232
911268867860296388117616227763443728384 455655203117582333369322235867038744576
227837986152508236339918178926177812480 113924185373112652997587619959418126336
56964688834985593912608075227873673216 28483642491707430663211170238019141632
14242470282961032185059151431050723328 7121559660034174519256358871545937920
3560942089293916472991571013783257088 1780552174285372843177481295896772608
890316651961893724929588542450958336 445178608390550514135218218476765184
222599445400077082902821082864025600 111304793302439454369016528244834304
55654931952420183643311257528827904 27828733626810320051057125467619328
13915000638705274140229311085412352 6957817232002694127465029718507520
3479067072326375592407701947154432 1739612764325702060541444517527552
869845996244108162439519030738944 434942805162682647304157901357056
217481306101655606694278143672320 108745604810984944868238668333056
54375278285571043194669132414976 27188877082824806977609465331712
13595057511432046178942182227968 6797838240725844434539815895040
3399073862867832889804270338048 1699614302686371781169316364288 849845836969413558718248779776

424942261297820613425919688704 212480802055467223746357493760 106245236731012070389877571584
53125036217145264453288198144 26563727034392246855818805248 13282467980105930742496755712
6641536221507869028542054400 3320919226481386342917865472 1660535171104419085782351872
830305364484072500052885504 415171571707967728607297536 207595230586949603594076160
103802337659957671442251776 51903530013220270543732736 25952945598230852683169792
12977063094925785047236608 6488826695368071876444160 3244560921636625614635008
1622354247794607645523968 811214017385451241865216 405625455436799330484224
202821951090436520017920 101415587231236687396864 50710099458627557392384
25356202650818385543168 12678677786161496195072 6339627123456899809280 3169957676916525760512
1585050896052300808192 792561476823169368064 396298752810094166016 198158383604301824000
99083695401778282496 49544099500702826496 24773175650258255872 12387150775082549248
6193856862517985280 3097069168747347968 1548604953117851648 774337660931014656
387186422651551744 193602007418798080 96805401755910144 48404899901210624 24203549462233088
12102324486930432 6051437121372160 3025855999639552 1512996719296512 756532719386624
378283539562496 189150359715840 94579474825216 47291884896256 23647016189952 11824044965888
5912290918400 2956279676928 1478206947328 739137028096 369585291264 184801034240 92404711424
46204452864 23103275008 11552161792 5776343040 2888302592 1444216832 722141184 361086976
180551680 90279936 45142016 22572032 11286528 5643520 2821888 1411008 705536 352784 176400 88204
44104 22053 11027 22055 44112 88228 176464 352944 705920 1411904 2823936 5648128 11296768 22594560
45191168 90386432 180781056 361578496 723189760 1446445056 2893021184 5786304512 11573133312
23147315200 46296727552 92597649408 185203687424 370424152064 740881858560 1481830825984
2963795869696 5927860174848 11856257220608 23713588183040 47429323849728 94862942666752
189734475268096 379486130405376 759006620549120 1518081960574976 3036301360103424
6072877598113792 12146304952041472 24293709415710720 48589617854676992 97183633755865088
194376063604752384 388769719395549184 777574623163187200 1555219615070552064
3110579967629459456 6221441410235629568 12443445770424680448 24888017440756203520
49778286681326092288 99561076962279555072 199131161123813851136 398280336646137184256
796596702089293332480 1593265461772624592896 3186675038733325041664 6373638307842801795072
12747853076437907013632 25496859074380420874240 50996023991770055442432
101996659669558538272768 204002542711153931321344 408023532166381572194304
816083957820910563491840 1632241702618115965190144 3264630979188821606793216
6529557106282822566412288 13059704508376003838476288 26120589608372725088256000
52243540399986884999118848 104491803166456639643451392 208993051065879018577330176
418004991597689515735515136 836047762127241988632739840 1672171082118209891588898816
3344493279963871611824635904 6689288791382646880942948352 13379182045675101076473249792
26759573017169816782121205760 53521563885978862822591823872 107047963475236184161882472448
214105598357029285357162594304 428230539527172404781120487424 856499764680572477695831572480
1713076900613600291658844340224 3426308543732111255852051070976
6852926572474043856772826923008 13706472114967730403683103408128
27414182169974746187641105940480 54830840220028063135832010129408
109666632200213267792763616755712 219343167920740818627726426505216
438706142882110203339851238998016 877451899845477538848499249971200
1754983027853469342034592043892736 3510124512031967212744371175686144
7020565936713991482839116527173632 14041765698728097080378981405949952
28084799048056422390159459515105280 56172133397313301239121912436621312
112349337397027515395849811686064128 224708815998856856626911596997771264
449437914407317364924247141246828544 898916393633842033189342176996229120
1797913916906098673060380142997602304 3595990093089026559484151864005492736
7192304704731711545695086884031561728 14385258446570739944843740080104275968
28771814967356113596594612784290856960 57546226083141494607003490816746323968
115097644463141524041635512129821868032 230205673520000117738528085252302176256
460432116227434374787570292489921232896 920905770829737028196168828950476226560

18418946184092106136343941458422199974656 3683955390317894341752901267566974992384
7368243087634734912474028486899020070912 14737150789267362282884508877328180314112
29475630806530509481641921561716640972800 58953920069052588795029650737553842634752
117913157050088317253550916703348806647808 235836947924142913834085063863179856052224
471695163496218386322136588639324197617664 943432862288301889952206099104577366261760
1886950795168334014520278041861012674576384 3774071731520128498272287771025741233258496
7548483745407177935008038916658914234728448 15097648055548197746943004582532692005879808
30196657240564079247739862663495111084605440 60396036740063526003187432323849676314902528
120797517997997787021790278641418260921188352 241605925031737044074411385270274338425143296
483233628134957028210484426515424310015819776 966510812412879936544292164980599886362705920
1933108737111691633335230953860702305387544576 3866391698795246787163755155520409676099354624
7733131846734220615314096806638829482847240192 15466960591755895312601366604473679226991542272
30935314980086698789149079191339398976577208320
61873417553323213906190850347462878998342664192
123752410292946060468167084624493920087061823488
247515970958491386247904937108124164354876637184
495054242662181303118951409934520977071259254784
990153086814759667484185891305587250865530470400
1980395376610313457460937925484265095177084862464
3960969159182215159907008136714711377246217568256
7922295130287606809784280844921785128276530823168
3961325971105391649877272708207073751030313254912
1980752188533489947431202496976627468961180549120
990420695757142034961884319924859031203602235392
495232648623769548104083695680702163963307098112
247627474684484039363612615699487406162406539264
123819312528541652337591691779311865171579764736
61912443857420642496688537854440013630978129920
30957615725285229412290614909612047338083188736
15479504760930068788118480446002043930338656256 7740100829608761435045826718599032095817859072
3870224639376244238016206607098521113233195008 1935199431974053879254749927448763089278730240
967643272129992819750698275674132810970431488 483843414136479349937010793811942038650748928
241932596103981144999336224893408835908141056 120971742569861307515083526440423326245453824
60488593543866021265249470217071117268418560 30245657901400694386478588606965285707055104
15123509515434189070166221052697506389950464 7562095040084015473546573900956184963186688
3781217661225468206005018637781808365699072 1890693901204464337618375162542762124902400
945389485898097286117120503097310033477632 472716010596981201712526712461619502252032
236368639122456880183246586687291993882624 118189636473211579755114908571887118319616
59097476692597359709303261900064119848960 29550067574294464770524534757092340269056
14775698401145124843198719282076310306816 7388181507571508650567585592803225239552
3694256907285227439767905772284147662848 1847211530392350277126009374083341352960
923647303571043417184032931012304437248 461844420972955847902530587491469099008
230932595080194993606522354738392989696 115471489836956031630889707865525714944
57738341066907283229259119180927467520 28870468607668275321536692214546038784
14435883340941454514221912419314171904 7218266189024385683837739365677662208
3609295353789022055282261260849119232 1804728806532925634322826419429703680
902404968085670120502261104217423872 451222766452438711921554499359997952
225621524431021181795989223305641984 112815832817911503815600598465642496
56410451710156208366603292639232000 28206493505678332412703143022821376
14103880578139280321052319863013376 7052257201719697217876534107308032
3526287057184877137613454141554688 1763222756754952833144320614727680
881650992458733548740957079339008 440845303269995340454876925657088

220432555155311953269637655822336 110221229337813118155918424408064
55113090548985129838509010452480 27557783214531850299529404350464
13779510577285567839902151737344 6890064773652605265019800649728
3445187129331213305044262715392 1722670935918061988789312552960 861374153585258662528246874112
430706419605743165330918735872 215362881209428499698857017344 107686276307992708366127333376
53845556005635583441413079040 26923986928637406349881245696 13462597927228510489527975936
6731601195069158902057664512 3365951713262031279675670528 1683051414494741554161254400
841563486179233734242336768 420800632555548345702023168 210409761010739912141438976
105209602871852825715933184 52607162619167847680573440 26304761901204641251590144
13152971246412679331446784 6576780771111519018549248 3288537959508349185687552
1644342766730469431050240 822208276853382134628352 411122585170764776865792
205570515957419243208704 102789869664728048992256 51397240675373238190080
25699773259191225942016 12850463090347916394496 6425519775550109908992 3212904002963130810368
1606524059075603333120 803298058334820630528 401667043565919797248 200842528982214639616
100425768090734690304 50215135845181030400 25108693822497357824 12554909861202100224
6277736405577760768 3139008940277235712 1569574838882795520 784822603813486592
392428894092787712 196223243139416064 98116019616219136 49060208831365120 24531203927310336
12266151719469056 6133350737641472 3066812807774208 1533475123363840 766771921420288
383403140579328 191710160224256 95859375079424 47931835023360 23966991253504 11984032497664
5992284684288 2996276559872 1498205388800 749136248832 374584901632 187300839424 93654614016
46829404160 23415750656 11708399616 5854461952 2927362048 1463746560 731906048 365969408
182992896 91500544 45752320 22877184 11439104 5719808 2860032 1430080 715072 357552 178784 89396
44700 22351 11176 22353 44708 89420 178848 357712 715456 1430976 2862080 5724416 11449344 22899712
45801472 91607040 183222272 366460928 732954624 1465974784 2932080640 5864423424 11729371136
23459790848 46921678848 93847552000 187703492608 375423762432 750881079296 1501829267456
3003792752640 6007853940736 12016244752384 24033563246592 48069273976832 96142842920960
192294275776512 384605731422208 769245822582784 1538560364642304 3077258168238080
6154791214383104 12310132184580096 24621363880787968 49244926784831488 98494251616174080
196997299325370368 394012190836785152 788059566045659136 1576189500835495936
3152519739159347200 6305320953295405056 12611204856544231424 25222535612995305472
50449323025804296192 100903149651235962880 201815306501726666752 403648627401962815488
807333283600944594944 1614738624795927117824 3229621364779930091520 6459530959936011894784
12919638380624327213056 25840429682753261273088 51683165208515736240128
103370942103049899868160 206751107578136654512128 413520661900347018575872
827078217288841456254976 1654230221553977750716416 3308608017060545177845760
6617511182026269708517376 13235612659862898122686464 26472405911346513656676352
52947173005934462135959552 105899068378351793917132800 211807581489669327124692992
423634052445270132830240768 847305883822403222822191104 1694687325508532359967801344
3389525766744516548582440960 6779353764943936754458558464 13559311992797680823504470016
27119832911414976276183646208 54242083674469181810716704768 108489003052216822138132234240
216987677510990561309662117888 433994697835094956686119534592 868028081296417581505829666816
1736133533845290499278840528896 3472421810195491671092043448320
6945153105400804687252811677696 13890925180821252064643072917504
27783088301681789509561044959232 55568652483442149779671888166912
111142256727041441080443372830720 222294416974397165203085938655232
444608640989422896490570263298048 889256896060102925149937298571264
1778593020282720114637468141092864 3557344496890468757950123370086400
7115005906430994573250620915974144 14230645638162103261201990183550976
28462558926924434751805477070307328 56927653155049325962413947547025408
113860376912499564842433881906872320 227730895029800955520079737439387648
455482072469205562710583422130061312 911004709757618428762014738762694656
1822090549153651464205725266530533376 3644343357584132141774842111071354880

7289011233721922710276467378163286016 14578671504551162274006501068367724544
29158641083316958254920134760817754112 58319878315063183923654534769800118272
116644948926984902674937600035929456640 233300282447686875005132261064517353472
466621334082807889320778644114351587328 933284206540484057262585532199336935424
1866651489830704671767227552339941392384 3733469133160882458018568080562417827840
7467270573320711145005362112889905741824 14935205760639314747947176129309951655936
29871740749274414411767256065680183656448 59746139954540398655280319745480928002048
119497596821063936974052254719202977382400 239005827466094153275087739894888197521408
478032922580120865204141940702740880556032 956108380456106847716216803231410732138496
1912301831503943930048299450114679406329856 3824773804191348329328330587533074696765440
7649887890749617597120124549673581161742336 15300456346233077071167175848562025859907584
30602273821933837896188205195553778792660992 61207269902803043300084117387967011731013632
122419984323476821615583648769652931753410560 244850857682695113261998125526743680089587712
489723493436873166585657907028362993344708608 979490543016712213294639126006477253020483584
1959068198319356186835924875912457038703099904 3918310621210575894165142999623919142730465280
7836969691564878829316872494845848416109461504 15674636281417211740606917980887717093515984896
31350666359409331645160181944167474709626093568
62704120311968479618213055853119030464440434688
125413815810236591892211495635806223019257364480
250838781993072449095993759130748770219267719168
501699864731343428815129053979770188800041418752
1003444330953083918876541179396085674323094798336
2006977864886961960245648501665261942092213518336
4014134135735512165476429289076705071076474880000
8028625083394200820923123149645772515937045446656
16057963790634754621786775442276269779442282266624
8029338707240553800863652292630497263505236819968
4014847759581865145416958432061429818644666253312
2007513082771726695201045358903805502768357048320
1003801142876260408846805750888448048107190484992
501922872183328735046544411162496672415101222912
250972586464263632834842973440384660388303601664
125491868418431449073206870649760492284528295936
62748721802365540864496127289664327187452395520
31375754697757678596194409627224204116320321536
15688574247166293380070377804808122319457222656 7844635572726873731021775398002071290377142272
3922492010935300386004180946800040710512836608 1961333117753581953248737097299522887918551040
980710115019756856747691860599512710290341888 490376835581361368435507586274631988310704128
245199306826422154248584621124753810738118656 122605097931081812139707724556095813660442624
61305271224476273577561569274907360975912960 30653996741705820542634638135883407560802304
15327678935586752148244245817156567316824064 7664179750160297012585586283185715426623488
3832260016263608975524524828896573597417472 1916215078723534722378128258100144740761600
958150074657632478496997050876001341407232 479096304976748797902464986350965156216832
239558786312340678278215723631964820865024 119784710068153478802599477044223531810816
59895013490068309233045546136232326594560 29948835973029939532395676875176443641856
14975082600512862224134290341118361993216 7487873607255377341035371122324251082752
3744102957127161785001798537044660584448 1872134555313317449742955756463597813760
936108816031527003492506122202432667648 468075177203197641056767183086533214208
234047973195315890183640652535925047296 117029178894516479919448856764291743744
58517185595687507373538693630310481920 29259890872058387393676479439237545984
14630594473136510550291806031659925504 7315621755121913701872686171850539008
3657973136837786064299734663935557632 1829067698057307638831563120972922880

914574413847861122756629454989033472 457307489333534213048738674745802752
228663885871568932359581310998544384 114337013538185379097396642312093696
57171042070293146007501314562457600 28586788685746801233152153984434176
14294028168173514731276825343819776 7147330996736814422988786847711232
3573823954693435740169580511756288 1786991205509232134422383799828480
893535216835873199379988671889408 446787415458565165774392721932288
223403611249596865929395553959936 111706757384955574485797373476864
55855854572556358003448484986880 27929165226317464381999141617664
13965201583178374881137020370944 6982910276599008785637234966528
3491609880804415065352979873792 1745882311654662868943671132160 872979841453559102605426163712
436509263539893385369508380672 218264303176503609718151839744 109136987291530263375774744576
54570911497404360946236784640 27286664674521795102293098496 13643936800170704865733902336
6822270631540256090160627712 3411286431497579873727152128 1705718773612515851186995200
852897165738120882755207168 426467472334991919958458368 213243180900461699269656576
106626312816713719280041984 53315517591598294462627840 26658939387419864642617344
13330059989520291026960384 6665325142665324866306048 3332810145285252109565952
1666478859618920892989440 833276323297607865597952 416656608392877642350592
208337527568475675951104 104173375470256265363456 52088993578137346375680
26045649710573280034816 13023401316038943440896 6511988888395623432192 3256138559385887571968
1628141337286981713920 814106697440509820928 407071363118764392448 203544688758636937216
101776847978945839104 50890675789286604800 25446463794550145024 12723794847228493824
6362178898590957568 3181230186783834112 1590685462136094720 795377915440136192
397706549906112512 198862071046078464 99435433569550336 49719915808030720 24861057415643136
12431078463635456 6215814109724672 3108044493815808 1554090966384640 777079842930688
388557101334528 194287140601856 97147865268224 48576080117760 24289113800704 12145093771264
6072815321088 3036541878272 1518338048000 759202578432 379618066432 189817421824 94912905216
47458549760 23730323456 11865686016 5933105152 2966683648 1483407360 741736448 370884608
185450496 92729344 46366720 23184384 11592704 5796608 2898432 1449280 724672 362352 181184 90596
45300 22651 11326 22653 45308 90620 181248 362512 725056 1450176 2900480 5801216 11602944 23206912
46415872 92835840 185679872 371376128 742785024 1485635584 2971402240 5943066624 11886657536
23774363648 47550824448 95105843200 190220075008 380456927232 760947408896 1521961926656
3044058071040 6088384577536 12177306025984 24355685793792 48713519071232 97431333109760
194871256154112 389759692177408 779553744093184 1559176207663104 3118489854279680
6237254586466304 12475058928746496 24951217369120768 49904633761497088 99813665569505280
199636127232032768 399289846650109952 798614877672308736 1597300124088795136
3194740985665945600 6389763446308601856 12780089842570625024 25561305585048092672
51124862969909870592 102254229539447111680 204517466278148964352 409052946954807410688
818141922706633785344 1636355903007305498624 3272855921202686853120 6546000072781525417984
13092576606315354259456 26186306134135315365888 52374918111279844425728
104754447908578116239360 209518119189193087254528 419054685122459884060672
838146263733067187224576 1676366314442429212655616 3352880202837448101724160
6706055553580075556274176 13412701402970509818200064 26826583397561737047703552
53655527978364908918013952 107315778323212687481241600 214641001379391114252910592
429300892224713707086675968 858639563381290371335061504 1717354684626306656993542144
3434860484980065142633922560 6870023201415033942561521664 13740650865739875199710396416
27482510657299365028595499008 54967439166237959315540410368 109939714035754377147779645440
219889099478065671328956940288 439797541769245176724709179392 879633769164718021583008956416
1759344909581891379433199108096 3518844561668693431400760606720
7037998608347208207870245994496 14076616186714059105877941551104
28154470313467403592030782226432 56311416507013377944611362701312
112627784774183897410322321899520 225265473068682077862843836792832
450550753177992721810086059573248 901141120437242575788968891121664

1802361469036999415915531326193664 3604881394399027360506249740288000
7210079701448111778362873656377344 14420793228196337671426495664357376
28842854106992903572254488031920128 57688243515186263603311969470251008
115381557632773440124229925753323520 230773256470348706083671825132290048
461566795350301063837767597515866112 923174155519809431016383089534304256
1846429440678033468714461968073752576 3693021140632896150792315514157793280
7386366799819450728311414184336162816 14773382636746218310076394680713478144
29548063347707070327059921985509261312 59098722843843408067934109219183132672
118202637984545350963496748934695485440 236415660562807771582250558862049411072
472852090313049682475015239709415702528 945745719000967643571058723389465165824
1891574514751671844384173934720197853184 3783315183002816803252460845322930749440
7566962673004579835473147642410931585024 15134589960007052128882747188352003342336
30270509148009889173638398183764287029248 60543676752011348179022603981649134747648
121092670416005836021536823191539390873600 242195974655977951370056876839561024503808
484413216959888461394080214592086534520832 968868969215642040096093351010102040068096
1937823009023014314808052545672062022189056 3875816159229489098847836778647839928483840
7751972600825899136159136931903111625179136 15504625766385640149245200613021086786781184
31010612662238964052344254724471900646408192 62023947583413295612396216445803255438508032
124053339684697326240207846885325419168399360 248117568405136122511246521758088654919565312
496256914881755185084154699491052943004663808 992557385906476250291632710931857152340393984
1985201884098884260829912045763216837342920704 3970577992769632042153117339325438740010106880
7941504434682991125292821174248887610668744704 15883705767653436332558815339693795482634551296
31768805331881780829063976661779631487863226368
63540398256913377986020645288343344020914700288
127086371700126388627826674506254850132205895680
254183893772852042567224116871646024445164781568
508390088290902615757589769461564697251835543552
1016824778072202292761462610359674691226683047936
2033738759125198708015491363592439975899390017536
4067655924211985661016115012931061138690827878400
8135668660347147812002494597354484651165751443456
16272050944540648603945518337693694049899694260224
32545529136774003167772094961356837594935771267072
16273478192233354563826576623663143545036077006848
8137095908039853771883552883323934146302134190080
4068726359981515130926908727408148260043114938368
2034452382971551687956020506577164723467581390848
1017270792976172905224293324725127658456802656256
508657697233284983235288198080836477589907308544
254339998989241756929214866899554562975706644480
127175574680920511120392817379345443578229817344
63590574933610071888089100654456802834303156224
31796681263379944107990896309620441939745701888
15899037529977426135968621146006241231169912832 7949867214132440108970897068601130746233487360
3975107831638083574978741782099570438441009152 1987641028104973547736017514949287751882637312
993864070195452653991332069424395142272385024 496953813169209267057327690687073204301725696
248487795620346103559494673330974418733629440 124249342328043786795162750659206117658198016
62127393422957260905289082326462512974790656 31065057840946314206498394661660983560241152
15533209485206998980176124080045355316543488 7766945024970420428551525414630109426483200
3883642653668670683507494394618770597347328 1941906397426065576369613040961243240726528
970995734008897905492739442306550591389696 485519134652381511400336182066239781208064
242770201150157035027151321489602133360640 121390417487061657177067275973042188058624

60697867199522398420279445600641654718464 30350262827756984126012626607381107703808
15175796027876384520942765207220694024192 7588230320937138489439608555375417098240
3794281313968042359203917253570243592192 1897223733733757736844015114726389317632
948653405241747147043035801333828419584 474347471808307712832032022652231090176
237184120497870926071273072318773985280 118597252545793997863265066655716212736
59301222421326266345446798576022716416 29651909284877766879630531912093663232
14826603679546200293268832268087984128 7413626358326758573361199290064568320
3706975438440208500043991223042572288 1853568848858518856703691400526430208
926824989248466731692693594765787136 463432777033837017516770744634179584
231726529721720334593597345942732800 115868335463261080214404659784187904
57936703032830996566005323298504704 28969619167015726512404158352457728
14485443408807977370902827527831552 7243038617054045742801787939717120
3621677764852051400076081057759232 1810918110588539964375634072829952
905498669375527114356613808390144 452769141728392123262705290182656
226394474384510344673551838085120 113202188952412313857875515539456
56603570356284727689487556018176 28303023118181649225018677133312
14152130529110467302646788128768 7076374749565054996392118845440
3538342117287438170730421813248 1769248429896174421632392101888 884662900574314878949786648576
442350793100271273541688623104 221185067956692553804241960960 110597369681624735418819805184
55301102692451596967759314944 27651760272045413113054363648 13826484598932513871114534912
6913544530921160592850944000 3456923381188032125072310272 1728537248457741976859574272
864306403160733945591496704 432172091046298451376603136 216095490256114964978728960
108052467494540352134578176 54028594930511610889895936 27015478056876522856251392
13508329324248620133777408 67544559810029489419714560 3377377478967334386270208
1688762526459962031341568 844418156718128434774016 422227525103137926938624
211122985923605818245120 105566104647821336510464 52785358166919881949184
26393832004964547821568 13197492463234577334272 6599034461993440378880 3299661346184796045312
1649902730686435950592 824987394140236939264 412511711468627951616 206264862933568716800
103136935066411728896 51570719333019549696 25786485566416617472 12893805733161730048
6447184341557575680 3223732908267143168 1611936822877749248 806003595810963456
403019390091526144 201518491138785280 100763643615903744 50384020831207424 25193109927231488
12597104719429632 6298827237621760 3149551057764352 1574844248358912 787456483917824
393745421828096 196881300848640 98444945391616 49224620179456 24613383831552 12307228786688
6153882828800 3077075632128 1538604924928 769336016896 384684785664 192350781440 96179585024
48091889664 24046993408 12024020992 6012272640 3006267392 1503199232 751632384 375832576
187924480 93966336 46985216 23493632 11747328 5873920 2937088 1468608 734336 367184 183600 91804
45904 22953 11477 22955 45912 91828 183664 367344 734720 1469504 2939136 5878528 11757568 23516160
47034368 94072832 188153856 376324096 752680960 1505427456 3010985984 6022234112 12044992512
24091033600 48184164352 96372523008 192753434624 385523646464 771080847360 1542228803584
3084591824896 6169452085248 12339441041408 24679955824640 49362059132928 98728413233152
197465416400896 394948012670976 789930385080320 1579929489637376 3159996418228224
6320267714363392 12641085184540672 25283269880709120 50568738784673792 101141875615858688
202292547324739584 404602686835523584 809240558043136000 1618551484830449664
3237243707149254656 6474768889275219968 12950100728503861248 25901327356914565120
51804906513642815488 103614316626913001472 207237640453080743936 414493295304670969856
829022619406360903680 1658117296406759735296 3316378708001595326464 6633045646379342364672
13266667753510988152832 26534488428526583152640 53071282700062379999232
106147177086143187386368 212303577544323229548544 424625601832720168648704
849288097153587756400640 1698649981283470351007744 3397447536519530378428416
6795190220944240109682688 13590970737698838925017088 27183122067018395261337600
54368605317278225345282048 108741933001039320335777792 217493310735044379961982976
435005510936020238504820736 870048800803903434171351040 1740173159471532782666121216

3480497434670517393979080704 6961297100795938445251837952 13923198664501684205091028992
27847606254822983039356764160 55697630361285195337062940672 111400096425848849190824706048
222809864258254615415047061504 445639071329623064896889421824 891316828285473797927369441280
1782711027823402932121920077824 3565576798151716536778202546176
7131463081313254418625129873408 14263545132646151527387709308928
28528328205331588435050317742080 57059132290741747630650433732608
114123216341640636782400463962112 228256336203595556607000120918016
456532479447819679298398627823616 913104572976896490765594027622400
1826288374116307245868781599195136 3652735204557643020412750286290944
7305787321765343098175874748383232 14612208468830800311052497848369152
29225684588261828851506492399943680 58453904477724114161815978206298112
116912879557849141241237943225417728 233835900320500108317687860076478464
467692083050603868305799667404242944 935424730920415039952447229311057920
1870930591479244686586590247627259904 3742023442235318586536572073264807936
7484371403024295599799927302550192128 14969391843155908053053420917141536768
29940081760526449813013974458365378560 59882759669482167039842214164895367168
119770711635822868907312958826119954432 239551807865362807469882978644898349056
479124384918159754250280079275113578496 958290308211187787121588402520860917760
1916663693172112131485233292982989357056 3833493539843697377454579561848513757184
7667319386686340983877385075462097600512 15335303387370574425691222054454335373312
30671936002736933767255347915968951091200 61346530461465437366256503446058462871552
122698377834914014396004622120358047121408 245407389493794308118992474697198336999424
490836046635521174891951410307361159512064 981714628566907467091835742440651290050560
1963514327725545168799537328533160522153984 3927198796634550806830806344370036928413696
7854737875636022552125076063347505625038848 15710156316005886981177078875909874786500608
31421673761479457716208011250249476645847040 62846069781894282940123729497358407437385728
125697584081659300895662872988435723166154752 251406057199060071822156573964309262915076096
502833892469603083705974803903494158995685376 1005711341082172047535272919756739584322437120
2011509794450275855317192463412981701307006976 4023193813472415231127678174624968467938279424
8046736076088557503241942844847947066525089792 16094169050464569088457058680891914394347241472
32189731897504046340860463344175869311288606720
64382251388157909009613618653135819667765460992
128770077962615450675012621235839801425907417088
257551306297830166661596010330815927032567824384
515124913340858863946333556379904502426641629184
1030294428172114789138950184196354301576295219200
2060678059325023700770466511265799196598614360064
4121534524611635646526065308277779580089276563456
8243425861146447783022395188047921533962648813568
16487565346139248545985319519080567815493489000448
32976557939971203051851697324130585126123360747520
65955970375327818023465511220200069242519486988288
32979412435356614971613813896069484116396126240768
16490419841524660465747436091019466805766254493696
8245566732685506722843982617002095776667222933504
4122961772304341606407123594247229075225659310080
2061570089132964925696127939996705131058853576704
1030829646056879524094347041434897862252438749184
515437123773638292670315056435721579487725355008
257729712259418411646728296076997113924615667712
128870431316008838479149531968066719052684328960
64438003251154235567467457948817440571530412032

32220395422152025947680074956800760808359329792
16110894609363467055813210469596400665476726784 8055795753825460568893191730396210463386894336
4028072101484593804939889112997110297017712640 2014123163028228662716591180398057681170989056
1007105137657080211481618902148780106916560896 503574346900023045802471107049265686623813632
251798062485752992932066381512070659894673408 125904475760747231481448604749754238238720000
62954960139308983248432009371736573265051648 31478841199122175378069858184298013705371648
15740101164294929565961855841363870389108736 7870390864514385721444391295289366962765824
3935365573440653329953927334948399365488640 1967767857312056899592829511126057624797184
983926463951893567104347677388957783425024 491984499623879342206140299607443377225728
246002883635905950430053380260203931369472 123006758729936114878518305358343087063040
61506037820959627271004960293292104220672 30754348138475598551375383953706332454912
15377838683235691733624143880383306399744 7689251648616792095780297891956723286016
3844791977807869162374261921860896686080 1922479065653671138429187448871715864576
961281071201703847835621968406491693056 480661304788286063228325106188562726912
240341036987860101269419614086939803648 120175710790788585462338337539799121920
60090451543823560144983434018064171008 30046523846126413779398849633114390528
15023910960170523743152991128598347776 7512279998638920298303278720319750144
3756302258596289362515030938170163200 1878232258936559287939211258090225664
939156694287486947310453523547684864 469598629553347125325650709025128448
234809455981475388498037328138207232 117409797593138607166624650881925120
58707434597769760042115318847373312 29354984949485108250459156126892032
14678126300042668239930326415048704 7339380062671391177315537383325696
3669848487660724117332955779563520 1835003471992876323004071433732096
917541350077695293670832488841216 458790482079476212919814630408192
229405144560052389502106508197888 114707524040183336272152850595840
57356237900170238896626223546368 28679356890124404828588010897408
14340297415081845104431455010816 7170458192550743897284452286464
3585383838780282621176588533760 1792769290642596646855475462144 896423330947525991561328328704
448231008286876829847459463168 224125175549995331957127380992 112067423478276124495262515200
56036129590777291505980669952 28019273721208260382165041152 14010241323513937505669873664
7005422893211872410128613376 3502862562333388033711144960 1751506839030419931178991616
875791198447072922751205376 437914488689467939956457472 218966689077699709268656128
109488066905332724279541760 54746394635907796962377728 27374377909574615892492288
13687779250597666651897856 6844184773204012678774784 3422239960554596015800320
1711193767253592846106624 855633777114943842156544 427835335301545630629888
213926891022809670090752 106968057197423262433280 53486334441720844910592
26744320142365029302272 13372736531934818074624 6686656496343560749056 3343472363359856230400
1671808239273966043136 835940148434001985536 417988088615510474752 209003051507009978368
104506029353132359680 52255266476379865088 26128759138096775168 13064942519001808896
6532752734477615104 3266517104727162880 1633328921107759104 816699644925968384
408367414649028608 204192503417536512 102100649755279360 51052523900895232 25527361462075392
12764230486851584 6382390121332736 3191332499619840 1595734969286656 797901844381696
398968102060032 199492640964608 99750615449600 49877455208448 24939801346048 12470437543936
6235487207424 3117877821440 1559006019584 779536564224 389785059328 194900918272 97454653440
48729423872 24365760512 12183404544 6091964416 3046113280 1523122176 761593856 380813312
190414848 95211520 47607808 23804928 11902976 5951744 2976000 1488064 744064 372048 186032 93020
46512 23257 11629 23259 46520 93044 186096 372208 744448 1488960 2978048 5956352 11913216 23827456
47656960 95318016 190644224 381304832 762642432 1525350400 3050831872 6101925888 12204376064
24409800704 48821698560 97647591424 195303571456 390623920128 781281394688 1562629898240
3125394014208 6251056463872 12502649798656 25006373339136 50014894161920 100034083291136
200076756516864 400170692902912 800375745544192 1600820210565120 3201777860083712
6403830598074368 12808210951962624 25617521415553024 51237241854361600 102478881755234304

204966559603490816 409950711393026048 819936607158140928 1639943583060459520
3280027903609274368 6560337282195259392 13121237514343940096 26243600928594722816
52489453657003130880 104983410913633632256 209975829026522005504 419969672451553492992
839975373700125949952 1680022804994289827840 3360189725176655511552 6720667680729462734848
13441911822211228893184 26884976565927064633344 53772258974863342960640
107549129635745113309184 215107482643527081394176 430233412031127872339968
860503717550403163783168 1721081222077101165772800 3442310018106792007958528
6884915184118763368742912 13770420664047885443137536 27542021919716488297578496
55086405022674411417763840 110177532411831692480741376 220364509556629124251910144
440747908579189727084675072 881533596090242411331059712 1763142750044210736985538560
3526436615815873302617915392 7053175463086650262529507328 14106955389083107839646367744
28215119703985830308467441664 56432657259610889875284295680 112870150222500238267267416064
225749971851557393567932481536 451519286516228621202660261888 903077258658684910538911121408
1806231888569825157345003438080 3612618519644560987224369266688
7225546524298943319517463314432 14451712018617529329172376190976
28904661977274344038619651506176 57811799834627258837789101260800
115628551429411659196677799018496 231267006379137601435554791030784
462553819798903768955507968049152 925147253679064670079812708073472
1850373735520643604497218960097280 3700905927366315737669625008095232
7402128767382688532689624191991808 14804891360065491180079996735586304
29611050370731210589561490174377984 59224636042662877637925973755166720
118454342687726668193457934323154944 236918826580255162222127842271952896
473857935570113976114679631795191808 947756435959435255570207158092955648
1895594001557285117822110105191055360 3791350262391399449007611788392398848
7583025043336457324742006732805373952 15166699123780231502937579777651900416
30334696321775096712782292179386105856 60671988791979460839378849606936821760
121349169880817456506386229710202863616 242708724355351982668029520413064167424
485438217898138104646573162811445215232 970917974171144487914174569593524191232
1941919025092025533070405627128315904000 3884004203683524180624924230139166851072
7768340714365994590218074412043403788288 15537346042729881638372600727616947748864
31076021313455548192618105262294175842304 62154701082902666216982018138708912373760
124314719077788472097455651505658946125824 248640071979543223521894533467800135008256
497301411607019005697755527848564755529728 994645358509903128703443977523058482085888
1989375787611536492022753798697974906224640 3978921716406533453277239284699665696555008
7958183715179987845017941944006763161321472 15917047995093817566962810637228389859065856
31835457119655318887779474772886506790977536 63673636498246005283266656542632467727646720
127352717514362745581948727078983843746676736 254716324064466961194728282145405504076120064
509454426200416862451118220265686641317773312 1018952408543799605025559752481124548966612992
2037991929373530970297766128861751630595358720 4076158083318925461088825505522508326514982912
8152664615781577963164237506643026783678496768 16306026129850610008301648004482073828654055424
32613446056276128180549641991356188179902234624
65229679705702072688991975947496457404992716800
130464934597703778033769335824561076900361928704
260941019568006821379109439508258477981476847616
521904339881212173381360414734789604324459675648
1043853281252821408009003900906124505371931312128
2087795765486436938510573944685339604189886545920
4175769936934462122006280175116860395271820935168
8351896685792100733982824921726083164327737556992
16704506995430554447906178986436891076223666487296
33410441238553814855693416258843231647583715721216
66823736972493041631148949089625362285440196935680

133653182935756907101822131323128522551425924857856
66829445963263865470673182233503160265985727922176
33416150229324638695217649402721029628129246707712
16708788738508672327549353844345239561632814727168
8354751181177512653744941493664982154600503050240
4177553996550344571857603032578672264192299368448
2088866201255966408421367659162426725542173605888
1044477702118380265456966901017758659494098763776
522261151804388663351624986227151978108555362304
261141726274793596987383260972712313235030671360
130576438323696431149477014415924318707891830784
65291006754998031902631199172746240399134162944
32646897174073924115261945568765160722161205248
16324145485324416139604145775578600622377664512 8162421191805935110788659383387310441837363200
4081384820474831075887622939492660286242947072 2040779522523347298190458093645832675783606272
1020433317404639529218552358772667604222869504 510238436773802704670937835361209435276967936
255130107422642822366299745668042534221250560 127570498229192146198565286827740175402008576
63787971373531440606990350410729541846695936 31895346816233404057349028703794497996193792
15948353972850543905601441101112112534519808 7974517268792192891264183925163488035471360
3987428775579556914863823649885459901841408 1993799458381508692047777668594587892973568
996942264486619463331821756123222917513216 498492399891242290319877338974575944269824
249256833769587424486921899943770214891520 124633733796776851906952565200126228824064
62319525354379995785222090214183675101184 31161091905185782808483948914152117895168
15581210566590783862178426360606199119872 7790937590294338160057439132068169646080
3895634948646642194512832541916619866112 1947900551073057654498472758899577454592
973991813911397105870264623420422488064 487016676143132692245646433695528124416
243518722665283415778080277840422502400 121764553629500242716668669416540471296
60884872963179388772148599956434845696 30443734555804328092981432602299727872
15222516315009480899944282613191016448 7611582676058398876698924462616084480
3805953597306028651712853809318330368 1903057928291428932538122693664309248
951569528964921769609909241334726656 475805046892064536475378567918649344
237912664650834094072901257584967680 118961402927817959954056615605305344
59483236765109436435831301209063424 29742886033154946447317147307737088
14872076841877587338359322005471232 7436355333588850726530035178536960
3718336123119453891940204677169152 1859247289722241210307695882534912
929663258942377737322644713242624 464851436511817434745720742608896
232435621776223000415059564298240 116222762648268641728629378646016
58113857204212891624864487571456 29058166542145731192707142909952
14529702241092508286491021017088 7265160605556075488314235289600
3632735045282948416691480035328 1816444893893929544612921212928 908261132573192440440051204096
454149909099710054286820900864 227084625956411944176808099840 113547148681484430605102874624
56775992192381444560900849664 28389205022010336909625131008 14195206973914975769399918592
7097905718412391541993635840 3549103974933647599643656192 1774627545330549714145247232
887351551597137814234333184 443694665264500385698021376 221856777365215932139438080
110933111049090835714932736 55468916707786852680073216 27735638945514143751340032
13868409768567430581321728 6934500032188894643486720 3467397590047036998156288
1733772581999813337284608 866923184488054087745536 433480038988100753424384
216749242866087231488000 108379233119062043131904 54191922402540235259904
27097114122774724476928 13549133522139665661952 6774854991445984542720 3387571610911068127232
1693857863049571991552 846964960321804959744 423500494559411961856 211759254478960721920
105884130839107731456 52944317219367550976 26473284509590618112 13237205204748730368
6618884077351075840 3309582776163893248 1654861756826124288 827466062785150976

4137506235786199040 2068841078823321600 1034464519876771840 517254250170941440 258638120201748480 12932455765901312 6466502760857600 3233388819382272 1616763129167872 808415924322304
404225142030336 202121160949760 101064875442176 50534585204736 25268366344192 12634720043008
6317628456960 3158948446208 1579541331968 789804220416 394918887424 197467832320 98738110464
49371152384 24686624768 12343836672 6172180480 3086221312 1543176192 771620864 385826816
192921600 96464896 48234496 24118272 12059648 6030080 3015168 1507648 753856 376944 188480 94244
47124 23563 11782 23565 47132 94268 188544 377104 754240 1508544 3017216 6034688 12069888 24140800
48283648 96571392 193150976 386318336 772669440 1545404416 3090939904 6182141952 12364808192
24730664960 49463427072 98931048448 197870485504 395757748224 791549050880 1583165210624
3166464638976 6333197713408 12666932297728 25334938337280 50672024158208 101348343283712
202705276502016 405427732873216 810889825484800 1621848370446336 3243834179846144
6487943237599232 12976436231012352 25953971973652480 51910142970560512 103824683987632128
207658164068286464 415333920322617344 830703025017323520 1661476418778824704
3323093575046004736 6646468625068720128 13293500200090861568 26588126300088565760
53178504399990816768 106361512399609004032 212732031998472749056 425482078395454980096
851000185587928924160 1702072428769895776256 3404288972727867408384 6808866175831886528512
13618308812416076480512 27237770546336759808000 54477846935682733309952
108960305557383894007808 217929834486804642791424 435878115717682995134464
871793124923513409372160 1743660036823321656950784 3487467647599232990314496
6975230443103645333454848 13951051182017649372561408 27903282955656016156426240
55808927094553467135459328 111622576555589803916132352 223254597844145347122692096
446528085154222172826238976 893093949240307302814187520 1786263456344340519951794176
3572678028416132868550426624 7145658288287169394394529792 14291921039484146103376412672
28585051004787906835927531520 57172519861215042930204475392 114349875425708544377107775488
228709422257974005787613200384 457438187329061845642021699584 914915060284351359417633996800
1829907491821158055102449188864 3659969726147226782739260768256
7320248937304274910547246317568 14641116844628192511231942197248
29283471629295670402738783518720 58569419138669911566027365285888
117143790037496964653154327068672 234297483595308212348507847131136
468614774231244990781414080249856 937269162543747113731624932474880
1874617553250008491800843408900096 3749393562825045512276873905700864
7499104038300148081904121987203072 14998841901900410278508992326008832
29998951454401048786419481355223040 60000438210002554031641956116856832
120005947022406020980889899046535168 240022035249613867796991771718713344
480064352908831387264407490688712704 960169270636870077869662875879997440
1920419670912154762421021540765138944 3841001601101138738205434659540566016
7682327720755935903137652475101708288 15365304478619188659728871262244569088
30731907031453011026364875148571443200 61466410211335289466544015545307496448
122938012719529113760716561586944212992 245886410032775297176690184166546866176
491793589252984733663894490318410612736 983628716880837745948817224607454986240
1967340510511412049139690937156177494016 3934847174522297212763494850194890031104
7870026656043540654495215652154850148352 15740717926084973766926883207839840468992
31482765080165732449726670222739961282560 62968188616323034731199148059600483254272
125941694144629209125889911347442087886848 251894022113224697578763053151366418530304
503809311874381953811492567215697322573824 1007661159044629024930918056257323616174080
2015407388680988284477701956166505174401024 4030984918545437038187135599636726232907776
8062310119457795014837734573880884234027008 16125300803649431906602395896976632004476928
32251962736766547567058645292382991081799680 64506647732468462641824997581625436309291008
129018739982807660299065409156969780909965312 258048369001356790628961646301377378402697216
516118516074196521319584948577630389970927616 1032280588291358922762493209105012046272921600
2064648288868649605771633042109526625207975936 4129470802309162732036559332018058315740217344
8259290053762052505059705159634126762128965632 16519277005811559092092583310464273785554993152

33039947808198026348131512603320588093704110080
66082683209545869024155717171425257232596467712
132170941605391370704096818272418676555569430528
264353033583382006719764404403973677291891851264
528728367911962544062670344526220002945289682944
1057501337314322149371623760488985302613591326720
2115091877609438421235813663851061198673206575104
4230362161180465087456759613448303584238460993536
8461081134284106664883783798388969542261017673728
169228758924145666309708096739762663832090226720768
33847179032521838579297251765494777159316836188160
67697212560429089078356620102928453308906437869568
135400134111629001996237473349734704598358406725632
270811686204799651671523412987225005157807875424256
135411552093170649675285939637490300559449468698624
67708630541970736757405086390684049269997499842560
33855742518678074338583601481311474130135132667904
16928584883185390149232329883640461812635757707264
8464649253515871564586429513312593280101974540288
4232503032719524027278347042402477826943035113472
2116340719340556136131739664074329506917541478400
1058214961160675129312152903473710050181782700032
529129781325536095279217987455127673452397330432
264576041035367312951179761586700160906951655424
132293595703983289131375264722918242543852322816
66149585445141460893580324326243202317114408960
33076186519145638610736508145513641681151328256
16538790157860273387341427063952841101872726016 8269743528073863734657300027574430681584893952
4135045988608795387821943261586220406116712448 2067610106590329454157618254692612735720488960
1033848609438130607202132439296057634191310848 516946082790548243662727875622904450261188608
258483930431015591862194765798890041713360896 129247409733378530946512796893163929148063744
64626427125624632980964105443441418719723520 32314574692280000244335906220150436432707584
16157967910873841999094879859290081752776704 8079324237803841938010903304252472644599808
4039832260085381438237183339429952206405632 2020001200634420953734457513366834045255680
1010043135613075594175161678509345993654272 505042835454470355741547300167637482340352
252532051551201457197756880540300983926784 126271342687583868262370055498391613341696
63138329799783503962930835363316367360000 31570494127887536897338321488718464024576
15785911677941660906605612647889372184576 7893288145969776682271032275709756178432
3946810226484361455619629113737413132288 1973488189991917285051871044809974087680
986785633370826921146963766375620804608 493413585872847599883996005173127282688
246717177530140869597255063579222081536 123363781061928969626256062285940260864
61684486679393752226942296391134740480 30843541413911509820378280819649675264
15422419744063071763642706721865990144 7711534390585194308548136516953571328
3855929454569426367637459836487073792 1928045856923127790500425707248680960
964063493280771198591060748126912512 482052029049989250965954321314742272
241036155729796451318189134283014144 120523148467299138576700553954328576
60264109534850025747153270383575040 30133322418025241102978131894992896
15067295034312734666189814299099136 7533964429806424390445281325350912
3767140671228240723897827750576128 1883649563776634626286507419238400
941864395969574445312050481594368 470952005025415788740423626784768
235485906033022177412411006386176 117747904776668230227305099689984
58876428268412685874202348093440 29439452074245628317376073170944

147203450071424568488254861475584 7360481988581049769481467854848
3680395736795435557275096317952 1840275239650173114904729354240 920176305451314225585955274752
460107495538770946859772936192 230063419175942390463284117504 115036545291249653748340883456
57520690497264056132519854080 28761554174451642695434633216 14381381550135628662304669696
7190993006522717988446011392 3595647618988810822869843968 1797899367358131325758341120
898987462610928620040880128 449512620771395788601294848 224765755118663633591074816
112387599925814686440751104 56196161146148778042982400 28099261164695106432794624
14050220878157911922049024 7025405586984135313850368 3512850367444657333338112
1756498970698623504875520 878286378837459171540992 439161636162803295322112
219590041453438502436864 109799632412737678606336 54902122049378052997120
27452213946193633345536 13726683433849120096256 6863629947300711759872 3431959088838431735808
1716051602013253795840 858061829803645861888 429048929300332412928 214533471849420947456
107271239524337844224 53637871561982607360 26820061680898146304 13410593790402494464
6705578370177957888 3352929922577334272 1676535330032844800 838302849388511232
419169016880300032 209593304533172224 104801050313097216 52402724179804160 26202461601529856
13101780556578816 6551165156196352 3275720017051648 1637928728002560 818998723739648
409516541739008 204766860804096 102387725369344 51196010168320 25599078825984 12800076283904
6400306577408 3200287506432 1600210862080 800138985472 400086269952 200051523584 100029956096
50017075200 25009586176 12505317376 6252920832 3126591488 1563361280 781713408 390873088
195444736 97726464 48865280 24433664 12217344 6108928 3054592 1527360 763712 381872 190944 95476
47740 23871 11936 23873 47748 95500 191008 382032 764096 1528256 3056640 6113536 12227584 24456192
48914432 97832960 195674112 391364608 782761984 1565589504 3131310080 6262882304 12526288896
25053626368 50109349888 100222894080 200454176768 400925130752 801883815936 1603834740736
3207803699200 6415875833856 12832288538624 25665650819072 51333449121792 102671193210880
205350976356352 410719132581888 821472624902144 1643013969281024 3286165377515520
6572605632937984 13145761021689856 26292621555007488 52587442133270528 105179282313052160
210367360719126528 420752313624297472 841539811620683776 1683149991985545216
3366440721459445760 6733162917895602176 13466888785744625664 26934903471396093952
53872058742605873152 107748621084839116800 215506249368932974592 431030513136375431168
862097055069769826304 1724266167733577580544 3448676450655231016960 6897641131686613745664
13795858724125530914816 27592870369755668676608 55188046582520551047168
110380704851059529482240 220770633074155913740288 441559712892385537032192
883156319272918493167616 1766386425522131824541696 3532920424996853325496320
7066135997898886003818496 14132862291608130713288704 28266905174836978837880832
56536171532915392498368512 113077065432313654641950720 226163575597593048574328832
452346040661117575729512448 904729860254098108620734464 1809535278371922131564888064
3619221672471296091776614400 7238745576397495840846905344 14478095615704798996281163776
28957400157229212621737033728 57917218166097654501823479808 115839272035473767520345784320
231688215477504452074089218048 463395773768122738214973734912 926830233162473144563538067456
1853737837577401625394257330176 3707630417659713923322877050880
7415570320329249191714478882816 14831759610678141073566407327744
29664757161395567527407713779712 59331990202869705815365225807872
118668932165896553151830048112640 237347767852107389345859289219072
474715342744843344776116964425728 949470299570943821721030700826624
1899019827304401907779654945603584 3798198110933832344234496979107840
7596713134517721745819368134017024 15194060094335557606339484619636736
30389387839271343442080465942478848 60781310979743143342963925291368448
121567692561887199603533837395558400 243145526328576225042279648416759808
486311335066756101754983244084805632 972663234952719506850814382672183296
1945407599543853620383324554349510656 3890977458364536454130040686709309440
7782279435282731334986864529439195136 15565207907672779523427295370919542784
31131713889560192753761723365921390592 62266023927549652921337711980007391232

1245372401519578406703039544563344002560 2490848648976327509958649699053464453312
498190498982699641302244061796009771008 996422536340267561225516367562653302784
19992928149430271679693089223066574127104 3986022452360016473870291422015683297280
7972377211718979176708808795796436680704 15945419037435850811354069495123013533696
31892167302867486538581042797306307411968 63786993061726542908907893208733175513088
127579303035436225481307401645707472404480 255169239894838730289598033747897187565568
510359747437610019233162528408758860644352 1020762030171085155774257978643446692315136
2041609130933900546164381800938751326683136 4083388403051261561560495289181218537472000
8167117088469444061584453952969868843155456 16334914741672730000095834655154601222733824
32671190612813143754045522808738929518313472 65345103484561655015798752614337313182318592
130695651486994045047012919222393534656020480 261402192009729560124856666432224885894807552
522826162090942060311374988839325404955148288 1045695880324850000746073289628402076241362944
2091478872935631761738793203156306685144858624 4183131970443127043970879654111618435613982720
8366612390029981128928345803821247001876496384 16733921678347416339829864598838514265050054656
33469237153269740843606075180069069052694233088
66941261899689298015104842324922219150576713728
133888098985678228685995068579412600391529922560
267787348343955722683560905017961524963812835328
535596997433109975990263345754195698289131651072
1071238596356617013226809762944936693301275262976
2142566395694028148946185668762963980048574447616
4285311197349644542877503623272109146989196738560
8570979206622465575725271818036580667762489163776
17142672037091284131391072779057886083093169700864
34286771321875274222663203844085221661322722148352
68576397139135960365088524260109342312918209789952
137158503269042744569701281664096482606381950566400
274328424519627136818451029615948561173854963105792
548679685002337568994998991807408314269892050157568
274351260482710432176547962191459753096037087051776
137181339232126039927798214239607674528564074512384
68593524111448431883661223691742736254554802749440
34298189303416921901711670131840817622413784121344
17149808275554813930796364208905133558775083433984
8575260949700583455368446675944929153171637403648
4287808880811879972669355623718645763477866545152
2143993643386734108827243954732413475184957194240
1072041423183764115659905048802752034315490557952
536043012337080588453094060119648665519251259392
268032656541139559538117797918960656940378619904
134021903456869412424844282889048490560565805056
67013739321584522540314833409308326325471150080
33508263457367169434103762687046203685329698816
16754828626971038799025054334719122103961911296 8377762762629246440499113662957571182629486592
4189055605886486740742850079277790656639008768 2094614915229175130618071663538397860981637120
1047351013757553445432359143718950196821884928 523697284950259662777841227834350731576475648
261859531510871301419751441904613182371004416 130935210273306385725291134946025499476885504
65470327395588560370353274469872203884134400 32736524827261963939030490733365829014913024
16368942978364823846442172115897778043879424 8184811771549332861684549432556320790151168
4092576026958126900074006403581876279181312 2046373084070793684652869045442796081643520
1023229077331261959634367444547327011848192 511635806313563538471150183186627991437312
255828536980748048562558322049796238475264 127919585402357163944770776253139240615936

63962451157170151804131195740690180997120 31982554806580860817938501677405370843136
15991942017288322866905702742232825593856 79963033156431076624210773228814828883072
3998317811321026945694651637323276484608 1999241982410250030089382306602905763840
999662529579993293665719397272086642688 499852033977430786143373820621360201728
249936401582432462726943971303338541056 124973393088074766191100516147998490624
62489292692466650509364523322163855360 31245944420447958961589394285164232704
15623621247331296334248263454623268864 7812135142219306593850914883332210688
3906229830386482510288849019676393472 1953196044831655861826120298843340800
976638587235035234253908043924242432 488339576027121268797377969213407232
244179929218362460233900958232346624 122095035211582143034556465928994816
61050052906991527976081226370908160 30526294104095992217442109888659456
15263780877348110223421803295932416 7632207351324112169061275823767552
3816262131987084613205824999784448 1908210294156056570940506043842560
954144761159285417639049793896448 477092187620271274903923282935808
238555997330449920494160834461696 119282950425382101768180013727744
59643951092769621644639805112320 29823213486424096202594801680384
14912225713231690791434850402304 7456422341625666740786149982208
3728365913317744042927437381632 1864260327911327357730899886080 932168849581891346999040540672
466103767604059507566315569152 233061555208586670816555433984 116535613307571793924976541696
58270224505425126220837683200 29136321178532177739593547776 14568765052175896184384126976
7284684757542851749485740032 3642493494498877703389708288 1821322305113164766018273280
910698931488445340170846208 455368355210154148666277888 227693622338042813623566336
113851533535504276456996864 56928127950993573051105280 28465244567117503936856064
14233212579369110674079744 7116901437589734689865728 3558598292747457021345792
1779372933350023348879360 889723360163159093542912 444880126825653256323072
222449286784863482937344 1112229255078450168856576 55616933382234298122240
27809619612621755908096 13905386267063181377536 6952981363907742400512 3476634797141947056128
1738389456165011456000 869230756879524691968 434633392838271827968 217325703618390654976
108667355408822697984 54335929504225034240 27169090652019359744 13585108275963101184
6792835612958261248 3396558543967485952 1698349640727920640 849210004736049152
424622594554068992 212320093370056704 106164444731539456 53084421389025280 26543310206140416
13272204858884096 6636377307348992 3318326092627968 1659231765790720 829650242633728
414842301186048 207429740527616 103719165231104 51861730099200 25931938791424 12966506266624
6483521568768 3241895002112 1621014609920 810540859392 405287206912 202651992064 101330190336
50667192320 25334644736 12667846656 6334185472 3167223808 1583677440 791871488 395952128
197984256 98996224 49500160 24751104 12376064 6188288 3094272 1547200 773632 386832 193424 96716
48360 24181 12091 24183 48368 96740 193488 386992 774016 1548096 3096320 6192896 12386304 24773632
49549312 99102720 198213632 396443648 792920064 1585905664 3171942400 6344146944 12688818176
25378684928 50759467008 101523128320 203054645248 406126067712 812285689856 1624638488576
3249411194880 6499090825216 12998718521344 25998510784512 51999169052672 104002633072640
208013856079872 416044892028928 832124143796224 1664317007069184 3328771453091840
6657817784090624 13316185323995136 26633470159618048 53269139342491648 106542676731494400
213094149556011008 426205891298066432 852446966968221696 1704964302680621056
3410069342849597440 6820420160675905536 13641403271305232384 27283932442517307392
54570116684848300032 109144736969323970560 218298481137902682112 436614976674314846208
873265982145648656384 1746604021885335240704 3493352158958746337280 6986992548293644386304
13974561557339592196096 27950276036183791239168 55902857915376796172288
111810327516772019732480 223629878405580894240768 447278203555235498033152
894593300598618415169536 1789260388173531668545536 3578668350299653013504000
7157631848504485379833856 14315853992819329465319424 28632888577259376341942272
57268138337760187506491392 114540999042003244658196480 229091442816972228606820352
458201775099875935794495488 916441329131614828750700544 1832958216126955571824820224

3666067547981362972296478720 7332437327417629601886633984 14665479117745066518360621056
29332167161309747665895948288 58666752174258724590141308928 117338340051795907696981442560
234686351510148732427360534528 469392045833411298921516367872 938822777293050265976623333376
1877722925838555868220427862016 3755600594182022408975218114560
7511510673373866163019161010176 15023640316767375016175771582464
30048518573574035412626442289152 60099513027226641585802682826752
120203977814610424692704962150400 240417859149535132427609117294592
480855525339698830939616620576768 961750664760654794048030013128704
1923580557683823852433653570207744 3847319571692676233542494228316160
7694956056035409524435362632433664 15390545937370933163571473616470016
30782359525342094556544443936145408 61567254351884645571891881278701568
123139579306170204061389749370224640 246289299817142233957991472366092288
492598882043888119586406891983470592 985238328906983542513661678469513216
1970557787452381691709019145944170496 3941277834181592596781429869898629120
7882880186916843620289642895817834496 15766409410941004094032852103676821504
31534116896096641894972836831435948032 63070829940622551203759938911036506112
126146852178103637235148408318402232320 252304088949924344125553877629462904832
504628947087282827561621877244242690048 1009299432549433933744271998459119140864
2018681941848604424730600484859505803264 4037530037196681963945313945601546649600
8075392381392310156858853842968163385344 16151449376782512771654159589466466942976
32304227981560810459181222985993214230528 64611114419113190750108253586106989150208
129227545750209521163708122400455099678720 258465725324385321654399475257392442114048
516952718296703201962765411427749369741312 1033947971889271521233463744681427710509056
2067981014370273277082793333014713363070976 4136132169924007023397318353333142610247680
8272604622214934985258100081273716988706816 16545889809163711847443126911762297513836544
33093140747795107448740107321954322100518912 66189003754525582405187921640768098346729472
132383452026921899825791257275255104984842240 264777793089585269682413342537948026552451072
529577364250653479426488341050771686270435328 1059198284644272838976299994051294638871937024
2118483681574477438199246612002091810406006784 4237141587720818396891786471803188686136279040
8474631624585363834770159439204387502921089024 16949960147458181751513491869604795267139239936
33901314091491271666973329721601631056872603648
67805415776132359661839351407987343158933454848
135616406738564351979464086745542848408243404800
271243963849727969270498941350222020997239799808
542510228444654469164139418418716690355985580032
1085065058379705999574561908273978677434983120896
2170219319740206121641689959421047948315990163456
4340617045442000488268512204588277083524028170240
8681590902807177466507288980668916540832152027136
17363895429460707912955107104322557829232495427584
34729218106614121785791272494614565153601373601792
69461290708613655491344661561168029297475512696832
138928290407998134822213556266213856575496556380160
277867998797537917323475578820183309112084174733312
555758833558159130005048090215877810146350473412608
1111563339042484850726290045582778004137065194717184
555804505484325720721241955366900193990714721304576
2779136707237045080396694439712056929564484226252280
1389625443526230778593589551294806444587697422991136
694841266716969508494415941366792212196576366428160
3474349058354118138460185535430906010496520106803200
1737245891561694367224145682013925480005079190732800

8686586269731648326090992981561989773809491640320
4343471540827412408030628776527176073796793663488
2171824973394500326507880531136678630344420753408
1085957088187647224500223337004884611895222337536
543000844839022142873253204220714954309117149184
271511572792110336748197369969493801335311564800
135761361582354801029884068914315062758032277504
67883468384327216842834726421941612424204386304
33943127988738516585363709193362846734696316928
16972260892656712374655027587877443628645220352 8486478895472083228314100289536731944971141120
4243413672307905134650343392567371037809836032 2121793948439884327571818320183188051567050752
1060940530362908043909232472041345292114591744 530492043252936962016277891995548279222829056
265256910662209951038969773985211956194181120 132633899848975710534900300986324886388473856
66319672183423222775157857490021897339928576 33161197221179295141432782243440675742810112
16581279175323489447643317870935201407827968 8290979870028665662285122310075032472125440
4145660076197793300374292842341232120168448 2072915108690626884803012264822474002137088
1036500089641178559709439054237165972094976 518271312468521838508685988031547471560704
259146290058227198581326224472255978536960 129578461941096738954154727464369110646784
64791889426539939308823171346305116012544 32397273941265754570284489480212838350848
16199301584630769743078696643636559347712 8099983099314331100507574273583349760000
4050157703156638664737900112674209923072 2025161928328055889611006544278372483072
1012622502538896223426531516109820002304 506332020456882251023779880040226881536
253176394822158195167147001012771880960 126593389707937632411202031002715160576
63299291002398083619415280749522190336 31650943575413675516614772998843400192
15826120824814154611760952811462852608 7913384930960735732607259561752002560
3956854724757197079667021358886289408 1978508492017013146515206468448288768
989294810827713876598451128726716416 494667687823460589969649511614644224
247343985116532120820036729432965120 123677063160666973327624351529304064
61841066881533943122615169171062784 30921801091367199790709081288736768
15461534370983714010055288995971072 7731084098141914062378018673786880
3865700505395985559864196424794112 1932929480860507044269691756347392
966504354511510654303642650148864 483271984296383893236219711062016
241645895668506229660309048524800 120827899594410256351254120759296
60416425677283698936176858628096 30209450778681134848363328438272
15105344359360210114319113781248 7552981664689926402228281671680
3776645574849873873648503226368 1888400158677392273091432808448 944238764964923804679307001856
472138725295575736406448799744 236079034054344785236622049280 118044352730450851135009849344
59024594216864654825854337024 29513506034251942042101874688 14757357480035778335638290432
7378980971472792825112821760 3689641601463848241203249152 1844896358595650034925043712
922485958229687974624231424 461261868580775465892970496 230640379023353472236912640
115324911878159605763670016 57664817122321237704441856 28833589152781336263524352
14417384872201026837413888 7208987584005692771532800 3604641365955436062179328
1802394469954012869296128 901234128465153853751296 450635510976650636427264
225326978860362172989440 112668101116199513882624 56336356401108970635264
28169331122059092164608 14085242021781849505792 7042909241267076464640 3521598735821614088192
1760871425504844972032 880471741549441449984 440253885173230206976 220135949785869844480
110072478492562292736 55038491046094831616 27520371422954258432 13760748661430550528
6880655805691985920 3440468640334348288 1720304688911351808 860187528827764736
430111356599926784 215064474392985600 107536635243003904 53770516644757504 26886357834006528
13443728672817152 6722139214315520 3361207046111232 1680672242532352 840370481004544
420202420371456 210109800120320 105059195027456 52531744997376 26266946240512 13134009991168
6567273431040 3283770933248 1641952575488 821009842176 410521698304 205269237760 102638813184

51321503744 25661800448 12831424512 6415974400 3208118272 1604124672 802095104 401063936
200540160 100274176 50139136 25070592 12535808 6268160 3134208 1567168 783616 391824 195920 97964
48984 24493 12247 24495 48992 97988 195984 391984 784000 1568064 3136256 6272768 12546048 25093120
50188288 100380672 200769536 401555456 803143680 1606352896 3212836864 6425935872 12852396032
25705840640 51413778432 102831751168 205671890944 411360559104 822754672640 1645576454144
3291287126016 6582842687488 13166222245888 26333518233600 52669183950848 105342662868992
210693915672576 421405011214336 842844382167040 1685757483810816 3371652406575104
6743579691057152 13487709137928192 26976517787484160 53955234598223872 107914867242958848
215838530578939904 431694653343924224 863424491059937280 1726919350864052224
3453979439216459776 6908240353409630208 13817043656772681728 27635213213452206080
55272678226718097408 110549860053063565312 221108727305381871616 442235469009273225216
884506966815565414400 1769085991225168756736 3538316097638413369344 7076920425652978450432
14154417312058260324352 28309987545621127495680 56622280934251468685312
113249173554521364758528 226507570481079584292864 453033587706232878137344
906104068900613175377920 1812281924777521188962304 3624711423507632054337536
7249717994920443461500928 14500026285651245628653568 29001233162923208668610560
58004827509087852159827968 116014377384658573964869632 232038199502282887220166656
464095288470497253021188096 928228355872857463204085760 1856532269609440840731590656
3713215654946333510110019584 7426733541347570677513715712 14854071545604948669614784512
29709352017029511968404275200 59421121885698253195157962752 118847079474674964907014750208
237703830355906846847427149824 475427003524927527761649598464 950892692676082723656889794560
1901862756604620783580960784384 3803880255714152239696283959296
7608069996438125824461292699648 15216758962895894339060034961408
30434755865831074058394969047040 60871987611740718877339736342528
121748926983638579275779069181952 243507757487591441593757331357696
487035322015811449271913048702976 974110258112880030712622869381120
1948299744388274325762839282712576 3896757945101577180200865653325824
7793832802853211417752105482452992 15588299431006536950204959316508672
31177866512613302129811415336222720 62358268326427060718425824078856192
124721607255255034354457634970533888 249453355715311894544127243566710784
498926993840227440758678434384707584 997894552499662184858204763271987200
1995870234637738976398105315549118464 3991902728552307166159602209108525056
7984129975658272759045987574237626368 15968908988423862371545541460516405248
31939116051062358449998215545115115520 63880828250553984313810696338394841088
127766848797966503455249923173118902272 255544082189650076565756907338896244736
511108933566734292442027936663109369856 1022259405508336863505084117296852500480
2044601887766410284252224722534972522496 4089369929032293682988562420952480088064
8179072165063533594945350793670030262272 16358808944124959647827153490870200696832
32718947116245704211527210788800681738240 65440552688482978254800229191721924165632
130886422288949096173092073611684969709568 261783478401864471673167377679852182175744
523588224451661502000301216272668849864704 1047218984199188121308535354371266670755840
2094523038990106477232936552394391283564544 4189216219163673423697604792092498451234816
8378772720694267785858672958792428670681088 16758226006122377448644272666799720877785088
33517813141712438651142398832029168828416000 67038348542360244809992504660917791802523648
134082141602591224635400423315554491896430592 268175172240923919301631674618546800375627776
536372122553330778664925005211969233916788736 1072787801249627437453173322373689734164643840
2145662714785186635152993268646882000991420416 4291499654142236790799279785092769067307106304
8583347757428200622585146065783548265262743552 17167392413143855327143465122763116791822548992
34336178622862618818233276227918274106239221760
68675144838875053964359244420620629257666691072
137355864864049740584503872770809420605709877248
274722880100698746480578513400755165392172744704

549468060946596023584298562519782979145851469824
1098980723383589108414880196476111255014714900480
2198050649747972339322326535825313103475453722624
4396279705457532923629785357396807393842955288576
8792916222838242337229835286285977161470006263808
17586546069522837654400199715556679070508203900928
35174519386738381268681457717082807636152790548480
70351893268862174457125032006104514262578346590208
140709495528495172753774297156086826505702224166912
281430409038531993186597060599929248972495510306816
562883654040147281731291053775369689867173144559616
1125812980006461154178775972701761763578710537011200
2251717303865255489789939675705568294846149569806336
1125904323858794335611163703003806531267439032795136
562974997892480463163678784077414457555901640343552
281498916927781879260887858326462824739041882144768
140755167454661763469968162307109210350066472058880
70380438222716293654746197725493504165306001522688
35191646359050852787254157148716201577789383507968
17596536803371779373567607717342825536462883127296
8798625213609066176754068430163775142015537250304
4399491012766121333362166500828068757899816468480
2199834709363854789173649393287124972395932155904
1099961956172324455833107768080107782920978038784
550003278831360758539695419758326539821994999808
275012789788279644581418477738299594091750490112
137511970080439454946494622798717959136251740160
68758772633369543801140003364143060613314117632
34380780113259680064516347664463570829251182592
17191086954917294114231346823427805675922653184 8595891926602374098102259907311912968609857536
4298120187873050569544423201454961549629194240 2149147206222457045018858224626983307476729856
1074617159254194402632752424263242920069431296 537330357698580141378037868106497093200248832
268676067885031540719849762040686363182891008 134343478460386505375340295014062089882828800
67174461489128620195377854503890499087106048 33588591874031993851542780750374976616398848
16794976501749838802698317124402351844622336 8397828533241840339812621936808607690522624
4199084407804380639138042655708019729367040 2099627274493920554184887171505867806736384
1049856172542825394400376507578862874394624 524949353919345255854154714702395922710528
262485310783638907254060587807680204111872 131247972303802593290521909132081223434240
65626644607892866477006762180161172406272 32814651531942218154376284897140866547712
16407990379969001535124594352100573446144 8204327496983446996530523127815356809216
4102329901991196612749374539790213447680 2051248027745334863616743757836374245376
1025665552247535710429400122888820883456 512853545311201994525214183429727322112
256437157249318066917864152707522101248 128223770921517568286560606850090270720
64114481609188051557094568673209745408 32058538878808659485454416960687177728
16029918476511646596180774792384741376 8015283756809481724817170552212946944
4007804137681570075771976854116761600 2003983198479199644567684216063524864
1002032164058807125624690002534334464 501036364439007214482768948518453248
250528323424305433076596447884869632 125269232314553629455904210755256320
62637151458477271186755098784038912 31319843379838863822779046095224832
15660555515219546026090271399215104 7830594670259830070395509875408896
3915455791454943563872942025605120 1957807123889986046274064556752896
978943176026250155305829050351616 489491395053753643737312911163392

2447556010471911049108556485754881223827522837526939765274207846406119385202195491774881350864076830598163951016744254681653444608
1529970094552801481747827628441676501599577738287538078629232643825234721391825049438293852160 19126947319483678609863281213449563860516004115986267546583044782123686133196333801726279682391158557132167337234839633921195627635598868253784408064005978379963158264194756981555229893108741610935602959613952 1494715883371527511606716006474738816483125412153272565763737091939883722436310466560 18686215278055871324786524169343485428346565234010357764671931608832597402813726722336060251745956094311137281168077349537806743607705605840622866013177200299212829204294921686603412799488
1460273775665366041205145673016640262320095588515843650979587068594455838720
1825563580510592066125824912818683743443452166144 456427788615795435634688
2282231176799345725931521141116170525985713684480 570603911060020705361922853134847450564211507214266250698005124481024 71334135793787139522563566850904877432832000
1783497510032754343936891784783813396135936 4459104063052075499522229642103518585159681114866087755566284805574555618759199948827873903993702842368 13937514946804842496696903894837913190434846602116779212801742400474583138304 871235421663657984
435635303017873408 217826447601958912 108917621847490560 54461009947000832 27231604485128192
13616351998377984 6808450877095936 3404362877501440 1702250158227456 851159438852096
425596899295232 212807039582208 106407814758400 53206054862848 26604101173248 13302587457536
6651562164224 3325915299840 1663024758784 831545933824 415789744128 207903260672 103955824640
51980009472 25991053312 12996050944 6498287616 3249274880 1624702976 812384256 406208512
203112448 101560320 50782208 25392128 12696576 6348544 3174400 1587264 793664 396848 198432 99220
49612 24807 12404 24809 49620 99244 198496 397008 794048 1588160 3176448 6353152 12706816 25414656
50831360 101666816 203341824 406700032 813432832 1626931200 3253993472 6508249088 13017022464
26035093504 52072284160 104148762624 208305913856 416628604928 833290764288 1666648637440
3333431492608 6667131420672 13334799712256 26670673166336 53343493816320 106691282599936
213391155134464 426799490138112 853633340014592 1707335399505920 3414808237965312
6829891353837568 13660332463489024 27321764438605824 54645727900467200 109295853847445504
218600503787913216 437218599761870848 874472383895830528 1749015136535838720
3498171010560032768 6996623496096776192 13993809942146973696 27988745784200790016
55979743368215265280 111963990336057901056 223936987871370543104 447891990141250568192
895820009079520100352 1791712075753078128640 3583568266694232113152 7167424763764615938048
14335425988281535299584 28672004898067677446144 57346315639144568586240
114697242964307564560384 229403709300651983896576 458825865345377677344768
917688624178902773792768 1835451035334100385792000 3671049644620790447996928
7342394437146760248819712 14685379170103879203291136 29371938931828475817885696
58746239046898386458378240 117497200460279642561970176 235003845653525024414367744
470026580772981527409590272 940090940477826011980890112 1880257438819377938285199360
3760665993366207705217236992 7521634218187319067728150528 15043872899284445450043654144
30088954724388505529262014464 60180327300416240316873441280 120365490304110939150445707264
240740652014778795334289063936 481500646842671424735373426688 963039979311570517604337451008
1926157329875596371475856097280 3852469402256103415486074585088
7705248289522028176040873951232 15411115549063699042219197464576
30823469038166683464713294053376 61649413956411937689976386355200
123303779672981016901052369207296 246617462866276316844303931408384
493254732773181199773006248804352 986549079627619531714809269583872
1973177387417753327767212083118080 3946513231160535184209611254136832
7893343374971127425769596684075008 15787320575242368966239941719752704
31575908801084966161881380142710784 63154352903370388782565753691832320
126313776409141690482737494196486144 252637694023085206800686962018615296
505295670455774065271797871288516608 1010631905730755433884443637079605248
2021344941099925474450583063164354560 4042852141476680162264557704338997248

80860288015070187512558985646985707 52 16172706640121354355965363441438294016

32346711354457342418837859506958893056 64696018857343952251489984262082396160

129397230011546439330608499020494012416 258804844616809948316474059033646465024

517630458421054035943462240052609810432 1035302455216976350507952724075853381632

2070687987183689258257961936092974284800 4141542127866851631000036848068483612672

8283416562732649490968299647902037311488 16567497739463191439873051199334214795264

33136324706922167795619006205728709935104 66275307869835905422983820025577980559360

132555932651654950509459255279397082497024 265122499127276180345901741015276407750656

530266265902484919345769942943517301014528 1060575067100834955999472807712963573055488

2121235204793400146614811459077785088163840 4242640550770260762461354605459286060433408

8485621383907442463386172585526003889078272 16971923332548726803699271920266871314579456

33945207794565137361252397338963469702004736 67893137848065642230212501674786393549701120

135791720214002019475840417343291695390785536 271594329463745508982511662674021207364337664

543210436998973958026684981322918047894208512 1086464430140913796176693274595587362119483392

2173015972567759352600033173090677256901099520 4346206169707382225693359593980359579126464512

8692760788558491492373305683558729288901459968 17386218475404437066719784358313478839099981824

34773830747383782297385914699018998200794087424

69550449087917380922664521362822077446776422400

139106473362134394501114426655212316983929339904

278224097096868054313799621169560958148611670016

556470494938934639250740778057394564658729320448

1112985591368266339747764627551334426040470601728

2226060385717326801988095397975759445526965125120

4452299177396241848961323081697700077945978093568

8904955166715660187892910734887762529676051873792

17810623957277673355726350612760249806920295120896

35622675162248052671333759511489949108976972988416

71248204819881517262429635594918797208226711470080

142502118630533858364383504333715392396998953926656

285015655242609364407815474955186380755088969826304

570054146448302024173727882485883953432360063598592

1140153964822770639063649630122790290709084375089152

2280399273497874459559686990547625349106897245962240

4560981234700415281984149441699340233591251483492352

2280581961202540822424462451151714884484354237530112

1140336652527437001928425090726879826086541366657024

570191162226801796322309477938951104965452807274496

285106999094942545840203205257231148443817465610240

142559208538242096759625835772493372202454263791616

71282458764506460299575034458185585091499897389056

35642656629945936109668575515062242040886331441152

17822041938819321034774816900515845768011357093888

8911377781332837007357673021750285257789774233600

4455867296628006748663968796621323815786934960128

2228022851294797496824550541183752501339491401728

1114056027137795809658558342028421547392757661696

557050314314096435452420706732483422057884811264

278536307529647483037781121225378035209695395840

139273728951123374174675944542257179695224193024

69639652068711503415230664235912670892800344064

34821219830930659871561678100348375968994295808

17411306813752784017754012041370208245794209792 87060018560201190498635925162831142535454635840

4353175152581923045425089505940562192097083392 2176674688576893282959191376869783628710674432
1088380900431412521602919000384643080686403584 544212228287189200863121156167197173508734976
2721170031793360704623914060710364033371340 80 1360639461075387702466111170292371099599503 36
680346953127047526310132655114780091256668 16 340187087858200600693604862541687316356792 32
170100349576438719116071698762992293542625 28 850535776118885689426704831275704644534272 0
42528490217778889163652558436822391067770 88 21265095814806746927984937654929774954414 08
10632973260362024637071798045724177187471 36 53166993066603379050755636319917334488678 4
2658455991569831745807614120560689152000 00 13292811649047472695387232125627557897830 4
6646671670122893330868196824225835017830 4 3323468757861025157021388792818945543372 8
1661800840330301824304339586762486788915 2 830933650865045535048992388557750403072 0
415483440782470078972907491867128705843 2 207750028066208695210659394727691105075 2
103879167870591175467432521760908928614 4 51941660854039001664767673078986152345 6
25971868886391207797909542638758920192 0 12986453672881457381717624369012382105 6
6493486451283655432240238709322652057 6 3246873033063291086810832617069556531 2
1623501420242377228750772939738893516 8 811783161976554457048064785471504384 0
405907806915960149860371550536781004 8 202962016421821535598355354168904908 8
101485064692831498133262466534709657 6 50744560587376114233673627992483430 4
25373294414168239700358011358806016 0 12687154267324211141939604360685158 4
6343830663782151216850101520983654 4 3172042096951098431365200430812364 8
1586084431005560627152675050566451 2 793073906767786019311374942863360 0
396552799016395862523206180221747 2 198284322324449357695362444505907 2
99146122570350392064560899450470 4 49575041989238052640720288323993 6
24788511346650454624580063461376 0 12394750849340941464399991380377 6
6197623012678327808254975515033 6 3098935300343092442154977669939 2
1549529547173510490091233791180 8 774795722087737379552489373696 0
387413335294359757029680925900 8 193714404772425412141558582476 8 96861070948835472884138351001 6
48432469755729119848748705382 4 24217202018520251627714117632 0 12109084579587971665526941286 4
6054784074957908758598411878 4 3027512930060915842216676556 8 1513816911321438652567073587 2
756938678806209692012904448 0 378484450975850028871136051 2 189249781274297605867909939 2
94628668530335098650125926 4 47316223211760697183148441 6 23659056079176922520616960 0
11830000276236748224829849 6 5915236256442517594675609 6 2957736187383330538468147 2
1478927123272701139799244 8 739493076426868505182208 0 369761295608693220232396 8
184888026501976093936844 8 92447702599802788878745 6 46225695974308765394534 4
23113770324358068174848 0 11557346330780876826214 4 5778903749691359782502 4
2889567166996140575948 8 1444841229573300630323 2 722449437824265486336 0 361239130430940328755 2
180626770974873957171 2 90316988367138874982 4 45160295623420385689 6 22581048531635666944 0
11290974625780570521 6 5645712492871653785 6 2822968836426511155 2 1411540713208597708 8
705798504101969920 0 352913325799820492 8 176463699774328012 8 88235368324372889 6
4411944338079088 64 2206060129969766 40 1103074045449994 24 551559012957552 64 275790501595054 08
13790074835566592 6895312295690240 3447793586798592 1723965512876032 862017116176384
431025737957376 215521458913280 107765024423936 53884659695616 26943403589632 13472238665728
6736387768320 3368328101888 1684231159808 842149134336 421091344384 210554060800 105281224704
52642709504 26322403328 13161725952 6581125120 3290693632 1645412352 822738944 411385856
205701120 102854656 51429376 25715712 12858368 6429440 3214848 1607488 803776 401904 200960 100484
50244 25123 12562 25125 50252 100508 201024 402064 804160 1608384 3216896 6434048 12868608
25738240 51478528 102961152 205930496 411877376 823787520 1647640576 3295412224 6591086592
13182697472 26366443520 52734984192 105474162688 210956713984 421930205184 843893964800
1687855038464 3375844294656 6751957024768 13504450920448 27009975582720 54022098649088
108048492265472 216105574465536 432228328800256 864491017338880 1729050754154496
3458238947262464 6916752772431872 13834055300677632 27669210112983040 55340619249221632
110685636544954368 221380069182930944 442777730551906304 885590645475901440 1771251659695980544
3542644056880316416 7085569588737343488 14171702127428108288 28344530154763059200

56691312109339803648 113387127818306977792 226783262835868696576 453584540070246875136
907205108937512714240 1814482275469063356416 3629108666126202568704 7258505562628556849152
14517587586009417121792 29036328093523441090560 58074962030056095875072
116154535746130619138048 232318294864298093051904 464655036472669895655424
929346966433487210414080 1858767719843269259034624 3717683013639128194482176
7435661175183435741790208 14871912646177230189232128 29745005883975177789767680
59492372951191790402142208 118989468268866450449498112 237988381270698640189423616
475995652007328758959702016 952029082946520475081113600 1904533723756766864485646336
3808418563240985557618130944 7617139357936874772529938432 15234883178783556859647229952
30470975283386728348469166080 60944368418412685955287744512 121893572540103830427274313728
243796816486764577887946276864 487612975786642989842687852544 975264637199513647818966302720
1950606645651482631905113800704 3901368033807875936344589991936
7803045552625573217757904764928 156066710075270789125653259091968
31214658090580863631581417308160 62431792061240298023712632864768
124868535882637737568524862226432 249746975285589758179248917446656
499513757611808082442896220880896 999067129304873297054589213736960
1998213486772260858446771971424256 3996585429869550245568731030749184
7993487772389157548487836237299712 15987609370078429211676420826202112
31976486390757086652754338355609600 63955508082714629764311670117629952
127916086767830172446229327048081408 255842314740462170727670627721805824
511704911890527993125765202694897664 1023450388600263289592378299892367360
2046981906838941185866452388789878784 4094126072954711585096296355590045696
8188576664463081596919375867200667648 16377802366033480047292318046442487808
32756902806281593801491768716967280640 65516401760992455016797802682099171328
131037995818843444861224135860527562752 262086376231403959377705332713713565696
524193521650242058065924787412744011776 1048428581675352394752877818796121784320
2096940240100441346747812125533511090176 4194046633700355807979737226949557223424
8388425574399657844927700405664184532992 16777515762797208147791852714858509238272
33556360753590201211456609236777298821120 67115379963171972254659026087675158331392
134236076838327084172809667403591438041088 268482787500620447672602565263665118838784
536986842649173453999171591440294723190784 1074016220594212025306276104706518417408000
2148117511780154285228418053064894776868864 4296405164743769039688567793433505437843456
8593150611854459017840598961474442643898368 17186981788442759912608124672163748824219648
34375324706353203579070102842757224721285120 68753371671641774665847912682373903588261888
137512187861154284347111239358466715467907072 275035264758050038725053306704371247518580736
550092307587583017511768269383618128202694656 1100228171318131915146859850716987522736455680
2200543454922195590540366325333477578135044096 4401261134416254701574025898465960221594353664
8802870717976236444134638292529930573837238272 17606438334239926970242449576255881408971538432
35214270465054762104431245134903803340537200640
70431328523259340536755182234591687726262648832
140868232232818313729295748398751537542901792768
281747614838235892770162264656639399266556575744
563517530421670316163466065031551446894619131904
1127079662333737693573215201499648190512250224640
2254248527648269509638996545872386974470524370944
4508675461258127264263125377490955135833096585216
9017707734439431018496515326474272645450288857088
18036129092725215016933559795933270038468769087488
36073685433143135993748177877835989572073920921600
72150225361671683907258472327610878134420607336448
144306159714114191654041177799099554249386745659392
288623737409770030987130821885954704459864553291776

5772703107826233573323585763474206008419112305295 36
115458629349141330538091101784586358552818670895 1040
230926393083515979219420976599377193874510191368 6016
461871054937498594725319499259163341286766081893 9904
923778647415930462023594090639144589649023562101 5552
461907592478431867298274591379981248362257480207 5648
230962930624449251792376068720195100950001589682 1760
115486032504841284967807420875199788859437219630 2848
577452998487289720197134036951510136219368222097408
288737917225186507777615484763510664070775173021696
144374667603364077728331975525633130015933117497344
72190188297067450783928104334755463998239324241920
36096521396226431351845110453347181494256044867584
18048974321959568655863084369658315494696213807104
9024843972902960817901806756321520121132202590208
4512600392413068653936035663906941247458149138432
2256389399187328449460583974826561217175098490880
1128239301084061285976575058849825905310561206272
564141951287229173611429065143185601016786583552
282082126016213852117285300430729124689146281984
141046638194406558714428034144932724434949636096
70526106690353095685106709037250443262663065600
35264447141751456006499700501017262153925656576
17632920469163182085223023241704651338259890176 8816808683725318083598098116450335799778476032
4408578566434522562292342306024172965213503488 2204376395503193041392817776911589015268884480
1102231753894562400819732200405545773965508608 551137655018764140471527756177648520148287488
275579716545123540266594706076262076656910336 137795302790432505148712767031849946619838464
68900373654151620082064090512784427455610880 34451547956543493794885898754821940800651264
17226454543005588774369876126625833936748544 8613567553869715325648401437920348736585728
4306953918118318132055932406263890252398592 2153562029650889300643832046783803068252160
1076823550121309767629848945217830505152512 538433042708587442468890933521879738089472
269227155178260000561428697217422111801344 134618894501113139944205963836952177278976
67312105706548139803848789532596649328640 33657382081269854817797298573358605008896
16829355654632819866835101190209442676736 8415010134315356162385776546869791424512
4207671220657151195677001249317430755328 2103918687078312155080557112599982899200
1052000881914024356161306800270625210368 526021210144446317391167522120629485568
263020989665940228350840822052973182976 131515687129828649003048941522815811584
65760439713343591915338736009572515840 32881517930886429664576500628868562944
16441408002550531685741816626475433984 8221028519828924269597691469258293248
4110676519191291348162237312639434752 2055419389234060280762814445324861440
1027750259436237443722255117165002752 513895412127722373531551505833787392
256957847268663012600987726542536704 128483994236732419218099850084089856
64244532419566666067852918448455680 32123533860383561263327955927433216
16062400755491894746364726315319296 8031517290396004430532737333460992
4015917101523030743941555754631168 2008037778924029636308371421265920
1004058503543271950322982482608128 502049058812264541245889627291648
251034432926446553665144006639616 125552216822338041835367159981670 4
62763559991768779937385598156800 31383017935923675348967698202624
15692127937981480364621298663424 7846373454000561527379374112768
3923334146950519143622404944691 2 1961748106005051054379205918720 980912738628753195323193556992
4904757121274904317283920773 12 2452475274703021328975936880 64 122628599438429524965495668736
61316717570853991741097246720 30659567711246610499723329536 15330388318533112564449017856

7665496390721459939518185472 3832899311088181798405931008 1916525213407816813526384640
9583003856357713639249018888 4791690822838171605433057728 2395939858748743195620802566
1198017153039200294262538244 599032188352014495357337760 2995279000922144217917030304
14976985300421079795236864 748878779811571925044428888 374454147301044930163507272
1872344523481519489024000 936209155228907163615232 468123024358527291359232
234070735551300500455424 117039979461668677615616 58522295573843552501760
29262300708426383097856 14631726814965494972416 7316151637858899197952 3658219934117525454848
1829182024652800655360 914627041123419291648 457331534960219127808 228674774679364304896
114341890939309522944 57173197269468446720 28587724534641065984 14294425217273954304
7147494083613687808 3573887779295199232 1787014258391777280 893542313567977472
446788748970033152 223403170578038784 111705983335530496 55855190691020800 27928694857138176
13964897184382976 6982723470098432 3491499174002688 1745818306478080 872943512977408
436488936357888 218253058113536 109130824024064 54567559495680 27284853489664 13642963615744
6821750243328 3411009339392 1705571778560 852819443712 426426499072 213221638144 106615013376
53309603840 26655850496 13328449536 6664486912 3332374528 1666252800 833159168 416595968
208306176 104157184 52080640 26041344 13021184 6510848 3255552 1627840 813952 406992 203504 101756
50880 25441 12721 25443 50888 101780 203568 407152 814336 1628736 3257600 6515456 13031424
26063872 52129792 104263680 208535552 417087488 834207744 1668481024 3337093120 6674448384
13349421056 26699890688 53401878528 106807951360 213624291328 427265359872 854564274176
1709195657216 3418525532160 6837319499776 13675175870464 27351425482752 54704998449152
109414291865600 218837173665792 437691527200768 875417414139904 1750903547756544
3501944534466560 7004163946840064 14008877649494016 28018854810615808 56039908644487168
112084215335485440 224177226763993088 448372045714030592 896779275800150016 1793628920344477696
3587398578177310720 7175078631331332096 14350720212616085504 28702566325139013632
57407384450091712512 114819272499810795520 229647552198876332032 459313118796262146048
918662266389543256064 1837396590373124440064 3674937295934324736000 7350162822244801183744
14700902105241905790976 29402957131988418428928 58808220106986050551808
117621051899990528491520 235251327172017911758848 470521101088109533069312
941079095664366485241856 1882231978305027808690176 3764611530562645293793280
7529518209030469940412416 15059626713871298586476544 30120434019363314584256512
60243229221968063991119872 120491180810418997627453440 240991806353803734545334272
482002502173538947671523328 964042783278940852504756224 1928161124421607619332931584
3856473364570667067312701440 7713248960596237791919079424 15427102384102282898425511936
30855413694024180426025730048 61713245239687590110400872448 123431326182653638737500569600
246872323771864194508398788608 493763990356842223083592876032 987566666339912114300776349696
1975210703932279564868733894656 3950576150369469802271830179840
7901461785748760949612385140736 15803542541517164589362219843584
31608323023073614558999338811392 63219121926225799878548475871232
126443195612608741278196548239360 252896294745531765598592289472512
505812396531692097281582964932608 1011664407144641326731962701840384
2023408042451796917801518947631104 4046974541228622364278224983162880
8094265995107301785906824142127104 16189165815514717686514396635856896
32379599281629663602430289974919168 64761733864459783663663573356249088
129528538331320480244933133525319680 259067217867442786325078240676282368
518154718144489224320580428603850752 1036350001108185751982008751710273536
2072781131854786110645713292425691136 4145724522986401434654818162861670400
8291773564526461296036419481743917056 16584196166160239445526405275528986624
33169690406535112597959943175140278272 66341976961499492609734151598445166592
132689146219857520047096833693219553280 265388677033432109749450728379097546752
530798123254298358809415578743511973888 1061637784883464996239859401457657708544
2123358646516666549721775290856582938624 4246883446532806213927663557595700920320
8494099200064558656823553066956471926784 16988863014127009771583558037443084025856

33979055256249804459040019881946448396288 67960768968491178749825847378013457481728
135926854848965497163143309984268036341760 271864343521897273653269850425018315440128
543749954691727105960506161763001116393472 1087542444679319329228945245351931203813376
2175169959950368893073756334355720349679616 4350510061084198255379244356015156583464960
8701360404535317449221952086637744935141376 17403401373804476775370830922490353406705664
34808163877076637304595515343410433886257152 69619050013088642116898737683680321918205952
139243544544048019249212889361079552127795200 278497978123837508529256606709596920838356992
557017734319157957120174869394069474842247168 1114079024781281794363673050737890216015560704
2228245161848495348973992725375282964693254144 4456664548268854218441278698549570994710773760
8913677545681435477869143892697152120070078464 17828051989650325037711460776590324501437218816
35657497775875558239369267535572689525468561408
71317783144900932806631227035929460096125370368
142641141476101498269047838001427082282627235840
285293433324802261849666443861990488746007461888
570609167394803054322474423442253625853520904192
1141262936280003169891231918321052548430053769216
2282615075540800462275029979515195690306131460096
4565408557043189169535192244776572567504310763520
9131173926009554829040649061045507508792717213696
18263061475865462638021827265075739765153625800704
36527550199423631235924712816120929025443634348032
73057954894232674391611542204180757041160034189312
146121618779236172622747317552239312062865599365120
292254655540013992924543101392234220086822260703232
584532147043111281207183135359979632095826645352448
1169109966012389153130560135870981648036017538596864
2338311275877111487693508002044008063760763572977664
4676805239458889338251791464692105662898984137523200
9353975854327111402233133850592390396552882258182144
18708682459472888255925369543601138934615592482635776
9354706605145776853692235693008748538062710224453632
4677535990277554789710893307108463804408812103794688
2338859338991110576287834383856276669893134547681280
1169475341421721878860111057079160718790931521732608
584760506673944234788152461115091551317647884812288
292391671318513765073124696845301371619915004379136
146201544650027706376086581566528483790503033176064
73103626820399265107805407355203140885524282081280
36553240657892338513783761963571019937898523787264
18277333952792522236832410124770234716517453266944
9139023788319437608386469633877479732042822320128
4569690300121307049178367102684921052913459003392
2284934353041447647081749694215551119902753423360
1142511778011120884787157918544320856674388672512
571278189750758973016720494990433076698700316672
285650245247978751819931015354352862530103148544
142830697810289008565750891606744593355428069376
71418136498294320610768137768156377722902282240
35710462045722068469330414866470229384045264896
17855927921148488316638380424431134953319694336 8928312409717971199305776707813577607308378112
4464330429430849120146181601705793868978454528 2232252327001356320319737424752399467151360000
1116169719643644040283192024325950999906746368 558106637893304960203257668137851133118906368

279064207982393950132459662056363383142219776 139537548509067710081645245021900599862493184
69771496513469222548530329507809754076938240 34887109386202295028119018252334604111314944
17444235257834989390986435875382165592080384 8722457911284415633956681312298514564251648
4361399096825668286210072343452973166231552 2180784619004564377720902015378344525168640
1090434844798147306168383929515101233610752 545238690047006211738158425670515102318592
272629978847469385196062443291739793915904 136320306335717832261522836874111018336256
68162811623850485962507226051176069857280 34082735039921027897126516832648315273216
17042032133958406406499710319854297808896 8521348373978149432218081111692218990592
4260840340488547830593153531728644538368 2130503246994010472538633253805589790720
1065293161871873514890344870873428656128 532667350123370896755686557422031208448
266344059655402518033100339703674044416 133177222124559793844178700348166242304
66591207210709164335903615422247731200 33296901679569215874858940335206170624
16649099876891924790883036479644237824 8324874456999620822168301395842695168
4162599487776639624447542275931635712 2081380873526734418905466926970961920
1040731001582574512793581357988052992 520385783200890908067214626245312512
260203032805247279868819286748299264 130106587005024552852015630186971136
65055828803712732884810808499896320 32529182052456594671806900953153536
16265224851528411450604198828179456 8132929338414262782652473589891072
4066623125532159920001423882846208 20333390790928594224338305485373440
1016735009545554244337949514661888 508387311813405688253373143318528
254203559427017127168885764653056 127106731473665705105542478823424
63555841616911423313321037660160 31779158748494997036935417954304
15890198344267141208605158539264 7945408657143391949371304050688
3972859071076606647220014415872 1986506906790758659877188403200 993292139021606998072184799232
496665412323917333102887698432 2483423775685155583584841498624 124176024487536250309119574016
62090430095407354412909199360 31046423973523291835629305856 15523816449671453232402006016
7762210456290630273494679552 3881256343872766965394178048 1940703729800109397020508160
970389643831917655671963648 485213711381890306416836608 242616300423910892498845696
121312872578438315894636544 60658797472460592769925120 30330579327851013796265984
15165879959735865603784704 7583235127773112154718208 3791765137839145753772032
1895956355895867715092480 948015071436081276649472 474025982462114347876352
237022214603094028713984 118515718987565441744896 59260165336791934566400
29631235589900574130176 14816194255702590488576 74083853558227446956032 3704336794301799333888
1852240454744937594880 926156256169487761408 463096142483253362688 231557078440881422336
115783042820068081664 57893773209847726080 28948012504830705664 14474569202368774144
7237566076161097728 3618923775568904192 1809532256528629760 904801312636403712
452418248504246272 226217920345145344 113113358219083776 565558878132797440 28280538578026496
14140819044827136 7070684400320512 3535479639113728 1767808539033600 883938629255168
441986494496768 221001837182976 110505213558784 55254754263040 27628450873344 13814762307584
6907649589248 3453959012352 1727046615040 863556861952 431795208192 215905992704 107957190656
53980692480 26991394816 13496221696 6748372992 3374317568 1687224320 843644928 421838848
210927616 105467904 52736000 26369024 13185024 6592768 3296512 1648320 824192 412112 206064 103036
51520 25761 12881 25763 51528 103060 206128 412272 824576 1649216 3298560 6597376 13195264
26391552 52785152 105574400 211156992 422330368 844693504 1689452544 3379036160 6758334464
13517193216 27035435008 54072967168 108150128640 216308645888 432634068992 865301692416
1730670493696 3461475205120 6923218845696 13846974562304 27695022866432 55392193216512
110788681400320 221585952735232 443189085339648 886412530417664 1772893780312064
3545924999577600 7092124877062144 14184799509938176 28370698531504128 56743596086263808
113491590219038720 226991976531099648 454001545248243712 908038274868576256 1816146918481330176
3632434574451015680 7265150623878742016 14530864197710905344 29062854295328653312
58127960390470991872 116260424380569354240 232529855960393449472 465077726319296380928
930191481435611725824 1860455020465261379584 3721054156118598615040 7442396542613348941824

14885369545979001307136 29771892013462609461248 59546089869934432616448
119096791425887292620800 238202806223811440017408 476424059191696589586432
952885011871540598276096 1905843810719376034758656 3811835195391341745930240
7623965538687862844686336 15248521373186084395024384 30498223337992886201352192
60998807859227207225311232 122002338084937284095836160 244014120902840307482099712
488047131271612093545054208 976132041475087144251817984 1952339640813900202827055104
3904830397355252234300948480 7809963026165408125895573504 15620530515240623566378500096
31242269956300861761931706368 62486957764240952782212825088 124978751231760364081124474880
249967173870077645195646599168 499953690553269124458088497152 999946066732765917049767591936
1999969504717987170366716379136 4000093751940885013267795148800
8000496988891591371604315078656 16001612947802825433346079719424
32004463835644936246967058563072 64011403551368443254483915374592
128027758862894028030067427246080 256065421246102339102334047485952
512150649532833244289066480959488 1024340913146923620746929733894144
2048761054456361505831453011738624 4097680565237751540338093111377920
8195678043125560138026560398557184 16391989911551234390753869148717056
32785247473702697010909235000639488 65573030248605850480621463407689728
131151131099612613878848913628200960 262312403404027053592909800882044928
524645089217657758856243549015375872 1049330743254522821053334992533323776
2098742616147460248788365774071791616 4197647491571749710940123126153871360
8395619501697157848607029408328318976 16791888040501632550667625128697790464
33585074155217898808242382881477885952 67172744458865065030299031011120381952
134350681214588664888226592518569984000 268711747022894399431710246029798408192
537444263233222938173934614044913696768 1074930064841314154968897472060461154304
2149943206432364867179851432062189830144 4300052566364202848843815840006914703360
8600437439727351926655857631778899492864 17201539493452596311248167167087939158016
34404408214900977538369238141236158660608 68811474885793524908484283896592878010368
137628266683570189480460183021426877399040 275267167191106658287903596499335997554688
550555602030145875229773653911636480622592 1101153739356156867767480229649201932271616
2202392549304043970150826302950261806596096 4404955239791548409533384293204239497297920
8810250761950017757530231961015910762807296 17621182088633877391987390671246685062037504
35243725306735438537828634840923097196920832 70490172872406244583364976678705648539533312
140985790262683224182145367351130205370449920 281982469561107918395121562689698227323666432
563986717193698776851904781354272087812866048 1128016990530363433827132874658295441956798464
2256121093346658627900912373216093416575729664 4512416411265180776295117994231191898475724800
9025181271674088593576822484060393927599980544 18051059441635631269126817959316808116497022976
36103512679846170702199981901025656755588169728
72209812952842157732292655766835394556364587008
144425201091983948120370695463238951203105669120
288861552556567161552312158785614226586964328448
577745405858332853727765853289501101535434637312
1155535413207062768701814778015547499793881235456
2311160029394919659896195698904185593033786392576
4622498464751427564777523683554552372959620628480
9245353741426031619525311938601467119703336943616
18491421106698416218991153020187658986974865260544
36984269461089538397863364326344767469086113267712
73971393417564488715488845224628433928444992028672
147948495825899801270501923593134665837435515043840
295908409633341250220052313474024927635962092060672
591839655229765795798201559523561047194106308067328
1183724982385698182312596984198144478232576864026624

2367541308623729546057581698698333724153882223837184
4735265304952125454979938858000756983685221439242240
9470895985313583635689428637209693038125356861620224
18942522721445832722837959116835744217760541689511936
37886506944528996348594121918504204718540739311566848
18943984223083163625756162801668460500780197622054912
9472357486950914538607632322042409321145012794163200
4736361431180123632168591621625294195949963388649472
2368272059442394997516683541114691865663710190108672
1184181701647364089474535635708368316676219342946304
592113686786765340095364750429695350260291795419136
296068261374924317726730841502603271091236959682560
148039839678232982702889653895179433526164010827776
74022774334501903271206943519528615753354770907136
37012814414943657595484530045733757371813768200192
18507120831318181777682794165851603433475075473408
9253917227582267378811661654418164090521633423360
4627137019752721934390963112955263232152864555008
2313657712857155089688047699350722209522456199168
1156873457918974606090306921111906401484240060416
578459029704685833668294996274225849103626010624
289240665224942182145718265996249248732565995520
144625907798770723728644516927692786456659492864
72315741492535178192214950428630474273517993984
36159264542842497260053821196707277659353120768
18080329169708702712000083589549659090973622272 90405130339980783969866282903728396761353420880
4520430741570902718986607392985424903391936512 2260302483071383119739950320392214984358100992
1130194797678657439993298472145858758510116864 565119176910811660058310892047805012420591616
282570477491147300059986274011340322793062400 141290683263444385045408550999389069687914496
70648063890657560030411982496553988989648896 35325393074796463769059844746706721567670272
17663377102132073761456849122568224320258048 8832028833432957819191887935891543928340480
4416184557899939378827675655249487848275968 2208177349541699924029703671276601866190848
1104131210066715079322784757464229904121856 552086872681290098315358839645079437574144
276054070164611328484662650279021961543680 138032351994288803905822940367752102150144
69018834453135971784657277797996611764224 34510746454563770808201542706058586226688
17256037841279777862037223256559433285632 8628351227638835159986837580044786728960
4314341767318890694477531765904928407552 2157253960409181904480822370893731725312
1078668518579459230861439429417499623424 539355028477163754741233836694066692096
269687898832298947025873979339691786240 134849141713008008340565520166175113216
67427167004933271584097025331252166656 33714881576681269498955645289708388352
16858089825447951602931388956895346688 8429369431277634228192477634468249600
4214846974915646327459630395244412928 2107504617096237770411510986627350528
1053792873367326188546603387816247296 526916719093266745943725641159409664
263468500751435198807074794205347840 131739320978118512321143383915495424
65872195790259712619374685364158464 32937365545730084539088839385284608
16469316598165156384245168044244992 8234975211732635249472958197923840
4117646062191346153411666186862592 2058902259258187341043426637381632
1029490743710350802690510090665984 514765178895803967429653431320576
257392492968216266757025908654080 128701198244265274899612550823936
64353075002211208210356073660416 32177775441144889485452935954432
16089506690592087432863917539328 8045062830305865061500683550720
4022686157657843203284704165888 2011420450081376937909533278208

1005748910666916137088357236736 502893798146571902610973917184 251456570479842868338884608000
125733120943199892686141128704 62868978323239175601419976704 31435698087439202429884694528
15718453506629408529529700352 7859528984769607922058526720 3929915608112255789676101632
1965033361919853809161469952 982554459891789861742444544 491296119411826409452077056
245657504438878944016465920 122833474585922341653446656 61419098476202605649330176
30710729829722020235968512 15355955210671368823635968 7678272753240863764643840
3839283950573021558734848 1919715762262805617573888 959894774619550227890176
479965834053848823496704 239992140398961266524160 120000681885499060649984
60002646785758744018944 30002476314383978856448 15001814617944292851712 7501195539348298137600
3750741884862224924672 1875443000025150390272 937757528809594159104 468896778803306561536
234457396600908021760 117233201900081381376 58618852749854375936 29310552274834030592
14655839087370436608 7328201018661928960 3664241246819319808 1832190992153837568
916130680449007616 458082932410548224 229050262298296320 114529529195659264 57266963621085184
28634581322170368 14317840416899072 7159195086356480 3579734982131712 1789936210542592
895002465009664 447518412374016 223767796121600 111888193028096 55946243997696 27974195740672
13987634741248 6994085806080 3497177120768 1748655669248 874361389056 437197471744 218607124480
109307756544 54655975424 27329036288 13665042432 6832783360 3416522752 1708326912 854196224
427114496 213565440 106786816 53395456 26698752 13349888 6675200 3337728 1668928 834496 417264
208640 104324 52164 26083 13042 26085 52172 104348 208704 417424 834880 1669824 3339776 6679808
13360128 26721280 53444608 106893312 213794816 427606016 855244800 1710555136 3421241344
6842744832 13686013952 27373076480 54748250112 109500694528 219009777664 438036332544
876106219520 1752279547904 3504693313536 7009655062528 14019846995968 28040767733760
56083682951168 112171660869632 224351911673856 448721003216896 897476366172160 1795021451821056
3590180342595584 7180635563098112 14361820882010112 28724741275648000 57451681574551552
114907761195614208 229824318484250624 459666229154545664 919367642681180160 1838805654106537984
3677752045701431296 7355785566379573248 14712134082712567808 29425394065331978240
58853039930477641728 117710583460582653952 235430174120420048896 470878362639349579776
941792754075718123520 1883657565745474174976 3767459246679024205824 7535206723734200123392
15070989908220703670272 30143132737946014187520 60288571318901242068992
120581754323820911525888 241172732019678677827584 482363910783431065206784
964764715055009549516800 1929603217086313937240064 3859354008125217550893056
7719003164155614454611968 15438596624121587614875648 30878373839863892641054720
61759108862969220104716288 123522940092421309854646272 247055324917808358999719936
494129539301548196580294656 988296857534959350322298880 1976669272933644614968016896
3953489661594741058582872064 7907281554644385774459420672 15815167572198578863506194432
31631544070216772356187095040 63265505992072773970723602432 126535847687424006458146029568
253081366781404929949689708544 506182076375923693966174715904 1012402838378075056065940029440
2024883048008605448399061254144 4049920838522121569332484898816
8100151162054064483733694578688 16200921294127771657604838719488
32403080528294828695484576563200 64808636936668228151518951374848
129622225633493597824137499246592 259254354787301478690474191486976
518528516615231523465346768961536 1037096647311720179099490309898240
2074272522785954622536574163746816 4148703501896937773748335415394304
8297723916443932604847045006589952 16596081658187979324394838364782592
33193430966976186878191173432770560 66389397235152830215185340271951872
132783865072706573347976667356725248 265577871350214972531165308339093504
531176025110033596732754563929473024 1062392615039274496806357022361518080
2124866359716963600294409833728180224 4249894978710756413952211245466648576
8500114475975171254631205646953873408 17000877989057659362715977605948899328
34003054052329952432339087835980103680 68008704253089172278492440920124817408
136022600803036879384613412336578854912 272055586199790828424483885665816150016
544131941587015796159481893316949180416 1088305421548899870939992030604532121600

21766939198475362991220405491503317647364353553993194545712728194074183198572544
8707440293388037654424614100131467231232 17415545200773967766785680103793074634752
3483241962954372044944426401464642961408069667497715079010730634335643413419917312
13934031234214116112476028651506796121292827869125850824860157650380348661816518246 4
5574037846644297618069740678862008158781441114850104624724640921881057598330602782720
22297852798411795164596279588485191476183044459740700865819502150987605000754179342336
89198216840985599427654385846089401268961281784032393293096176245780391843274379021516 8
356820089953296072787694613352952146532761607136674024959458206524662966744988345224396 8
1427389250170598991459086733286186751958712322854887390698612683226481746446751669745090 56
5709992562112054767069580052642259671145512961142042068565376833537239322478203200560168960
22841712494166854273211252688559089337824706564568516723405234375135543785510822932889206784
913738189595419579125767406661965599642694451218275460690195845664488521124435332254150950912
3655231517696659949292338823126270503089602560 0
7310741794708301531373946842730949110698029875 2
1462204110804656632832643207841871443043370926 08
2924519725335305918780994094275106127894271754 24
5849262458122597143793403545732938739403603312 64
1169897093114916490004963780583133044603732623360
2339883389210627102502493704039356682653489168384
4679945184382842449990119693824894552199026180096
9360247180688861389950503959142151478182148046848
18721207985224075759841537061269027703932487467008
37443843218140857479564132408507504903001357680640
74890540931667126878890381388953908796275480854528
149786790854105077597304995921785615573096492695552
299584999689751802873658458131326827107284047364096
599192835342586901105413848838164846136750218674176
1198431342611340392927021562827352076117864685240320
2396954029075013967286430855956748919924457866264576
4794090745854694297437637172517587375226372724097024
9588546867118721320604825266243353821207659431329792
19177824485056108092668752374903065783925146828931072
38357110471749547088255708434638847850869949590405120
76717143946773755982347824238943128267779211045896192
38360033475024208894092115804304280416909261455491072
19180747488330769898505159744568498349964458694017024
9590739119574717674982130793492428245737143330144256
4795552247492025200355840857350303658246028656640000
2397867467598345781610308158977196596811742824103936
1198979405725339481521347944639620682250235659943936
5995125388257530361187709048953215330472999539179520
2997676873944181657384339187354163624847410389319680
1498895526879799067087411925115859792229160504524800
749476308393753652741327128277318886017307907194880
374752426673803885969474146998353937960017781063680
187383349575365472784142364929024216455690804264960
93695242906914501291773828179435731965686358999040
46849405513073133095738236947179677851763657932800
23425594786344507772794779902320744860342068183040
11713243408076224498860220665525825397401153699840
5856844711490097555661525689945639182315636654080
2928533859471041430946470523564182832965348229120

146432268159851704203108910107777303738643906560
73218921673075668429447147018672732914510200832
36610854633112742378669919491728406979849224192
18306124214843825271308132737060223751221673984 9153410556565639676640652864128122006259367936
4576879502854683358813619679863066068453949440 2288526863713273439653456463831035566889107456
1144306987999602599950051543865269049775620096 572175272071284240036687427907510158053343232
286098525071383590049174541941192895609438208 143054707053562530040002684964315356096102400
71530075785716632527709049479017132193742848 35766399022326000017708378237938293169717248
17883880075896841885781115868184010121281536 8942280320315341881354021308699436828852224
4471310301341131409908742341653434298531840 2235740221262295939570237014478575091318784
1117912645927013087093051429065216516685824 558977590611439102200492175445572743856128
279499429129685830427229318179268614684672 139755031476826054877106274317875428720640
69880174194404597270298944773058275049472 34941416325198083551022376193589417869312
17471372776596934233447640000324849106944 8736018695297413345692045951927494639616
4368175501148179787330135951846282362880 2184170827323826450907124463864408702976
1092126952036781504074590475902838112256 546084245205824891347809359936735936512
273052507196629515329161740961026408448 136531445895173292492209400976842424320
68268319096015913659918965736585822208 34135457622222590536866615492375216128
17068377848218612121886874058228760576 8534513442662964487670220185134956544
4267418980608311457198501670577766400 2133790619942570335280946624294027264
1066935874790492470981321206649585664 533488219804849887161084550576078848
266754251107226769415754248913682432 133382196156014297625483111269662720
66693633379207605271544549041242112 33348084340204030865173771223826432
16674675995402129547287633963515904 8337654910351121830994191157559296
4168985911500589444172282666680320 2084572183912808986423734877290496
1042325706037661625380664210620416 521182660059459378774730491297792
260601233550043972429564438642688 130305568535179127735881815818240
65155260147668134628490706157568 32578868013873352694520252203008
16290052976956319037397575663616 8145335973487980863767512612864
4072822729248901104418118696960 2036488735876905888476240543744
1018283053564680612371710869504 509160869595454140252650733568 254590106204283987159723016192
127299888805420452096560332800 63652362254349455306629578752 31827390052994342282489495552
15914299489406978455832100864 7957451976158392885209726976 3978877103806648271251701760
1989514109767050049949270016 994794833815387982136344576 497416306373625469649027072
248717597919778474114940928 124363521326372106702684160 62184121846427488173948928
31093241514834461498277888 15547211053227589454790656 7773900674518974080221184
3887097911212076716523520 1943622742582333196468224 971848264779314017337344
485942579133730718220288 242980512938902213885952 121494868155469534330880
60749739920743980859392 30376022881876597276672 15188587901690602061824 7594582181221452742656
3797435205798802227200 1898789660493439041536 949430859043738484736 474733443920378724352
237375729159444103168 118692368179349422080 59348435889488396288 29675343844651040768
14838234872278941696 7419398911116181504 3709840193046446080 1854990465267400704
927530417005789184 463782800688939008 231900196437491712 115954496265256960 57979447155884032
28990823089569792 14495961300598784 7248255528206336 3624265203056640 1812201321005056
906135020240896 453084689989632 226550934929408 113279762432000 56642028699648 28322088091648
14161580916736 7081058893824 3540663664640 1770398941184 885233025024 442633289728 221325033472
110666711040 55335452672 27668774912 13834911744 6917718016 3458990080 1729560576 864813056
432422912 216219648 108113920 54059008 27030528 13515776 6758144 3379200 1689664 844864 422448
211232 105620 52812 26407 13204 26409 52820 105644 211296 422608 845248 1690560 3381248 6762752
13526016 27053056 54108160 108220416 216449024 432914432 865861632 1731788800 3463708672
6927679488 13855883264 27712815104 55427727360 110859649024 221727686656 443472150528
886977855488 1774022819840 3548179857408 7096628150272 14193793171456 28388660084736

56779467653120 113563230273536 227135050481664 454287280832512 908608921403392 1817286562283520
3634710563520512 7269696004947968 14539941765709824 29080983043047424 58164165109350400
116332728265211904 232674252623446016 465366097432936448 930767379237961728 1861605127220101120
3723350991928557568 7446983458833825792 14894529867621072896 29790185635148988416
59582623070111662080 119169749739850694656 238348506678956130304 476715027756421742592
953466084309862449152 1907004226213762826240 3814152567615601508352 7628593365607354728448
15257763191967012880384 30516679305438632607744 61035664453886478909440
122075940593791385206784 244161104559619625189376 488340655863312959930368
976718205214773338963968 1953510197405841516134400 3907167968764272708681728
7814631085433724770189312 15629852466677808246030336 31260885524976333903364096
62524132233194102629335040 125052986832871074903883776 250115418398707889098194944
500249726263347256777244672 1000537231458557470716198912 2001150020780840855755816960
4002451157289133540158472192 8005204546033170737610620928 16011013554976148789808594944
32023236035771912208791896064 64048889923183053675933204480 128102615549644565868565233664
256214902505846048770528116736 512449147824805931607851532288 1024936981275839531349293662208
2049951333804134398965768519680 4100057410113179470465899429888
8200424305236180286000523640832 16401467580492003262138496843776
32804173101023291904551892811776 65610822082125154569653583872000
131226595924407450660406764240896 262463095369129184363012721475584
524945997778886934810423828938752 1049931609639031001789644429852672
2099942447440576267916882403655680 4200043351206181064508951895212032
8400403615062419186368277966225408 16801441055424952487437304284053504
33604149761450133204276105271312384 67210834824100722867355203949035520
134426740250602358652316394710892544 268863621706006543139844763047428096
537747525821616737950113473346142208 1075535616462440779241074841194856448
2151152362563296165163845471394856960 4302466984403421543691082520800002048
8605258487360501514108948197620580352 17211166011828319881671462707282313216
34423630097871273470250058038646931456 68849856344171814354314381325458472960
137704904985202163536257293147246166016 275420194564121396727771647287150772224
550861158315676932766057416559618424832 1101763855006222144153143077089870610432
2203610786762180845548342642121008742400 4407387727023834805580798260124552527872
8815107761046615840129822472014175141888 17630880136091124138196096847558490456064
35263089500178033192265097502177261256704 70528837456347636216276002618475083202560
141062991824678412096043620465191287783424 282136617473323103519070471386864818323456
564294502594578765692107403686694122160128 1128631540485022648692147729199317215346688
2257348151561775532000161302050492372746240 4514866444307011533232054291404700629598208
9030073170980944004927571957416833027407872 18060826906695729886782070664048529591238656
36123014942859143527417994826526786255323136 72248752144653654562543696649913026656337920
144502948807178044140502807293544961604059136 289016786650097558311836442574527739790884864
578055351371678056685334541123931112747302912 1156154258886321993493992394197613491825672192
2312395630058575747234631412294729516313477120 4624965484689015014962556072388464097951219712
9250279418521757070911698640374938326550970368 18501255735330968223796570271945896914399002624
37003905267236844611539486526283834351392129024
74010598127623505550971665017351749747972505600
148026771441546643757728713964271661586321506304
296064693255692552827028195787679647353396002816
592151687256583636277197927293631943068297986048
1184347976003564333800678926023809182859607932928
2368785154987922790093923994920708959165239787520
4737748715937433825172980275587599105222527418368
9475854243798044140316225122667560584229150523392
18952422111442441260572979388319845916026492420096

37906271470577588481027017062609141327189367586816
75815397436540588881816150697157181644651500666880
151636503863852001603156534538192161269848532320256
303284425709245650885361535364139918500788126613504
606591687381574597128820003303791028923758377172992
1213229046689315784973833871758604441691881002237952
2426549437230964751380055473819253651072490500259840
4853281562166595865624886408242596837522437992087552
9706928499742524456979323737693372745799789967310848
19414587750303714365417749317803103633109407900893184
38830637002244759633753702320438923549238471734329344
77664197007764181073343812010543279664516255333744640
155334240022077685758360438760417424461111134397661184
77670043014313504685016626749874144796594879063916544
38836483008794083245426517059769788681317095464501248
19418972255215707074172360372301252482168375698522112
9709851503017186262815731107358805311839101832396800
4855108439213259494272641014283492191297007907766272
2427645563458962928568708237443790863337232449667072
1213868453655648055000547983872917815512980472725504
606957062790907322858370924511970099678672360308736
303489949376995309108233928543740645800427242127360
151750683679268478393641197415748120880759152050176
75878196335019651116582715279812959430652341518336
37940525415202531518172415925875929210462553505792
18970976331447618739026737105922689352799468126208
9485844977646985859483633124453707050183829749760
4743100894785081174726948847973034711983962718208
2371639650373334709856040566859607949438005280768
1185864426677064416174303354866349271442014601216
592954514083730738710293213151447284082513281024
296488407414464634666717374434859966222009630720
148249778893531949989144071146998145201381310464
74127677039915791322464727538283153645878902784
37065232316532803825178709751533617345533575168
18533313056553855994562527866962828934063849472 9267004977420655038267850429079424597680455680
4633676713282191039627218462338717364164493312 2316925468927027280060255855068861214744379392
1158506290606479520153451239484181873703256064 5792749233747227001383872757169666570017161216
289648350723102820100024465845921101591347200 144829619879422145065427646916679459087056896
72417532198646440040421530455199183689220096 36210127228790903774064618726029318917455872
18105744179129293763959236112229522995150848 9053212371931567820443081430722193265786880
4526776327149244379453272402664812516999168 2263473234166352424342502044984264200552448
1131779152379041329479183944318061071302656 565910843837453223393558433071995021164544
282966055742692891023762446992479753338880 141488344783329585175372838724480998047744
70746830847656362419432226976361059713024 35374744651823966125589017295240810201088
17688036939909875520730960551150545272832 8844350776953883989333706227340342722560
4422341541976415109150966089552706404352 2211253847737944111817539532717620723712
1105668462243840334529798010329444122624 552855000309354306575413127150038941696
276437884748394222942963624567677911040 138224134671055646299110342780168175616
69114663483957090563369436638248697856 34558629816193178988591850943206653952
17279963945203906347749491783644479488 8640306491155611600601529047842816000
4320315504854635013664156101931696128 2160238882065732113513773839970992128

1080160005852073360097734814488068096 540100285335640331719291354495320064
270060283872621991694857650873303040 135035212538711908765034812249473024
67520141570556410841320399531147264 33761338435878433650061696468779008
16881303043239330939731596585992192 8440968434269722527216172468797440
4220642673459889792283273322299392 2110400564892459160479230205100032
1055239896527486712408411874525184 527639755304371922288604323250176
263829781172500244186501354618880 131919842346407263614350273806336
65962397053282202567724935151616 32982436466680386664137366700032
16491837203359836022206132912128 82462228086689739356171791237120
4123268785849780350620258009088 2061711764177345511577310199808
1030894567714900423922245697536 515466626670564046027918147584 257742984741838940047356723200
128876328074197928540377186304 64440581888738193528538005504 32221499870188711393443708928
16111354398004163011309207552 8055979430456985162948280320 4028140830955944410120978432
2014145973341698119383908352 1007110765602712016853663744 5035742722267287487007686656
251796580866609482794270720 125903012799787611042349056 62953867583135240343781376
31478114383188337583194112 15739647487404527497248768 7870118891607443101450240
3935207019756311227138048 1967677296854450451775488 983875541915372644990976
491956217701760032047104 245987332222916870799360 122998277797476862787584
61501444741747645087744 30751875292378429390848 15376514106941518118912 7688545283846910771200
3844416757111531241472 1922280436149803548672 961176246871920738304 480606137834469851136
240312076116489666560 120160541657872203776 60082522628749787136 30042387214281736192
15021756557094289408 7511159753523855360 3755720614250283008 1877930675869319168
939000522306748416 469517853339418624 234767722762731520 117388259427876864 58696328737193984
29349263880224768 14675181695926272 7337865725870080 3669070301888512 1834603870420992
917336294948864 458685327343616 229351253606400 114679921770496 57342108368896 28672127926272
14336600834048 7168568852480 3584418643968 1792276430848 896171769856 448102662144 224059719680
112034054144 56019124224 28010610688 14005829632 7003176960 3501719552 1750925312 875495424
437764096 218890240 109449216 54726656 27364352 13682688 6841600 3420928 1710528 855296 427664
213840 106924 53464 26733 13367 26735 53472 106948 213904 427824 855680 1711424 3422976 6846208
13692928 27386880 54775808 109555712 219119616 438255616 876544000 1753153536 3506438144
7013138432 14026801152 28054650880 56111398912 112226992128 224462372864 448941522944
897916600320 1795900309504 3591934836736 7184138108928 14368813088768 28738699919360
57479547322368 114963389612032 229935369158656 459887918186496 919810196111360 1839689111699456
3679515662352384 7359306202611712 14719162161037312 29439423833702400 58881046690660352
117766491427831808 235541778948685824 471101150083416064 942237484538920960 1884545337822019584
3769231413132394496 7538744301241499648 15078051552436420608 30157229004779683840
60316709809373052928 120637923218373476352 241284853636001693696 482587721670512869376
965211472138044702720 1930495001870127333376 3861134118928330522624 7722556468232812756992
15445689397217928937472 30892531715940464721920 61787369274890143137792
123579350235798713663488 247167923843634282102784 494354294431342273757184
988745482350831966617600 1977564751677958771441664 3955277077308507219296256
7910849302522193791418368 15822288900854746288488448 31645758393330209988280320
63293877969901854799167488 126592478306286579243548672 253194401345538897777524736
506407692157009274135904256 1012853163245881505433518080 2025781884355488925190455296
4051714884438429679027748864 8103732000331763015349174272 16208068463573333345285701632
32417345852966281319746109440 64837109557571791897841631232 129679054818422042312382087168
259367781043401001658161823744 518754904899915837383118946304 1037548495426059342899828490240
2075174362104574022066838175744 4150503466714058716668038742016
8301316418437938778404802265088 16603251806895520246947054092288
33207741553830325874169007308800 66417958987739222508887812866048
132840869735635586538875222228992 265691642991585456119949637451776
531403093023799478324297660891136 1062845800128856088817392093757440

2125770828420226441972377731465216 4251700113165481412619942550831104
8503717138981019882590259277463552 1700806810326215387988126690652979 2
3401740385712453598916403051626496 0 6803734301544952843713105443894067 2
1360797566332999697918680956907028 48 2721696544714017654189481650070487 04
5443595913524071825083202772653834 24 1088759747524021668357488449033338 880
2177600624686457943396672687071821 824 4355363508649745100156736952153931 776
8711051535853148627040257060328439 808 1742275210881361410753408043269803 2128
3484680229184186192197529348947836 9280 6969620073211299125776485222712134 8608
1393975937610845173431582349505719 17312 2788055721158861043415735308938022 74816
5576319134192063479936611837729214 30016 1115305365213280974608350611516476 620800
2230693807176298506458757710974220 763136 4461553767852070127401628397830976 569344
8923439842703086483771482747427023 224832 1784754429940406542547941739838418 6621952
3569641782680391576683173860382865 3588480 7139549410959940136540928482177786 7866112
1427963051311819423943101848717968 57110528 2856032440863301641156036002000759 56977664
5712277558205928868851736613131163 99468544 1142498046937050891078280244452161 769963520
2285081164465832016772426332556181 481979904 4570332470115124502776584352416078 848065536
9141005222597169944016632079439589 464342528 1828269100992818176496019090809404 2465107968
3656674314932404728377423531461781 2003061760 7313620855758346207525617762609507 8151815168
1462778616330376591659277692459090 64595013632 2925666123018167883626863664792559 45772793856
5851550026751165167870343889333875 24711120896 1170353561493198913697392089816526 315753308160
2340794235272329587641430803532555 164168749056 4681762695116522695776154854864115 393661763584
9363873839376772432538896205326240 917972058112 1872844457704099894705096540184850 2097241178112
3745828295065690605804827678608904 471707648000
7491935349446362844398924553696217 0479341207552
1498442821752268895437638750034925 03049058910208
2996997147230530443990985178661213 30278870810624
5994217301913046194213385714505153 08919247601664
1198888061873006300088960214337575 914561507164160
2397865326726806722670486571548242 4225690382499 84
4795909059415201690326105428842666 0320301243432 96
9592174930753579870622475429177694 4378443443732 48
1918506348535351272118548000134011 3623256880119 808
3837155421839973140225201828864967 6741650142986 240
7674596293218487472426615314923825 2473573051465 728
1534976348551405732880565394423543 0292769163391 7952
3070066876918227942551615451724642 0181647432980 8896
6140362113467288838684200229204395 9555513078356 3776
1228118094619624358453033910991901 5749546258150 19520
2456327533091581898338455552285847 9175979801258 22976
4912837753887830159541686565175785 3705734172432 13824
9826040883184993044812924051559749 8119017484695 63392
1965281251718865154108494994553585 7765313324905 398272
3930708653601463398508810357590443 1813646305743 339520
7861709607530392977601261452147429 6193331923351 764992
1572400381571571831636980437822794 5751874247043 3701888
3144917683274130135507417170432206 4530164218832 7747584
1572517301702558303870436732609411 8778289971789 4045696
7862878808840257699935824400013602 6457489170812 108800
3931585554583861940259732568490072 9511764241338 597376
1965865852373797515275776468486672 2897391948635 570176
9829694637278320302108433263641540 5194508883009 20832
4915030006343826513918992092424859 7951029011420 28288

24576063470242464383918837765144746652401790667980 80
12288488454382898099121357534082597164644537812910 08
61444725868222820031416480927964105015440901459148 8
30723504732265574783613087092757612103829556926873 6
15362323265209869775758966860766585849969331562086 4
76814470821434760798556950875771828240119423303680
38408662658410086359159533723855363615196094398464
19205044953051396159520296004912406555166238572544
96028792884488745697304125739485656513672149729 28
48016180501860255298503385727204640125756553297 92
24008982280738068874177354292333225997338515865 60
12004937155273005049551507860532065965899377541 12
60026915850884878310071692874487594665647485747 2
30014572962702365686192923223157429750899041894 4
15007843999981146108675000004535531084487170457 6
75042007593055546871267691987461736467624099840
37522397593102681599580191976122908756406173696
1876189569483879488176326897925747463950014873 6 9381296296563124481868220985226747450398605312
4690822372853425761427403740412378790523568128 2345498298712644640960348494105691927923916800
1172792705499288200603497559002597230293024768 5864181308211270403634104354761742483120455 68
2932199544463049902125360457255249407387893 76 1466154217410232301216834368564813786607779 84
7331043312944698256854942542510014347608064 0 3665657769419117503812856621097979881088614 4
1832896941182942939599120985470476294186598 4 9164824988281635636459068301959813239144448
4582582635324278287461265838283622503677952 2291376388253869378346498762793669193891840
1145730729422799806481182303222763567972352 5728866323593324618945576125243462694993 92
2864539500036325102742620367186553775063 04 1432322919137993948006226335875688101314 56
7161880441289126723205712440790496575488 0 3581073143444141853190146601101276322201 6
1790603033121860172388718490903652178329 6 8953347472608247090911818406283330977792
4476839889803596659940022179024200531968 2238503021651534887212067577453367787520
1119293049200635722227062032697317654528 5596672937877520004240451383339757076 48
2798440314875930698672796301596462940 16 1399272080406550697612683455761523671 04
6996620016875680229444843803624079360 0 3498439815859303485413135164220270182 4
1749284811640383428051924213314250342 4 8746748576755575566986404222591827968
4373536547654616996856593689306202112 2186849403465723105109992633658245120
1093465266552068855895844211331694592 5467529156856380796183460529171333 12
2733865990476208656443850000842096 64 1366983701262113457397984868549263 36
6835172036430612932870223683387392 0 3417712783275329289375261512014233 6
1708919774167676056157705591167385 6 8544915783488437338138902131638272
4272616348069247197744638153719808 2136387402197137863209912620810240
1068233315179826063773753082380288 5341364646305415979712749271777 28
2670781358355850820278366565826 56 1335440196779496825350179247882 24
6677448571905341202805876064256 0 3338848079956599139430427944550 4
1669485936980263838728958928486 4 8347739169911140538713519423488
4174024327460480941891122102272 2087089534982695807212742246400
10435834531175755717399617208 32 52181106937190161993677615923 2 26091520609250772700178572902 4
1304624387495323220175916892 16 6523363722640539026714525696 0 3261802753902230976274733465 6
1630961823242096219596102041 6 8155111347665384755274186752 4077706789560144206283931648
203892895264379801746538496 0 1019502255253761965894402048 509770017092812461528055808
254894453279371970054455296 127451949006168854672441344 637283356863258621588275 20
318653484347836484907171 84 159332645132021829510103 04 796692740450627082833100 8
39836112762057250905784 32 19918794250791573834956 80 99597660602772611085107 2
4980067497579367649771 52 2490125982510052372643 84 1245109108115210460200 96

62257761248769736704000 31130033545889475198976 15565593233697041022976 7783084847224672223232
3891686538800411967488 1945915326994243911680 972993692294140919808 486514860545579941888
243266437472044711936 121637722335649726464 60821112967638548480 30411682383726116864
15206404141816479744 7603483545884950528 3801882510430830592 1901011623959592960
950540996351885312 475288090361987072 237652841274015744 118830818683518976 59417608365015040
29709903694135296 14855501602881536 7428025679347712 3714150278627328 1857143858790400
928606289133568 464320324435968 232168752152576 116088671043584 58046483005440 29024315244544
14512694493184 7256615682048 3628442058752 1814288138240 907177623552 453605588992 226811183104
113409785856 56706990080 28354543616 14177796096 7089160192 3544711168 1772421120 886243328
443138048 221577216 110792704 55398400 27700224 13850624 6925568 3462912 1731520 865792 432912
216464 108236 54120 27061 13531 27063 54128 108260 216528 433072 866176 1732416 3464960 6930176
13860864 27722752 55447552 110899200 221806592 443629568 887291904 1774649344 3549429760
7099121664 14198767616 28398583808 56799264768 113602723840 227213836288 454444449792
908922454016 1817912016896 3635958251520 7272184938496 14544906747904 29090887237632
58183921958912 116372138885120 232752867704832 465522915278848 931080190296064 1862229100068864
3724595639091200 7449466156089344 14899482067992576 29800063647612928 59602326318481408
119209050683473920 238426897459970048 476871387105984512 953777958584057856 1907626285912293376
3815393309312942080 7631068093602594816 15262699137158610944 30526524174224064512
61055300148261814272 122115103896150999040 244239214991556739072 488496444381622960128
977028917560264884224 1954129892714567696384 3908403900617211248640 7817096031610574209024
15634768523973451841536 31270689969451510530048 62543685781912234754048
125091983249842896896000 250193189871722648567808 500404826487519006687232
1000846546463185432477696 2001766879902665703161856 4003681333757921082736640
8007657815421021518299136 16015905926652401742249984 32032992444925520895803392
64068346073092476614213632 128141414512667822873640960 256292273758301385037709312
512603436982534248656273408 1025244652896931454474256384 2050564863657588823271931904
4101280843042629475190702080 8202863917540162607675080704 16406332297990132529937514496
32813873521799879689049735168 65630164895238988636448882688 131265165493756435789596590080
262540002394069788612590829568 525099347601253411291976957952 1050237380828734490717544513536
2100552132909924317702270222336 4201259008324759307938902835200
8402827501659339960946530451456 16806273973338322612030510465024
33613785886715930604335920054272 67230047653510431969221638356992
134465047067178005459542873210880 268939997654670293961284939415552
537899802349969154006968264818688 1075839218781195440182733301612544
2151757665724905144703060147175424 4303673787774838818081307382251520
8607664488199734693512988940304384 17215962801699583501726726232211456
34433193253999395232854949167628288 68868921809199246924512891741667328
137742914220799406766631770296156160 275495969646400639368475514217955328
551012221702404930407374975687196672 1102065008224017164155597845876965376
2204211146086448934992891480759074816 4408584551449727083349174539528437760
8817493621453112593425132235077451776 17635636280013542040303830782196056064
35272570634241717787514794188474417152 70547737416912702988843853625113444352
141100667130683940805316237746556108800 282211718855084951265889536485770657792
564444206897604041842293194956858195968 1128929952170076362305614633884350152704
2257942981089889281853285755709967826944 4516052115679251678190684487302470696960
9032436538357449585349594926370011480064 18065537690712791628635641756270163132416
36132404609421368173144187319600606609408 72267467674834306178034182253321773907968
144540252261651752019559979734884669194240 289091138347269783366103189926251581145088
5782035443432472125386172840765467647803392 1156449623980809368080278603356864266633216
2312984318553348970776423050365586475319296 4626138778290158410784577880348888834744320
9252617838947237760032618950677209437700096 18505916242628317396992164650569282411823104
37013193614724318547838182799568291896492032 74029109488384004603384072595996037938675712

148063663494638744222183559185710984168734720 296138216025018958475197946358859784920236032
592298210121520857012057548692595203006005248 1184639976386007594147438409334941672343076864
2369367065057946948541523442569385877348286464 4738908354687757417576340132937776820020838400
9478165158519241876139266761473563770690207744 18957027215325937834251706514143147802677477376
37915448227226783832449759010678336127949078528
75833684047603383992792209986140753301086404608
151672943281506400641369803901849668692549304320
303357036935612066594310375662835661565851598848
606736374616422663811762287043943971493209178112
1213517350723242388869807645524433239709430317056
2427123904427278900232181433921957072864884555776
4854426214816146045449495153590095332621816954880
9709209241555468580869254878672553039027729596416
19419132106957290141679038900329830825623650566144
38839691461607286243239136086629111146383683878912
77682237418599984406240388745197121283040133251072
155370183827970792652005010634272040546625797488640
310751785637483232983058487556299677054342656950272
621526407238049761324213907688110546030867437846528
1243098486402266113364621680527243475906099123585024
2486288316656865408161631091356531719500926742953984
4972759321018397179188037643317152974379310477475840
9945884017446127084105626207842485019513534938087424
19892498785710919619670354258101328180536897842446336
39786459073059170142258912201035372644093451617435648
79575841149393002090354231771736177854226215099957248
159157528305335327792381278282803220840531053930086400
318326748623769302808108186044268171945219355320516608
636676881273735900062907631045859804418753205561720832
318350132649966597254799445001591632473533850241204224
159180912331532622239072537240126681368845548850774016
79593379169040972925372675989728773250462086290472960
39798151086157817365604541679697102908250699077779456
19899806293897574134261372682264909595635177505161216
9950268522358119792860237262340633868572502735716352
4975316948883726259294894091774406469663708359426048
2487749818294196311079834776189248002520582675496960
1243920581073264746256111253245646385104655585640448
621983126499715668486152559198334384474509916766208
311002981231399481922124745886922788198346020356096
155507199606470564800586606087339192079718541164544
77756454298620694320055419615608495030132036075520
38879654397003053119908768093773697010202400784384
19440540822347879539894913189871573252669391765504
9720627223097116259917721166428149000118791569408
4860492017510146374943992868960255686951443628032
2430335211735867309964562577353218436921745735680
1215212207358330716228564360113154515183884828672
607628404424363888737423715774849905953448394752
303825352584781209680282625746561277157477187584
151918251478690237495926696802848800669115088896
75961913332494935075856040366208481379745792000

37982350462822375701874366165496281212467019776
18991872129698641932910356073944160867530571776 9496284513993048007441764532570090564413816832
4748316481568387524214175514084050347531173888 2374245353070125522353734380941527706427719680
1187166232678028641300190502420515119544926208 593604894410497260711756907185133192937996288
296813336240990100386709281580004413051764736 148412112638365785208770054783721114817265664
74208778578118260112092734388720011554324480 37105750418526813809900220692789732850008064
18553555773997248781877037095609729961426944 9277118169365545329401981922412296748924928
4638729225866233133932722648509864258568192 2319449683524846801582227167906790071336960
1159767377058288518099046505779324006694912 579904956177076817703489713802626488860672
289963111912504688178728087357795487186944 144986872868235483752855658907138864971776
72496094890109311708173637067689993175040 36249376673050440769959722340905276932096
18125352950523112842916313073982778638336 9063008782260502650426382488756459405312
4531670544629724439697304220260764745728 2265918349064598777090708598071649894400
1133000712907167667166382543006458707968 566521125641017972893705393488546234368
283270947414226056102109757736931557376 141640666003971562878683409364794998784
70822929150415048853155969930562109440 35412762649422158133485117589363359744
17707030361818395920196125106722832384 8853839699462856386824845709381992448
4427082109008257406775814432701284352 2213622184142543310069603005355786240
1106851656890478958375649397180465152 553446110854843130858248645841518592
276733196632223391264336296546402304 138371668918512608549774135086022656
69188369760456760733690060949422080 34595452530828608596246527177916416
17298360090714418412824011940560896 8649496958007266263762380146081792
4324906935328661660556377160941568 2162532695826845094615782124421120
1081305961994679679476687834185728 540672788037968405822742303080448
270346297539298485953570344534016 135178100529806384497884768763904
67591526144981763009492182630400 33797001012530166885020990439424
16899119476284726132647944781824 8449869223152184411392697171968
4225089354081002878230710976512 2112622048292956775382536683520
1056349709772706055824858939392 528194197699466861979224768512 264106770256290348023010033664
132058220831423632528203841536 66031528267351045522451333120 33016973059495137390400372736
16509090992657376009787539456 8254847727783591662187446272 4127574979619247659740561408
2063863047673349744193699840 1031969302768537829258559488 516003540850200393210134528
258011215158065935895494656 129010329945515837592961024 64507526155999353619087360
32254943669620394220847104 16128062130620555816075264 8064326213215457260863488
4032310680560318306844672 2016229127256453991628800 1008151457116374414917632
504094175302260917010432 252056311023167313281024 126032767197602084028416
63018689441810255708160 31510497642409734701056 15755825281957170774016 7878200871354737098752
3939244550865444405248 1969694333026760130560 984883195310399029248 492459612053708996608
246238813226109239296 123123910212681990144 61564206906154680320 30783229352984182784
15392177626445512704 7696370288199467008 3848325881588088832 1924233309538222080
962151839141199872 481093511756644352 240555551971344384 120282174032183296 60143286039347200
30072742531301376 15036921021464576 7518735388639232 3759505133273088 1879821286113280
939945002795008 469989681266688 235003430567936 117506010251264 58755152609280 29378650046464
14689861894144 7345199382528 3672733908992 1836434063360 918250586112 459142070272 229579423744
114793906176 57399050240 28700573696 14350811136 7175667712 3587964928 1794048000 897056768
448544768 224280576 112144384 56074240 28038144 14019584 7010048 3505152 1752640 876352 438192
219104 109556 54780 27391 13696 27393 54788 109580 219168 438352 876736 1753536 3507200 7014656
14029824 28060672 56123392 112250880 224509952 449036288 898105344 1796276224 3592683520
7185629184 14371782656 28744613888 57491324928 114986844160 229982076928 459980931072
919995416576 1840057942016 3680250101760 7360768638976 14722074148864 29445222039552
58892591562752 117789478092800 235587546120192 471192272109568 942418903957504 1884906527391744
3769950493736960 7540175865380864 15080901486575616 30162902484779008 60328003992813568

120660406032138240 241329608157298688 482676808500641792 965388801373372416 1930847971490922496
3861836680470200320 7723954835917111296 15448472621787643904 30898071143482130432
61798394086777946112 123601291773183262720 247211590745621266432 494441195889752014848
988918420576522993664 1977908898747083915264 3955961912682243686400 7912212055740639084544
15825000572233581592576 31651154065971770032128 63304613974952753758208
126613839635923934904320 253236902643884724584448 506492252031843158720512
1013021397551833736544256 2026116582079962311294976 4052380738112514299002880
8105056624130207950831616 16210703544070774607314944 32422587679762266625933312
64847536542765968074473472 129699795452014805794160640 259409035636995350878748672
518836960739922180338352128 1037711700411707317838413824 2075498958687140550000246784
4151149033101732928647331840 8302600297658369514588340224 16605805058226546343764033536
33212819042272707316702773248 66428055936184643891754958848 132860947575647746300208742400
265731566557852409633815134208 531482475928818653334425567232 1063003637483864974802441732096
2126084646220185285872064659456 4252324034945281244278491709440
8504957554900383833625708199936 17010534079820410357388865961984
34022306099680106095052631048192 68047088079438782950655060344832
136099127919034707422409717186560 272208159358383697887018627366912
544436125757395961858435640721408 1088911865596049055885668053417984
2177902959354612376108929650786304 4355964375034253280893046389473280
8712245662718563619136466954747904 17425125150737241352973682261098496
34851517952074710935348861225402368 69705571205349878329500715857215488
139416213013100669576607418527252480 278842567231003164988426810680147968
557705416871609981647277568611581952 1115451398562427266635403031725735936
2230983926763269139952501852456615936 4462130112803367493268395282923520000
8924584744160393413263573721867616256 17849818525428103679980713755776385024
35700935125070841066868560135635075072 71404466398570949547551385519434760192
142814125094000433922731301535198740480 285638634781717937500719664063055921152
571298038750870014311953450111428722688 1142637615876608307244935144193491206144
2285358308502953171731926776328249933824 4570882770505379457947966528539034910720
9142097848009705144864159008843139907584 18284860310017302747664769921216419987456
36571049848030390411202443649493120319488 73144758152052350654150694913106801328128
146294833216087840971793005054454724034560 292600300256141961270569240565391690825728
585221868160216481195104942043747867164672 1170486271616298079698142805913424705355776
2341057613824326394012151455478707352764416 4682285368832113257256034598261130589634560
9364911020031147452975532571129692947480576 18730502604796136782877991891474249431384064
37462366339059957319609837281378225935613952 74927454937055282146927381559615906016919552
149860354391981299309270177112950720325222400 299731597819704068649371182213339257233211392
599484973710891077360404020401554147631955968 1199013503564748034844131352752859561594978304
2398114119415427829934909329405221655852089344 4796402463402719180363111906609448377028444160
9593153375949165401712810308816906884705419264 19187003650185784885398793608829834030707900416
38375401096946477934743933200051708584009924608
76753589787042772197380558364887498213208096768
153512754760385177050546500659343158516792688640
307036659893369619412663769177822641214338367488
614095620531937769448469074073917930790182715392
1228235842554272600143221219584381158303377391616
2456560888089339322779008582041852910052778704896
4913300182140266890543149449829887006997605253120
9826957176203710271056563471152136387779306192896
19654627976253773522053656085288997523126803759104
39310683200200253003988370456547444541389990264832
78624220895785917927738857485033788073052746022912

157254150782342659695001948113945374126651023032320
314519719546226967069052362515646344214393108037632
629062275055537229496201657606803880350968340021248
1258170222037241049708597180364630144546300927934464
2516431787926815280849582091031305056781330351652864
5033046263558296924563939642666699648940117694873600
10066457902525926574857430206541578368635149372882944
20133646555870518601173962255499514878780126712037376
40268754613378368105266128195831746040579909356617728
80540432230031398016368663761328924647199130578321408
161086710466612119644410142261988714426476884886814720
322185112946322886512165914002639159117111017233973248
644393609918843067471023086962601778762536529388634112
1288833987890080723835428691839850478581702048618643456
644440377971237656364405604877248699819165519230009344
322231880998717475405548431917286080173740007075348480
1611121786505908061314447030697973905218948627267846144
80563816256228692463059922718652385175513625499009024
40283369629751677134448165044158908870776468682047488
20142415565694504018683184364495812576898062307295232
10071573158256584735071143103456085359203945136783360
5035969266832958730400347012332132214979429559959552
2518075977268812546632561236468110875178443275763712
1259083660560572864032474483385077821433585885773824
629564666243369727374334174268050102638975066832896
314793751103226511366215553421780647280578595389440
157402584542384079522632009854768121620834828681216
78704146766577451681078121499322959800690179833856
39353500630981431800420119035630929395481472663552
19677463939337068880150588660800189445308927705088
9839088781591710930045558901892457096438559539200
4919722796757443710007911736692409735111327612928
2459950601359515977496522011219295461001687728128
1230019902170155049994544077046193027223855824896
615032251830276055620413574241369161973433892864
307527276287737293121777554979820905167469936640
153769213330168279216674161419478614674111463424
76887394258233955936229772674523388382243979264
38445090925691886132061232319653734713716113408
19223242361133397148003789151022887618155118592 9611969629710425614988481071109453939726090240
4806159039427076327987533783353732035187310592 2403166631999469924240413515576368550255788032
1201626872142700842243530069737935541458960384 600835214142833361183426690843843403895013376
300428496107158150622544173409359518530273280 150219692571449810326687500698398667556519936
75112568544660272671051457346058787923951616 375557645401797820089379582171459121034821632
18779503265632751921616717834944424053833728 9390091915183296899271822292079643795128320
4695216098775108918867642833343537781669888 2347693119979284694049687260323626832887808
1173889095285507464332776551987742387470336 586965815290686290820354736906835679248384
293493541469309424737160598909900082380800 146752087646637852032071914683191162568704
73378702279310495847781764955716141973504 36690680367651032839763786284918351331328
18346004797823408877818345045989315837952 9173334705910650667877398474759728005120
4586833506454798448422812213262399045632 2293499829977135781453462594572467044352
1146791453363436169347759541256867282944 573416495869152223984393892613750521856

286718632528293181647454007299533701120 143364508561005125651355534146096070656
71684850428931830239492032321212645376 35843723288680548826653148784688627712
17922510681447591266780140704385466368 8961579859277454060116853508213309440
4480952188915556243421818332116942848 2240557224096192728392604955063615488
1120319176867303667537150372034379776 560179870843255485438999133268475904
280100076626429568554711540259880960 140055108915615697194961756942761984
70030089759008305056283871877791744 35016312530104380757543432642101248
17508790090352304493472464672653312 8754711957826209304086606512128000
4377514435238133180718490343964672 2188836445781580854696838715932672
1094457836972047559517216129941504 547248725526652345843006450958336
273634266283640455963702418472960 136822084901977369502950805733376
68413518331067255512025201115136 34207997105572913136287499681792
17104617522806099258281199403008 8552618246412870974209324482560
4276463865711346159639024631808 2138309304108128416086693511168
1069193337680291876176937353216 534616011653259772155263975424 267317677233186803111029637120
133663674319871860072213643264 66834255011575159294456233984 33418336431607194276402823168
16709772678713404452788764672 8355188570811605883688058880 4177745401133254770490867712
2088948258430353299568852992 1044511908147039606946136064 522274843539451282053922816
261146866502691380317388800 130578155617828559803908096 65291438992155714724560896
32646900087698574773583872 16324040339659646092443648 8162315317735002399047680
4081305232820090875936768 2040726403386340276174848 1020400095181317557190656
510218494334732488146944 255118470539403098849280 127563846955719976812544
63784229320869202100224 31893267581939207897088 15947210251721907372032 7973893356237105397760
3987090793306628554752 1993617454247352205312 996844755920695066624 498440392358857015296
249229203378683248640 124619105288968994816 62311804444298182656 31157028122055933952
15579077010981388288 7789819980467404800 3895050727722057728 1947595732605206528
973833050674692096 486934117523390464 243475854854717440 121742325473869824 60873361760190464
30437780391723008 15219439951675392 7609994853744640 3805134865825792 1902636152389632
951352435933184 475693397835776 237855288852480 118931939393536 59468117180416 29735132332032
14868103036928 7434319953920 3717294194688 1858714206208 929390657536 464712105984 232364441600
116186415104 58095304704 29048700928 14524874752 7262699520 3631480832 1815805952 907935744
453984256 227000320 113504256 56754176 28378112 14189568 7095040 3547648 1773888 886976 443504
221760 110884 55444 27723 13862 27725 55452 110908 221824 443664 887360 1774784 3549696 7099648
14199808 28400640 56803328 113610752 227229696 454475776 908984320 1818034176 3636199424
7272660992 14545846272 29092741120 58187579392 116379353088 232767094784 465550966784
931135488000 1862338084864 3724810387456 7449889210368 14900315291648 29801704325120
59605556133888 119215407235072 238439404404736 476895988678656 953826337095680 1907721393668096
3815580226289664 7631435330486272 15263420416786432 30527940345200640 61058079713656832
122120557473824768 244249911040671744 488517414267387904 977070012906864640 1954210394557906944
3908561526604169216 7817404528185049088 15635372006323519488 31271869912553881600
62545991624921448448 125096486849470267392 250201980898195275776 500421976194900033536
1000879981186819031040 2001832019967675990016 4003808155123427835904 8007904540623007383552
16016385541998318190592 32033924005501243228160 64070153854011700150272
128144919394041827688448 256299062160120510152704 512616571064314729857024
1025270035616776878817280 2050613858209848595841024 4101375290372286868094976
8203045728649753089015808 16406681753109864883683328 32814544097840447178670080
65631449378922329179947008 131267621124327528005107712 262544686981620795300642816
525108263429173069182140416 1050254305790209095525990400 2100584169444144105375399936
4201319454615740039397638144 8402941140686383736088952832 16806486744282574786765258752
33614182414384764202705223680 67230782680408757663759859712 134466401064095973844218544128
268942473534748864721834737664 537904289882611563510464774144 1075847265391450795154520145920
2151771902035356926576221487104 4303698546575624525686805364736

8607706578161070396442335510528 17216032126341783483022120583168
34433302192722852346319140290560 68869080265524275453188078829568
137743112291205692427475754156032 275496128102725667897150701305856
551012063246079901878699788599296 1102063740573416935926196349173760
2204206709309348136189986242297856 4408571874943724801055159572496384
8817460662537506659460693320794112 17635555150375127433622134993190912
35272377951350483096645766689587200 70547291203901422652094526785585152
141099653010203758221795040383991808 282209447225209342278802054393626624
564439176860022336228028056038539264 1128918918539251975796904006579650560
2257918966716918558275503802164445184 4516000192710666329914399182339178496
9032324903974991086555581520698933248 18065298845057299026564729353439019008
36131895764329231760036591330960343040 72266387677087730933887447910085296128
144537967651033996695403426316499812352 289086319895785063046063913625658064896
578193408979004265402641949236633010176 1156428356332876809426312142443899781120
2312939789415490176094680772829067083776 4626045732330453466673474521540669210624
9252423771659853162315174994846408507392 18505512157317598782566801893222957187072
37012353542630982481006507593506194718720 74027365541253534793758822801132950126592
148060047994490209251009260830507021631488 296130729812946697829001752117496286019584
592282727273825954311969965147957057552384 1184607989843517025931872852121843086131200
2369301050278764286479611547895544114315264 4738772241740989042190954783094804112736256
9477884765848899022845372940797039993683968 18956450096431639922617672630808943523790848
37914261322330963599089198760047614120427520 75831244903597294705886104516954682386546688
151667934325065324427187623027628273064476672 303346757685872118885206074042694362711719936
606715293443227177832073804060264358588973056 1213474143029420235787470920070279983509012480
2427035398344772231821588464040062499680157696 4854245021261407984136470175879130064684580864
9708838491666543009259526847356270260017692672 19418373881620540100492226685908560781332447232
38838141559815988364930799354209162085259018240
77679070712781793057754290673202405215706284032
155363716611863218771293965275972972521789063168
310738583596325702854158698411082269224331116544
621499467937849936331458932540437186810168213504
1243043537366096933909200936517419670343348387840
2486176277712987990310968015907929934132720697344
4972530961387564225607068317562041055157489238016
9945418734698304941184401206616444484099074162688
19891551093242962862309331556217613715766339698688
39784529434178631684499721398404676926669062144000
79571913363742675288761559368748252843610889781248
159149535718256174417047351881374303667767310548992
318310489418053996513143170050504203296625683070976
636643814799191288384383272676519598515433490087936
1273333301524549167484960410504061580875231228067840
2546757946901431516402308551310167929439190951919616
5093698581507529395669392563224425394255838895407104
10187762538424391517068336047657029859266591773949952
20376255827667448485595773937730417860043011514171392
40753973156972227874109751560293552003105678960885760
81510869317219117554055910490252536572250669786857472
163027584640987558719784635719835938276579963303886848
326066861295073764662914900918333606817317174068117504
652157106616344823772521060793990674162948843056922624
1304360981285084236438424639502628269382526675955220480

2608815498674957650663614314834550380878311331593191424
1304454517389873414225189675331922111495784655637970944
652250642721134001559286096623284516276206822739673088
326137013373665648002988677790303988402260658830180352
163074352693382147613167153634482859333208953145262080
81540099349965735612419984186906862232643788437716992
40771511176620198709128195778286147399341550151401472
20386486339128764806023199731559431841180603041972224
10193608544973715128741150786987894991345215504121856
5096986960191523927235350854098037031050064743628800
2548584823948095145050063157351063283213760867598336
1274338083900214163241225443826554025451244681691136
637191877913190376978709654488788204647804464791552
318607356938136836168403293532149698284993294368768
159309387459839241923725879909952647123042178170880
79657548225305032881625056526915222551793854578688
39830201360345222400693586549427060771033310035968
19915814304018964180287322417698255133084846391296
9958263963932658580113925780341489940326518882304
4979310387927917535042095175916926157055307284480
2489744396944752890013613730831553671973677563904
1244916799962773506253089936852322132709850742784
622480700726585283749686504144433714716431351808
311251500735891907186414019931353181538968666112
155631325554245586248992393895244752859860828160
77818450370272609452388888912406457475118661632
38910618981711212890140790438595269260153454592
19456006389143060527043568210493654891373789184 9728351643715257304508370600844837576335425536
4864350046429492172747478548221423853491978240 2432262135500677846620385898010214459408121856
1216174623893304803433516260954858496035127296 608109090018135341778419786452304881183096832
304065434044809140920040721213590257174315008 152038161540275305475435774600514036878540800
76021803029073020245425594297116472584962048 38012262644004193876566650646987963365326848
19006811886735938815210252072708845219086336 9503746225734890346068589410961854377754624
4752043254050905642266026392784643072983040 2376106697617183055748879040044179478544384
1188095884104456645182372441848018710298624 594069209700160881245152681836973840662528
297045238674046719949559571374969163087872 148527936249006499638271400915725702922240
74266626580494819650881508071983412150272 37134642518243194741313657843051986419712
18567985873119489828593280825056133382144 9284325243558691143264866364293136777216
4642328775278818686116546158029103431680 2321247464389145900300329566955819237376
1160665270569441228771193027448543379456 580353404472154753696110635709588570112
290187086829794446503312378847452725248 145098735711755758079284719920055582720
72551964004307146453456625208192401408 36277280076368206933635445228178505728
18139289075291420320271288926130405376 9069969056199368586862427619085778944
4535146787376513506794605387553177600 2267654523326671360078998482781732864
1133867826482542983380347135893438464 566954195650875143360597515198005248
283487239030239397515510731224645632 141748690117520611675361352425144320
70876880359960762296483669618982912 35439707830580609377643331512696832
17720487740590418803522414107951104 8860560782945266459111581229776896
4430438847797661758230977702789120 2215298652061345143453082395344896
1107688940111929703895337969647616 553864277096593418032067370811392
276942042068610992058232878399488 138475972794462637550216035696640
69240462277309889535657816096768 34621469078694230148103807172608

17311353509366757764189353148416 8655986239693200227163401355264
4328147862351510786116063068160 2164151302428210729325212729344
1082114336840333032796196962304 541076511233280350464893779968 270547927023197092265844539392
135278799214877004649621094400 67641817459077731583159959552 33822117655358480420754685952
16911663290589047524964696064 8456133876749427419776024576 4228218054102165538534850560
2114184584914808683590844416 1057130071389267298957131776 528583925160565128059420672
264301407313248303320137728 132155426023107021305282560 66080074194794945475248128
33041217689018190148927488 16521199140319453780115456 8260894718064906242883584
4130594932985042797854720 2065371253468816237133824 1032722520222555537670144
516379706855351478386688 258199076799712593969152 129104150085874724372480
64554380885946575880192 32278343364477894787072 16139748142991250817024 8070162301871777120256
4035225266123964416000 2017684690656020135936 1008878374125029031936 504457201461023997952
252237607929766739968 126123307564510740480 63063905582069055488 31533078690941370368
15767102295424106496 7883832622688763904 3942057048832737280 1971098893160546304
985584630952361984 492809907662225408 246413749924134912 123211273008578560 61607835527544832
30805017275400192 15403058393513984 7701804074663936 3851039476285440 1925588457619456
962828588548096 481431474143232 240724327006208 120366458470400 60185376718848 30093762101248
15047417921536 7523977396224 3762122915840 1881128566784 940597837824 470315696128 235166236672
117587312640 58795753472 29398925312 14699986944 7350255616 3675258880 1837694976 918880256
459456512 229736448 114872320 57438208 28720128 14360576 7180544 3590400 1795264 897664 448848
224432 112220 56112 28057 14029 28059 56120 112244 224496 449008 898048 1796160 3592448 7185152
14370816 28742656 57487360 114978816 229965824 459948032 919928832 1839923200 3679977472
7360217088 14720958464 29442965504 58888028160 117780250624 235568889856 471154556928
942342668288 1884752445440 3769639108608 7539546652672 15079630176256 30160334094336
60322815672320 120649926311936 241308442558464 482634064986112 965302489710592 1930673698897920
3861484836749312 7723244551405568 15447038858625024 30895177228877824 61792553481011200
123589505008533504 247187806110089216 494393204406222848 988821593184534528 1977713555113246720
3955567847714848768 7911417170406408192 15823397290766237696 31647920481439318016
63298092762692321280 126600689125012013056 253210385449278767104 506438785297067016192
1012913599391152996352 2025899256376343920640 4051942627940763697152 8104173486257679106048
16208923433267661635584 32418999788039930118144 64840305419089073930240
129685222524196575248384 259379668420430005272576 518777783584933720096768
1037592460658014859296768 2075258708292324556800000 4150664990537238790012928
8301625128979656932851712 16603840553769672571355136 33208861699160062554013696
66420084581561559930634240 132844891529605989506482176 265699227792177718303391744
531417345050286915187638272 1062872469032436787536986112 2125820495928599489397391360
4251792107584650807441620992 8503886446624205272176918528 17008377356158217858941190144
34017963638136050347057086464 68038345127911329952463585280 136081525959101118421625995264
272172723324759153876649639936 544364789462632141820094578688 1088768264551491951773779755008
2177613900355439239814740705280 4355382543215789152163843801088
8711074571441399649396412383232 17422768112902441988930274328576
34846774165844169358135447781376 69696024211766909476820693811200
139397000183690960474740984119296 278803903887696203991681161232384
557627614816020974067760708452352 1115294843713299080304318188879872
2230668915589112424946229921710080 4461496287503253378567646931320832
8923309487656563814485668038443008 17847252800613241743672084428488704
35695773251826711716745665560182784 71394081804853879892294324526776320
142793234212108672702194635866374144 285596609629019171239601245358391296
571213501667641994149626437968068608 1142467568154491291640100770438709248
2285016265947397189961897329882562560 4570194791171623593287186237775413248
9140714100896905613301155631571402752 18282077238901128080055877575183958016
36565452552016889867018887774450221056 73133501252463047147852040797065052160

146272194801784629123332612090459324416 292554774197286327901922285173577089024
585130317582006795114358692332471058432 1170302173538881868849745628635575877632
2340687423827500294941547745212419276800 4681541001154473704367208466307373596672
9363414309307893637702642884379817279488 1872749323261367973334173767228977473126 4
37456315693223144382556379151639829807104 74915289842437858596858565917400220303360
149835896596858856857208747063041561985024 299682427017683993041400724582565366726656
599386121683300544736767910078095218966528 1198814778662466206781468741982119408959488
2397714627916662648178803327616096759971840 4795599397016785765589338342535909404049408
9591539076400492469642140059679250576310272 19183758717534826816211206868573364689043456
38368878564537337386276267235576456450932736 76740479388010042280260241468012367047557120
153486403293890819575935896929743642386497536 306983695623523109182702621846925101355761664
613989169318529158427066899668725835877056512 1228021894780024196977457111287202938085179392
2456130901845980154201560846473908408832491520 4912436028263823828896414940746821882989248512
9825220505671374698779416377091653896627027968 19651137909630203479532005745379328054551117824
39303669615835315123010357473150696631696359424
78610126824820446573913406911085474308580966400
157225828835940525803612197751739110707538427904
314462808044480316918795163362614545595829846016
628947916834159164460731862443501739553165672448
1257940435158715390167746796323548775829343305728
2515970073298224902828059735520188145104710533120
5032118552558038050641251756786557477101468909568
10064593917039252591252768085065477327987033505792
20129901457924858162446065313115679403542258384896
40261230163542422284773188912200808302220899516416
80525314822470256489308494396340515594714564526080
161056338635711336818141221936558829169974660038656
322124095252964321315330910160873254301040382050304
644271026469011937988758752897257700524262888046592
1288587724864190466693711370945537784892890023985152
2577266793580714114819810472193120337474508543754240
5154716274866094592504396404990330210326474079076352
10309797925141521910738343731188839491407862141288448
20620326601101709272935789304794037124325552248848384
41242114703840749448789782294420790531670760430239744
82487152410956160703415971958507013629380832725565440
164980150828461645018504758656344892390840289181302784
329971993670021937260355146791351515045837825822949376
659967371366241168967401552540026490619990146566586368
1319981510784876926828185622994699902296609282974547968
2640056557674543031443136281818693646706476545631846400
5280300187558664418459802635295974977639469050629193728
2640243629884121387016666353477281330932992504997347328
1320168582994455282401715694653287586523125242340048896
660107675523424935647549106283967253789877116090712064
330065529774811115047120182620645357159095805505699840
165038610893954881135232906049653543711626526483021824
82522228450252102373452860394492204421852575106596864
41262575726763382089644633882078818493945943485841408
20632018614200356496281418783455767388482799709192192
10316374682509510973870260312936062764996313837731840
5158370028959421849799905617072120917875613910433792

2579276358332044106332340538838105226626535451000832
1289683851092188643882364134570074997157631973392384
644864761509177617299278999860548690500998110642176
322443798736130456328687966218029941211590117294080
161227608358836052003868216252892768586340589633536
80616658674803437921696224698385283283443060310016
40309756585094424920729170635162091136857912901632
20155591916393565440305114460565770315997147824128
10078152770119959210122821801775247531782669598720
5039254791021567850046543186633804952783382642688
2519716598491578047515837736189993069837715243008
1259902900736186085004201939531541831641869582336
629973751113291573125242505484043564182440771584
314998025929245051874192020601158106271973376000
157504588150922158592881394230147215226363183104
78755081668610895624333389079857688658369839104
39378934630880355976113040522320884851779043328
19690164213727632070029693252356462687186583552 9845430556007543076001433121776241474241822720
4922889502575635058494009808687125802445176832 2461531863573749289493651528243065433884721152
1230809487929840524870149076071283983273426944 615426522036403202496736194010517624802246656
307724150053943071279198924992696628983889920 153867519544842270655014876490067222783328256
76936482031356502835215145241893065537355776 38469602145145935171461426119376259841523712
19235481637306809462657639808902993457184768 9618081101020325669792283279058928496803840
4809210691693623304127873326833180132507648 2404690416438541886679802507068448008306688
1202387743515136060647834175360152975179776 601215139405500588977883548593040973103104
300618203526716573815925004753002729308160 150314418675341426571454117604742486032384
75159867793662283117472866416491803705344 37581263124826926474609337015306182197248
18791296176411355695241120411183231270912 9395980395204624076588786157356685721600
4698156351101785152778506054560877903872 2349161252300629133631309515221706473472
1174622164525182845436683001581486997504 587331851450025562028855622776060379136
293676310318729850669684872380688629760 146843347456223460162470966686673534976
73424269876540997495049748591501377536 36713433012485132454432006919832993792
18357365543349883080669569771957649408 9179007290228599967061568041999400960
4589665904391129196894175599009988608 2294914081833979205128783588510138368
1147497605736196905905239688757641216 573769085277702104623043791630106624
286894683843652878146733869440696320 143452412524227351990972921533169664
71728741563314132454289454172995584 35865638432257294456546223789703168
17933453041428761342973860246454272 8967043433364437728837304299028480
4483680173007247393093839237414912 2241919314666137960884513162657792
1120999271414326112611053353304064 560519442747791622389925062639616
280269624894210094237161724313600 140139764207262188639680458653696
70072357983709665080390027575296 35037416931894117920469912911872
17519327435966701650372406018048 8759973202993172170254927790080
4380141344001496757661826285568 2190148043253203715098094338048
1095112707252829525682637766656 547575696439528596908114182144 273797519626321215487454740480
136903595516439066260426194944 68454215609858762388562509824 34228316730748995823455961088
17114762828284305226315333632 8557683645597056270451343360 4278992938525979963872509952
2139572027126715896259674112 1069823792495220905291546624 534930785713541931226628096
267474837589736704903741440 133742141161351222097084416 66873431763917045871149056
33437896473579240346877952 16719538532599978879090688 8360064414205168792371200
4180179781055174072598528 2090163677503881874505728 1045118732240088356356096
522577812864117887729664 261298129804095798640640 130653676588066326708224

65329144137042377048064 32665724990025795371008 16333438955765201108992 8167007708258752266240
4083647969317451988992 2041896042252763922432 1020984049923400925184 510510039360209944576
255264026879359713280 127636517039307227136 63820510319467298816 31911381059640492032
15956253479773667328 7978408214863544320 3989344844920127488 1994742791204241408
997406579974209536 498720882173149184 249369237179596800 124689016636309504 62346707341410304
31174453182332928 15587776346980352 7794163051397120 3897218964652032 1948678201802752
974373460639744 487203910189056 243610545029120 121809567481856 60906931224576 30454539354112
15227806547968 7614171709440 3807220072448 1903677145088 951872126976 475952840704 237984808960
118996598784 59500396544 29751246848 14876147712 7438336000 3719299072 1859715072 929890304
464961536 232488960 116248576 58126336 29064192 14532608 7266560 3633408 1816768 908416 454224
227120 113564 56784 28393 14197 28395 56792 113588 227184 454384 908800 1817664 3635456 7271168
14542848 29086720 58175488 116355072 232718336 465453056 930938880 1861943296 3724017664
7448297472 14897119232 29795287040 59592671232 119189536768 238387462144 476791701504
953616957440 1907301023744 3814736265216 7629740965888 15260018802688 30521111347200
61044370178048 122093035323392 244194660581376 488406501031936 976847361802240 1953763443081216
3907664325115904 7815603528138752 15631756812091392 31264613135810560 62531425294876672
125067248636264448 250143293365551104 500304178917146624 1000643542206382080
2001357453156941824 4002855643802238976 8005992762581188608 16012548475115798528
32026222850138439680 64054697500090564608 128113898599808499712 256236804398871740416
512491623196252962816 1025019275189524889600 2050110607973087707136 4100365331134251270144
8201018892644654252032 16402614246041611927552 32806381413587830702080 65615068670184875098112
131234749026388177584128 262478721424813209944064 524975889593700129439744
1049988672675547677982720 2100051132327390194171904 4200249838607370064756736
8400794825119919482339328 16802179946050197670330368 33605540483721112751964160
67213442150683660326535168 134431606667850190298284032 268872658068666119886995456
537764205603263718354845696 1075566190138390393871400960 2151207938140506702066221056
4302566992008465232779280384 8605436215471834122852237312 17211476893853475560291827712
34424162713526565749758361600 68850743278692360757866135552 137706322260663180032431095808
275422315927883277098259841024 550863974668880388263314980864 1101766634963988444660220559360
2203610641180432225587622313984 4407376024865775123709607018496
8815061534741371592487938818048 17630742039502385875113327198208
35262722019044057130501553520640 70527919918166685021552905289728
141060791596490511564205407076352 282131486713295306170610007146496
564282780467219178425618400280576 1128605175015695489020033572536320
2257289578193905242377660689022976 4514737612712839013430508465946624
9029792138075735084211391107694592 18060218101451584283123530566991872
36121703853503396795648557837189120 72245943008207250050100109080788992
144496956618815413017806204974399488 289004054442432651870824383574441984
578028391294468955412072714400169984 1156097347408145214164993323302912000
2312275824454705035011682435610968064 4624713908186239283386756449232224256
9249752334926136993500296054485024768 18500153706959590840454158421011202048
37001605488133815387815449466104709120 74005807124696898189445164180374028288
148016806546252331206518858857077276672 296043997686221732068294778706812993536
592108764559877603447103679398942867456 1184259067494623485515235602768519495680
2368601211738983528272527693478306512896 4737368576977440171029168362839148068864
9475069460953826571026562677443366223872 18950803535905545599989577258416872620032
37902936299806876115852058323894025584640 75808531055605322063449924261908611858432
151622379023193783790391463752058345095168 303255391870353846907766157960598932946944
606532051388640252469498776834162351407104 1213106638073145622246930475494253673840640
2426298346738021479109726794640365289734144 4852766834659503427451185276584446463574016
9705873951685927793365833927776324695359488 19412428468105697463658594604767512927141888
38826218065679078681171042707964752927129600 77655158390293524870049792412788959999950848

155315761298457784755514998819296828291284992 310642411632657039541860825626031473165336576
621306601336797019145383307226938579496206336 1242656758816559918414089926403628425323479040
2485400629919051597074826476706759383309090816 4970975484409966714642946201212523831942447104
9942299417963660470272478898023057794533425152 19885295734214775022518130787242135850363912192
39771985265004458208982607556876312223321948160
79546758123158732745857907078536705491832143872
159099091432617098147501198086641573074040782848
318209333237833461606573164032419470328834555904
636440967220865453836287863783111589019175092224
1272926535932127968918858799002768474761362145280
2545942274845050060330283740878627542968748212224
5092062955651688365645699767503436272829544267776
10184482723226553221261664106499234919443184222208
20369679070299459422463857355983194586454559817728
40740785388291624804808772997935838668045502382080
81484425271968661529379662567810576326363770257408
162974559534708146898283558279498950633273071501312
325960537050957941475615582846753497227637204975616
651943910064999178309328098269018186377456533897216
1303933492056164947334850061689058756599277315686400
2607958327964663076102087853680162280887283127156736
5216099343633992515068951167964414097152023245881344
10432564062677317755867453257137007265058960474898432
20865858876173300963194008356690372671627748916068352
41733179253983932829306220398213461626275153764679680
834692815112425274644448848166092355818589619394445312
166944409029034378540570511071515576769257862519062528
333900510071167404304486651621692883802672972498468864
667824404168532103055664562200709228133660439917625344
1335695576389458795004711642316065377323949869676625920
2671484688883706767796188320461424596761157719036002304
5343156449976991891165906712581436877748831397437505536
10686687044373140493478873568480049123950694713606012928
5343530594396148602312966855898612246201863316168507392
2671858833302863478943248463778599965214189637767004160
1335976184703826328365006749803946903663723808724877312
668011476378110458629194633859296912360176399283126272
334017430202153876537942946408310186444245447101906944
167014561107626261880644287943485958354201347281125376
83510203557087792746158551341408411743139985505648640
41756563280181227275997479355536922154589648685367296
20879012390909279089457841520184819218804652308955136
10439871570863972270458471681300588680157240137613312
5220118473136652498094011301254383875456077060374528
2610150580420659430479393380929236705416767025971200
1305120962136496305955890555615640736552747760877568
652583317031331448336042210383331560198556004384768
326303076497207371847069571479421376060369064165376
163157247239374509763059018883588486010730063069184
81581478115072666801291626013733141995637797027840
40792166305229039360526871292836020492955281260544
20396796776460872660203964789402734994045832003584

10198755200153612820072246966193729870807011688448
5099556006038394655021255768843046122295553687552
2549867205999991450003194027294613654593800765440
1274978204490392786247880085083852124019912343552
637511402990394923747081578260198710371462152192
318766851867796727185111556989235679366484066304
159389001120197996248341162424186001773618528256
79697288153248814452063273176877081931997511680
39850037873199315389977982570830581488592879616
19925715834887111776962164276611311005593501696 9963206366587282929467668633903665633445281792
4981777407865504985227127564750837882046906368 2490975816218684252860210406274921473685585920
1245531464252308006553428515087212003173859328 622787510197636943338375913518481634752462848
311404644134559941700018784746678633958998016 155707766585150705865424806367058225270882304
77856605551510720440420110180388566781132800 38929663905223043974063908588624010463412224
19465512517345363863958881043526868768129024 9733096541039602870442903896370866152275968
4866718411703261904453183635489148960243712 2433344276443361186842457661396432422174720
1216764673517545710729161752524145182113792 608403604406705414018547337175037076570112
304212436027318986336256899044000781041664 152111534925642632831620064750241511899136
76058425918812886247555839989241316638720 38030542187402228039650823801680938663936
19015935707699006477761863804370609504256 9508300160848449467849157853950374838272
4754316233923697848408691902857722462208 2377241193711585481446402439370128752640
1188662135230661019344229463655698137088 594351836802764648982628853813165948928
297186302995099394146571487899241414656 148598343794408231900914274445949927424
74301768045633383364271402471139573760 37152182097031325389042833859652091904
18576740085622979547974983241867198464 9288694561365148200714274776954175488
4644509539959403313720528966487375872 23222335899618116263541960272248832000
1161208514628265435111828030626988032 580624539723736369226337962564780032
290322411066670010448380954908033024 145166276135735918141796464266838016
72585673369068415529701225539829760 36294104335134435994252109473120256
18147685992867332111826803088162816 9074159909083723113263775719882752
4537238410866890085307074947842048 2268698433595959306991131017871360
1134388830879236785664362280910848 567214222480246958916579526443008
283617014760437762500488956215296 141813459140376022771344074604544
70909205450266582146221835550720 35455840665172576453385816899584
17728539302605930916830358011904 8864579136312786803483903787008
4432444310661304074276314284032 2216299526583107373405338337280
1108188448917781354836259766272 554113567272004511484925181952 277066455042559172775860240384
138538063224558044904628944896 69271449463918251710663884800 34636933657778740484506648576
17319071291799177556840677376 8659837877354492435714015232 4330070054404698046503845888
2165110585066074937575342080 1082593071464900425949380608 541315425198381691555545088
270667157332156585068199936 135338301032561162179313664 67671511699522015912263680
33836936441381725367435264 16919058516501221389369344 8459824406155790047510528
4230059777030484700168192 2115103675491537188290560 1057588731233916013248512
528812812361031716175872 264415629552552712863744 132212426462294783819776
66108519074156605603840 33055412458582909648896 16528282690043758247936 8264429575398030835712
4132358902887091273728 2066251509037583564800 1033161783315810746368 516598906056414855168
258308460227462168576 129158733713358454784 64581618656492912640 32291935228153298944
16146530564030070784 8073546756991746048 4036914115984228352 2018527426736291840
1009298897740234752 504667041056161792 252342316621103104 126175556357062656 63089977201786880
31546088112521216 15773593812074496 7887071783944192 3943673330925568 1971905384939520
985987052208128 493010705973248 246513942921216 123261266427904 61632780697600 30817464090624
15409268916224 7704902893568 3852585664512 1926359941120 963213524992 481623539712 240820158464

120414273536 60209233920 30105665536 15053357056 7526940672 3763601408 1881866240 940965888
470499328 235257856 117633024 58818560 29410304 14705664 7353088 3676672 1838400 919232 459632
229824 114916 57460 28731 14366 28733 57468 114940 229888 459792 919616 1839296 3678720 7357696
14715904 29432832 58867712 117739520 235487232 470990848 942014464 1884094464 3768320000
7536902144 15074328576 30149705728 60301508608 120607211520 241222811648 482462400512
964958355456 1929983819776 3860101857280 7720472150016 15441481170944 30884036083712
61770219651072 123544734269440 247098058473472 494213296816128 988460953370624 1976990626217984
3954118691389440 7908512260685824 15817574277185536 31636248065998848 63274695155253248
126553788357017600 253116372807057408 506250337800159232 1012535859972407296
2025142088688992256 4050424914866339840 8101131304709390336 16202825559372201984
32406777018651246592 64815805837116178432 129636115273859727360 259281237746974195712
518580489892457873408 1037197008581934710784 2074466074757907349504 4149076264703890554880
8298440759783932821504 16597457980320169066496 33196068882144944979968 66394443607299103653888
132793498900616634695680 265596221173270124167168 531210889090613957885952
1062458671669375334875136 2124991130315045507956736 4250129834582680692326400
8500554817070540737478656 17001699929951440180609024 34004580451523597772521472
68011522086288630367649792 136027766539060130380513280 272064977811086000051453952
544148845088103478683762688 1088335469108069914529234944 2176746496079865743381889024
4353644107887183315410616320 8707590447229270288114909184 17415785357368347890817171456
34832779640556310410809049088 69667977132751850079967510528 139340789968782158676633845760
278691251344121234386665340928 557401845501356302840125980672 1114842376628940273813842558976
2229762124510335883894866313216 4459678991525582440324095016960
8919667468060986225716914814976 17839953906141615141571279192064
35681145752322515663417457508352 71364767384723602087384713265152
142734486529604345695869023027200 285478876579522974433937239048192
570977560199674514952272864083968 1141994734480606162073342500143104
2284068697123726588484278544236544 4568295850572481705643744176373760
9136908613795020468637862528548864 18274451052890155051976473408700416
36550169756380538333354443520606208 73102874813961533125511880447623168
146210820230323979168629747708067840 292431781665449784172471469041778688
584883845740503220015366885334843392 1169808256300213743371581665172258816
2339697642238842093424859119349661696 4679557543754513400213109816709611520
9359439606062685227153002789439799296 18719528249232687307759571890920751104
37440354572680008322426276405923807232 74883305293789284058666818060012224512
149771802884437102944962166616353669120 299553990362591275545181394225365778432
599128749912616690400876910436048437248 1198299038200101659422782064842730635264
2396681153149939876087620617626728792064 4793528459799352866659354211135992627200
9587389226597651962286934374037055340544 19175443067193196382510320651604250853376
38352215362382177680893545110268782051328 76707089180755925193532897834658124791808
153419495273494990050557410897557370961920 306849624370956259428098052251596984680448
613720516389845077510162565416158454874112 1227483568075555272328258052658245880774656
2455052206742840779272381948968349703602176 4910274554669142027776495585240415291310080
9820889391705204994016454545088262350831616 19642459348144251864959835839391388238086144
39286279825756187483773525177212503549018112 78575281910447742475254757351284461243727872
157156008338766219965924928696287830778839040 314322905713273909962680685380013478140444672
628667589498030759987023026734902589446422528 1257378735139027400097369365419556445223911424
2514844582563986560441385354738615423109955584 5029863389699836641376063957276235911544176640
10060075228543400323738714410150481953736884224
20120847355374254729450601811496984168770830336
40243088507323417622847549605386008860135784448
80488964607796651573587791175556098765459816448
160983504401892935802960966280680359621296128000

321978159176385136917492700420497043423345246208
643978619097968804458126936559266735208196472832
1288001839686334670162536944555078767139404906496
2576092882353463462817640031983248127724833734656
5152364170668515170620412349712677442341715312640
10305085153260206831211089270917717258467526311936
20610883930366766642362707684820159264503243997184
41223195108426239244606473655609768024142870740992
82449244712237890408975063883158435038558506975232
164904198415246604657474360910194668057662544936960
329819814812034856993997188108144932076416151846912
659662465587153009346091308791801056075014427639808
1319370603100472609408376482734624495994393103171584
2638832550053278400249140695771293759677514702127104
5277847787811223163363056852146677054732486395822080
10556060951031779052455664625501533180219886774779904
21112852652882223556370431093419424501949601515831296
42227166807401778015659065871671565286918858964205568
84457256618078217837154539113008563139877029793497088
168920359242705759285981892965347991411832683317166080
337852410498510165795309415409357713087822614094675968
675728205023217626037310089776038886703959723110039552
1351503178098829840968002697466724694464548436061454336
2703099892302448859722770430762743231042354851805659136
5406386856814476075019070933184074146311225662976819200
10813147858048108861185202009685323661075483244684640256
21627044004934531144664524306004998059057030326831284224
10813896146886422283479322296319674397981547082146643968
5407135145652789497313191219818424883217289500438822912
2703661108931183926443360645738506283721902729902161920
1351877322517986552115062840783900062917580354792456192
675962045285190570504222679349273491987104672316915712
337992714655693932475456969153298476257709583618801664
169002203334396289849401299315980103260933415539572736
84504024670472806730537057027655484196506019634872320
42253473836873734268186732198660458381272665749979136
21127467669255532585552467941746587332146160841261056
10564099210037099018505784892081472736827994403766272
5282232292723215872117667906644825903791454193451008
2641207490213941117491221683624457719584455592509440
1320649417033137149461804706963251243636592044146688
660347544479651870088999286057136813740478146019328
330185190221367582723548109316324002831330134982656
165098304101454615201298287802039799396210598477824
82552006546112719520411260472958798688378064732160
41277430520749065720086688522448848839325415112704
206394288884220885839983873404209149167230898929664
10320071254033619409962201273596936957399545151488
5160214032978397949966232922544649665591820419072
2580196219469993097475682604145415426241934131200
1290142711225393609984124373509253009843979026432
645093656357895335615203722472899153283495493632

32255797855154693311917262909558590082250073702 4
16128456446207309921537169847736111250162686361 6
80645069824186365935578541203464637296001679360
40323928708668091131735616584124359170594963456
20162661252621499647840981283258199846594543616
1008167907545447686490707713722711005394580275 2 5041013762299101952946831816412560092297166848
2520593993435482736720062532105782578810716160 1260340552860707248483354578002642555736424448
6301920545018365643033389449761969110337454 08 3151069162866597521825003004755362720996392 96
1575558902661200611106665564231487044341202944 7878217358953567306104048911260297631629312 0
3939244792423552028437409805473121523099238 4 1969690452685160201911397577658047115191910 4
9848792545792721948020451262897667344171008 4924566414079821443241957318752549556191232
2462368277631640956236844503028132720148480 1231226674111685595426355173339995331100672
6156346047037753563671440475829621510635 52 3078279361758539575105552542479633182883 84
1539192849999101184187692423522227805224 96 7696230095594662904113042879023195095040 0
3848247970596909943643811820217625581977 6 1924190446698244217615551100461826808217 6
9621284540490167317045981454074204127232 4810808423744556773007103702919637106688
2405487288622014943745608339401086074880 1202785182685875750493832413671176798208
6014133605303720145574303288209052794 88 3007170648589030769339722254031110799 36
1503637247263100732946146431978847600 64 7518445851158430406112158684710699008 0
3759352733000678573746792604763580006 4 1879741270211070972218752933585905254 4
939903086960901328782054782395010252 8 469967769408133585727366548998533939 2
2349919976679082535318528533997813760 1175000553158748571000112161501478912
5875205589889779371704800280020254 72 2937704206992907944204519876266557 44
1468902809520463101278319806261493 76 7344767577722361152271898371948544 0
3672510553921203399076098856294809 6 1836318659490613111008124263307673 6
918191021010312261239099549233971 2 459111356137658983487068483407052 8
2295636008850809181772935960985600 1147857618506661723055264752467968
5739486162939594276120307622215 68 2869842116672939968482145741045 76
1434970575938041399452068835491 84 7175100467698064073315324002304 0
3587674027852960574685151913574 4 1793898910928444556356320912998 4
896980403965204412685032934604 8 448505676233093273595952706355 2
2242605752417921704246944727040 1121341561835188520257062961152 5606901237307080941953267793 92
2803547332719109641310610391 04 1401822023392339405822293442 56 7009351902125619954946408448 0
350479684364477144039067484 16 175245886811336645165407272 96 876259657202173591556404019 2
43814494017383197864288583 68 21908002587328858075378483 20 1095437908298305860930633728
5477378436150844090461716 48 2738783665405079438135132 16 1369439056367368415519703 04
684743140016098555985920 00 342383375924256452105994 24 171197590920231813109514 24
85601746939167700083015 68 42802349209109746805637 12 21401912474317821784883 20
10701325172040385083473 92 5350847053460929637253 12 2675515760450833366384 64
133780399708560095707136 66892505697289261547520 33447405770149237620736
16724279345826922233856 8362427903289612828672 4181358066832882270208 2090751091010479063040
1045411574302258495488 522723801549638729728 261370907974074105856 130689957586664423424
65347230593145896960 32674741196479791104 16337933548193316864 8169248249073369088
4084764862025039872 2042452799756697600 1021261584250437632 510648384311263232
255332988248653824 127670892170838016 63837645108674560 31919922065965056 15960510788796416
7980530272305152 3990402575106048 1995270007029760 997669363253248 498851861495808
249434520682496 124721555308544 62362925137920 31182536310784 15591805026304 7796170948608
3898219692032 1949176954880 974622031872 487327793152 243672285184 121840336896 60922265600
30462181376 15231614976 7616069632 3808165888 1904148480 952107008 476069888 238043136 119025664
59514880 29758464 14879744 7440128 3720192 1860160 930112 465072 232544 116276 58140 29071 14536
29073 58148 116300 232608 465232 930496 1861056 3722240 7444736 14889984 29780992 59564032
119132160 238272512 476561408 953155584 1906376704 3812884480 7626031104 15252586496 30506221568

61014540288 122033274880 244074938368 488166653952 976366862336 1952800833536 3905735884800
7811740205056 15624017281024 31249108303872 62500364091392 125005023150080 250018636234752
500054452338688 1000143264415744 2000355248308224 4000847935569920 8001970749046784
16004491253907456 32010082019442688 64022363062140928 128049124170792960 256107044434608128
512231681055260672 1024498546482610176 2049067461709398016 4098275660907151360
8196832796791013376 16394228543535448064 32789582986977738752 65581417773769162752
131167339147165696000 262343685493586132992 524705385385681747968 1049446799568382459904
2098965656730802847744 4198075428649681551360 8396439087675514814464 16793454636103333052416
33588062193711272951808 67178430230431759597568 134361472146881946583040
268732167665800747941888 537482782075675205435392 1075002457639497829974016
2150078702255290498154496 4300304978463170672721920 8600905104831520698269696
17202400505473400102191104 34405981602567517615685632 68814324388376470053978112
137633371143235809753169920 275276187019437358796767232 550571263504806196174389248
1101180305941475349510488064 2202436169746676613344395264 4405023455220805055335628800
8810349141896513767964934144 17621302746702834850517221376 35243814419225284330209148928
70490046690089797918767710208 140984929083458054354234245120 281979529573473025741866139648
563978401960059885550527578112 1127995489546347439234645753856 2256068350345150214736472702976
4512291443195211102007307796480 9024892371400243549083340374016
18050403712820129788304130310144 36102045365679544956883159744512
72206566611437660674316117737472 144418084983032462869731831971840
288846073486379208781662856937472 577711954013386983647724099862528
1155463522108031099464244971700224 2311006272378576463266083487350784
4622171001082181455207354062602240 9244658914814419967765082301005824
18489951654928954050230912953614336 36981170960458136329863322610434048
73964877222116729118529638627278848 147934825046634371154665264067379200
295879791298070568144542501760401408 591779865005744787959508950772088832
1183600294830696879259865796046749696 2367281719299808365201427381098643456
4734725697876445943766246340207575040 9469775914306550314259275836435726336
18940200865720417481972117984912605184 37881699805655468670851368593907515392
75765995759740204755517002435979640832 151537183816338944338662535368288501760
303084752226394958332582131729235443712 606190273640224055975678385443787767808
1212422085655316390572385014858209296384 2424927248060369338386826517657686114304
4850020649620211791257766011197907271680 9700373606239369811483757974160884629504
19401411826476632080903967851851909431296 38804152880949049077680839510764099207168
77610964217889667987107486635648759103488 155227245347762475637706588499538639585280
310465124519491230602396407455559521927168 620951516686915019858759275824083529367552
1241945568669695157025451473474096029761536 2483976207931120548666768790600050001575936
4968122557045701566565269268503815887257600 9936585396458324071594001911615063542726656
19873851357650490020114930572444990621876224 39749063844768663794083714643319708316598272
79500849948472695095875136283498870778888192 159007144414816125207165686560716649849159680
318025177865373720445162201108871116281085952 636072133802230380951986058192617865727705088
1272187823747426642027295428334986997786476544 2544462759780785044301237480569476528235085824
5089099744133433609095768208937958121794437120 10178547937410594259178122913473926374237405184
20357792773108642600329418818143873009771872256
40716979342792193364605183618679786542137868288
81436746278734203057103059202143654129463984128
162879067743768038769991502333855470349304463360
325769285860135342851553772526847264879361916928
651560872465469216326249080771967178120229814272
1303166346421335493898781232980479652963471589376
2606421895823465110290128608834049899372967100416
5213022197608518465565389503414280985637982044160

10426401207140213421101043578320924345060059774976
20853516038126779822142616299626573437688310923264
41708459323946265604166290885222596370513004593152
83419773143277943128094698342384091731298774679552
166845255277326710095713629828645981443143080345600
333701928536195067870475725945047558847377222664192
667426693035473431099048384465606309616936569274368
1334899057997113452914290634082235003078237386440704
2669889459846560087260968998466514773845203268665344
5339961607397786537386713457537119083067863528898560
10680288590204905800502977836282417236890641040932864
21361307931228477052465057514981192615291110048137216
42724077364094285007848318714795101513601876028817408
85451077731463231821533044799255635593243063922720768
170908001469475787254738904337842136318564751575613440
341827694952050221732823438154346002901286750611570688
683678773930297737912338135266015466330887996143828992
1367404315912990064718058788446677853718404982129033216
2734902167930769307222882612722649549550067943940816896
5469991408071116970019295297103886783326651847247134720
10940356960561390651185650737524948935106335613225271296
21881462209961094724665421761684248607118735063912546304
43764420997598816293919084096637198688049597802749100032
21882958787637721569253662334952950080930862738836553728
10941853538238017495773891310793650408918463288149278720
5471113841328587103460475727055412888685747603440140288
2735650456769082729517002899357000286456131781402820608
1367871996436935953651883967593147064284694880542785536
683959382244665271272633242753896992670661935192080384
341991383135431282859662250855610226599488215056384000
171001537574264965041503940167135978431822731258363904
85503691790407144326588377453233421781950677494267904
42753307396840903066212392411449427173994994679676928
21377384449239116984565298048141071728507325306109952
10689057600028891218012199945278714935008576636190720
5344711487719111971870875433243447002881745309663232
2672447087711889167367825446923768269129601150615552
1336269215782111174400106588612906518409164823199744
668157443854138882558150226881964451126764535545856
334090139908611088958123579728737821524473329745920
167050778945076368318586023008246708742782195859456
83528243967923596079055128076062253361663863422976
41765549231654503999408622324000576175968314458112
20883488239673604979644840304985012835552348602368
10442100931759978979792684723984868791560269987840
5221228871841577734881474647738615582672182837248
2610703638901582989933303466742398384782115340288
1305396420941188556212934804807744489114069630976
652720511215792808729608938122144892918540795904
326371405980495669676375236920208770640023388160
163191278176547467493973002389672547410388189184
81598426681423550074879193159620354750382342144

40800607137286683201385942562202217897785294848
20401000466930795682666144272297129210189709312
10200848682609124882319658631746574735743385600 5100598565876425961653122563672292433195958272
2550386395224144741073207905735648749260111872 1275236753755038250659927264817575640961122304
637640154949002065391625288383663453646094336 318830966510242502726643472179269543405813760
159420927772991986378737150083353679994290176 79713186145431360697076282038536294142836736
39857954202183364102391994517697874144264192 19929657665825523928122924008063800608555008
9965169115279682902524925378639332072488960 4982754698823301920494194376623381920350208
2491462420003381194862963031963548902227968 1245773745297555714739414437807703422140416
622908140296710416023673679816816196583424 311464703972321487338820070364890341048320
155737668898143883329016504106862919024664 77871492905063511498196632819463706640384
38937075680527540664971220216792133664768 19469202454261662790422062011926207004672
9734933534129777624179256957728173588480 4867632920564361926573741454746621837312
2433899537031917520528927215314578440192 1216991306890827038885491851627922980864
608516422632847658753260047799278370816 304268595910140899031887084892297625600
152139490251928984343572072942478032896 76072341274393759585600301719403626496
38037468711411513499707283483784118272 19019383392813073603307208053933211648
9510016214960195228380387182987182080 4755170366756926827553585169503879168
2377666313016878020458488373757083648 1188873721327646313570092081381113856
594457143073426808455469987941842944 297238712741515230062946967596564480
148624426973158527949079470611103744 74314748787779720433342728711962624
37158642044490088446072861059186688 18579954847545158337737178881196032
9290294336422636222621896361639936 4645305624536346641784668896100352
2322732040430687585229927992000512 1161405634296600924783760767975424
580722624188929028476278769975296 290371215614778797280338577981440
145190559567546540161268885487616 72597755663851840841184240992256
36300115771965205800867019620352 18150676856002245590570959372288
9075647913010944140354204467200 4537978699010382742711464624128
2269066720757646707622913507328 1134572046005051021945047351296 567305365815639345039318974464
283662354314376589553057136640 141836012860466753293227393024 70920424281872605904963108864
35461421066755917581656260608 17731314996287766105415483392 8865959729598786710001418240
4433130980526845183647547392 2216641048127148506147192832 1108358302995437210235305984
554198040963650083698507776 277108465214790781139681280 138558954973878260215054336
69281838670180564930134016 34642099926710999876370432 17321640259165858643836928
8661115277488108674744320 4330705212696644013785088 2165426393324616845099008
1082750090150455841652736 541393491819301630377984 270705969281687669964800
135357596326862262370304 67681104006440344879104 33841704924724779286528
16921428923114693066752 8461002691933498245120 4230645461154824978432 2115394788171450417152
1057733422882744172544 528884725839881568256 264451370119195525120 132230188659225133056
66117346129426251776 33059798964619968512 16530462432263405568 8265512691108413440
4132897083042562048 2066518910265458688 1033294639504818176 516664911938453504
258341252062248960 129175024077635584 64589711062073344 32295955042664448 16148527277146112
8074538516480000 4037406697193472 2018772068073472 1009420393775104 504727376756736
252372278312960 126190434123776 63097364545536 31549756014592 15775414878208 7887975874560
3944122155008 1972128186368 986097647616 493065601024 246541189120 123274788864 61639491584
30820794368 15410921472 7705722880 3852992512 1926561792 963313664 481673216 240844800 120426496
60215296 30108672 15054848 7527680 3763968 1882048 941056 470544 235280 117644 58824 29413 14707
29415 58832 117668 235344 470704 941440 1882944 3766016 7532288 15065088 30131200 60264448
120532992 241074176 482164736 964362240 1928790016 3857711104 7715684352 15431892992 30864834560
61731766272 123467726848 246943842304 493904461824 987842478080 1975752065024 3951638347776
7903545131008 15807627132928 31616328007680 63234803499008 126473901965312 252956393865216
505929967599616 1011894294937600 2023857309351936 4047852057657344 8095978993221632

16192507742257152 32386114996142080 64774429015539712 129553256077590528 259115308248203264
518248208682450944 1036531601736990720 2073133572218159104 4146407881924673536
8293097238826057728 16586757427605536768 33174640755117916160 66351533310049517568
132707570219726405632 265424147638707552256 530866309675924586496 1061768648148868136960
2123609353891774201856 4247362822971624259584 8495013876319400230912 16990604213391103885312
33982361348286814617600 67967028539582842929152 135938668765184113246208
271886560902405081268224 543791568548883872088064 1087620030585915163279360
2175313848148125164765184 4350775270248840005943296 8701845688402859364712448
17404281672616077435076608 34809743936852872281456640 69621849056947179385520128
139248420480377228416253952 278506285693720196122935296 557031460853371870826725376
1114100700638606698815160320 2228276959140939311953739776 4456705034009330452554317824
8913712299473564562402312192 17828029061856936439391977472 35657267049533487507958661120
71316951950706204274266734592 142638739604690867065232293888 285287150615938651163862237184
570593644044991136394519773184 1141225973716209940922630144000 2282529318684875218112441483264
4565213379874661108759245357056 9130736244759143562587215495168
18262091459537929815311880552448 36525420859115145010898660229120
73053317598308860782347118706688 146111586956774863085793833910272
292233077433864009213786860814336 584485961908356584511972107616256
1169011537897970301192740987207680 2338102303958454866723075518365696
4676363064241938262121338124632064 9353043041133933581593050425065472
18706719907567981277886849201733632 37414707465736190785175195106672640
74831950232672838029153383619756032 149668971067746588975912754052333568
299348083340295003787037481730310144 598716449090193659244498910711906304
1197473462999594621829845715926384640 2395028055637603850341387220857913344
4790218370552036914046166019726114816 9580761259657732254819115195472805888
19162171556422781363091796702986764288 38325641187060196433090726030055833600
76653878522549660279995717308276277248 153312949341957855387619965112881774592
306636283277632780430496991218421989376 613293335742699700171508104422160859136
1226628209860267678964044452814955479040 2453339496470271915170145393571178479616
4906845146440016944824403763024892002304 9814022599878980118617033477814854090752
19628709813755852695170518859159848353792 39258748855507490306213941525379977052160
78520156167006550444173690664880514793472 157045629245996240551838996558002150965248
314101892315958760430661223572486544687104 628225052279850079515288908057937574887424
1256492639855565276338510737941804120801280 2513070350302860787292887319535466183655424
5026310841789182043817506326374648251416576 10052961965945285026098476027356728271044608
20106604496624411929123878803928320078512128 40214570122716507612101611106286367229870080
80431862504368382731910929209432188605431808 160869169526607500479237272412583285502246912
321749228088956470989305372812604387587260416 643520234249395882040272401600084408340054016
1287084024641757644203868115149920083011174400 2574255161569447048654382854199342698684481536
5148684547710757617802058956197690462693228544 10297717544565242276590704407993391056034988032
20596131987417938635154581807182802373367037952
41193657771410785434255509596757645269328199680
82390103135971387196403711158299371583844646912
164785781458242407048592806246166905258065788928
329582713289084079408756380351470134696884568064
659187727323366689440654296421212917755275116544
1318420056137130440127591664278971132233562193920
2636929315255055002747749471431032857913148309504
5274037036471698250480631228608246902718344462336
10548430884866572990931527028708856179220784611328
21097575393579498961803583200402437106009760595968
42196578034851703883488224686774323707155903938560

84396010565088819686738565945487546404584573370368
168797730120948463213001365034852890789714677727232
337606878223438574105051196357461377540520417427456
675236592409960443568199325290433947003222958800896
1350518856746087477852592515731890277850810165493760
2701129057344508137137572761765825323390348826771456
5402440802393682637139920984135740182158154645110784
10805246980196698000009392889479659435071223273357312
21611224711212061451477887621375677011652274512986112
43223910924061453805873978927584070306324204958515200
86450744851397569417584365224833573178687721782116352
172907335709344462446841545188998011489454067294404608
345826363431787572117028719856657753243065382049153024
691676110889772438680748698670638967014445259018993664
1383398989831939466254879915255924855085519507879362560
2766891515768668110296524866341143552284296995441475584
5533970103746914576166579804340874788795109950248452096
11068314351912985863480219751998924946043251819227906048
22137376992664285149254559790632200628992567475917815808
44276250563005197143097360154533102731797262626759639040
88555494281363647975371201455603608411218780603367292928
44279243718358450832273841301070505679421517976607653888
22140370148017538838431040937169603576616822825765830656
11070559218427926130362580611901977156761443331613917184
5535466681423541420754820377609576262607237625172459520
2767826876816559888164175224634081973416876792268980224
1383960206460674532975470130231687907765067385975865344
692003487256534560934426324073167414410848187908620288
346013435641365927690558791515245437469581341414653952
173012563827232287456952210496953583866869294437498880
86509204916890805534312512618142224499473959083835392
43256063960082733670074459993903828532756635474460672
21628762730860032286496331839368272407888145703501824
10814746740839348868977716840892315274698986834886656
5407556058124340797353633881050247172726950409011200
2703869372914503580109204670827168354052203700289536
1351980358383418380770796200564606560870466098036736
676013015154792485743495032857814472357415172964352
338017925558937890550795982716662832139798648455168
169014671770239769114922224502209214050444855214080
84510190380505296477223228823043506015495193100288
42256522437945354198492672697491202502883979296768
21128974842819030079186865491730325999010181021696
10564844233332691529563697317357525373289186197504
5282600522627934009766980944424943873536640942080
2641389464294761127376056615085562530214344392704
1320739333637777624934311378979326561830184157184
660391967564087343090297225207935929276598059008
330207134154642936856719380463104288819052019712
165109142263621101084145074161120306499902504960
82557358724960366869965229045344234295139500032
41280073159055091598928960505064157670163873792

20640733477814999881437653243728099096378998784
10320715188051226981705413117462059678838030336 51605318185974770113459998065300349047432806400
25803530215846702659196465271645199850337730566 12902200669353010130831465755320112588479528966
64513181153913344660323494374088126258950963632 32257679480530819333244829985787844787752140888
16129384192052483168163956392265813223014400000 80649643219197783348527488958188520260763648888
40326182739066575428117597977523987203227648 20163771934267129590985725737976857138036736
10082226249500485733956326243595860337229824 50412832659337033362098948091016460527206400
25207267035585819027208132482026809684131844 12604058870751560686683395459272694552330244
63022421118551059298813623387659921312972888 31512273941672157582105134739478184932147222
15756668662034392757401728892563204603904000 78786001766163533618754452076936583708672
39394330111077551725250129845528572198912 19697829669536668320561516826294426271744
98492471417672803892489843649122832220166 49247897243831133091086051583386766540800
24624779389412932117963590671106058485766 12312805078455148845192077775259366850566
61566102311019158157011801074828522291222 30784089614881286044031606636680105164888
15392564037126496504778656367972974592000 76965416334061749937707547088029483008
38484006241245508675760906168097046528 19242652157730071191334019396089675776
96216505974186940223937928540654141444 48109875579861762245602880050429952000
24055749086315027189618397915266416644 12028280191349586628217677902658928644
60143429197708298308130784238423244888 30072728719334331737586589481775923222
15036871419907257160553893422170112000 75186892400736742261572460517261312
37594713850968599360187726961836032 18797990750784413794794611832520704
93993122880422639547476800920616966 46998146003461605060490271339315200
23499865283355945173621071109160966 11750328782490543908498503274332166
58753624616515576150932354970419222 29377802660289216379686096784588888
14689396506160322341953008041984000 73449458410880182470314838458368
36725967145479376615432318353408 18363602542759330997853608738816
91821107563894868439955291504644 45912101206996540945321269657600
22956824316022823835332446781444 11478799014273688599002129367044 57395929352679826401690176716888
28698931816995604904184853299222 14349949478825648303762309120000 71752165245767470777160957952
35877291548703350017755185152 17939250237261482323464945664 89699273500856448190261493766
44851147907702742381599129600 22426329532488630334033756166 11213542555562944738633973766
56069601724407871551255347222 28035745335500509704670412888 14018344904398541816856576000
70094085705234143906889728 35048223444237789364748288 17524702017929253388025856
87626461568698060468387844 43814706523874926998323200 21908091131700411881226244
10954414500731680131645444 54773917178065771613388888 27387880926236571284275222
13694401631720128380928000 68474314001609855598592 34238309922309534646272
17119731421907070746624 85601539413296870850566 42802210858529193984000 21401826005204976271366
10701273290572677775366 53508167892714337075222 26754984666282642636888 13377942693104058368000
66891965265333977088 33447108532573831168 16724117216240336896 83623400830968791044
41813107790367948800 20907257582625751044 10453980635033763844 52271662393773260888
26136710806188851222 13068795207745536000 65346175061983232 32674187042619392 16337643277123584
8169096516468736 4084685697187840 2042411568070656 1021240143773696 510637251756032
255327215812608 127667902873600 63836098920448 31919123202048 15960098471936 7980317671424
3990293053440 1995213635584 997640372224 498836963328 249426870272 124717629440 62360911872
31181504512 15591276544 7795900416 3898081280 1949106176 974585856 487309312 243662848 121835520
60919808 30460928 15230976 7615744 3808000 1904064 952064 476048 238032 119020 59512 29757 14879
29759 59520 119044 238096 476208 952448 1904960 3810048 7620352 15241216 30483456 60968960
121942016 243892224 487800832 975634432 1951334400 3902799872 7805861888 15612248064 31225544704
62453186560 124910567424 249829523456 499675824128 999385202688 1998837514240 3997809246208
7995886927872 15992310726656 31985695195136 63973537873920 127951370715136 255911331364864
511839842598912 1023714044936192 2047496809349120 4095131057651712 8190536993210368
16381623742234624 32764346996097024 65530893015449600 131066184077410304 262141164247842816

524299920681730048 1048635025735548928 2097340420215275520 4194821577918906368
8389924630814523392 16780412211582468096 33561950323071778816 67126152445957242880
134256808491541856256 268522624182338453504 537063262763186388992 1074162554323391741952
2148397166240821411840 4296938447669718679552 8594165125715589070848 17188906712183481565184
34378966345871569977344 68760238534752353648640 137525088755523134685184
275059400883083124146176 550137248510239957843968 1100311390508627334791168
2200696567993549507788800 4401540709939688691990528 8803376567784556736806912
17607343431379472179265536 35215867454379661769834496 70434096092000758362275840
140872914550484386369765376 281755273833934512029958144 563529437133800502640771072
1127096653199463962443251712 2254268864262653839209922560 4508688844252759507066683392
9017679919960422671427043328 18035964302830652657441439744 36073137531480919944057585664
72148692914601069146464583680 144302221532480596809627992064 2886141114471518110652653633536
577247571756150055372102565888 1154533829138527778877795729408 2309145029529510894022772654080
4618444801563932460579907698688 9237199088137686266228540178432
18475017146295015222594529918976 36951272232629315825463958962176
73905020345337202411477716172800 147814992450831546344055028842496
295639888421977375730309250678784 591299583884583317545016887345152
1182638781850423767258830546665472 2365356791863361798855254637281280
4730872040051752126385696362463232 9462060992753561310121766900727808
18924755810807236734944282153058304 37850779272214701699290061009321984
75704093845629859857383115425054720 151413258293660632632372217662930944
302836657792123091099956408951504896 605693597993849833870336765154295808
1211427760806906971081521424811163648 2422936651252228548844738638627471360
4846035561781286311052868855265230848 9692395642116231048832520866551037952
19385440321339778951118608045143228416 38772178716894191609144348714368761856
77546953582217650632102962676902133760 155099099461293836091834455850133487616
310208583516304741838925972692925415424 620437936220043622988366067371167711232
1240917410814955524597760378712969183232 2481917898379647606437577245367205888000
4964001950258768327359267466616946819072 9928336207516482883686760884998963724288
19857337029030858225309973673528067620864 39716003286057501366492851154116415586304
79434665028106572564731509922353391861760 158874646968196284792954635072947905101824
317759927760358848912892500602378052960256 635541123168650256479751462117720591433728
1271124781633165630267435846061370153893888 2542334633858061495150737535774598249840640
5084839408899583459533206758852912383787008 10170019100166087857529876892313256535785472
20340718765066017591986680533841376607993856 40682798659599718937827214566112480288833536
81368319578134805383362136129084414723358720 162742083674140345782139686251887737738100736
325495056384022161595110200491213292058968064 651011890839527263251882056957302217283469312
1302067337822020406627087425864355700898004992 2604221787929972573500821475628213934458142720
5208617800431808667494936199055432934240550912 10417584050007344375976458893708875999129632768
20835864998302142833926090778613772259556327424
41673123793179193831798527539619585041706778624
83349035179508203991489747044023251128601804800
166703645545316040638764878017614664347580104704
333418441463231346589100523894365652875913199616
666859183671661223801342583507003954113332379648
1333762968833719508848968238450553204949676720128
2667615140648233140190502619774197003345377361920
5335408687258054525366137525294575193582802567168
10671174186439285540702539622081512760949700820992
21343061996724924061345608387147750269467593015296
42687551241142554082572275060264950034071568777216
85377956977670520084906666692468799058415903047680

170761622946111864009337566528815396097377337081856
341534663873765375697723599345386388155845736136704
683092163710614046753544131266283968233873596219392
1366229999347394684223282127683590320312111440330752
2732551342547122549878951985669225408312951376445440
5465285372798911462622679431942540352003359744458752
10930936121007155650974909785093259774761633472053248
21862602992832976753408921412602877691033094910377984
43726667487303284409736046510038471665085845753298944
87456257977881230625308500389742375896211003371683840
174918361962311784862289815518815616924500630473539584
349848415937722216947925260516292964113158508407422976
699720215901641728342541779989909388754631511735533568
1399487199855678045578466077894465698565892013312442368
2799067935816145268943697191618225239245042006307635200
5598322943841868893460924454895038162716599971980771328
11197020032102894498068909053107251693886231862692544512
22394788353044102418431938392848854124678527562847092736
44791073283764831681452117358966409723169182800618192896
89585139722882917052080715864470222393962620951084400640
179176265756472341482514394022015250683173752601864830976
89591126033589424430433678157545028289211131650780430336
44797059594471339059805079652041215618417693500314222592
22399278086073982952196660112654958546114910587619115008
11200013187456148187245390199644654641510487212540559360
5600193665937652449196225171480915004981759565635780608
2800190369073615402384877621569751344604137762500640768
1400141952589202290085821328699522593358697871091695616
700094360320798439489601923307084757207663430466535424
350058872173497866968146591132204108867988962693611520
175035282093298257095746110305432919566073105076977664
87520564049923790353709462522381892349075864403574784
43761743526599226079772934946023662457557588134330368
21881602514118278491345569315428189370288622033436672
10941166632468471971402335578922273755899224999854080
5470766003938902348565943250065226413327069491494912
2735474345821784355715359355334657974352263241531392
1367782844837058768573873542818351371020495868657664
683914258381612679645033703984686877432430058274816
341968547172347987501565318280099034677306091110400
170989982576944817590306892283927315319198576541696
85497845783857820714915562713902556649872053764096
42750350139621616317338839642920727820072409628672
21375888693657161138609948964445088657604396187648
10688301158751757059275239053714906702586293780480
5344328985337466774622751812603634538185194733568
2672253695649527509803942049174907862538621288448
1336171449315160816148254096023999227992322605056
668108025402778938697268583730272262357667282944
334065163073988734660205059724272455359586631680
167038156723293999858879137917043897701698109440
83521865954796816320836648860636275930273153024

41762326773973316324364670412710178487730700288
20881860285274112244155508197551109505162412032
10441278591780783163064340594373564883229736960 5220813520462255102025463544985787506939133952
2610493872517059311259378396392396286131699712 1305290492401495535753012510145949409396916224
652667024272230707938167911047850337863991296 326334440117185682399991478351136298551476224 0
16317764510379914701537280574940040104876441 6 8159154481083494101539410987155965467007385 6
407971335348851542615509084342095544078827 52 2039924733217641900770238096631964074036428 8
101999639484551304423146538577672521383936 00 510015211541102569038905861618734195330252 8
25501611282972430798103951517455289187041 28 1275123099444486657213130497698693430378496
637582817370175887260531709762311200702464 318802042509054222957249085337637843107840
159406338166510251142116157897060042932224 79705827539246695402803886562650582155264
39854242997619132617274847088385571422208 19927786112807458766573875447722925883392
9964225363402675612255163675626533027840 4982278835200810920611694813695801556992
2491222494350142017547903894789168300032 1245652785549939287394980191365217910784
622847161962403783008004217667925835776 311433965574918961159259169826621358080
155722175084318015407258115409639899136 77863683690588275117443322952984559616
38933139919508771265628794100574584832 19467218996861702486267963362328444928
9733934016984509669860764837184798720 4867129267769084048293773996602687488
2433645763522956630828582787306487808 1216863446580685618755139288155815936
608452005699946461047993591329193984 304236144054775056359208769290240000
152123142629788441097210371457941504 76064106616094677007408179135381504
38033320958647566733105586270896128 19017294304623897481253541487050752
9508964064962005797977144919326720 4754640488806031427663759547564032
2377399472565529978169473317732352 1188739350364022121253533430841344
594389482222639626711165101408256 297204644631634096397781743697920
148607274075974189719990468345856 74306112918065665620545032421376
37154294399072118190547415334912 18577766169555701785411157229568
9289192569787672237774303395840 4644751027398746791421514088448
2322452884951828731977938239488 1161226512810214203412255971737 6 580651906864184851128075157504
290335624838649342597435228160 145172648122603129815416438784 72588741912940794166057631744
36295579882290011712203522048 18148394404054813170689114112 9074499433482310242638233600
4537400832468606949965955072 2268775974098029389306396672 1134425765980877651814907904
567231772456370304488308736 283625330961150891534581760 141817387847058315412504576
70911055106770592528859136 35456708145006013675732992 17728944368313365543518208
8864767332061862124584960 4432531239983520738705408 2216339406968055207559168
1108206596972175022882816 554121745230161220993024 277070095987117465272320
138539659679577160024064 69272135682797793705984 34637220762903503699968
17319186842204055273472 8659881651478179348480 4330084940927165530112 2165114528057620692992
1082593292825829310464 541314660811424137216 270666337604966809600 135337672402110775296
67671088000869072896 33836669900341379072 16918897900124110848 8459730425038766080
4230005950007738368 2115073343748046848 1057571856246112256 528803520309100544
264410556247572480 132209676170297344 66107037108404224 33054618065829888 16527858788728832
8264204272271360 4132239575089152 2066188507021312 1033128613249024 516581486493696
258299333181440 129153961558016 64579128262656 32290637873152 16145855807488 8073196339200
4036732387328 2018433302528 1009250205696 504641880064 252329328640 126168858624 63086526464
31544311808 15772680192 7886602240 3943432192 1971781632 985923584 492978176 246497280 123252736
61628416 30815232 15408128 7704320 3852288 1926208 963136 481584 240800 120404 60204 30103 15052
30105 60212 120428 240864 481744 963520 1927104 3854336 7708928 15418368 30837760 61677568
123359232 246726656 493469696 986972160 1974009856 3948150784 7896563712 15793651712 31588352000
63178801152 126361796608 252731981824 505480740864 1010995036160 2022057181184 4044248580096
8088765595648 16178068062208 32357209866240 64716567216128 129437429399552 258883448733696
517784077336576 1035602514411520 2071273748299776 4142684935553024 8285644749012992

16571839253839872 33144778019307520 66291755061870592 132587908170252288 265184612433526784
530386817053097984 1060808818478284800 2121688005700747264 4243516748889849856
8487314972756410368 16975192895466242048 33951511690839326720 67905275181492338688
135815053962612047872 271639115124478836736 543296244647467155456 1086628518091953274880
2173329093777944477696 4346802302743964811264 8693892835864081334272 17388362132480466092032
34777877186465539031040 69558060215940291756032 139120732117899010899968
278250687607834876575744 556519821959743462703104 1113076537407634344509440
2226226861791563527225344 4452601297535716730863616 8905497742976612814553088
17811585781763584334757888 35624352155147886080819200 71251065493537206984245248
142506853353557283613704192 285023151440080306517835776 570065192346092091616526336
1140168163624047140394762240 2280411885111820195112943616 4560974885951092218872725504
9122252003357088095039127552 18245108469623983504665608192 36491425865067581638505922560
72985269581774392535361257472 145975374866827243587421339648 291960421140211404208240328704
583940185093536642483275956224 1167919055813300953100142510080 2335915482879057242467466215424
4671985708263025157469294821376 9344280901535871660007314423808
18689180773091386010152078409728 37379599486222057400579055943680
74761674852522685561707910135808 149528301465202512644515416768512
299066506450719308331230026530816 598152819942067182746858439049216
1196345253965391497662513650073600 2392769736093297259662620844097536
4785697928511623048000428776095744 9571712769673303153351231727992832
19144059364646720421403211807588352 38289386379893669072207920318382080
76581308060987794603218834043174912 153167686724376502124043654899171328
306345514653554830083299283423985664 612711311716713311837022514099257344
1225463188252633927014892922701086720 2451007506143682460711481634407317504
4902177271564194134786354846824923136 9804679061682046696299492849670422528
19610007160471410246052552011381997568 39221312395157454199012236646846300160
78445220938744175811838738541857210368 156895634174346886451306007580043640832
313801652942410842557869076152745721856 627624075072255824426252274290808324096
1255289688519379927473532792552250408960 2510662453788496412189122073045768339456
5021491061076465938862357121974071721984 10043314429151878106692940195713213530112
20087293472301648671322332294956567232512 40175916172599082258517568396973414809600
80354490801189734348780944408067390308352 160714298514362608361053504044375901995008
321439230852691496049090238545234046746624 642899729353315550752146938003432579006464
1285841994002496218812226797832794129039360 2571769058596722672240319439317446200131584
5143708258376905813712370565938608284368896 10287756799120732565888204506484648336949248
20576194162975307008703335762184160210321408 41153749455418297771260525022798047493488640
82310221169771963050228757042455549132668928 164625886857414661115872928078630006556721152
329262662750570792262576684144697829696208896 658547103572624524586815024264271292557950976
1317137763288214929296953360478293851446968320 2634362638862361618840553344856090235556069376
5268899502296586758174399937511185536436404224 10538147453736900557335386370620381203521339392
21076991805761255196643945732436782668339740672
42155377408097418557234237447265605859273605120
84313542409344653442361166859315292763735457792
168632660004988939540507717648198747617847410688
337276470382577144392586203155533819416447811584
674575241510352819408313942029340287194401603584
1349195084511102700062910955495225871111815168000
2698479372002999522618388053863542335669654257664
5397137149967587290221908393473265858231356358656
10794631111858351070414081358438894090246808403968
21589975847563055120768691859862512928061808181248
43181378942818816201418442005694475351259999109120

86365612381023044322599000583327849692792763711488
172736933752816912484722234310533497366131058409472
345485285487175472648492934908822590693353178791936
690993406937434240655082802393156373308888481529856
1382032485801035072026359469937335130462141210951680
2764156315454403325485106670176715028613010917687296
5528495318613473013834988800957519592603478826942464
11057356012636278753399528523123218255961871637020672
22115442776091222958258158888662794653433571240312832
44232347053819776819434521462158305589886798413168640
88467617110914215444705450293982043745812908691423232
176941080228377754501083715327294952623704441113018368
353893852469854156225513060133251635511566129686380544
707811088965905606897717379223826731551446754293448704
1415668945984205802688817276362300384159522498428272640
2831431428073200783164399588553894610432302976539295744
5663049928355979921902329248766376905091121912444092416
11326474001131116554951718640849929178635275743619186688
22653696291100546532197557568334209094176615324700377088
45308889159877719908983355709937119662165358324324761600
90620771475108693507143192566411642271954971998497538048
181247529260923894392639347425898090439158454696691105792
362507031143260803541984619437945792668813930792774270976
181259501882336909149345272012047702229655476096083165184
90632744096521708263849117152561254062451993397889597440
45317868625937480976512799149549328505038124373868806144
22659682601807053910550519861409014989425126024396406784
11330215445322683666422320074021682863165594930929205248
5665294794870920188784690108669429115809313424830103552
2832740933540249272179110090164008400017914692097802240
1416417234822519224982937562996651121065586335890276352
708232001437456906938160040455649021061107662865825792
354127692731827100692425649706486240794711078893256704
177069692372462873957885639592573985529434163176800256
88537769189506098784779227165952425330756393453486080
44270346096390380295307817267808928948397852659286016
22135903799013855599113010476320822615708754295914496
11068317274916260525286056159368590378609291131092992
5534341325162796625507803540288384724682102557114368
2767262006433731494186289500446237130029779774341120
1383676675143032337809338615374140948859254135062528
691861173534599464262766240262581666351809191477248
345942004748841379810431586419046429136995657711616
172976711365191513744740026353401012549043359842304
86491210177981168792132129748639405264794445414400
43247032336683290355947123160289152127533605453824
21624229792187998157914090723129300811334994100224
10812471708017175568927309933057012779451592736768
5406414259970176029448787252274687576617844211712
2703296332965882137216959769010434381754946027520
1351692767973338129854762955941762487600484974592
675868684731867595550523013689153892161748467712

337945492738533063086832274703713270261627224064
168978321555566164199201521281424797221190107136
84491948370932898427493452605496479655783301120
42247367982041357377693072285140280350485774336
21124380889308132770819709133766160436539949056
10562538893797793426396441062481090348918505472 5281443671470760233691513779039550239783518208
2640808948021311877092403513419277652553891840 1320448030153621818669525068659390092608012288
660245793148293849396424190304570679469539328 330133785609888394729042923139723156317536256
165072337322814932379936875563580486450151424 82538890920342833697676144778649697370767360
41270806589639100602691925887754575758229504 20636083859553392178272889693092151415537664
10318382212143617027599908221153507475980288 5159361247255268983031685797880469622095872
2579765694219364726131708742592092753100800 1289925382405547480373787293121975347576832
644983958850706298840860107473952159301632 322502613249319428747413284193458322407424
161256623536642854037198257324970282582016 80630970224312996850344936276605701980160
40316814340152283341045371945363131334656 20159071784074034128459137876211705839616
10079868199035963293197794889870923005952 5040100253017454761083010420817996546048
2520133203258463937783561698350265794560 1260108140004100247512809093145766658048
630074839189484263066918668558200209408 315047804188459201188716395271758544896
157529094391088135421986728132208492544 78767143343973335124807629314268856320
39384869746201301269310947281216733184 19693083910207967488109039952649519104
9846866473657642170781303132345335808 4923595496105650298754043144182956032
2461878877691239756058717361096622080 1230980003664827181370206575050883072
615510284242017242355527234776727552 307765283325810447012975591014006784
153887712265306136424093782319824896 76946391433853524670849884566323200
38474463367526990564826438986366976 19237865509063609397113967844786176
9619249667181861755907358098194432 4809783289915959406628866136997888
2404970873120493967652026612449280 1202525050641504115994810078199808
601282332361380624081803425087488 300651069701004595083100905537536
150330486610659439062650049265664 75167719185408290291874822881280
37585097532743430526212310564864 18793167736391357953243604844544
9396893353205500321690527203328 4698601419107660833379625992192
2349378080806285752956994191360 1174727726029370544612087693312 587383205827799106372839145472
293701274320456470219817222144 146855472863506693626607435776 73430154283392576071653130240
36716286067515902665001271296 18358747496667758647087988736 9179675979788782980837670912
4589989105621843319065673728 2295070110674647573856256000 1147572834269186744089837568
573805306600524850625773568 286912098033228164603314176 143460771383096951946870784
71732746874789910796042240 35867554029015672809324544 17934367310318195110313984
8967478803064276907982848 4483886975484728130404352 2242017274718658903408640
1121045530847476870807552 560541212167812144955392 280279829455942927253504
140144526413989891014656 70074569050004159201280 35038437446506686447616
17519795184005646647296 8760185822378975035392 4380237026377563373568 2190190570782819614720
1095131314188428771328 547583671492723867648 273800842945616674816 136904925072435707904
68454714336031539200 34228483067922612224 17114804483914727424 8557683716934074368
4278982595955392512 2139561667721873920 1069816017733025792 534925601052557312
267471596619300864 133740196356161536 66872297201336320 33437248112295936 16719173811961856
8359861783887872 4180068330897408 2090102884925440 1045085802201088 522560080969728
261288630419456 130648610177024 65326452572160 32664300027904 16332686884864 8166611877888
4083440156672 2041787187200 1020927148032 510480351232 255248564224 127628476416 63816335360
31909216256 15955132416 7977828352 3989045248 1994588160 997326848 498679808 249348096 124678144
62341120 31171584 15586304 7793408 3896832 1948480 974272 487152 243584 121796 60900 30451 15226
30453 60908 121820 243648 487312 974656 1949376 3898880 7798016 15596544 31194112 62390272
124784640 249577472 499171328 998375424 1996816384 3993763840 7987789824 15976103936 31953256448

63908610048 127821414400 255651217408 511319212032 1022671978496 2045411065856 4090956349440
8182181134336 16364899139584 32730872020992 65463891525632 130932078018560 261872745971712
523762671812608 1047559703363584 2095188126203904 4190513691361280 8381302260629504
16763154277072896 33527408065773568 67057015154802688 134118428356116480 268245652805255168
536508897796554752 1073052979965198336 2146176328674574336 4292493394837504000
8585268264651718656 17171099479256858624 34343324858420559872 68688901516654804992
137382306632936980480 274773620465128701952 549565255328766885888 1099166539454552735744
2198405136503143399424 4396954388194362654720 8794197006764877021184 17588970474282057465856
35179093870068721778688 70360493583146657251328 140725598852311741890560
281460421076660338556928 562939288897394386665472 1125915471282936192434176
2251904729542167223074816 4503957033036924122562560 9008209213979027597950976
18017008723768413901553664 36035198039157545214410752 72072757261556525251428352
144150236889595920148070400 288309918512157579586568192 576638726490246637753991168
1153315231912356232669691904 2306706021688438379662802944 4613563159104328587972444160
9227428549663560833238564864 18455461562236928981064482816 36912132050293472591303671808
73826681952226174440956755968 147658199607730807398612336640 295326070622018531830622322688
590671484057150897728039944192 1181381653740529463589670486016 2362840678733514263446522167296
4725836099971939199427406725120 9451981684953699743923538231296
18904582339927042177984526024704 37810402619893369736243951173632
75623281119865310233037700595712 151251513999887761987174997688320
302512931520089807016549188370432 605045670080808180117496762728448
1210130954242873492403790297432064 2420341136648261249145174138814464
4840840729621551026965535365529600 9681998371893159111281444906860544
19364630569086432337263638165323776 38730528788773092903928773033852928
77463592878746642266660539474116608 154932256359894197450927065761054720
309874653924590220737066105147752448 619769590258784093144556157546790912
1239579745336775489629960209596153856 2479240620311965585941616208197451776
4958643499900760385246623994405191680 9917611518355179197220031144830959616
19835872073817675247893628601703071744 39673042221849984202694389827488448512
79348680592129235819203044903141507072 158702553481117006466034620302612234240
317415491555951082587326301597882908672 634851752299336304485166725181082697728
1269745042973540887591361694332799156224 2539573162696818332424779876606865833984
5079312478893109779333672729096266711040 10158957264785165787635571409957603508224
20318579143568224033207594723445347188736 40638487515132232982288093253950974722048
81279633486256035796321994122022510133248 162564583884495211256135603472286141644800
325139801592956701839254437401054526046208 650300870833845962332475335715073537605632
1300644276963557041972883593256076046237696 2601373624518884431856163303016401003452825 6
5202917390221149106354997747631735953162240 10406175062809219151173458869870903674535936
20813030690352280179273844488956670885494784 41627422510172244112401542476343068843835392
83257567279279855732510791949545591833362432 166520579076430446480436997892810091958108160
333052047188602362991704823773058000498982912 666125872448687666045071303520991634163499008
1332295301040341212213465918991734534658064384 2664677714366614184673578461882971601978261504
5329529653305091889840450171564948269280788480 10659407755753910820667486838727906669210107904
21319512409795275723308146668651833599717277696
42640418616165459610562639319695707722028679168
85283624825480735549017970604175496489245605888
170572824837261103753821325137919155068867706880
341156800047121472819213418134974634318488403968
682335900839441476261568371988221916998482788352
1364716403169280013769419815412989130719977537536
2729522009319354150031405773699068854885978996736
5459222424600296545047943833144318896664005836800

10918801661123769580066152237781000167112107360256
21838316946093892140072833618546725081792406093824
43678061139880490240026725523062899658721194934272
87358976775146392399815567618064698307715155361792
174723662541063608639155368380007194595975841710080
349458743063668864957359203047769985153042745393152
698940322090421025272815338671051162228267614732288
1397926316107008641261824542493124708300899477356544
2795943976066350463956036815288294184290527450497024
5592070639837367290776849091180677903958511892561920
11184506655084067307283249103569534878671937768259584
22369744060986800066025600049555427898853703502790656
44740949623610931034969403783943572080727062938124288
89484822250496523875775214937552576727493437741334528
178975490507542371363223244614436018587065499212840960
357962673028183389949792118707533767438288245886025728
715948730082564074346275496372390995404890986692739072
1431944228217522737585933510659428911866410963226853376
2863981992539834652958632057148151665846079906136457216
5728151057289247661490794185954891015918675771638415360
11456676258997652034128648515226957400290383462007832576
22914100806833617490551417317088265537486830761477668864
45829698191343861825691075207445232548785789197879345152
91662389538040977340558631561427868045195833745606705152
183330765386788462059470225415930541985640178190909440000
366673503394989938875646375418010695761777377781210939392
733370952032805907264704600008320615104548798361205997568
366697448637815968389058224590309919342771420579995058176
183354710629614491572882074588229765566634220989693558784
91680348470160499475617518440652285730941365844694794240
45841670812756876582396999793594844339282810597271404544
22921583695216751713492620183431772906547469136097705984
11461165992027532567893370235033061821726766486779854848
5730770068223344639520215189175118595089899202755428352
2865478570216461497546872630416853139658207581060464640
1432786053160625337666818833123073490885732780371607552
716416410606509963280100675518860205971180885106491392
358219897316353628863395967238091833249747690013589504
179115794664726138043370798358376781756952468736966656
89560820335637730827521806548853823444515546233569280
44781871669456196316679106959259628005277429049327616
22391666585546763609798655322046172144148542490935296
11196198668182714530628878582231265142829185228603392
5598282021796023628179214751719722106792049605869568
2799232354750344995521995106161905821084753298718720
1399661849301339088477191418231975294386740897251328
699853760613752839596692641691498839115552572571648
349938298288418067477394787133505015518867348258816
174974858134979857578221626710630305739979205115904
87490283562875340708872929927254051860262368051200
43746569029130376314317523249596475425267566772224
21873998138411541137099290767782962460201974759424

10937355881128947058519909955383843603885083066368
5468856346526061774245087263438102988834589376512
2734517376243825009615109774592142087863318609920
1367303289612309566053837958732616340654671265792
683673945551353313650060515084580818688841613312
341848123148275922136601025401426733525173796864
170929636760437593724085896630281528852963393536
85467605973368613189935640279924845471669944320
42735196783259214758914166122354463258429095936
21368295289917061461430256052373251890511609856
10684496094102257771701714521784636075904335872 5342422271622992406344150508691323103276433408
2671298248097427963418721878245164084300349440 1335692680191679861832684251072333308481241088
667868118167322870978003781511042287406153728 333944948119402905519832718742958960285843456
166977918577572187775331773365198388434305024 83491681547721461395373593679458648362844160
41747201903328414451540650338159051254267904 20874281516398049102697251918294389163556864
10437481040565945489812089333754626349989888 5218910661466433214137776354181029059100672
2609540401324946841684754020742372471603200 1304812735958338538150309932197115206828032
652427635627101827729121427011522088927232 326222445163751719319154394396224328722024
163117542730741736259263587209362764988416 81561429821362437961377601218801943183360
40782044138677003896561704416461251936256 20391686683336394406217304111760766140416
10196175648667143432076878007645453156352 5098253977833044830522551979705261621248
2549210065666258972503332477793898332160 1274646571207997764872694482867582926848
637344054791433021746861363419108343808 318683411989433580528687742702212612096
159346398291575325091972401847435526144 79675795294216929959800466171882373120
39839195721323098686807365710023491584 19920246897768866196857249167052898304
9960447967438091525155407739547025408 4980386242995874975941095447783800832
2490274251136352094652243512897044480 1245177690387383350666969650951094272
622260912760329532700390877272683315 2 311314705006449489337166359989059584
155662423105625657586189166807351296 77833746854013285251897576810086400
38918141077606870855350285108248576 19459704364103549542375890905726976
9730169094701831828538319628664832 4865243003675944442944346902233088
2432700730000486485809766995066880 1216389979081500375073680269508608
608214796581378753621238520741888 304117301811003659852818453364736
152063602665658971447508823179264 76034277212908056484304209838080
38018376546493313622427004043264 19009807243266299501350951583744
9505213106642971095744200572928 4752761295826396220406462676992
2376458019165653446470412533760 1188267695209054391368796864512 594153190417641029751193731072
297086266615377431908994514944 148547969010967174471196082176 74276402357122816493947453440
37139410104381022876148432896 18570309515100318752661569536 9285456989005063033624461312
4642879610229983345459068928 2321515362978717587052953600 1160795460421221750688186368
580416619676542353924947968 290217754571236916252901376 145113599652101327771664384
72559161009292098708439040 36280761096266766765522944 18140970843943742088413184
9070780569877050397032448 4535537858891114874929152 2267842716421852275671040
1133958251699073556938752 566997572593610488020992 283508009668842098786304
141758616520439476781056 70881614103228952084480 35441959973119082889216
17721556447311844868096 8861066454032074145792 4430677342204112928768 2215410728696094392320
1107741393145066160128 553888710971042562048 276953362684776022016 138481184942015381504
69242844270821376000 34622548035317530624 17311836967612186624 8656199958782803968
4328240716879757312 2164190727184056320 1082130547964116992 541082866168102912
270550229177073664 135279512635047936 67641955340779520 33822077182017536 16911588346822656
8456069051318272 4228171964612608 2114154701783040 1057111710629888 528573035184128
264295107526656 132151848730624 66078071848960 33040109666304 16520591704064 8260564287488

4130416361472 2065275289600 1032671199232 516352376832 258184577024 129096482816 64550338560
32276217856 16138633216 8069578752 4034920448 2017525760 1008795648 504414208 252215296 126111744
63057920 31529984 15765504 7883008 3941632 1970880 985472 492752 246384 123196 61600 30801 15401
30803 61608 123220 246448 492912 985856 1971776 3943680 7887616 15775744 31552512 63107072
126218240 252444672 504905728 1009844224 2019753984 4039639040 8079540224 16159604736 32320258048
64642613248 129289420800 258587230208 517191237632 1034416029696 2068899168256 4137932554240
8276133543936 16552803958784 33106681659392 66215510802432 132435316572160 264879223078912
529775626027008 1059585611792384 2119239943061504 4238617325076480 8477509528059904
16955568811933696 33912237135495168 67826673294245888 135657744635002880 271324285363027968
542666162912100352 1085367510196289536 2170805389136756736 4341751515761868800
8683784506500448256 17368131962954317824 34737389825815478272 69477031451444641792
138958566502516654080 277926140204288049152 555870294807085580288 1111776618411190124544
2223625294416418177024 4447394704020912209920 8895077638417976131584 17790731737588255686656
35582616396681118220288 71167538636371450134528 142339688958761327656960
284688601289559510089728 569395649323192729731072 1138828192134532878565376
2277730171245360595337216 4555607916443310867087360 9111510980791801087000576
18223612257393960879652864 36448405106408639170609152 72899171396058713163825152
145803065158600295972864000 291615575050166331236155392 583250039566264141053165568
1166537858064391239268040704 2333151273992508392859500544 4666453663712468614365839360
9333209558879840886025355264 18667023580669489086638063616 37335256087158592802450833408
74672930025956414863251079168 149350695755191288243200983040 298711062916939493519799615488
597441468646992821106394529792 1194921622920213310346379657216 2389920617092881956959940509696
4779995976690674586454243409920 9560301438391170517977211600896
19121221846801983726091872763904 38243681633643252832458644652032
76489839147365076425467087552512 152984630054887294372033771601920
305979163630088871786266736197632 611978134300806309656931858382848
1223995882682869751482660488740864 2448070993528253767302914521432064
4896300443381536063281016130764800 9792917799413129183912406437330944
19586469424126372482525561226264576 39174206498852973194452619155734528
78350948298906402847708231717879808 156706967200213718613022450248581120
313424075605229263061256874122805248 626868433620062177792937695496896512
1253777432059331658926723285496365056 2507635993757077924535142359997874176
5015434246790985062433676298006036480 10031193012135628551594135752032649216
20063035061378573956641837816106450944 40127368196971781620190808256295206912
80257332542372830654195881760755023872 160519857381604196136020294017839267840
321050099356925461927297649028336975872 642120967901285063165109420041990832128
1284283474177438404951247084054615425024 2568650025104613367144550656050498371584
5137466203708699848773214287983531786240 10275264714416345926514654527732133658624
20551194042830584310965760958994407489536 41103717313656953537804425725049095323648
82210093083305476907354659064218751336448 164425503078594093478200933356678624051200
328861639981154466283385097169839490859008 657744547610241491220736655252643467231232
1315531630516348099749406232331215905488896 2631148331624426434114678308314289753030656
5262466804432313337461088303932295390167040 10525273891231547613385639982472022548545536
21051228347196937103698206714158908633513984 42103817823861557961250266926747544339873792
84210357906658483430208240850354542825439232 168426160331187701875831895694427993942261760
336863209698116873782494619376293804467290112 673748197467716687626650894727463242100113408
1347539951078399255376625101404677750531293184 2695167014442730270999896826708858033724719104
5390508253457324062493086901216721132773703680 10781364956058375165972760298031452396195938304
21563426810404204413918693587258925053688938496
43128247417383316991783733156909890629972000768
86259282427916450311460158278603862305132249088
172524140042132533278705700486775886700640993280

345059430456864331868982168832688097582034976768
690141161658927194361105873383648843525575933952
1380326924808251449968494818203842983774163828736
2760743052597297022429555779280776560994351579136
5521664511156182289844243844307734308880751001600
11043685834235541069658752260107830991545597689856
22088085292317435119258033663200386730659386753024
44177597832327576198397125612370222956455156252672
88358050160040564316556367796679344903183077998592
176721809310851952472636968737236487786911686983680
353455036603245552624322403762228571534914435940352
706932909169574400606741740099968334992010995826688
1413911490265315391929677345350959053828386239545344
2827914324382963965291742421003962875345500974874624
5655601133647059429344826030261201528606845894131720
11312388048350521312626071526432209642891831865769984
22625506847519708076711244895280777427293491697811456
45252475196676747056340693475394271137606639328165888
90507873396628155918517794320453974841252590521417728
181021592799805635448708403380238814814583804773007360
362054877612709918120762436239139359893324857006358528
724133139251617130688216131435602180314964208933404672
1448313046555628850269814780785851281686557407708184576
2896719629216046878326394597400996405486372795099119616
5793626330641672112226319266460580495199261549563740160
11587626805702500935599698676238336358851555017858482176
23176001900243315293493517639111023454609173873178968064
46353500378163257431575275851490748383030475421281943552
92709993911679768552327032849518899713685206192411901952
185425974134066044483007027992112605322618923084519833600
370863920889545103722719980570374822435734867568431726592
741751787021916236958851810313048868452463777935647571968
1483551464529484532944527318970696184066915641468863381504
741799677507568295985675508657647315614451863533215809536
370911811375197162749543678914973269597722953165999964160
185461891998305088753124801750561440694109987282696011776
92733939154505798065738882021818123294679248991196020736
46368466154929525877457681584177763121151752170522017792
23184981366303076361022961078723232297481939922723012608
11592864827570694891658540682678791517194001880092508160
5796619485994925801402800412997983442823516899411755008
2898403279102252078488165242328285563525016429388627968
1449248407603520628137465139078789702819137204535689216
724647587827957608515423828496718311937883097188532224
362335485927077451481057543727020886233098796054609920
181173588970088049352201586602841308248628021757476864
90589717488318686481937200671086086690353322743824384
45296320245796674143886804020375759628196317304455168
22648890873717002523402503852604237955607986618499072
11324810812267833987430802847510298048558907292385280
5662588093838583356580176884359238559656910637760512
2831385390771624859722476172481664047517183814664192

14157383673119790205774319513918544076029561552240 64
7078920196190728056468129082714383957236602015580 16
3539574277910780505024549204234747938229211627520 00
1769844228863098490907516933556151948920061123624 96
8849506593854033646513796324974649643627582167449 6
4424896021696287419245003991084269771327429358387 2
2212519373232779007616554909840607360420533816524 8
1106295367808707152805303912069539917588676476928 0
5531655245005124009011651846093880774835430227968
2765916825483356126998392065920030980863739035648
1383003014232075124745479104396560787154881478656
6915238078612360929958810879165530419389467197 44
3457730543032173118095113118174128451502263500 80
1728921023379082885605410398382745846654896701 44
8644883876210396060816321188392137337793308262 4
4322581317762688846802795192435272721156066508 8
2161360348710089831598714895337238386707739443 2
1080715019269417619898016097228420206418722816 0 5403749320918951619983373733941106097417879552
2701961772745407570238333490870055581371072512 1351024442515669665242490057384779057016602624
6755339993293177726829066846672651616738344 96 3377778887004003563722841703210703974196838 40
1688943888680709132015574991542541070012252 16 8444991669297082410848645657398650764630425 6
4222631947595309580809708178542298089599795 2 2111384030271038978097546764192635398442188 8
1055726043372211582895119719557060876042240 0 5278800358044518383707330285089020264316928
2639485249613989426469530986196368074211328 1319785160102859830542698414924113008132096
6599138476993624739253156683750209895792 64 3299675576736475162896410646439927375462 40
1649890957488068978083121475502374901514 24 8249720633039501873590188138923930576486 4
4124993239319329428382384450167993322700 8 2062563081059453959984837415437010678579 2
1031314771229621602889241302895012347904 0 5156740009647581128930319490357596782592
2578453081573527121707216233120065912832 1289268079161631839474636360530666717184
6446548087682500590478323022506502389 76 3223377889778420991791732121179835596 80
1611740867857795844172151365553209999 36 8058963954131905962242183352582511001 6
4029611784487416351811804938699486003 2 2014870795954439861251259100553858252 8
1007467849832585773298307865878986752 0 5037501508439758079854930907405221888
2518831883858293646609161242707755008 1259456506748354126645428515856449536
6297485357837807149931382051795107 84 3148844090966921833317810762153984 00
1574472751507470045834965249205207 04 7872617287657395875055125586667110 4
3936435408888720760467712463654092 8 1968281086974371791703931066987315 2
9841722347521916015870029510737920 4921019630085986536610201843269632
2460589043205507532642694465585152 1230334135684010898490144004767744
61518687488263401532947038837 1456 30760334096163129070693438717 9520
1538066222409727868745667900866 56 7690578700056496419783319329177 6
3845413144032176747919149577011 2 1922768469018052642973319744716 8
9614151830100084559935323504640 4807230657554952952502024142848
2403692700029931812518193266688 1201885035641193574392687230976 6009618606337106212631389143 04
3004906017234122276649671065 60 1502501365649845723491823779 84 7512748613413151543294060134 4
3756495199288537234564500684 8 1878308045935249348740985651 2 9391842461131150400998604800
469607234629302702914614067 2 234811173101023942889648947 2 117409364443698267160995430 4
58706571168442281438583193 6 29354230057517714648334336 0 14677587265407144288688537 6
7339029751027715626604953 6 3669632934675929554432819 2 1834875496919000647781580 8
9174672632500182591733760 4587483890202680972279808 2293815732077635324346368
1146944759526965081276416 5734908265075562501898 24 2867546366258149798707 20
1433819299989259173232 64 7169327084247217235558 4 3584778834274069302476 8

17924470632122649935872 8962523546437476679680 4481405888406814195712 2240775001797445025792
1120423529695741476864 560229779246380220416 280123896822444851200 140066452010849796096
70035477805238583296 35018864802526134272 17509995351216488448 8755279150584954880
4377780312780832768 2188960525134594048 1094515446939385856 547275315655737344
273646453920890880 136827625006956544 68416011526733824 34209105274994688 17105102393311232
8552826074562560 4276550476234752 2138343957594112 1069206338535424 534620349136896
267318764503040 133663677218816 66833986093056 33418066788352 16709570265088 8355053568000
4177661001728 2088897609728 1044482359296 522257956864 261137367040 130572877824 65288536064
32645316608 16323182592 8161853440 4081057792 2040594432 1020329984 510181376 255098880 127553536
63778816 31890432 15945728 7973120 3986688 1993408 996736 498384 249200 124604 62304 31153 15577
31155 62312 124628 249264 498544 997120 1994304 3988736 7977728 15955968 31912960 63827968
127660032 255328256 510672896 1021378560 2042822656 4085776384 8171814912 16344154112 32689356800
65380810752 130765815808 261540020224 523096817664 1046227189760 2092521488384 4185177194496
8370622824448 16741782519808 33484638781440 66971425046528 133947145060352 267902880055296
535822939979776 1071680239697920 2143429198872576 4286995836698624 8574266551304192
17149082858422272 34299265228472320 68600729480200192 137205857006911488 274420510106845184
548858612399734784 1097752409171558400 2195575187087294464 4391291111662944256
8782863698302599168 17566290346558619648 35133706593024081920 70269664985861849088
140543833571351068672 281096674341956878336 562211363082423238656 1124458754961865441280
2248989567517768810496 4498123250223613476864 8996534730823378665472 17993645922399060754432
35988444766302728355840 71979195375614670405632 143963002437247768199168
287935228246532391174144 575888903237138491899904 1151814699962424402903040
2303703186901143644012544 4607553947754876964438016 9215403043414933281701888
18431396382640225269055488 36863973356901167949414400 73730307897043770721435648
147465338160570411088084992 294940121054106561466597376 589899131574144601514049536
1179836042080152160189808640 2359747642024030234703036416 4719646399775512298052911104
9439595031005928253399498752 18879794524921663821386350592 37760797975662942271947407360
75524013802965113802244227072 151052863309208686121187278848 302115398024974289275772207104
604250138863062412618339713024 1208538963352352493370270023680 2417155297957160323007721242624
4834465338419231318549804875776 9669240161848283982168334532608
19339099293716210654474118627328 38679436527471706689223136378880
77361348935021984138996071006208 154727649630201109799091738509312
309465202780716502640382670012416 618950212602061571365163726012416
1237940039285380274899124224000000 2475959306733274814135841991950336
4952077069791578156946871071801344 9904471052233213371244116319404032
19809575929766540857188980990410752 39620419510133309943779458684026880
79243374321467076346361910774464512 158491819245335065610329808361750528
316993779695471957055871590349144064 634007841800547565782167127949574144
1268056248420302434905182150401720320 2536193626479019476492060089808584704
5072549512234868166347511757627457536 10145423543023394759421806671275491328
20291496123154106372297179654592135168 40584290320522846451501491933266575360
81171176789474960316817249114697760768 162347545875808455461263028725724741632
324705476345333980577783118444107923456 649431721878102100466080358873532727296
1298904982131072479553188961717699215360 2597893041011881516348434411376665952256
5195952235523236147180981798635866947584 10392236778045418523330189549036803981312
20785138170088729504596831001603748134912 41571605568173243925066565810267776614400
83145869592338057681878939234656113917952 166297056096659255027249493697553349214208
332604746017284789381482217851588941185024 665230759682502137416930896616142367883264
1330504054660869392141794715058213706792960 2661093179913469018899455273768285355638784
5322356501010398507030642234840286595383296 10645053284387717952524747844288004958978048
21290787133509277781976422437790873454379008 42582935396486239317806698374011473981603840
85168593051907846143321103744882402108899328 170342630621686427302057621483483712509181952

340696150279114324634946070954405241601130496 681414078629711589331553797883686116367794176
1362871713402389058786430907717123499066654720 2725830539090709877819508439333749530795442176
5451835302753283276132310126466504126915149824 10904019054650293593251206748531018384478830592

21808735007588041268475586488258057030254723072

43618863811750990700897518958908154583103569920

87240515216651797729687729882600390211395387392

174486605619603228115160843694768942513167269888

348984361611805721541892455248674209207087529984

697991023968809973706926446215621066775681040384

1396026649428017008660135963867787430274374041600

2792142501836828139812838070608665453994772004864

5584463409635244524610808426963512094881591853056

11169283631193665539191881425419386563547279392768

22339280886233684058324291993823497874662750158848

44679989020160074076529642273616445244461883064320

89362832535705560072821401119171789479196531621888

178731374062181943985167035382221376938938594230272

357474166105905535649382537052198349838968250433536

714971168174894366656862006679907891600118624813056

1429988008275955324029917878510838167044601497518080

2860067360404243829492223487323721101777931490820096

5720317408513154021849222435251531738933319973208064

11441000192435640769427995791711242548621553929551872

22882731135689946990315093425838843238752935825375232

45766923773017224883548390536510402760525527583293440

91536770549309111572933188442686238087090367031672832

183079387105167546757539191624703341306259357793517568

366170466223433740738424012728068412876675963047378944

732364316473064775923539284413460286281666421015445504

1464775400998524140740461086741567493619961831872266240

2929644338101837459267687209312428829353181643427282944

5859475748413253274108904490283445342932879246220066816

11719325641245663259364869123884066054318790411171135488

23439399571329639941023858534402482845543644659804274688

46880295720335906726635957642073667164899416994532556800

93763584596025067142448396430684737277423089338913128448

187533155502756641663249755154444280450094689377522286592

375078283626926298083205434895038172690686400154436632576

750180512496678625679822718962375568962366843107657383936

1500408915479009310386469136269349585086721771812883005440

3000913611929322738826585669227896064497419714820902486016

1500504696450313428440116532958546479410697943008019480576

750276293467982743733470115651572463286343014302793859072

375150119355404386623440982411935843433668528550788988928

187581045988408700690073453499042727612082774975090524160

93793516149557604034213207896058766753665642837393276928

46898254652455428861694844521298084850644949093620645888

23449875615066027853141542547283393162228538384272326656

11725311951952170637717831416958871949567301110867165184

5862843048185663674432445780138023659010166514799083520

2931515060197621015002987925898305671618341237082292224

1465804298151205096394876480863799756865799608382521344

732925533101799842644129499389223338961214299111948288
366474458563998568545410379173273399744764397016317952
183243075288548607884378004325967565004460822238330880
91624460647548965748025409532649215068269722984251392
45813691825411813776930908451157323817154517424668672
22907576663524572339924556067995020050087086678605824
11454153707171618895691828955205689095798457322438656
5727259541290475810710689938206934083276685652787200
2863721114497571086787732699405511809327071322177536
1431906229174952134110060214853778288507899908980736
715975950550559362413127040002400336176132078436352
357999393256821328885611986288955764049157101191168
179005405619181488282330226288355680005124081582080
89505557304976156060927229716116738992834806284288
44754205900180783990344673144027818991553785888768
22377816573936744975112865714998634243345084317696
11189265098891548977526697428991679495456637845504
5594810955407362733748481000242020934620366766080
2797494680684475489366806642994101060756207304704
1398791941832634805929686392933595827101115613184
699418271661515933587984732185070561912063787008
349720286203357232105563133951671605136784883712
174865718287978248708566950905403964658768936960
87435646737138940682176167417486063374572716032
43719217165144378505034429691135072209880481792
21860305480859643334490387836763556366237302784
10930501189573548708231780413979788313767182336 5465424819358637874609183454788899222207856640
2732799521965250697551238351293952143766061056 1366443317125591228898942487596727338214096896
683243436634278554511132899773239302272581632 341632607352880747286397277874057467719057408
170821748194311108658614052930747642150912000 85413596356090921837014733462233275221147648
42708159307513144672361220229546364683419648 21354760218490414213107536863988045878132736
10677720391612128045017231806601454707277824 5339030336989524491740347590604443237744640
2669600239086492480486039638954079560925184 1334842654839111357550952741302968751489024
667442595067488237429442831564448861257728 333731931357710398041704646238706673385472
166871282590838338684343938347594458071040 83438299751410739173917776787917789724672
41720479103701154502831792201019175206912 20860904165848469709352348004039727775744
10430784389923181083644399953784933974016 5215558348461063656306312952775002030080
2607862250980268385395212964328768536576 1303972663865002471318634726135018029056
652007101119935374969831485052825894912 326013935153684757140172803519071387648
163012159873700913397714932255864913920 81508676085279724112671731376097067008
40755636116854495763242998312130838528 20378467095534564735075065468106571776
10189558066320940794264315890073862144 5094941292437299610495549523047219200
2547551775857064411929470550528753664 1273816452747739509305583169766948864
636928508783473406323215532134760448 318474395596538528996819739693023232
159242268400670177416015856659333120 79623669501535545166810921736077312
39813102401368000812806957571244032 19907185025984114521104227137224704
9953909425642114317902487744413696 4977113169146085687626430960107520
2488635812735557108150809024004096 1244357520449035686244201283977216
622198567265146409206499027976192 311109187152887487645448706981888
155559545336600885343823949987840 77782248548379013432461773242368
38892362214228792096505785745408 19446800077134038738390342434816
9723709523576840714263895998464 4862009504293331029666310389760

2431082123399120851100336390144 1215579747325788093683758792704 607809216476007880908674695168
303914279644560857487734996992 151961975525555887260566323200 75983405614418672888632573952
37992911733028951073490993152 18997060329424282851332849664 9498832396167045082960101376
4749567313810974370126888960 2374859214769213099386863616 1187467386316469506855141376
593752582624166232008425472 296885736045048855294640128 148447590389007297292533760
74226156377745083468873728 37114258780493259145740288 18557719686056988278521856
9279154990933673492086784 4639725069419426422456320 2319963821686008049434624
1160005054331151443820544 580020973909649431461888 290019710326861570506752
145014466849449212641280 72509539267733820014592 36255922555371516854272
18128537738438061850624 9064557099595182637056 4532422664985667174400 2266283390086871515136
1133177723840454721536 566606876318736842752 283312445358623162368 141660726278938951680
70832614939283161088 35417433369548423168 17709279634727632896 8854921292340527104
4427601383658618880 2213871060573487104 1106970714658832384 553502949515460608
276760270850752512 138384533471887360 69194465759199232 34598332391227392 17299715951427584
8650132853620736 4325203865763840 2162670652358656 1081369685917696 540702022828032
270359601348608 135184095641600 67594195304448 33798171394048 16899622567936 8450079719424
4225174077440 2112654147584 1056360628224 528197091328 264106934272 132057661440 66030927872
33016512512 16508780544 8254652416 4127457280 2063794176 1031929856 515981312 257998848 129003520
64503808 32252928 16126976 8063744 4032000 2016064 1008064 504048 252032 126020 63012 31507 15754
31509 63020 126044 252096 504208 1008448 2016960 4034048 8068352 16137216 32275456 64552960
129110016 258228224 516472832 1032978432 2066022400 4132175872 8264613888 16529752064 33060552704
66123202560 132250599424 264509587456 529035952128 1058105458688 2116278026240 4232690270208
8465648975872 16931834822656 33864743387136 67731634257920 135467563483136 270943716900864
541904613670912 1083843587080192 2167755893637120 4335649226227712 8671573330362368
17343696416538624 34688492344705024 69379183712665600 138762765471842304 277534327036706816
555086246259458048 1110207676891004928 2220485722526187520 4441112182540730368
8882505840058171392 17765574630069764096 35532275160046370816 71066802119906426880
142138107839440224256 284285222878135189504 568588460154779860992 1137212949106578685952
2274497955807195299840 4549140026802466455552 9098568283981084622848 18197713028714472669184
36396578978933552185344 72795463800876318064640 145595539287771063517184
291200301947578981810176 582419050639231673171968 1164874994766610765447168
2329823776509516369100800 4659795126971622414614528 9319885401848424182054912
18640361099507207069761536 37281902790635131550826496 74566166764511697924259840
149137055895506265493733376 298283556523978270277894144 596586002513888019136643072
1193209783959638995434995712 2386495125783003905193410560 4773141367293459639033659392
9546584966041822935360995328 19093774394993453185309343744 38188757715806520999793393664
76379933283252271257936199680 152764702269783001032571224064 305539075946122919098540097536
611097494705359672263875493888 1222233675036947012661341585408 2444544721326349361589864366080
4889244185157609395714091122688 9778797855325040136496907026432
19558214680669722963131263614976 39117667301378731306537426354176
78237810482836033373624650956800 156480572725829208268348898410496
312971048971972699578896989814784 625961904984573965242192365617152
1251963424050405062653181503209472 2504006076263324389643956550369280
5008170608851677307963100188639232 10016658130353411673276574553079808
20033950086006937461253897457762304 40069167822614103151909291618729984
80140870946428662762621576643870720 160286812495258238442849140100562944
320583766195318302720910253826768896 641187814800240257112244454904823808
1282416194419687817565336804312219648 2564913518477790241812369397629583360
5129989296232409696988130373269454848 10260303111018477820703043902559485952
20521255259144272494859654117160124416 41043808592503178696626440858402553856
82090213333435624807067146964969717760 164185618963729784441762824426268655616
328381622521176638538782709845195751424 656784014229787416388079541675708383232

1313609566834443111397187327322050527232 2627302210418622780036431142585368576000
5254770574336718674556975261053272195072 10509873455672383578082176473871614476288
21020411525342659614100804851273369124864 42042152278681104144074513509607018594304
84086963013353778119894834633334597877760 168179242938690695903281284494910317133824
336369119701347671133545799446302877024256 672759507050627900921058059805570239561728
1345561549397120919150049041437069450149888 2691208169385972072915963926525996842352640
5382586479955404615063659540355709568811008 10765513242277730168590782455318850905833472
21531707049289302214108491659852565348089856 43064775228046288182070836818134857769025536
86132272715027943871849380633129169683742720 172269989947926622759114175259977247658868736
344550868931594715549059178507392311900504064 689123515934672371159780012989660256966541312
1378290588012310622442883337929071780264148992 2756668288310553005132413299757646093190430720
5513510801192969530758119847314297251705126912 11027370051529666102502826190226604634058784768
22055437001346786286978825371649229529414631424
44112267799268480737903996725690499581423386624
88227323191686777803700685416165080208035020800
176460221569673188263186754761898322506446536704
352931593511945641837944277382932969193646063616
705885487769089814299030090484138586748798107648
1411815577028576689844343252404822470220608176128
2823720357037947502181252647682735533887240273920
5647619120037483249347637581111652254666528391168
11295595051998142988665539733715666883117152468992
22591903727842638957271608610416058513802496311296
45185234703377983874424275506801566522741375369216
90373323902141379668610667585542032035755516231680
180752356795053583176745568314961862052056563449856
361516131571648814032539602917679320065204188872704
723055099106380923423176138410869832052590501691392
1446155870138928437562546141972762047949545251274752
2892403084130190056557480014247568863587818998333440
5784988855965046475979735489099227262553094988234752
11570343087339425677689021899406633595861103959605248
23141416925497516806837145641229625333232035885481984
46284295352632364516592494967291966949483727703506944
92571513708539390839021397304249366465006767272099840
185148873423628105289715609347829598062092158274371584
370309438860354857802776848174320926388341564009086976
740642261746907010052244955305965313304997622938861568
1481331291546208608997872428526577547666624235719098368
2962756119197206395782509892882448937446506451120947200
5925699310603991147138549857423485559119528861607395328
11851772765627139005424159858164146486692089641945792512
23704293820092591433142440002962643710290243121353588736
47410084217861809710873120579193988894392613917631184896
94823161591076873110922722304925380736409483185110384640
189652309492860253600198406902925567368067477069916798976
379316591607133521957102738392000746526631975539225657344
758657128457093073427617325956300716634257993877235433472
1517362147399838205882058350257199880430504073352039104512
3034820075770980529817764097203596655184984317899214684160
6069831713484569295742822987785587099017920978188702318592
3035011637713588765925058890581990443832936660289487634432

1517553709342446441989353143635593669078456415742312054784
758800799914049250508088420990096058120222250669940146176
379412372578437640010750135081197640850608146734362132480
189712172599925327383728029833673626320552584066877095936
948590794553159173810404960633742161079005473832865628l6
474310363053345855351084886049558095277624013665672888 32
237162664415056061898483645891122555007872645207456481 28
118585073651719598060712424378733031188466641791038259 20
592944075479555825860915129059523924364984804891741388 8
296481391350256830709134068112691346393818200414145740 8
148245372480367874243905285847810365302571999191210393 6
741250246428036665666217688196375287041174490876739584
370636815227116980056454473576849373784744492898713600
185324253620107813639900051527755552024450870179528704
926650498133285686257864331335432085782647469548503O4
463339864083016152158114202516043205721520294099681 28
231677239549694730593648119682185184275858426712555 52
115842273528940692554119569053174382845478353187635 20
579229636415170099057075391326280867765137465094963 2
289623952592818367671776468693344910651441582125875 2
144816543489025842907507620861774693710157215852134 4
724105553408212509895635036884384660472968203206656
362064194685647902626865984729947926197575163576320
181037806333594775152957225508851761079333112774656
905217576621827994962407293263647795299393218805 76
452623060787841057080014229491518392601060436869 12
226318666632384058339412406175606443776212132167 68
113162901435423794069408848802726845625947022950 40
565832347773277794845557472588252346818939899084 8
282925094184718309672035350581435232754072341708 8
141467007241398860960645982434372146049337366937 6
707357336952192835426371447890133378608192815104
353689818848695683024756491804203013484849397760
176850484610647474168163629831669668832801193984
884280298984735534119745068806189154615888445 44
442154087458116848699335994227014982533885460 48
221084012711932965169399727025467693879913349 12
11054549084740375299456572846871394824644198400 5527448766942051170221579671234702477646364672
27638114957569573453574364595168537714853150 72 13819493040214445528020415417081781520737239 04
69099643008220521646268242682896470920239513 6 34550910407684407826217204140192017118396416 0
17275999655629277414650143469467899388336537 6 86382720537081754580958424344198951087374336
431927213980085610443330656705292026165329 92 215970412637381223990934595844794648446894 08
1079886091423598213801019316684716419055616 0 5399600598301451538236828270727297979383808
269988536974245600373427997901550693174476 8 134998522016709311917507291133368243689881 6
675013877731479118241502916579805703962624 337517572689705838447734688746385094737920
168764103256836058887358959601433668747264 843847100844095992754252874148373950627 84
421936842702005845535855475144789778759 68 210975067490981847347292256607696291102 72
105490856815480385963328387821498846412 80 527470899427349241265053236695747736371 2
263743757388648276356732267142000062033 92 131876032531810966040468957968063686246 4
659400931846488969512858911825635311616 329710850516961554411686516905476096000
164860617555339312033471788949067268096 824329049260989234305501597226982440 96
412177505372640954221822124854314270 72 206095243057393645645446725547568660 48

10305086671423340708999119433399009280 5152705594988499567862951294709792768
2576433927132664390613171436360040448 1288257528385539498647433612682592256
644149046602373400994140753592582144 322084664505988526332282350421934080
161047402855395176083747162023788544 80526236728898044500676574418305024
40264386015049250479739783912357888 20132826832824739354570640307781632
10066730329062426734635694329692160 5033523620856241895993034252746752
2516841038590635212334110670323712 1258460133376574738335852107137024
629249873728915935252324439556096 314634840384772250668361412771840
157322371952543266855280302882816 78663661856350204188189949689856
39333068868214387474369873969152 19667153404126836427322386546688
9833886187073239558729918054400 4917097836041530451899321417728
2458626289273220562216841904128 1229351830262837949242011549696 614695257944532808687801073664
307357300378823321377298186240 153683485892690119205347917824 76844160797984288861023371264
38423289324811759059686391808 19212249125315686844430548992 9606426794112747079508951040
4803364512783825368401313792 2401757814255638598524076032 1200916686059682256423747584
600477232495772606792728576 300248060980852042686791680 150128752856908890988609536
75066737611695880316911616 37534549397468657569759232 18767864994544687490531328
9384227645177523098091520 4692261396541351225458688 2346204485246970450935808
1173139136111632644571136 586588014799890031837184 293303230771981870694400
146656227072009362735104 73330419379013895061504 36666362611011554377728
18333757766258080612352 9167167113505192017920 4583727671940671864832 2291935893564373860352
1146003975579205894144 573020002188112429056 286519008293310955520 143264007746282848256
71634255672955109376 35818253736384397312 17909689818145619968 8955126384049520640
4477703929513115648 2238922333500735488 1119496351122456576 559765767747272704
279891679966658560 139950238029840384 69977318038175744 34989758530715648 17495429021171712
8747989388492800 4374132133199872 2187134786076672 1093601752776704 546818056257536
273417618063360 136713103998976 68358699483136 34180423483392 17090748612608 8545642741760
4272955588608 2136544903168 1068306006016 534169780224 267093278720 133550833664 66777513984
33389805568 16695427072 8347975680 4174118912 2087124992 1043595264 521814016 260915200 130461696
65232896 32617472 16309248 8154880 4077568 2038848 1019456 509744 254880 127444 63724 31863 15932
31865 63732 127468 254944 509904 1019840 2039744 4079616 8159488 16319488 32640000 65282048
130568192 261144576 522305536 1044643840 2089353216 4178837504 8357937152 16716398592 33433845760
66869788672 133743771648 267495931904 535008641024 1070050836480 2140168781824 4280471781376
8561211998208 17122960867328 34246995476480 68496138436608 136996571840512 274001733615616
548020647100416 1096075653939200 2192220027355136 4384577493663744 8769429865234432
17539409486282752 35079918484193280 70162035991642112 140328470029795328 280665736152612864
561349064491270144 1122733313354629120 2245536995453435904 4491214728395227136
8982710931767164928 17965984813487751168 35933095526882344960 71868442853578375168
143741389306784120832 287491785812822982656 575001586024155447296 1150039200845329858560
2300150459284697645056 4600445033757471145984 9201178297891094003712 18402933056534491430912
36807019034573589708800 73616343912156393111552 147237299510331213611008
294483822392699281997824 588986091529472273547264 1178009076547091966197760
2356091940070478770601984 4712331454093547217616896 9424958056092273788059648
18850506407994906281771008 37702193407610529974845440 75406747998462494772297728
150818218363407859189809152 301645881459781457670045696 603310652385494393920946176
1206659083702851745003601920 2413393725269429404330622976 4826938566266310637308084224
9654179363987524931909844992 19308963190884857178407043072 38619135307589328985988792320
77240688466817887230326996992 154486212636914232977352818688 308982096680385382988103286784
617983536173884600043001872384 1236005757973996868219594342400 2472088887200449072706369880064
4944332516905808817947102150656 9888974518821438980962929082368
19778568007662520652063307726848 39558373955364326684401514577920
79119223790807224129352827404288 158243399341771589779805251305472

3164967022038574626018096956047 36 6330132114483434912880177771970 56
126606603697794411474483232636928 0 253221130211840249382725819668889 6
506458106056183351632970348127846 4 1012947903377372409000978113835827 2
2025959189284756229472031062831923 2 4052045143629535281884211795984384 0
8104343817379116209648722932609843 2 16209194694998323711058044546501836 8
32419403510476830005637286455567974 4 64840835261914025178316967636264550 4
129685727005748780690718724722786304 0 259371456697533902204960702834608701 44
518775359878360965435553214493202841 6 1037583171612087773543784744588463308 8
2075231246934907232432926120381041868 8 4150592301291277835556565503170314240 0
8301444217425482412494557531157089484 8 16603407664536818307751968111947100979 2
33207853788445343581029642323160045977 6 66415784495634101093110696844851779993 6
132839722828755030048324218086766936064 0 265687753332483715820854084967660624281 6
531392122014914743090119467523574752870 4 1062817474729724109077061530223656514355 2
2125701410859237463947768250800327045939 2 4251535744518053419482826882306682126336 0
8503337334635263822140234526025420321587 2 17007206360468841610629630574874952781004 8
34015476103334311153957584195398129837670 4 68033078971461878173311814482092708226662 4
136070411472510268077416921146778313555968 0 272149330004193559616420426658742421317222 4
544315674126733166156014022047856431045017 6 1088665376490158426158374381556456038911180 8
2177398809453701040009441438034398431464652 8 4354933373185417045540426822591176957021388 0
8710139689601877661579307151509484554996940 8 17420823830990828824700155702390859939132211 2
34842736565555804652483394203525501536541081 6 69687650938259903311132954004538566389635481 6
139379657490816394634598239204052259412377600 0 278768026210225965293861140798054772090968473 6
557553474877638282637051606376010050714363494 4 1115141794669649269372761862311821114493580083 2
2230353279168043946942841023743244255116866355 2
4460845937993578710280316645725692562493145088 0
8921970635302139053349902487929793229505114931 2
17844498789234241372278343368816402668047879372 8
35690112615728409275713763523546437754171057766 4
71382455305976671613741680618920140344492713574 4
142769370760993049352111668381494810361286623232 0
285547661820065510953479951050298680067175638630 4
571113164236289846405473130675215478823556061593 6
1142262009664897341807972718499667195025521691852 8
2284595381714429981609998351297806864807862521036 8
4569333488198130559208102531192558679129363316736 0
9138952425934802310392416719579007257286003182796 8
18278475750946687004737256753545794312626559464243 2
36558093300047538777379360135867148221362225125785 6
73118470196403407090568413529285415634942662646169 6
146241507585423473252756213573673069654321750081536 0
292492149556080264648751200177550616077516349741465 6
585002567882627165583979946415510185692778398639718 4
1170041673306187603740914984951838278461048195593011 2
2340156421694241752627740154145312371073079187813171 2
4680458993552216595547300676773896370448123968880640 0
9361210287431899371678242090514335997500179124269875 2
18723005175518731104523765654961758508208220621556940 8
37447179552347326931382094257789690042832165989148262 4
74896697507314383307433314411311726138495781470365286 4
149798071819868225504204880614088144382654461924868096 0
299605497250215368787086264811105672976634721818011238 4
599229701721388573131525536788070114375921039572572569 6

11984968178846928173777570879078577655971452710182453248
23970684646532169769849262044791506048848969257826910208
47942865870740966384286764662851713571510066190577827840
95888724896835186457750010472240830090644387731003670528
191783436104376880293852983237556466076537286161703370752
383578844830166775344411891061262543943571593722798800896
767181634903159580202235631294824311468137230244381720576
1534411160291971219431294960934247070098262546086331678720
3068918101555246556916237318557691034520501263367799832576
6138027765053101349939769430493775857688954869125872615424
12276438653991419172094128447744339292673814423032291131392
6138410888938317822154359017250563434984859553906418515968
3069301225440463029130826905314478611816405948148345733120
1534698503205883573592237151001837753070191059671741104128
767373196845767816309530424673218100116089572634654670848
383698571044296922911471136922758661848541807716719394816
191855271832854968834088530754454136819519414558055727104
95930629071780738106220746523764471357383962628875878400
47966811113566995897698613835150937152504108989361946624
23984153845621811371143427204209819313158118332142977024
11992451067230062396718773745422085025032091084802490368
5996412605824609553932916944369630196742561501766746112
2998299839017093954753223508014108940484538730566123520
1499196687560941566269994271921701391298898355124436992
749621727806668077581688394918174156177763672482906112
374822555916432686014189826937748808353039083701796864
187417123964765666618767728208205269308598165581070336
93711484985657495115220271473768067220338394655621120
46857203994466078460528339421716749893188853260353536
23429332748051704681723271553274733088104254596448256
11715031749435185066591186697845545614807041281359872
5857698562422258896160368809526862342780977632247808
2928940625063462629512572135065475939079217311907840
1464515984457897905472479932683760353383972903845888
732280828192032248094336898917391368614168575868928
366151832077557771726216915746451280268175349907456
183081625029549709702632691017103438114633205940224
91543667010160266771078462080490618047589368463360
45773260752772839345420289326214758518931066978304
22887344000232772652650673806092104007033724862464
11444028812039562816295601474538414377300958117888
5722192811981369653132933023015388375542526902272
2861185608971478949059032654380784781217287372800
14306374059761365357775799398626937687331655647232
715341003733266798511041235031741492027333804032
357681652239232664567091385375007070194419892224
178846401305915964939331076617071697187586441216
89425988246107798797558230273319929638981468160
44714387919628807562725461119052005342084857856
22357890858101857863335903550722022932339490816
11179293878194655972654538270959021596818276352 5589821163669191506820562383278515863733403648
2794997694120527513656927815538760464528834560 1397542403203229636951787219719131498595483648

698792979673097758537555265834441382463275008 349407378872290349299608460904658507814404096
174709133954015909665219644446048162198585344 87357289235943322340317529219883535244984320
43680005747439344924012618108371494695337984 21840683438453514338933235803400610884091904
10920682001593678107930081276307737210257408 5460511141980299523196772325457584489234432
2730340641581879996214252006380650186670080 1365212856086805115415058925016254064361472
682627695691335116361495923421091517693952 341324481669633837507731192167028001603584
170667557746800058417357211311755122180096 85336437329391599040424413269998121779200
42669547892691584436085110442059341234176 21335438560343684675979007124559810789376
10668051587170788566957729514044975480832 5334191947084867397962977732905022783488
2667179050292170256223545354393778913280 1333631063520953406732800921167523217408
6668376300947910842676914582569078489088 333428535067672490993714352277197684736
166719459830694780324485706634928062464 83362326063776657576057118565628641280
41682461106102962494935691906896625664 20841879590158798100921412265489465344
10421264313633057477187489288765308928 5210794416093357951957136222392942592
2605478337685093582660263900201615360 1302779733661754094670979844603379712
651410149240480699005913869552975872 325715215825042175338168908402130944
162862678514922000586690441013886976 81433874558661456752148213913354240
40718204929930956605475603659882496 20359736290265592417438550181543936
10180185057782853266069649266573312 5090250985216455161710011721187328
2545204720770741845192599404544000 1272641974466628054765096474247168
636340794273942593466946623111168 318180300657285579775672504549376
159095102088799931408935848771584 79550026924478536465017722634240
39776251402278553612783760441344 19888744671158919496529329782784
9944681820589281093333389672448 4972495652799551219201057226752
2486325197652230945867709808640 1243201284452343141067445501952 621619985039285404600518049792
310819663926199619333656674304 155414667666378268183527161856 77709751684828363350112993280
38856084768233796304231202816 19428646847026705466702954496 9714625654968256390645153792
4857463943211580023969415168 2428807529469515926308126720 1214441543666620920315772928
607239661299241938738741248 303629275382586708659798016 151819360057776223975112704
75912041212129546810163200 37957201197685490816385024 18979190894653104113844224
9489890595231731409747968 4745092871568455381286912 2372620222760522528849920
1186347004868408683528192 593191949178278051315712 296605197961175880433664
148307210666606367604736 74155911176312397496320 37079108509660805595136
18540130715582706221056 9270353588167504822272 4635320909271828267008 2317732512229952061440
1158902284911994994688 579469156854506979328 289743585626508230656 144876296412881485824
72440400006254428160 36221325903034056704 18111225901470449664 9055894425711935488
4528087950344323072 2264114343916339200 1132092356330258432 566063770351173632
283040681268609024 141524738680815616 70764568363663360 35383383693459456 17692241602543616
8846395679178752 4423335278542848 2211736358748160 1105902539112448 552968449425408
276492814647296 138250702290944 69127498629120 34564823056384 17282948399104 8641742635008
4321005535232 2160569876480 1080318492672 540176023552 270096400384 135052394496 67528294400
33765195776 16883122176 8441823232 4221042688 2110586880 1055326208 527679488 263847936 131928064
65966080 32984064 16492544 8246528 4123392 2061760 1030912 515472 257744 128876 64440 32221 16111
32223 64448 128900 257808 515632 1031296 2062656 4125440 8251136 16502784 33006592 66015232
132034560 264077312 528171008 1056374784 2112815104 4225761280 8451784704 16904093696 33809235968
67620569088 135245332480 270499053568 541014884352 1082063323136 2164193755136 4328521728000
8657311891456 17315160653824 34631395049472 69264937582592 138534170132480 277076930199552
554171040268288 1108376440274944 2216821600026624 4433780639006720 8867836155920384
17736222067654656 35473543646937088 70949286317129728 141902970680770560 283814737454563328
567647067095171072 1135329318562430976 2270729005869039616 4541598749226434560
9083478973429579776 18167520896812580864 36336167693532004352 72674587186877693952
145353677973382758400 290716363146020257792 581450740690549997568 1162937510178118959104

2325947077950275846144 4652038271088627548160 9304364772553406808064 18609306005859117039616
37219764933222840926208 74441835709454895546368 148888283104928218480640
297785789581893291737088 595590025907860293025792 1191216945303868005154816
2382507677584030848516096 4765162929120651373445120 9530621006146482099716096
19061832308103322905083904 38124845207827363221471232 76252051598896161265549312
152508825564275192176312320 305027095861516123643052032 610073081188963725866958848
1220183941309790408895627264 2440443440483306732114673664 4881037996694065292876185600
9762378224843034243046047744 19525360912595875800679448576 39051930751011366230533603328
78106279353661961719416619008 156217394410602381955532062720 312444460227761680944461774848
624908263268637195955718848512 1249855212163502060045028294656 2499787795579459456357237784576
4999730333663829585248837959680 9999770152337480515566400700416
20000159274694603721270250962944 40001556489428492822815401050112
80005588858935556406180600348672 160016129478028254333460797194240
320042162476370791709120787382272 640104131993370149502639960752128
1280247878067997431174076693479424 2560574984298509126685746930909184
5121308424922046782046680949719040 10242933762494150621443736075239424
20486501350288415357588220502081536 40974270351177058944577937707368448
81951076003554574347958868821147648 163907222609510061613523724455116800
327824586423821949062259422535876608 655669455257247549794942792323039232
1311379475333702402930733479148650496 2622840080305819412543162747302445056
5245842419888468038449717072615178240 10492009358330594503626217301250932736
20984667753768505860706000914543017984 41970633581751645428319134453168340992
83943863311932558270452534154501292032 167892918920723651368533598805331804160
335796222435164372392324258603322048512 671613214057762884095162639191960977408
1343267966490394046811353522354555715584 2686619009730524650864763532650378952704
5373404172960522416213640041183292948480 10747140652919991061395506034131655983104
21494945919837874580727463971793452138496 42991221067671534077327831750647184621568
85985100591334637986401471115414929932288 171975518094652415636294557459070981242880
343961670013271110599572345374624205242368 687944607674474779853111151662212895997952
1375931750644814677014155225150354763022336 2751948571881359588644176293952567468097536
5504067284946179646520084275208850820300800 11008474852259280231503631925025133408813056
22017630269252402339934190599265130354049024 44036621667972488433722234696959987780943872
88075965594880344375152176390779429707579392 176157375707631423765719766775277767706542080
352325640451004317562270361537993351995850752 704673058973491575186202379050862337157234688
1409389674089949030495728070051475940645535744 2818866460465829821238102764002454413953204224
5637907145503523162969498775803913893230673920 11276162740150773366925584047205837917109878784
22553022378589000815824341085607696095516819456
45107438553752909795595028153607432713627762688
90217664700655635919082748271998946472443772928
180440904587610904493950880473566055035264040960
360892959547821074299472528806268434251281072128
721808219840840679222086593330809516864068124672
1443661041172078419690456258098164330451148210176
2887411285324950961873478659069419254348320342016
5775000976611490168732089603885019695588688527360
11550358765146156827434443779262401764961472741376
23101431154138666634809416701509528277491136856064
46204289555970039229499891688988506050118656458752
92411433607325490378761899949915911090510078410752
184828576205421804597048033043709620161565687808000
369668570392385256873144532375174836284222437588992
739359976747853809104385997325860864490626999123968

14787656254218742089249658598027441128256182461399O4
2957622594696081599282319449907532993339964988063744
5915427877096829561429414360419155522057386967695360
11831221129602991848588379642046490114869687918526464
23663173010024649148635861126509338371249203803324416
47327807521686629200189925937851393025518063539191808
94658538046647920206216259245368218617075438943469568
189322922099845164024105333230067302366229501617111040
378657536212788975271556295938796334996616250694565888
757338456451775244989803850834916130521546996309819392
1514723680955945078872990219584479182099722982461014016
3029540898016679335532745474998252206312703944604778496
6059268868242937026639021021655092096851923848575057920
12118911880905030764425102186627359562156879615881117696
24238572050648374951144324659889069861219823069224239104
48478640678973376746876889893046841196251773813372485632
96960274513300007182930260932631085340127802976592986112
193926535337306521744213484158336976575504116652882001920
387865043296026058245132892902823564941505254705156063232
775754031834878146003677634977946353464004552209096245248
1551555954155408351034178968300491154089997190015760728064
3103207689282120820122005333290179202503970551226657931264
6206606940506849876351305459958752193655893444843588812800
12413597004898916224917200506674291964607691574467723526144
24827960257568265394263580186862159083807192518496538853376
12414363252669349169346379680187867119199500944028815327232
6207373188277282820780484633472327348247702814404680613888
3103782375109945528443889713425360568447827578397476782080
1551939078040624823248768555057278731385901874796306628608
775993484263138441137796126700938589273944980196937433088
388008714752982235325603987936618906427469511497860775936
194010343687197625041154956261384259108983266448626417664
97008164998952066209753959277229532502115888574161223680
48505579077152659949465220211883467724870071962004619264
24253537827414643397026730392576084599341099818464313344
12127143058126478409660425339605217668123581827963158528
6063758601272817560403742741461196518288306873347080192
3031972836741197957988636406559892101257411416356290560
1516033186422993567887700721194592971685334698019520512
758039977237694078390541619554619946370981843930447872
379031680631945686418616439255971703449648169425567744
189521686322522166820981034367316716856902708442955776
94763766164535745216326924553323790994490666086563840
47383344583905203511081665961494611780264988975824896
23692403042771267206999348231636640316423222454183936
11846566896794966329222951833279001108657607521022771 2
5923466136102149527479534626999095078665494596681728
2961824411903407945172155043801592307021475794124800
1480957877877870563302271387051818537355102144954368
740501774902018577009232626101420460599733196423168
370262305432550936183664779338465826260957660184576
185136861707046291931356622813110711111024361078784

9257128534890855788544042797849425454578494603264O
46287069922146984902601272275216576768028855762944
23144248584919845431241165280593013131582619254784
11572481104383099205590847211788868939575405314048
5786418958153137847780555891640615656679750500352
2893298682057363046382844088693398421785899171840
1446693942519078584437705115783244507615961546752
723369272004737822841994093609894902169486753792
361695786374968176732567814664083775265496367104
180853468373783721022069291261610049723124678656
904295217800416768389273375955589105906750586880
45216154686595746583410014780186593475969417216
22608774241585327373678180381289316999281770496
11304735569936390727825676686242668630289416192 5652542009540058884406131590920339380468973568
2826358117055961202449712419359672222896619520 1413222614670946481348179521629587377779376128
706633085406956180735751416789669322055221248 353327431739219560398706536382272477610377216
176669160387480515214768682184855147096571904 88337302452675625115092048089287027693977600
44170012355805496311399877543073240919834624 22085686742636590032626865520751483996340224
11043183653685215954776896134983173766381568 5521761968026068446620179754795302767296512
2760966054604764457925955721049509325701120 1380525562598247346270910782350683633876992
690284048947056231789421852088306302451712 345152658297494395221694156500635393982464
172581646060730337274338693478558818369536 86293481486356738468915154353399969873920
43148069971174154150330480983760265281536 21574699599584969533101692395410272813056
10787682106791430995519072149470206492672 5394007206895188612243649050617638289408
2697086680197330863363881013250086666240 1348584878473533710302968750595677093888
674313208424200994461998497283155427328 337166988805817566886256309634236153856
168588686699767318270756685313447297024 84296939498312926549192607904888258560
42149767823371096981503436576526434304 21075532948792865344205284600304369664
10538090992950091098829425456172761088 5269207755751874762778104306096668672
2634685007514351988070747942053478400 1317383068576383297376221865529311232
658711816697795300358534880015941632 329366049553699476014479413633613824
164688095379250650924845693629628416 82346582990825781921225840221224960
41174559146013119190014416813817856 20587913398306673709707956758511616
10294273611803393912204352555057152 5147295262226725484777363365429248
2573726859275877006726275226664960 1286903043719195635531934385307648
643471328900226383850365578641408 321745567970427474967381982314496
160877735745370879004790587654144 80441343752764010262945092075520
40221909816421290511747445161984 20111573878230287946011172143104
10056096424124965318074310852608 5028202954567393331571517816832
2514178848536152002052940103680 1257128109894303669160060649472 628583397760265668646825623552
314301370286689751356810461184 157155520846623334195104055296 78580178274950896355901440000
39291298063295062807125426176 19646253494557338718150066176 9823428978733573016368709632
4911865605094238336831193088 2456008360410845082739015680 1228041959137285498531217408
614039869034574227846463488 307029379250252853213659136 153519411991609296252043264
76762067179046082948628480 38382214181143758885617664 19191697386382238148460544
9596143841096298427056128 4798219494500738889940992 2399183534226664283176960
1199628660601479560691712 599832777044813489897472 299925611894443599724544
149967417633240227250176 74986014659629327319040 37494160251319270506496
18747656586411938676736 9374116523582121050112 4687202376979136380928 2343673246083606118400
1171872651838822023168 585954340317920493568 292986177358214987776 146497592278734864384
73251047939181117440 36626649869497401344 18313887884702121984 9157225417327771648
4578753446152241152 2289447091820298240 1144758730282237952 572396957327163392

286207274756603904 143108035424813056 71556216735662080 35779207879458816 17890153695543296
8945351725678592 4472813301792768 2236475370373120 1118272044924928 559153202331648
279585191100416 139796890517504 69900592742400 34951370113024 17476221927424 8738379399168
4369323917312 2184729067520 1092398088192 546215821312 273116299264 136562343936 68283269120
34142683136 17071865856 8536195072 4268228608 2134179840 1067122688 533577728 266797056 133402624
66703360 33352704 16676864 8338688 4169472 2084800 1042432 521232 260624 130316 65160 32581 16291
32583 65168 130340 260688 521392 1042816 2085696 4171520 8343296 16687104 33375232 66752512
133509120 267026432 534069248 1068171264 2136408064 4272947200 8546156544 17092837376 34186723328
68375543808 136755281920 273518952448 547054682112 1094142918656 2188352946176 4376840110080
8753948655616 17508434182144 35017942106112 70038031695872 140080358359040 280169306652672
560355793174528 1120745946087424 2241560611651584 4483258662256640 8966792202420224
17934134160654336 35869367832936448 71740934689128448 143486267424768000 286981330942558208
573980254071160832 1147995692514410496 2296061753772998656 4592264245034352640
9184809965045415936 18370182880044253184 36741491659995348992 73485235119804383232
146974973839236136960 293958954877727014912 587935924153963511808 1175907877104945987584
2351887811803929903104 4703919738795935662080 9408127707968023035904 18816831876688349495296
37634816674881305837568 75271939192771825369088 150548490071562078126080
301106203515161011027968 602230853774395731607552 1204498601036938882318336
2409070989050172602843136 4818289552052934882099200 9636874252011049117024256
19274338799832456939700224 38549858191285631290703872 77102077565812697404014592
154208877498108264453242880 308427199729182268196913152 616873288924296014974681088
1233784356780454987111071744 2467644271424635888545562624 4935439658576723605737963520
9871181548608350868769603584 19742967560126509052126560256 39487144046072632733427826688
78976705943784494725205065728 157958247590847447967108956160 315926166588251812967615561728
631871675989617460002026422272 1263782037605462588137643442176 2527641446463380512542468079616
5055437635431671697619298549760 10111184755873164740307321880576
20222988481765972170752093323264 40447214903571229721779085770752
80896905687221030204107969789952 161798763134599201929315536076800
323607429789512686900830265147392 647234666619653939886058916282368
1294508947320565011940914604539904 2589097122803644288219422753030144
5178352701932317105114032593960960 10357022316514691267578439363723264
20714678458329496649857627079049216 41430624567259221529116750861303808
82863784435718899517036495129018368 165732639473838711951678977070858240
331475420152479249738569927767359488 662971122714562151147563802786004992
1325982810248331605635975500074582016 2652046750135077817953646789154308096
5304255759546984849270685156318904320 10608836037647628125268153468658384896
21218321112402573103989873249357922304 42437940299019779914886879122798149632
84878476746468827243588023493760909312 169762145789796189314804577483851038720
339534676173309448284866215960360517632 679090121534053035880246553906037915648
1358221781442974350381521351782709592064 2716526639635685258005099191506686705664
5433219432770843630494311358895908454400 10866771172540633489956848669556886994944
21734206959079159437850149242643914162176 43469743146154103791573202292348108668928
86942144748299777414892212198816778027008 173889606408582694493276039625874677432320
347789846641131668313535309708231597621248 695600960930195895281037080329427680755712
1391244457156256907870007082484784332537856 2782573984904244050355880008621426607128576
5565318110991948569943491704546569098362880 11130976504350818078350446783700569964937216
22262633573435478033627820316616003466297344 44526628276338639821109494131661734005440512
89055978811612647149926695260182922156572672 178117402141096029315268804514084752604528640
356245693317933528661368437015607321791823872 712513164707349997384398530006090276749180928
1425069885557665874892120371961931819829428224 2850226883401263510030887367823366172320989184
5700627991374390540555067983445737409966243840 11401604431892508122096722462489484950581018624
22803905762072470326166617916174990162459099136

4560920532071984881627958181474202084751232204S
91221198234589513960451855594268122740212891648
182447971655478660576689095118104407570802278400
364907093683556586464948958095345139322357547008
729836488112311703553039451908962927006221074432
1459717577715020468352361975254471150735454109696
2919524358410835059197290093382032894916932141056
5839227122783258363379712472510246976725912125440
11678811057489693216729689516512856327235919937536
23358335738825739413399908176010437402040031248384
46718098725344184786680874637990324299216445243392
93439051946073781493123865847919547588705655980032
186883812882918386825771964839716893157956842946560
373779043747378421330592395967189382277004747866112
747580923457840138019281724509889956476191619678208
1495207518841846866754757314170802296796747487248384
2990506381536026914941902358643649361282223470280704
5981195450776720192748580177891388257941903932129280
11962756276962773111226711276990955586638721847394304
23926243304744211673912524396398269314787271661060096
47853948111125754250743252477629254912594199254663168
95710819225526170307322912324923942391227710374412288
191427484457601664226318639389178749914534044478996480
382866660928301975675982908257019230093225336418336768
765756705882801245798657075471361920714765167757361152
1531560179817997080490696668857370762486159325356097536
3063213895740783338768158373544035367085576630394945536
6126614863691145033109846818746658418397669220155392000
12253603871801446777366753780810492205248370359041785856
24507956032441206977027627848255335147402804555545575424
49017408642559040798643496269779371768617736786015158272
98037810440471335286463473686096146484859728921878331392
196081607191649177951279909665267098864967968543452692480
392175187004711370659265743916683809520432958486297444352
784374319252248770831943337005666842621859959771379007488
1568796528990149600690710372355932132405708005140326252544
3137688838951603319435068141401061159135392181475788980224
6275569239845814874977431076180516106918736705341850910720
12551521603576846222169451739117819791133378095464247721984
25103809454924125388768082651749214736858565560489587245056
50209151405389116666394523650525579782900749860101358092288
25105341950464991277626440998776365046042184299611770847232
12553054099117712111027810086144970100316996834586431324160
6276718611501464291621199836450878838806450759683488612352
3138455086722036263864247314914636313727201551036880781312
1569275433846670190958947355801916604025588861116008628224
784661662166161124992885527073257525593788473356788432896
392342803704493577253148688122778374587391258077786275840
196177388162953296004927306354463993188944139738589167616
98091687236829901691640134323769399542096325219142598656
49047340196091577690408307735153401244860290284495306752
24524418386884102267498274154211051359336208979709657088

1226258333786120784489619722042270104812113640858583040061314787411401822780216286818699382082870841636584161283065832906674880316797579376764262946256800061511958528153296322138983474729217220629677839418502902059735449676650499472111466809277736210571265762082900521936486438326418937365598126973431053151805907457175007002624019163794069337731424653997000508989466936449876518502495821893349963318929106392372210379900721561247678464479124081766189903674713998709379062333804365563822082395693483912816063519480177788531125820004624446259211978832794973413043326951810150834699854937105367040598959908519137288452825136567950688530492554425139229948908864480196236965134131417982103411912679096321497491115150176402564450571721921489014959881846784748768393538171496640322218436471936429662064869376374395614750627395999209575505991564175922094407680187203516366084521839129020896873580068506578190336936046126784276728393266270203756890245260545884164680373358690654237954437179615729400739941004083223402580417299624169712715041063371751267896393728117016470205729885748266220920240482494180438835205851001916248082532398443331758205311601069785088292559016110483538869178780875219324924655881420814628396820428147555921769758126419213462913679367314421417666059084192300236245936090346516643843657322212559022195211857796714331286980788224001828716858142507424163782737652847264394159063049143863050027518753608182884742644426489620070445720709046712501931987260406105262655042224128228610514216437050479668031942486515888181739521143087415996557956496998809272233592505761792057156113045546533029782872941601730278530744322857892764563258411735790270979589046588669952142898993842459508599121844743954578962540134471451674728378048305727087969464852797823385635726926267763171155946626783476208057188352017864007585668659079514854791109994857732505689322760187278662905281980952409428434354176446627412231070152064948439746344412900229122233205117628734948017434873653208418143436811166365870510595678550637742873473858928640558335307643875830850705055874045281357004827917616088111093888693911230220843488378881395923339701419811742628483336971145445376697982937498642464525280702581450058235904349002102573287511589623581747207271874560174506368198626895458303406101844757315584872558425553050175608975106650429393469444362925050564829369632165913958175001804821815289866822039306097281473321015181312109079772404099658820168666884255776768005454154773704456055492546320095323881472272716046360196458498832964798892946227213636217701758505711151930679650984919046818316542753594248681106559678661263363409262117313967820893123889765915033601704682981625569258722847249846249717768523674522970773034995662774047709593642619670689068498881885446494320852992213104843816415662943962895592015790081065556670937444157392492793562136576053279456139640500003258555458209710082664053936620439606844623561915629568133206753312942710676315967546038681666605404897431720505200378498147942433303716569196042836121386611638272016652365344838112709821291987101286483264362025391020007909453341917184416334486632957382333562233741639682081735815694798323137886003868467210408995991124048673039804195143680520465645188705286519508918547251226024074541060406969351381366865921301243341134277480636365840318464

650641477607767306402581306146816 325330642324197936243489846067200
162670272922256109642844519530496 81337612341206625581972058013696
40670044110642598171260928131072 20335641025340941775767913627648
10168129997680292232952681594880 5084219741345056789010703187968
2542187241924983730772532789248 1271132306588719533519856992256 6355585496107473600826723794944
317802419460293717446759546880 158906045433425317240078598144 79455440568351887878388711424
39728929209995558568369061888 19865069067907586598771884032 9932836765408696956679618560
4966569498431800306986647552 2483360307079626067816742912 1241717932471675991070081024
620877855701769474115895296 310448372583850476348375040 155228908658408107819401216
77616815512445488732307456 38809588347843461777457152 19405384469732089594380288
9702987382771224150016000 4851641265338201751420928 2425894419645395713916928
1212984103310845276061696 606510498399496347582464 303264472571785028567040
151636847971910941671424 75820729828964684529664 37911517835986949111808
18956335378745777979392 9478455919749040701440 4739372075062596206592 2369758095125336031232
1184915076359686979584 592475552578352971776 296246783488431226880 148127895343842983936
74066199471735177216 37034225635774431232 18517675767840636928 9259119358897029120
4629700416936869888 2314920577212612608 1157495472978395136 578765328675241984
289391460430643200 144700128261832704 72352263154171904 36177231088713728 18089165300170752
9044857527992320 4522566202949632 2261351820951552 1130710270214144 565372314976256
282694747422720 141351668678656 70677981822976 35340064653312 17670569197568 8835553034240
4417910734848 2209022476288 1104544792576 552289173504 276152975360 138080681984 69042438144
34522267648 17261658112 8631091200 4315676672 2157903872 1078984704 539508736 269762560 134885376
67444736 33723392 16862208 8431360 4215808 2107968 1054016 527024 263520 131764 65884 32943 16472
32945 65892 131788 263584 527184 1054400 2108864 4217856 8435968 16872448 33745920 67493888
134991872 269991936 540000256 1080033280 2160132096 4320395264 8641052672 17282629632 34566307840
69134712832 138273619968 276555628544 553128034304 1106289623040 2212646354944 4425426927616
8851122290688 17702781452288 35406636646400 70815420776448 141635136520192 283278862974976
566574905819136 1133184171376640 2266437062230016 4533011563413504 9066298004733952
18133145765281792 36267391042191360 72536981107638272 145078360261787648 290165516616597504
580348625419239424 1160732435210567680 2321535239165313024 4643211215818981376
9286703906614673408 18573970763182768128 37149067426272378880 74300386652358443008
148605276904344256512 297219561007943254016 594457136414395990016 1188950301625810944000
2377972660845659815936 4756089436879395487744 9512467104134942687232 19025510669022188797952
38052174259548984442880 76106654362107182579712 152217920410232792547328
304445064192502439870464 608908575129078589292544 1217854043746304597688320
2435781874468904033583104 4871711322890397743579136 9743717793685974839984128
19488025883182308385619968 38977232357985334182543360 77956825899212103187693568
155918374164907076020600832 311846193062779891331629056 623711275591491261244112896
1247460330114845479649935360 2494996218093416873623289856 4990143551914285575893417984
9980589335283474809080512512 19961783133476756932748378112 39924775192773128494671462400
79851968237185486247692337152 159708772177649431012083499008 319427215761855779057564647424
638873774336825392181924593664 1277786234299878452497439784960 2555649839852212241262060765184
5111454422209335155058483920896 10223218329428491655185692622848
20447055628876626000508834807808 40895349197792537381292568739840
81793174275663645523134935728128 163591300311484432567369467953152
327192504143283148176938128900096 654404815327194862438274643787776
1308849244735646857045346059550720 26177777717633807978428285663051776
5235713891592644485531758414004224 10471744695835346028413891003809792
20944123216970806171528530359222272 41889514084541840572458557421649920
83781563470284137603720108249710592 167568197542969188125046203312242688
335146536290740202085304380250128384 670313354991084055841032707751542784
1340667274801375415022913310005657600 2681415679241165436727522409016459264

5362993617759160086818436396043206656 10726311754071978600363655948106989568
21453272545251274054180878208255131648 42907843164717181815268889040592568320
85818282477863631044352043329349746688 171641757252585796916332617155028713472
3432938999098888663487922295302715867136 686608567385211466286358712590748614656
1373258673145291211193745669152130990080 2746600423040318979629547826245529501696
5493366999580111073743208628373594046464 10987066306159168376454643208512258179072
21974797226316229210845738320554656530432 43950923680628243337564380448169593405440
87904505817248056506874568510459747500032 175814328546479252677240752249160616378368
351639290916924784681464734954803475513344 703299849481782128016895930822571436539904
1406642234259429373341724783471071844106240 2813369539110588981299315410594001630265344
5626909219404638431830362508491719144636416 11254158721176197802124188391590870057484288
22508998007086237481175303532396603651391488 45019357143640158716204460563222934375628800
90041436546215684940116628123305322896949248 180088317610302104895648670240329554085281792
360187524256345679822128168468096924753330176 720396826584174299705917992911069482672193536
1440837209311314479535159297771890231675453440 2881761530908560719316965219443282996013039616
5763697286388984959127223686685571057350344704 11527743021921696959241033868969152245349220352
23056182942130848000455240729134324751995502592
46113759680836604164856827440660690026585128960
92230306954823024657606346846105461098358505472
184466189095945681970998077621779084287093506048
368943528564490629253566923102694492754940002304
737909357874179789130275381923661633871385985024
1475863317238756639506833835283868564465783930880
2951815837458307401506233813440827722377591783424
5903810080878203047997599912627836631647231410176
11807976973679582585965464396748035637078558507008
23616667571205518151871457936480796021725308387328
47234762390103742263623974158931041538586999521280
94472379275592896447010064889800982067446764535808
188950467541956616733544362923479762115439060058112
377912353065454881146137192134715120191969182089216
755847542093993057650371316844941432306120488124416
1511740756114152706016936498840905248456605224140800
3023572856080638593466260727983855264601938944065536
6047328399865943549797296916571800064581334879698944
12095022175141219825324144754351779199917583742533632
24190775101101105102107391351119916541344995451338752
48383011703839541107132986387072549365709646835220480
96768946410953744020102380143810531297458605535526912
193543738828456811651877575026951927726995834801225728
387099169670012270527100779532565585718148917062795264
774221723366221835500892818022454631964612329046278144
1548490214784838259895168153959556184985853647933931520
3097073965674465697577101343748406212084965275550613504
6194335003558509750727732759155400108396446510466727936
12389044151536176212602525661627975585245924939664457728
24778836591910665847499171609890301907397913716790919168
49559169761497958539586583793049305288607955108505845760
99121332678349170768349648732636013524840165566859706368
198248651667404848915052259758346832944928841833415442432
396509275956222712586810444102843277680354705066222944256
793042497155271454687032737377985778941703452931230007296

1586132884796194968400889173100570005045394991460028252160
3172361550563694054855425742890336904414766154115192979456
6344914663069996345818146279159067597477484650620658909184
12690212450025209163850882145074922772250873986021863718912
25381191147820851272130943463663420699093557341604819238912
50763914791182568433120245274353991707370733422331822080000
101530894573446868643957207242762284033108704322908011364352
50766979782264300210836961968408292325737970900576189284352
25384256138902583049847660157717721317460794819849186443264
12692511193336507997138419665615648236026302094705139122176
6346447158610862234676504626186217906661103389742842511360
3173319360276735235391899709782305847654527866066557730816
1586707570624019676722773553235751370989252018630847102976
793377730554835867874798625790174909075620052114207670272
396700837898830948694105237481237066328307047456495894528
198356405260121981725405581033693339059402034427943976960
99181195785414244551879271663384072477325272563820003328
49592094470383749120527876404960737712474763956834009088
24796795524030187982558058489114719593143445815879008256
12398771906434250702426089387874535165024754826670505984
6199573025426703706786574765595855266738893372700753920
3099880048818141031180052418627221475482704666033127424
1549986792461465104483408727228257658797981322857938944
775016780256929846688395622571452289927305156349657088
387520082141563570567543440764387875227809825635172352
193765887077331108895444535121524802745983536547758080
96885866541940216253558674930427833939031080138964992
48444394772607439029697541150046633252535196002025472
242222928137122384966307872417439674767777425967284224
121111829443970525208883487129928016454643626966777856
6056097409689928967306519025568097762699270474956800
3028140048697297665085647243086093649038363733262336
1514115696274815423259017486694069208363546114523136
757080684100491006987605675922545796103955181207552
378551760031787151172851304249028494013068652576768
189281589006664399425949885268392044987079857274880
94643648998717611632737059206134921483812694130688
47323251747051511776249587889036910237042729811968
23662339497372108868065323087503179866089556279296
11831526560609230924002926115243952306828873826304
5915941686266203706986595343368157340306484756480
2958060046113895975985863814557169263599266299904
1479074624547345049239214978715129928522645110784
739559613018871055242749025075837612622828535808
369790956882034792932945280397055130492167258112
184901053627317029122258024128095727336460124160
92453314406808330889021704028831944713418309632
46228050999979073608457197996808012879303278592
23114722398276990886201771989600026700948701184
11557709648282222484087472490398023481122881536 5779029048712974762537029492998016805885706240
2889601636642419141515161370398510935604985856 1444844374464175450880903997149006734133559296
722443965303570665502113654549379000232312832 361232871687526802781887655262127316698923008

180621880361634136406359241624782566640844800 903136624397524357108873278092507374661140 48
451581923493439016092975174030550958059028 48 225797767394057926815756854507424114393743 36
112902286520698172792513060999786374878986 24 564528446721836910885738473729303462805504 0
282272730420091478904455821229837525608038 4 141140618739632251183021202797511659906662 4
705724361346093814569072474900522785046528 352872814497013186611519467906743635279872
176441724160489732969251349181612939018240 882235205362364363163714822049270301982 72
441130894961140030740586449095237954437 12 220572093620548939949657743582920378941 44
110289369880263932264511131309110890332 16 551463464751266972770966954133807955968 0
275740040050607142109689125861030730137 6 137874173862790398916947387327578741145 6
689391638501386133895251058623210586112 344706203844410136602882590304263733248
172358294219063603129069825648461086720 861817432579610689783491780723951534 08
430921697031951681960817216602798817 28 215467338887049009514944271421810933 76
107736914629061089024739967271111229 44 538700799072988366460038994156584960 0
269358512500335643898189075978806886 4 134683312732088552283179327439660646 4
673436846070046413086320584449589248 336728564239825032378372265850437632
168369352722313429106792119738040320 841872116623571710121990532754309 12
420948734817788137355010233409208 32 210480705661895209824512600220631 04
105243521957448175485760041868328 96 526233455419743730296318918131712 0
263124650526123291581918813460889 6 131566286671187359007839083927961 6
657851240396565361123593805627392 328935523718596963603996095807488
164472713619455623323097644400640 822388326898063824220986204487 68
411206542849424765913242093486 08 205609461124908809857995542364 16
102807825412552618379685018992 64 514054601313254159151861334016 0
257035037781872613202648786534 4 128521387453559073414683453030 4 642626280080909201140212563968
321322811447011517603503931392 160666241426784217318450790400 803355385650313379175748075 52
401689782083352835879621099 52 200850935670774491085684080 64 100428490149936282115778805 76
502157562322426593443577856 0 251086336947585888154130841 6 125546946366979239793236377 6
627753621300827677547036672 313886255383379578063945728 156947850058172658677186560
784762862123277641612001 28 392393236977845994919034 88 196202521447026584516034 56
981042122025650857862758 4 490535818408084396572672 0 245275287901671682106982 4
122641333299650582963814 4 613225113242326624370688 306621779993200166961152
153315501682618510868480 766600566843184691281 92 383311812636638414110 72
191661670925842241290 24 958337177666826377625 6 479183000352220774400 0 239598705935514179993 6
119802955847458986393 6 599032793635804413952 299525404017156947968 149767205608205844480
748858546039166074 88 374440532018651463 68 187225895508859944 96 936157625041970790 4
468092886269820928 0 234053480009328230 4 117030258441872998 4 585168884395409408
292593238290726912 146301017191874560 731527076191928 32 365774533212241 92 182892764164259 84
914491308611993 6 457259398201344 0 228636571048345 6 114321721498009 6 571625787359232
285821483614208 142915036774400 714596658708 48 357309066772 48 178659902095 36 893326354022 4
446676598784 0 223345010278 4 111675860582 4 558396080128 279206428672 139607408640 69805801472
34903949312 17452498944 8726511616 4363386880 2181758976 1090912256 545472512 272744448 136376320
68190208 34096128 17048576 8524544 4262400 2131264 1065664 532848 266432 133220 66612 33307 16654
33309 66620 133244 266496 533008 1066048 2132160 4264448 8529152 17058816 34118656 68239360
136482816 272973824 545964032 1091960832 2183987200 4368105472 8736473088 17473470464 34947989504
69898076160 139800346624 279609081856 559234940928 1118503436288 2237073981440 4474282180608
8948832796672 17898202464256 35797478670336 71597104824320 143198504615936 286405599166464
572828378202112 1145691116142592 2291450951761920 4583039342477312 9166353562861568
18333256881537024 36667613274701824 73337425572659200 146679249191829504 293367294476681216
586752181139406848 1173539546650902528 2347149462045982720 4694439661580320768
9389160798137352192 18777884546228125696 37558894992363094016 75120041784539873280
150244587168707117056 300498181536668975104 601014377471847432192 1202064783740713828352
2404201625075465584640 4808547365339007025152 9617382961054165762048 19235342382860634947584

38471837687225876742144 76945981217460967178240 153896574120940361744384
307802371613917578264576 615623189971908866080768 1231283273431965151264768
2462640333840225140736000 4925428241633039957884928 9851151631171259268595712
19702893558152877242843136 39406967707926471896989696 78816296599094378616586240
157637315564671626878386176 315284075862308993047199744 630587041190549464675254272
1261211861312961886512218112 2522499280489649687347855360 5045149676706751203342548992
10090601584868406063978774528 20181807632646619442544902144 40364824191112853514264510464
80732066233864936286878433280 161468968171008331090455691264 322947607748573579214309031936
645914558310260992495413362688 1291867802246749653124417323008 2583812975745954642516015841280
5167780693996819957566394073088 10335870873003461260201512927232
20672360716026565210540475416576 41345959372092415801355849957376
82694394624263402363261498163200 165393741008683946247622592823296
330797385537682175537444378640384 661614578115992917159287143268352
1323268770313242966487371058511872 2646616768789000197312335660974080
5293391993903028923299858409848832 10587100900456114903950090995499008
21174835626212343922600930342600704 42350938903024916074603357388406784
84704413107250288608009708183224320 169413896816901490133625403179270144
338837934838604806102462779984183296 677696152086813263875349507219652608
1355432868992833831091546908941877248 2710946867624082268864789606888898560
5422055994524993751092970791788085248 10844436507603645928912724739596746752
21689522052314608711279015791234646016 43380342178843851129465164206551597056
86763280506116969672744593661267804160 173531753309092474173117717818864828416
347073891211902018001492496630388097024 694168551611238175313499115246093074432
1388378641597344629248026474462819909632 2776840359944425815738109436866907340800
5553846873388324745960331849616349724672 11108026053775595720888889650997769535488
22216716721549083899714231205525679243264 44434762671093952715301366218111638831104
88872183798179475262348540050343838351360 177749684508342090188188695328928798081024
355510002840650459703360621114339838918656 711041273329233478060687703141644163350528
1422125081954332073429308328109217297727488 2844335234500394381474482499870292537507840
5688840610184249232180696687044300959121408 11378021502735419402824856748696033686454272
22756723570204680682576640246606930909331456 45514808269877045119007133991643588891508736
91032338798689457745721974980146631928709120 182070122115249650506859363954012172148801536
364151133266240771044549555895462160880369664 728324044603964482150760767765799954926272512
1456691645350894844424844847481351176183611392 2913470402987721449096336318862204885029355520
5827115030547306418685965885523414835382976512 11654578510238339878358518266644839801414483968
23309853918764133838690209524485699864126029824
46621101634103175841326765031363440250846183424
93244990861356168010546222027510961546880614400
186495556909011968676877827984590085184137723904
373002264190623202665326423828316494549028438016
746026829126444935953794383374905637459562856448
1492098259743286933153871838186356571642137673728
2984285722467367988800309819245803736730299269120
5968749850896324222585751924237788660352646381568
11937856513715824935141768419967939694489388449792
23876426651278002850224065982920604136546968272896
47754280550248711660329190251810657768230319292416
95511415595882835240420497075560214526733404078080
191028540182536494320365227294998227034012339142656
382068498346614636319778920877520500291157402583040
764159832656312567997654774331015291980413604462592
1528365337238791726711503413813052967805191456817152

3056822018329916634855394557928150703299111409418240
6113826724364499632575564576460390941975679810404352
12228018824138331990880680074128960954706273603944448
24456768399095329433220461990674280050922375174160384
48914998299827989769359127666181276384864406280863744
97832919602930641344554662702027985335768124426813440
195671685212410606300782140143386835803614872583798784
391355062437919859824909909765435401871386992627941376
782733508902037014096511078488194264271088480176570368
1565513785856468617086404674891035449598805950194515968
3131121107817726411959574385611364741310869880071782400
6262429287845031179492678842881317166848255719509065728
12525232720109219070132417829079809702149543357749133312
25051213729056751562558955944793970141205150552960270336
50103924035790129969706152462856641756222428780844548096
100210841226933513628588786072250686460069112911537111040
200427668764573534635530534437576178815386736522770251776
400867310150560084027766993461301969421270494444932562944
801758565543946197568945836094903162423535031688649244672
1603565021573544454164715370534404772009058148974866726912
3207225824118393026383078137758006438342092469144869928960
6414643210179394288873451068894406665332137280680012808192
12829669544244005049961491724545600907960179246140571516928
25660105336258443044352162622604776970512167861842234834944
51321743168057751977562683592236704250207954462806653272064
102646551327197235732842083878527709118783146403857673748480
205299232636557935021117601145164019474300767764204081905664
102652681309360699288275517266636310355517621360346408157184
51327873150221215532996116980345305486942429419295387680768
25664702822881040710927237663686227898063024079208785641472
12832734535325736827678208418599901526327416724384938721280
6416558829605476649946399002678344551811660704582742310912
3208375195774042443026846898028369170229806523486507630592
1604235488372673280540247147358783032276891347340822052864
802141689429162669783535422851690739719439716469195145216
401082817335994349648473636011994981650216879633989632000
200547394978703682202589780299072296720356950516690845696
100276690644705094790471371296073551307802730608193437696
50139841900029174239823926221305477127713492979020726272
25070669238852900542206083397287089300762810326972366848
12535708763845606982250101841960720018834437082217185280
6268041454132381846698580992638947693643734500474093568
3134114263170980101136055532148767688935125229919797248
1567103899637884639461410283989030765524191604801273856
783575333845139614177396400951838843290410297321324544
391799358935668454312043829954581151909362396121006080
195905525474383550767694729716621441086759821790674944
97955685740466437189683772227976153109419222760423424
48979304371870549497760089798820792837729267312754688
24490382936753940200339146741826754560374461622648832
12245556843786302825899124292121556350942144794460160
6122961109597817775814337606664867710848529388797952

3061571898651242069339556533634478623112993190182912

1530831621251787625385972131968261695400860842983424

765438646588977108051082998559642039622612545437696

382730741276030201704589965567576615772397334691840

191371079628785924691819215927666105866744198332416

95688394309778374265671724535771951923644864659456

47845624402581893092716920553855425456958815076352

23923525825137299526298989419912437476047598911488

11962119724491826253119759281448581111807895142400

5981238268207501371545011926470471742795995414528

2990708337084544808265072106108326464844021628928

1495398770032669465378819124490708529145022775296

747721685761533263312551097963626912934017368064

373871993253365896967846316840949780647761674240

186941571812982581139708542350043052414257332224

93473573499641106897746963139805607252316913664

46738180546395461612819827552294844148752580608

23369787171485184888383086767343442335673352192

11685242034886319485178129879269731298485207040 5842795242015023263082358187433870714566868992

2921484733293443391787825717616437889945567232 1460785922789687576017236170757970211303849984

730414739466326728070279741353860738817458176 365218258768904834065970698664368185991495680

182614573902323152048400763325903001287131136 91310009210096943531908088659810954789257216

45656365734516155519807897828335204467474432 22828863431991919636830875663382465770160128

11414771998362880756878901206298664653291520 5707556140364900847671182290453048210751488

2853863140774180658451456988878382047428608 1426974105682955446533661416265119994740736

713508320489410281920797169045524482883584 356764794068671420287381814979244484198400

178387713946318849807182522717863363477504 89196515429150994735337068973052242427904

44599586942571282283541438293586401558528 22300458085283533599707171050323340951552

11150561349640713028821811476926740561920 5575446828319829628895018714345905324032

2787806490909651371689565845114220183552 1393944783829693964465811166527743852544

696993161102281121543419705249188806656 348506965144857630426966913617252843520

174258674869287350041111987304955641856 87131933583072942434370258900642430976

43567264865751104924092262074403520512 21784281469982869315499697349242912768

10892465253545093084476631830642032640 5446394886049375755601707493331304448

2723278572663102484482549535670796288 1361679851150758545582122662337970176

680860207984982924461485278420271104 340440245197293288065954612835778560

170225193201047556950583293230710784 85115131901724234934094640021766144

42558833601462345696448816714088448 21280050626031286962925156708646912

10640342225665700538812952530124800 5320329569157878798081663352963072

2660244012741453663378425220431872 1330161620451983963858009382191104

665100617266620548013403077083136 332560212153624557048900731535360

166285057836969420045549962264576 83145004798563280783324779380736

41573740339320925771937288814592 20787489139680105576106093969408

10394054054849874133121771765760 5197181769929847739095248273408

2598668256217379205814805331968 1299372813734917271040993263616 649705749680572469587291930624

324862546246843151827043614720 162436108826700034430220632064 81220472264989246473459728384

40611445058314237865904570368 20306326992066926247539638272 10153465727488366781063495680

5076883979471635219178586112 2538517547599543523912712192 1269296552731634719118065664

634667165831748838139887616 317343027648840158360371200 158676236190902948825399296

79340479278692909235306496 39671420230967172028956672 19836300411293944720130048

9918445353552151712890880 4959370250728665532858368 2479758912340627604635648

1239916349658461221421056 619976621573304320262144 309997534158689014906880

155003378765362934841344 77503995225690681114624 38753150534349947404288
19377151727927277125632 9688864094339790274560 4844576162357970993152 2422360138773023424512
1211216098183530676224 605626063490274820096 302822038944392151040 151415523071823446016
75710013335725408256 37856132567769546752 18928629233838194688 9464596091895808000
4732438783436259328 2366289760462307328 1183180064603242496 591607624487665664
295812608336855040 147910702214938624 73957550130724864 36979874576990208 18490487044308992
9245518400061440 4622896638984192 2311517038968832 1155792879222784 577913619480576
288965399674880 144486994804736 72245644886016 36123896184832 18062484963328 9031510917120
4515889676288 2258011947008 1129039527936 564536541184 282276659200 141142523904 70573359104
35287728128 17644388352 8822456320 4411359232 2205745152 1102905344 551469056 275742720 137875456
68939776 34470912 17235968 8618240 4309248 2154688 1077376 538704 269360 134684 67344 33673 16837
33675 67352 134708 269424 538864 1077760 2155584 4311296 8622848 17246208 34493440 68988928
137981952 275972096 551960576 1103953920 2207973376 4416077824 8832417792 17665359872 35331768320
70665633792 141335461888 282679312384 565375401984 1130784358400 2261635825664 4523405869056
9047080173568 18094697218048 36190468177920 72383083839488 144770462646272 289549515227136
579116210323456 1158266780385280 2316602280247296 4633341999448064 9266958876803072
18534467509420032 37070034530467840 74142268084191232 148288934214893568 296586664522809344
593190921231663104 1186417026835415040 2372904422415007744 4745949582318370816
9492180639613452288 18984924229180325888 37970974358267494400 75944200516348674048
151892904632324718592 303794816463904178176 607607647326317838336 1215251323449654640640
2430574704493347209216 4861293524174770274304 9722875278725692260352 19446327018203687944192
38893806957911982735360 77789919758833179164672 155584451203684785717248
311178125779406426210304 622374698302886561972224 1244786290093920543047680
2489646367164135924301824 4979440308280861525016576 9959175764466902402859008
19918941824744163511369728 39839064241109044434042880 79680489665459523690692608
159365701697401917026598912 318740848127769573343625216 637500585721470625268105216
1275038950374804207697920000 2550153458613334329719259136 5100458032954120488085356544
10201218297363144633464389632 20403041057636096581516132352 40807291041091807792206970880
8161699933822844842763354112 163238835570924148202225532928 326487342548405213437848715264
652994027909924260942492729344 1306026741446076190018576056320 2612130854144607716304333307904
5224416450794126105143029006336 10449142386598073555354782793728
20898903743215789800847015149568 41799045426470864981968929423360
83600566733020300724487657095168 167206085226197742970074910687232
334422073972709768982349014368256 668863954986048104049096414724096
1337767524053353340266989601423360 2675614276269220944871572746797056
5351387008863470418418332581494784 10703090930376997894187039338790912
21406815686054109903074827029184512 42814899022708448035551150761574400
85632333346617352529905294929559552 171269737295635617977416576671940608
342549615796073061790045126969524224 685119514001749775250514201190334464
1370279592822706853841876296883240960 2740640315283828314365448382771625984
5481442889844485842094288343553540096 10963210298242630110915359843127656448
21927069633592577075284285998296465408 43855437341399787857475704620675235840
87713470831228843128765674489515081728 175432133959316221085159879475359383552
350874652512349511825576819943377207296 701770074212133162961667761872071294976
1403581686799134604544363767714776350720 2807246450348005766330784023370820222976
5614659054195484647145681022624175489024 11229650415389915523259587997013421064192
2245996544477723504455627897556982300672 44921260117551231924784159602174244945920
89845178691094033681314126818469050580992 179695674294171207026119868865179222540288
359401982412308693379222968186840687837184 718825232472549945412412397286645861187584
1437693000240965008132757716399220693401600 2875471071073660250881381276450299328856064
5751112283330780970994494240204314541817856 11502564849028482880452451855016060851847168
23005810262790807637831830459246985240117248 46012981655049299029517514416923697553080320

92028685569033965566742735830706849251852288 184062815655938666148900885655132606795087872
368136520347618802328632599297703030172942336 736294818766720544718926854570281693511417856
1472633193676406969561177021090314653353902080 2945353499638745699369000666080131839369936896
5890881223849354919231294579959268744064139264 11782110896842436879449175655516547618776809472
23564918691972327840871524302229115498850680832
47131231180519563845689394586850271520295485440
94265249954188944019271481138484624085779218432
188536075094677520694328346206537410261934931968
377083300561954306700227460272211144704622854144
754188901869107144023596456262694937770751688704
15084222405228611349293475983961935172264515338240
3016934013438016821079518110796960937975054598144
6034046432837621887144168507340103062842157039616
12068449677598420264258601586172568499468409765888
24137612979043193508457732315329861746505010905088
48276653205779092976796522916629172988146404556800
96556160906943597873355162405197244966565574606848
193118030804658019586234557954272287913676680200192
386247479590857686851517582196300171788444422373376
772517795144798669061132096968111535499070968692736
1545081262215763928838458059087245454842506185277440
30900253868283861039109303848476535677373740866338816
6180690424272388441083383157557160890124938724245504
12361746223954109607896317236322500851004791431626752
24724223198726884667251736315061359843519410829524992
49449907899091100237421676314955435970058477591592960
98902738801456862280679759999576304506156267048271872
197811323609463048173032334738483474144391157826715648
395634339232024743569410298955628678552939563113775104
791292062490246781585511856868580817634193621148237824
1582630893032888152064406231651808556325016232137850880
3165355322170565481915577499132910954763290443958452224
6330897716550709319404685069924409593753096847282405376
12662169577520575349956430283165994555959225613295812608
25325087443879464122206980852966339848824515064053628928
50651671465435555089002202279201381171461157803031265280
101306336086224363867180885704940165290546570955910545408
202618658483155235112714733702955136476341652611517120512
405249289587723484982135391992059884743180326622426300416
810522524418272999477682633156418993067354696043636719616
1621092939322198057982188964657436433296697477684841676800
3242281659615700234018025326004069760917371126564819828736
6484754881174008704143345445386533310482694595519912607744
12969892886233233880501280477529854198261293875820371116032
25940552020236900705431740128573283551114397121201834033152
51882636536014667299721838604173717411412412981525851668480
103768338063111066377160393902401735441192063441296070541312
207542806108385596309754221192912072119118601839080875491328
415097872181098119730375309162041346711706153591139219800064
207555066072712523420621087969129274592587551752058344308736
103780598027437993488027260678618937914661013354273539358720
51891831509259862632871988686336619266514125416258953281536

25946682002400364260865173516681884787848872077690568441856
12973724125085398602647176345097729971220340723625830121472
6487053624485307537430882965927258774258122704203188011008
3243622593213957886769088879652826281453037523296730480640
1621859187092631002411368138171011587888506847245933477888
810953538789141530719095918257805017525247466421750857728
405488742015983780116253883715052120553120754610267488256
202750357318698397436479904150600866171808888004829773824
101378171814702452407416433221837836033528699352262901760
50690582485027853048296457184187619490576477351055458304
25346039531352239946442348878728160482194302512989732864
12673393910095276684368234582681255609550183175225868288
6336884027257216697757647362999215489001607546978435072
3168535549733397526665588717328901586614061753171968000
1584314542919093352226176876579097714363659866427359232
792180655485743970559779697246872317710144428134367232
396102019755970632503235478102097889119229461527527424
198056855884534639863290553790379809691693354493935616
99031350945541981737481684264855337411885989112053760
49517136974408321771659045817260384988962650488569856
24759299238022826337288624751046550635991153210556416
12380014994420745894373863296731454388750490588413952
6190190184915039310051707108969816729752702285774848
3095186436309852836458241284786953132565079638671360
1547638890081093008945314507544498950126904067227648
773842281003629799830754186347760666985634157559808
386932558483356547594425559461635929453908140752896
193471988232449097636737012874695762707499601362944
96738848611609960738130623009286780344022566174720
48370851553497686328946369790612839667147665833984
24186139400595196144413714038291144581142024290304
120934265122220774562177121590637934664355107831808
6046891662071975526073693081065148519069601759232
3023535034016781885529412683405664852980824801280
1511812118498788004010989413139377723213424361472
755928359994592532628636242287961509968218161152
377975330369895531625888889003117079164862070784
188993240371247398468729828431126701672807530496
94499407778773515562257606180347431881592012800
47251097685961665945075149072565756463390130176
23626245741268287054510747527478898492992126976
11813471319777870568241960259337459377144594432 5906909884460798804614273377467734753896562688
2953542054516331162553783312633369909610414080 1476814583401131461400214968266436221136273408
738429069772048670761769140108093743733669888 369225423921765805411715398041484688449601536
184618156478753637721273113014461252516184064 92311800498312186368344263504090080403783680
46157261378623776938025985250474767274737664 23079311254045730345939919374452247173791744
11539995909389786111433423061833555355107328 5770168095878353524948443218220493561659392
2885169118530906997090087452762104722882560 1442627094561318615852976648206981332467712
721334814928591866580454785016455151747072 360678041288262212617210622964709818630144
180344337556114245972096926710596030693376 90174827234048692817794270969418576035840
45088742845020131324770039291769568362496 22545036036507958120321471549414924353536
11272850325252925289128961726472532262912 5636591316125935759048593839118801174528

281837873481270443676635340750066810880 1409230905781220497004204947720967815168
704636222078044387812616595845800787968 352328495632739263561565358915558834176
1761694401132281666084112099541086371884 880873162050433507180198702252189286640
440449561767363090659170677366917693344 220231271254754713864121001803870371844
110118880812913941199328332462140948848 5506106299922526273329808201117335552
2753134279599677743346599889563811840 1376607704619046175014147839284477952
688324134719126739177497866893524992 344172208564365195423960907072405504
172091174884583510629586440349024256 86048122743492211773596213580922880
43025329022346334116199603493666816 21513298336473281172800550098436096
10756966080886697643750649225019392 5378641496768377350550511700410368
2689399976546702939612849394155520 1344739602354608601975221469052928
672389608217932867072009120514048 336204707629280716578203753250816
168107305574797499810201473122304 84056128667477320665650534809600
42029302273777945713100166529024 21015270106908615546687532826624
10507944538464129118412491194368 5254127011736975231740607987712
2627140877120942952137485189120 1313609124186699144202333192192 656823904906463406167961894912
328421623859788620117378596864 164215647633172768575388123136 82110241668225613546043473920
41056329759932421402196443136 20528769342876018015685574656 10264686902892912665136463872
5132494567173908161215070208 2566322841450679994930954240 1283199199657202954627186688
641618489294532955894448128 320818689380232217237651456 160414067056598978264039424
80209394711540923954626560 40105778947391179388616704 20053529269505948399960064
10027059782658153552805888 5013677465281666452815872 2506912519617128064614400
12534903153296711451410432 626765023392429435256832 313391735068251572404224
156700479220144213590016 78352545453081320488960 39177425648045267091456
19589289284774936969216 9794932872763620196352 4897610551569885954048 2448877333378980904960
1224474695486509416448 612255362141764190208 306136688270136836096 153072847734695788544
76538675667161579520 38270463733487632384 19135794816697237504 9568178883325329408
4784230179151020032 2392185458319687680 1196127913531932672 598081548952010752
299049570569027584 149529183331024896 74766790688768000 37384494856011776 18692797183819776
9346673469816832 4673474173861888 2336805806407680 1168437262942208 584235811340288
292126495604736 146067542769664 73035918868480 36519033176064 18260053458944 9130295164928
4565281800192 2282708008960 1141387558912 570710556672 285363666944 142686027776 71345111040
35673604096 17837326336 8918925312 4459593728 2229862400 1114963968 557498368 278757376 139382784
69693440 34847744 17424384 8712448 4356352 2178240 1089152 544592 272304 136156 68080 34041 17021
34043 68088 136180 272368 544752 1089536 2179136 4358400 8717056 17434624 34870272 69742592
139489280 278986752 557989888 1116012544 2232090624 4464312320 8928886784 17858297856 35717644288
71437385728 142878965760 285766320128 571549417472 1143132389376 2286331887616 4572797992960
9145864421376 18292265713664 36585605169152 73173357821952 146351010611200 292710611156992
585438402183168 1170911164104704 2341891047686144 4683919534325760 9368113946558464
18736777648930816 37474654809489408 74951508642234368 149907415330979840 299823626754981888
599664845696008192 1199364875764105216 2398800120272388096 4797740978033131520
9595763431042973696 19192089812039368704 38385305523985580032 76772862847784845312
153550229295197061120 307109465789648863232 614236945977807208448 1228509920752633380864
2457091899099304689664 4914327913386685235200 9828944057149522182144 19658464575051347787776
39318082071607302422528 78638469986223818539008 157281551658466064465920
314572326688968983707648 629163100122011676966912 1258363093732170773037056
2516799974440636384280576 5033747522833862444974080 10067790193572904242774016
20136170682956167191199744 40273521957533051793702912 80549405098307538410012672
161103532563097946465239040 322216509859161632220905472 644451909184254743022665728
1288941597300372443207041024 2577958752464470800737501184 5156068620656393430121840640
10312439472767690517537357824 20625483408445188349662068736 41252175742709991328498843648
82506769337059211915347099648 165018374377396882347393024000 330046420161350681728183697408

660112183135815197523162693632 1320263051897858063179915984896 2640603475048171462627013165056
5281361692601253597788388720640 10563032870212328540645502222336
21126684710444299771428454006784 42254607360927884923131807137792
84511690601934340606813412524032 169028332964025822734726421544960
338066569448365928511652036083712 676152945937360423107702458155008
1352345505955977978384201688285184 2704770240074470221105996920520704
5409698936473968970887180928942080 10819714785597994999124736033685504
21640063396496104112950220418973696 43281394443592436455301937541152768
86565324188385329369406868488716288 173135718979171571656419723790254080
346281579163144969148051421206151168 692583440735893589966526789663588352
1385207446290994483273901473829748736 2770496022220403573229498736664641536
5541154303717636359822389051339571200 11082633125988931146371561258699718656
22165915289085179146196688829440589824 44333128652384991999300510282963484672
88668853453199251412415285814091579392 177342899203257037652459102124512378880
354696183000231144960175265241683197952 709413135187896429230864652468683276288
1418867808750661137082757548908000313344 2837818694251058831407571585757268148224
5675803542001590777299256147397071339520 11351939391002127783566738246559212765184
22704543396002148025069928396648565702656 45410416020000080966012760600357411749888
90823490495991731763771328814835384188928 181652297903966603191034272857911889756160
363315229631899485709051776172306022268928 726651726911731530072070013257576530051072
1453345989119328177452072948341082031128576 2906777048830386589520011740334022004310016
5813724238844233648271755167971759892725760 11627788760055388235006973710550951553662976
23256258084844618346940874170316766643748864 46513877299156920447735601839063260360343552
93030476857249208403178910674985974866378752 186066398232369151821773235343690858024140800
372143685500479773674377298674819532631048192 744309149072442487410416253324514698427629568
1488661854287850854944155818598780663186325504 2977410820861633470134958261097063859034783744
5954995866295130460763209769993132783393832960 11910340181733987962513006035584275697436196864
23821377261755430006999185062364571656169455616
47644148320085768177944716107121183834933035008
95291084233321352683782124179026448715054317568
190587743652942338023349632287621059520485130240
381186637678483941358270032434378443221723250688
762395576102166413339681600587029534804952481792
1524835753694729887925646272610604366332916924416
3049760710370253898343858688094299326111857770496
6099699826702096041672849661934779839115763384320
12199756465327368573315963895361922052015622455296
24400226554501090126572456933708568851599436283904
48801880356694886213025972153386587198335255314432
97606615208775184345814060878712073386943276122112
195218939408321192531152354901301944754432083230720
390449296798184032741353176090359485469955228434432
780921429559451360840803284756230162862092580814848
1561888531045069312397800434663482709568549409521664
3123868405942471806227988599629010186825827314827264
6247919499589609975320752659862109909029111621222400
12496204374588552676371056240932398888813137225580544
24993139499995770804201214324281155919136102417432576
49987740501628872511320632333395028121291860767408128
99978404006532406828477672036455488808623033399902208
199962654019614137268628158812241842749324690529976320
399937000052326921760601947103145415762806628520296448

799897384130851137967895153163614292053927751961280512
1599841536314096864829172824241875505164484493763936256
31997766087329829074451106843130448524442226967210622976
6399740289675544170463751440284677389110969893786746880
12799854723770245052074563023886530146674971706304495616
25600457736378803526443246334407411030256007250070994944
51202412050434233897474733242083523534324142175065997312
102407817256221721484125947630704450016272539699980009472
204821620823149950346604857554483705927793590099656048640
409655214267712915449915639695117023646084201598704156672
819334373778251860413243128562533270873162445996192432128
1638716638042155779853309955469664988908312977589953101824
3277529057055615677760267307628526872140602126375042678784
6555249676053839591627829408635447532929156595140358307840
13110882475992895655470248404027682643154217875061262516224
26222531199756224255369675981568940440900245119683616833536
52446594895053314399597710310165031190984108978489417269248
104896254781188360576912137314384363000335455435223201742848
209798639544540184709257708016877327237405385826935137894400
419609539053407296529382282809971856948279721566847744606208
839243598035468447280498299172378118843497342959650426847232
419634058982061150751116016362406261895217621392802682241024
209823159473194038930991441569311732184343285652890075529216
104914644727678751243212437478710166710538880304689404968960
52458854859380241510464577086382233664453058891466886086656
26230193677460553699661467716704691986818338815294534844416
13115479962615493322045323445109133570705074092427813322752
6557931543250354897129956515932960574000489388604179611648
3279061552596481566618625654655677181324220865497226280960
1639578666783892842336136525672437037824098518346181378048
819813278634772450681480112008517742493043301971874807808
409918611938799240097445980590408483037018672385329463296
204965292280106127427075952588279047413757846892360761344
102485639295406317402714457440676926654503178796028395520
51244316225379785545945469293607164801063717072938205184
25622906401528206195266854933437933137437922373931106304
12811827345183259808780487610036141937171993105696555008
6406100744801208259963773876676658652812512512213778432
3203143908505393307768651974167623168519514235789639680
1601618722305091242777708504998458505316386107736195072
800832745178742915835545511456552713186507548788785152
400428064602470105141118385206938086857411021854736384
200219878307784376182232007342799908560784134657540096
100112862157166849896952411041065386846431379193856000
50057892580220755851394409205365409706235345529470976
25029677040929043377156306445099062994627500731006976
12515203895873854414307704143757710568068664348639232
6257784635641593570018627532482944819411789165887488
3128983661673129966441701496543517177394623078727680
1564537502762731573937044613422780972541675787255808
782291587344449082326619239286901678193020017573888
391157211653766188842358085931206435057601070759936

195584314817653918260703276109481015509346066366464
97795011904212371050113754626679406744945798676480
48898933199798891484937935599309152867609282084864
24450180223745798722409496942639301181372832415744
12225446923796075851175013042812012964470511894528
6112901867859626170572638807152187669127303790592
3056540136910607207778885546449184428009675816960
1528314669945700665135725844661137510727849869312
764179635718048863191004458048841403725430915072
382100968231623696907072996883557026043468447744
191056059302111481109321882371346675112110718976
95530817244205556882553633150457418601243607040
47766802418677686605223162557620749823215927296
23884098107626297384584754270006395172905025536
11942397502956875733278963630601207717101043712 5971372976050301387132775063099608923874787328
2985773600311082453813034155449306994599526400 1492930356298507107029840389674404763630829568
746486956220736493576581850812078014980947968 373254367146109716819121753393476824073240576
186632628090925593424976290690457320328003584 93319036304398164220195852342088114309693440
46660879281666765863951779669473784227692544 23331120205567224808902816583951755650269184
11665900385150533342914871666583309593346048 5833120333758727140689167520595370680778752
2916645237471093804960449603949543282442240 1458365154031412019788157723800700612247552
729203844663638568548045322813314791636992 364612556155785563601005891863139638575104
182311594989875921463994561159810940665856 91158455950929530563743088194026031022080
45580557203460550197744447904073295855616 22790943215728167556808675855566788100096
11395803914863030007372563879548464136192 5698068110930988118170394915656767111168
2849117132215230616327253945769651077120 1424600104482483586784655216855459299328
712320821428675932702841730413046530048 356170795308055036006677926199181705216
178090589950886052830967493595920072704 89047891123872293829298012046124646400
44525243636150780621556138647144628224 22263270855182707164231635635613466624
11131959946145012008842600973827309568 5566142232349335217784692064923942912
2783152245813082215574041821467115520 1391616687725748411127868805236129792
695828626272477857234358349869350912 347924454341040754452391148560318464
173967297772921290143801561092980736 86986184187661101530703773952901120
43494359744430778994753383679655936 21747813697515503612077440191430656
10874223761407808863389094271516672 5437270337028932960369734223659008
2718714396676980744522460655779840 1359396812419747504430027099865088
679718213250502318299411935920128 339869010145565442191905160953856
169939456832939862617052176973824 84972204296548502069075886735360
42487340088313536414812842491904 21244289014176410897543870808064
10622453992098026793840660185088 5311381738553924069454692483072
26555768240529417370994527436800 1327922805890936353630854316032 663980745758582010882222456832
332000044285847922474508877824 166004857846202419753953263616 83004846774740439135326044160
41503632313189834196837728256 20752420619504724413006217216 10376512541207265863796785152
5188407386331084760545230848 2594279251029268294596034560 1297177404446497104459726848
648607591689180030810718208 324313240577555754695786496 162161342655260746993106944
81083032510871808319160320 40542696847056621570883584 20271938719338669491093504
10136264507574514098372608 5068279827739846725599232 2534213700846218201006080
1267143743911256519606272 633590318699701969354752 316804382721887839453184
158406803046962347114496 79205707366490387251200 39604006604749800472576
19802579763127203659776 9901578111939753541632 4950933171157952626688 2475538643173014241280
1237805350383526084608 618920689590272524288 309469351994391003136 154739179596822872064
77371841598225121280 38687046699019403264 19344086299463122944 9672324624708272128

4836303049842491392 2418221893665423360 1209146131204800512 604590657788444672
302304124987244544 151156460540133376 75580429293322240 37791314158288896 18896206834958336
9448378295386112 4724326586646528 2362232012800000 1181150366138368 590592362938368
295304771403776 147656680669184 73830487818240 36916317650944 18458695696384 9229616283648
4614942359552 2307538288640 1153802698752 576918126592 288467451904 144237920256 72121057280
36061577216 18031312896 9015918592 4508090368 2254110720 1127088128 563560448 281788416 140898304
70451200 35226624 17613824 8807168 4403712 2201920 1100992 550512 275264 137636 68820 34411 17206
34413 68828 137660 275328 550672 1101376 2202816 4405760 8811776 17624064 35249152 70500352
141004800 282017792 564051968 1128136704 2256338944 4512808960 9025880064 18052284416 36105617408
72213331968 144430858240 288870105088 577756987392 1155547529216 2311162167296 4622458552320
9245185540096 18490907951104 36982889644032 73967926771712 147940148510720 295888886956032
591794953781248 1183624267300864 2367317254078464 4734771947110400 9469818772127744
18940187300069376 37881474111766528 75765147246788608 151534692540088320 303078181173198848
606173954532442112 1212383093436973056 2424836555618123776 4849813848724602880
9699909172425916416 19400381294805254144 38801888489517350912 77606028778848387072
155216561157324144640 310442129513903030272 620902273426315542528 1241840575649650049024
2483753208893338025984 4967650532974751907840 9935589296325655527424 19871755053403614478336
39744663028311835803648 79491631899632885301248 158987875485284197990400
317984974342605250756608 635988395429284211064832 1272013684346715841232896
2544101155669726520672256 5088349885292042717757440 10176994918489264788340736
20354580132788888282333184 40710340857198493975969792 81423042897638422774546432
162850808161759715194306560 325711061056485169679040512 651441011578901817938935808
1302919802089666593039581184 2605915162043059100402581504 5211981439813570029452001280
10424265111082043716197679104 20849134685073894746982711296 41699478295967404123140128768
83401374443574037504629669888 166807584590426533525958164480 333624840587409984085313978368
667269023987933802237423255552 1334576733602095272608437108736 2669230838456645881484055412736
5338616419418202435502473216000 10677542323846226216073671213056
21355703617712095122284791988224 42712645175463475624844483100672
85427766231005522010238764449792 170860484222168185541577125396480
341730871964650654125353443786752 683481550969929874335105273561088
1367002716021116880839007319097344 2734084660204748026015608182145024
5468327776734524580706403452190720 10936972466119106218763181080182784
21874578757538326552227110511968256 43750425165676881333855717727141888
87503385632554219126514428860694528 175011841867509351170634844534210560
350033824939820528176481662694064128 700087932289244708023387272639414272
1400216429397696719387622439781400576 2800513988433808045456940668567945216
5601190236144445304277272915146178560 11202704990842549035281328986312933376
22406059018792414924016224284667019264 44813416111799463554939581193416343552
89629428372028194523693427634997297152 179264049040914923875015385766323814400
358538482675546917405287832525306068992 717097734538527974121089787035929018368
1434237007451924226863207818042491797504 2868557091653585010968472124026251116544
5737280336806643136421057223935037276160 11474892980612232501810340399635144638464
22950450575222357461557132702800429449216 45902230378440499838987169212661139243008
91807119212872569509720146039442839175168 183619555337728278682931907307126799728640
367249744499422836692847045070735842213888 734520756646778232039660551054436169940992
1469084048589421581387254023934801310908416 2938253167770573397390373891521460563869696
5876676476724607264012479470346637011845120 11753693235816135466488422315300705791901696
23508067036366112809903771379816275120226304 47017495202199909373661396258062277313298432
94037712663335186255030499512984008772288512 188080869844541107525476413019686925835960320
376172628724823685081783654026811668254687232 752367035521130310225228964028498969674907648
1504777627185226500573781240006749205680881664 3009642366656384761394209103913000944023896064
6019458957884633043281711455625006953372057600 12039266364912993127550009406848024037392646144

240792296281134403370731918048920683360823541763
481598530528017888380927295921761771947588321283
963224936987533940040781511491364354347059118083
192650562583806420663941686227841032959788318720
385312275540212106639454140314818390100329627648
770646851825622743902049816347909428562165235712
1541338305141642549050382704132364153847342432256
3082765813264079220593331551137818901140708786176
6165710032489746686171795388021818989173465415680
12331776876902669862313855347536000352131026518016
24664267377651692704568239838056725451830244409344
49329962002996091369017537962082900398796871565312
98662778501377594657797192496104699787866508623872
197331265599352601315511861813608719755627854823424
394673949968593673989285702559929991073648158441472
789370735900270643336668337695371174069478440828928
1578787143726707877389530540541764731983321129549824
3157766563130574893621144881138557423165537075488358
6315551395031616423528767308337523799868819850133504
12631393276041661196304897087958655068131310985805824
25263517302901987844068896018333668277772449937883136
50528496107441306591055995721500052838564555808309248
101059915218157274987948398812665538243168423481704448
202125676442863873587569612364661941618415470693580800
404263044898826394398484854207985613500988188847505408
808549473823850083243660967373294687530290872615698432
1617145715700094755380704452661236296117210735072772096
3234384967504978688548173941151766434347679449828294656
6468957007219535732669877953962120552921874859022090240
12938288158858228176486816051241416474296781636775182336
25877324606554769775267752389117183685499627111012368384
51756145790786166395123745351503068844811381896948744192
103515284736925586479423971849543540637247019143745503232
207036555784557680337200905992161887169742548987187036160
414085084190528375431107736570473386129982119373766131712
828194113623882780375627322313245995840958281546316382208
1656436117733417619778078342971090438843904648690201001984
3312968016438139357609804082631377772011785468575538479104
6626127594818886951326902958641149332671523279541349908480
13252638313522990374868395504039086242638951243863245717504
26506042874816413694165970181591747639869711857287583236096
53013618245173693277190298710210645588923042453697350074368
106030301481429118332097314114475591796213322385639067353088
212066732945021700219628061617059784829161119727766869114880
424145725854370327550122990010336772131791189368511207047168
848315971637394509321979713573107949210520278562977351729152
1696680983132096727087426894251084708314916356777864578727936
848365011494702217765447180677976759104396078214887226998784
424194765711678035993590457115205582025666989020421082316800
212103512838002481552228661945711392249567969466699275567104
106054821410082972553831047666909996743151222211594004987904
53028943200582352165773882180482148680759229844919186096128

26515237848061609027316120263754649494971424292020684849152
13258002047916020985872649718634112324781616830790888325120
6629192585900618729043619652695449951038760757785717112832
3314692073921613482575457223036921869843356550087995031552
1657393927446458800314552309863059382083666360641565753344
828720908966055429670688004103828914622827223119566995456
414372427104440729592049926638064069101910632959175557120
207192199862926872174377925612106840446203827179283808256
103599093086816689776365443952590823170726168939489918976
51801043121084971732770962549564113059175212144668966912
25901269849380799288679601561416407266493669909796487168
12951009069109556355486860924025379001699866873629245440
6475691606764356533316960533671277185076449396180123648
3237939339486967444445245302664932434651482677772812288
1619016437795878311116005169247113138382370328727781376
809531602924136450004693843580880029719499659284578304
404777493475166872225692551269101745123907077102632960
202394592744132759724519090373881737694032162281488384
101200219375341041668095952556606301413055393005830144
50601571189307851736966179963135866989547352435458048
25301516345472591319942191823984291636283504184000512
12651123548145628385700646833200324888896666075136000
6325744461777480555715098877204251979825790029135872
3162963574741073459289937168904170757601623510351872
1581527459296703320361162449603107762645176003067904
790786565611434955538678157377065073244770125479936
395404700787259125448387544976288132583476124712960
197708059384400386563718005632021864272283593342976
98856884187585605201621119387949831126414562164736
49429869341485508560691617979944365058343663828992
24715648294589107260286338132956907276740023287808
12358180959217730120113433637970816012154107330560
6179268885570453305041849104731589192969101508608
3089723645766020775013490695238885189930574675968
1544906424373407448753028419055987891688299298816
772475512931902254999655745246266594205655629824
386248906838550392811398640482269621283580805120
193130028605574829061484704170702972732166897664
96567801895937230858635044050135567411271696384
48285294744543523593263868007459824228229971968
24143344270559215878605106994925932375412047872
12072020584423334980289139993060976318354554880 6036184516783531010637863244329493224501542912
30181793706776972655565578246064249144912904192 1509133241481814512906112434981875838787518464
754588398812390196514717873465813552559292416 377305088441936568288189764720344592862412800
188657988738839019159510296353891204722589696 94331716628354877087462855173805056506986496
47167219443645122297585281085332255326339072 23584290286556403025719567291880991199592448
11792485425645122451323247020547927368007680 5896412854006021694893355197577679568109568
2948291497594741082062543442440697726107648 1474188284093235658339204643046277834080256
737115409694550387823568782436103402553344 3685568338671241473238767621674533944033280
184289486247603876282875426065508093394944 92147401579793507973183520646874607386624
46075030017892538902464664130497584037888 23038179622944161909168783968778932191232
11519422118471027183552617936154536181760 5759877212734986706260421943959803133952

2880021683117229910372267459921169088512 1440052379933483233807161973931218305024
7200469591541757562140951089509260328 96 3600338641708049477623046154681214566 40
1800221243822610087087808382303899484 16 9001365833955977176820468436335958425 6
4500812724399451959100947480576209715 2 2250471265910457664895830371492220108 8
1125268084810594675120593501348167680 0 562650268332980258896635908475112652 8
2813332471303315901164875331380707328 1406706800470865253923285560192925696
7033736826450362786320667273477488 64 3516969825273199651512453372995174 40
1758535618660608954932286554625802 24 8792931623423090420541732113770086 4
4396592576771568033211015727205580 8 2198359670915795428075582698763059 2
1099211526722903419772828766961664 0 5496216089939545627539330922708992
2748187273132287078107259005304832 1374133250647400671222426274627584
687086432364328901695611523301376 343553119702478733890004954644480
171781511611396508466102073819136 85893231685776824993600835158016
42947853782927697877075316703232 21474545861483491628675107913728
10737582415751567159406278737920 5368945950380694252237501759488
2684550346442802462385932075008 1342313858847628899326556635136 671176272236928283730073616384
335597807525021058898434457600 167803739465788987965916053504 83904287584533723241307439104
41953352718086476249828425728 20977280821953045439501565952 10488942642431426377044459520
5244622436943165017169068032 2622386776335308422907953152 1311231167099517168615686144
655634473015690062888697856 327826681240810770734776320 163918062986888255012601856
81961392676685562328907776 40981876929963498575757312 20491528760792107993530368
10246059528301233349591040 5123177338103206351208448 2561662456027898013810688
1280868121502096426008576 640452507495121922555904 320235477119597816053760
160122350245817335414784 80063480965917881401344 40032893404463547547648
20017023162984077197312 10008799811868190310400 5004544021122171011072 2502344068155123433472
1251208062874580680704 625622045835799822336 312820030117154652160 156414518658204696576
78209511128916033536 39105881464364859392 19555350368213585100 8 9777033316044636160
4888657395510673408 2444399066499514368 1222234717621846016 611134950996967424
305576271591505920 152792533842264064 76398465944387584 38200332483821568 19100715997724672
9550632876769280 4775453877338112 2387795658145792 1193932188811264 596983274274816
298500227072000 149254408503296 74629351735296 37315749609472 18658411675648 9329474273280
4664871354368 2332502786048 1166284947456 583159250944 291588014080 145798201344 72901197824
36451647488 18226348032 9113436160 4556849152 2278490112 1139277824 569655296 284835840 142422016
71213056 35607552 17804288 8902400 4451328 2225728 1112896 556464 278240 139124 69564 34783 17392
34785 69572 139148 278304 556624 1113280 2226624 4453376 8907008 17814528 35630080 71262208
142528512 285065216 570146816 1140326400 2280718336 4561567744 9123397632 18247319552 36495687680
72993472512 145991139328 291990667264 583998111744 1168029777920 2336126664704 4672387547136
9345043529728 18690623930368 37382321602560 74766790688768 149537876344832 299084342624256
598185865117696 1196406089973760 2392880899424256 4785899237801984 9572073353510912
19144696462835712 38290492437299200 76583183897853952 153170765842219008 306350327777460224
612718247740964864 1225471679854018560 2451013728452214784 4902168194392784896
9804617863762280448 19609798677477982208 39220723254862807040 78443698309539299328
156891900218705969152 313792807636666679296 627603629671842840576 1255243288140704645120
2510558633875447218176 5021261382938970292224 10042810996254092296192 20086198453260488015872
40173549828025582878720 80349405499060379451392 160703422684139186290688
321416068740315227357184 642850584224704164265984 1285738061937555747635200
2571549910851406333476864 5143247395655402343366656 10286789939215984039559168
20574170174242326784770048 41149520940105370980843520 82301403063452176784293888
164607528493387223213801472 329224501719740185718030336 658467892905411850016915456
1316973564742686657195540480 2634022687349099228714500096 5268196490425650286075838464
10536695212306204229445353472 21073994887522215773478060032 42149198700864046176130826240
84300815253367321610611064832 168606466210013101737920954368 337222603826583120509239558144

6744645504662800750852744151041348967786558787818304139427840 2698012944370030972875460050944
5396180631244972618285282492416 10792670747499766581639289765888
21585960465019175853416029093888 43173158870077637087106957312000
86348793620233844934763712872448 172702539000624831390627022241792
345414981521563945823453237477376 690849770083756457731304860942336
1381739154248770047631406493859840 2763557536660054359600406531670016
5527273529645137247876000151240704 11054863971940331553102374478282752
22110361769180777220905497308168192 44221991188961782671212491319541760
88446517679124021801227976045494272 176898105960648956520061938903810048
353806353126099738875335851433263104 707632988661803129421095650117812224
1415306542142813562183039194738196480 2830694213924041731047774178481537024
5661550687124912675458939934973362176 11323425892803483777644663025967300608
22647500822714284408742892363975753728 45296299719643202524392917352033812480
90595195587715672462600099952232235008 181195583472289879752828730400793690112
362401551538296829160914521794245820416 724823872264027797632343165573808521216
1449689282902923873885714575118250803200 2899461642555584305013485638177769127936
5799089438610641724511084252238073298944 11598511184220229677990394456241216684032
23197686982438351813917240816012573540352 46396703192872488543707385439085427425280
92796064841736546919160578492291415539712 185597446595456233501812772212823952457728
371205527014878746330608774882130147672064 742432321677690051315184010677224780857344
1484907178651245219938300943180378532741120 2969899427894220674492467730012615007535104
5939968996971901818216667147328945899175936 11880278276310724574896797669265323566563328
23761237117355291026720522087745510669549568 47523835364178265807294897673920748411944960
95050392987291899122297502344700950969581568 190106230492454533260010418683120810230546432
380223350020650536550851665353679437043859456 760468478112784013163364986682234507253252096
1520980512368533906450053285314220280837570560 3042048137022999573146753194527943094337273856
6084270498617862666786799636854891253998813184 12168889446379452374560185769307792638646157312
24338475791046358831093544529811605538589376512
48678345378667625826133435042015251599772876800
97359478350485067980159562048814584244734001152
194724531887269768616104508027197330579844497408
389460214147138802543779783913530985340441985024
778942729039476135710701103545334619042389950464
1557930059569349332667685278527214534807791861760
3115949322119492787827936699927519663061607645184
6232077050200573820641005685601220513015263133696
12464509123243241312522759426948033998146219540487
24929735448495001242445081028374331547197435281408
49860898144682708444771220342718112589531253309440
99724650784750828809304557257375124169335272112128
199455010560272481458133347658628046319216075210752
398921439102086610595315161605011688599523212394496
797865714167256516548727255785534569121228548734976
1595777100260679623813648376722091522086821345361920
3191645544373692429059684483746227811862371186507776
6383473776452051220984144428096545159102199364583424
12767312928313435167697839777401269388959312712302592
25535356607445535786854781397218896919428453390876672
51072174716528402476627766479270510121876562714296320
102147272436331466759091940328206452809792437293678592
204300390879212257129856695395743770751663498317529088
408612473771523161483059020270149271767484244095401984

817248331569243617412809299497622004063282983111491584
1634543431190881823719001116909890929183194956064358400
3269180398486552825224767269649075700479647891811467264
6538547869182684006023064610956739085185811742988435456
13077469882784524723193189365230653538824655404707872768
26155688054407362868680499017095657814555374646877749248
52312872686491352581949238607460017102922876968679505920
104628738528335958853074958361457437153470009287207026688
209263463367378425084502879015989680202188529274110083072
418538899356169864925711682618128972194874079947612225536
837101743955165759364835214408557167970742202694008569856
1674251378395983577756494127161712783103472490985585377280
3348598537763271273566635651012622460530921153166307229696
6697388637469150783240566095403638709709794648722887409664
13395160398823518038695721777564064996715493982226320719872
26791087045417469021820622728641705148022797334013733240832
53583706586375803932499603804310560605229213407149650083840
107170478163833339642715924302675421828825664292543667372032
214347086309830142840865281993459444894385803541576069152768
428706432583987212792597430763136092262240556996129607122944
857437385096628279806928595078706589471419013818214151880704
1714923810050564268057324657262281988836713827288338179031040
3429945699815743953001584248734301597461179253880496108601344
1715021889765179684944259591472019608624465426592157929570304
857535464811243696693863529288444209259170613122033902419968
428779992369948775457798631420439307103054256473994420027392
214396126167137851284332749098328254788261603193485944422400
107201128074650657419883091243218428012498039074987339415552
53602096532866194598799903968636364315432638276615853309952
26801814514203530243829131157831757312308128507869018456064
13401290380986981594129155165672666233449968938715055128576
6700836752436099033171872376214726905372936811747800514560
3350514157189353634639583584796560347010444577069036732416
1675304969080328876346615490742878620667210374132086603776
837676429782990467686719594543738533914599229864827420672
418850187512908248600065721858018878747796636331805769728
209431080067160631678385823222084245269146828865598914560
104718533188933569528369392757579525582197669782647472128
52360763172143411608772936952058464264910962566247743488
26181129874910019226680588762663582869361545120585875456
13090939081874166324487354524648966803133804479023939584
6545656613146661517817207333983071085793418198877470720
3272921842678119936695368702820829385009967079121485824
1636507689391454557241066869325061613561612529402118144
818277228721924573067224693619854267309120759621746688
409150306374060933756957976288588863918717627271217152
204580999193579790490151802883625297091437437365780480
102293422600064557050912308811478081111758030547976192
51148172801669609428374358090571756838898671206531072
25574817151653470165646280887702236560959163569537024
12787773951236067808552691365059297351234495767904256
63940696633227002671411211431337382109947048755200000

3197126175513683315002948301868913873186080933543936
1598608759683008248217668016085479320437404714663936
799327215804587419466930940618250852140884481277952
399675025883835357412513936596881022031533302611968
199843221932688502545781201442318308996312182292480
99924465461729663192652717293098053488428856639488
49963659978557537556207416932518476239350811066368
24982543613125121758044237609243962867243596906496
12491628618485737368992383376114343807405894139904
6245992715204456929481323973803353090594994913280
3123085560583022587233228129774767138743521378304
1561587381781908354862897136323928866094772649984
780815991636152708054590103880237081408892305408
390419146190676561933886581979925486488519914291 2
195215148281637442325218293829195594532976066560
97610361733968537490501838879381878311676280832
48806574663559176909197265422082979678432264192
24403984230067042536571805702237510100513193984
12202340564177248309272489346716765180905127936 6101344506660487675129537921157387655776829440
3050759365616175597811415584478196360550547456 1525423238951053679029031104188849446606340096
762733397547009779576177208069300356468703232 381377587809246359818919432022087994817118208
190694238422493914924875130004762905699942400 95349841470182324970145271999240906995662848
47676281864558846238926489498050180570677248 23838821497013264996390171498239953821761536
11919751030873553436658549123727408679092224 5960045656620237187561006249167420223651840
2980107898901848828396368968235568053878784 1490096484746789531506117405943712997965824
745069510021327324407025163884820984496128 372545388834629941530495812398892735004672
186278011329298110428739521427687488880640 93141664120640625046115568327964305129472
46572161288316097438930687971042432909312 23286745258155941177401795889051356626944
11643704936076916817669123896290748399616 5822018621537931523318674924027909242880
2911092387518702318901393949955222142976 1455587732134219438071725218948244832256
727814635254543858346376731459439296512 363917702220988998828445426722378088448
181964043407353034241851243857518264320 90984617852105784534739887176923742208
45493607000267525974277076212544176128 22747452537241079840592104418313240576
11374050787174198347022835365177196544 5687187652863928386874809260598886400
2843674956070378800119100419304587264 1421878042854396703400398104154865664
710959303836802003370622999328718848 355489793123202827520523473290002432
177749967164002326677867723457822720 88877518883201619797736855135322112
44440027092201038128269924270866432 22220647371400633178835710487035904
11110640598350373646768229419319296 5555478755500215352059301797560320
2777818605912621940367244442730496 1388948917037568102352418993340416
694494265559412617260607882657792 347257036300020591672503134322688
173633469910167437357351163658240 86819210835162289439225380077568
43410843357620430099887589163008 21706040648829857740081244143616
10853329809424750215109346852864 5426819647217285780089035816960
2713487194861098226311699103744 1356782283056776781289440149504 678410484341502224711515373568
339214913577308029389155336192 169612292491932473211276492800 84808564097605465863987658752
42405490974622347561168535552 21203349950220981095171620864 10601977206565394204879486976
5301139719010148931086581760 2650645417368800379866710016 1325360487616263147095064576
662699133274063052128387072 331359011369997265354620928 165684228051481502322524160
82844475208982185983868928 41423418196111810403237888 20712299393866263907270656
10356444844838311306461184 5178369996371745329643520 2589258785162167503028224
1294666286069231170617344 647351589778689294860288 323685018261381502205952

161847120816709178490880 80925866251363802939392 40464086047186508316672
20232619484345557581824 10116597972548930502656 5058443101462541107200 2529293608325308481536
1264682832959673204736 632359430878346084352 316188722638427783168 158098864918841262080
79051684259234316288 39526968029524000768 19764046964715421696 9882304957334421504
4941293216155566080 2470716976821960704 1235393672783069184 617714428577579008
308866010381811712 154437403237416960 77220900641964032 38611549832609792 19306324672118784
9653437213966336 4826856045936640 2413496742445056 1206782730960896 603408545349632
301712862609408 150860726272000 75432510619648 37717329051648 18859201396736 9429869133824
4715068784640 2357601501184 1178834305024 589433929728 294725353472 147366871040 73685532672
36843814912 18422431744 9211478016 4605870080 2303000576 1151533056 575782912 287899648 143953920
71979008 35990528 17995776 8998144 4499200 2249664 1124864 562448 281232 140620 70312 35157 17579
35159 70320 140644 281296 562608 1125248 2250560 4501248 9002752 18006016 36013056 72028160
144060416 288129024 576274432 1152581632 2305228800 4610588672 9221439488 18443403264 36887855104
73777807360 147559809024 295128006656 590272790528 1180579135488 2361225379840 4722584977408
9445438390272 18891413651456 37783901044736 75569949573120 151144194113536 302296978161664
604611136192512 1209256632123392 2418581983723520 4837301406400512 9674877690707968
19350305137229824 38701709786087424 77405618595430400 154815635237371904 309640066567766016
619297725321576448 1238630635015241728 2477331638774661120 4954804015037677568
9909889505052065792 19820341960057552896 39641809820021948416 79285871439857582080
158576246479342534656 317161500157939810304 634341014714389102592 1268718058225797169152
2537508174045632266240 5075160463279340388352 10150609156934832488448 20301794774621968400384
40604742470748543647744 81211790784506300989440 162428193255031029366784
324865609882098913509376 649749666508271536570368 1299536226504690492243968
2599146239985675822694400 5198440053923941321801728 10397175255753061996429312
20794940807316482698510336 41591062206253682808324096 83184485595748800439255040
166373693557980470523723776 332756831848926680337874944 665532553163784839256604672
1331102885259432635674918912 2662281328382591185673256960 5324713772492634199993352192
10649729776440172057280380928 21300064015790151429148114944 42601336957399917487470936064
85205091766439064233291284480 170415019236156586983281393664 340839709878870090999960436736
681698762570854016066716172288 1363436210767935700267022942208 2726949792788326736801227079680
5454054328081564146136816549888 10908418141172949637342357880832
21817455252365541964822165323776 43636148444770369309919229771776
87274772769619309380388257792000 174554497299395760281876112080896
349118898119105803605951417155584 698257603278840173296301220298752
1396554820638937478761399212572672 2793188869440389221860391969095680
5586536195205806972395971026092032 11173389303061671002142316227985408
22347412431423456118985380807573504 44696092513447140467372258318352384
89394720328094737393547510043115520 178794511258590387704701006899052544
357599163721982601244613987423748096 715218609853568854159651922098782208
1430477784526345011660151738700136448 2861036698691104630001999266405416960
5722235656659038473367390110821122048 11444795831871735373461563377662820352
22890240700850787600376693067366793216 45781779475916208907660518758815891456
91566155100261685229135302765796392960 183137502497381905285899136027922006016
366285389588480880227055333048502452224 732591548364395899764624788082321784832
1465224635103660078150277820135277330432 2930532346957056713542612128211822182400
5861230847413586541569337232306179407872 11722794001826119312106900416377428901888
23446252617650131082150252736284997976064 46893834463296047080173409279630276296704
93790327382583663992092626173381113282560 187585971677150467647676867575003347943424
375182577178267214622336965606488938643456 750386422004466987898640392125942362800128
1500815379304799093105213706077813696626688 3001715829201328420826293255807485335306240
6003601799586117310884318198918686554718208 12007543881539155560232099772444804877647872
24015768327812152997391126294104473291718656 48032897785091989748636106086638673656283136

96068517829119347004979919170136801458257920 192142480176109429025375252333992511207899136

384295849387960328081581332655422838998564864 768613476847403596224824321285721311162662912

1537270509837773072572971954521193888656392192 3074628131961477905392590532941890309974917120

6149430488494819331278474313682785685274099712 12299209426133365703543535122963581501196730368

24599115750554185489060243237123183263690522624

49199625297683279142066832456638407049975169024

98402038188516374612026356878060895145138585600

196809651563332381879838097685689952380653666304

393630453499264029071246963230516228942060322816

787283207743726588765635462179305106245626626048

1574611016977850238777553995795155509214265212928

3149311236936494600047674134463401611874554347520

6298800879834577445080480554672984410641156538368

12597958571592331380131225680838331195066408763392

25196630767031015740202980504661387137701008900096

50394688781754737440287019295292223770538400546816

100792232058894886800336155162523346531349566586880

201590173108560597440196543468924491043244664160256

403191764198662842559441553225604578047580390293504

806406364360408980476980039026720348017342904532992

1612858400646984551670153943204463079879050056957952

3225808145146302284772695616710970927446828609699840

6451798977997270932410166694026031390271114210967552

12903963331403874590549884309260241851297142405070848

25808657413626414632558870460936841844104112776413184

51618776328890160168035944606706399971227881485369344

103240475661054982141908296583078232508495074835824640

206486797328659287895489407905487330149068773401821184

412985286670417223014324445289636390562294794263986176

825993957367031740475340149536596241652904083448659968

1652034682786458069844062816987839404362437156738695168

3304162901677705317474890669804972650838132293160140800

6608512875564988990523311411268532985902780545685782528

13217399895549134692193682965854241340258593010102566912

26435548079936582806681486218342833417423249857667137536

52872592737549792457951213009954368308658627390258282496

105748178630452838605078907166446139564941510130364579840

211502343571612184588510776625967085025131530960425189376

423016659764637383933727477838083781840760083320242438144

846057264772100797380866804848466787262514209439268995072

1692162420029853653788557308041532021687016504476106227712

3384420621031011425630762012772260937698009180147348930560

6769032804004631087368818818922915664043970702684970811392

13538448731894478646952227224602618905383846090150487523328

27077663711559390238333636227188129653595015498620066847744

54156859918659646365525625592464776239902621838846317297664

108316784828401024508767967878983853098172481155937001799680

2166339699638965512572969369146076307433079437268362738008064

433291659242257952256805605068369817339627824449702944833536

866607838413169758735344943689174039626193548725360827301888

1733264716683647225914157354483216889146262897102631529873408

3466627513081909868715249643176171398080277393509082810286080

6933451185593050571204369154771818035736057985625805121650688
3466823672511140702489119511595646637655780592116722311364608
1733460876112878059688027222902692128721766095710271030951936
866754957985092884065747345003780469307820947681090453110784
433389738956873369143740539278107437127379423753522695372800
216700999460600148127303703027162319800424186833250082095104
108353564721381805841368568207635460518579330894869408251904
54178314856231768809542642450844880568473284186556887728128
27089923675886317349200500398936015438828451462839535665152
13545344961828375146814839786224795296710130416200313733120
6772864042856795809514714686490791437003017550490429816832
3386527802399702022811004739934592612825484946440351383552
1693311791685503070432326068311894753574730558817743929344
846679841085577564729574883328246600368359322207656083456
423351893164201797121493366250272911974676682503220101120
211681932892807405939099645418211261882586851951306080256
105843959601756956658726303855643033888917681325501054976
52923476378555105173951392501090218418270968337674534912
26462486478115866009269816537179459946041548006299271168
13231617383477089715781968411906905341473805921880637440
6615995763948123213464514277612040354963418920305819648
3308091418078850784519022174635314019594967439835660288
1654092477091819981152893605232303930854112709759205376
827069622572107285023138061573475425955370849800290304
413546503299152289734914660265399443241842672360488960
206779097656125468479130144872030586752999959910416384
103392471831337396045401479805680725942539291820294144
51697697417306028925618943587673079254289301842690048
25849579459471679914268573636252897768654478887616512
12925155105145172682863837739334627955082153426944000
6462760240277252704296694330271403512918533705039872
3231471463990959533580734895437746524147995348303872
1615781403921646357506561312869895645918361922043904
807913537923906474111377589010459014881363084967936
403968186943494884734737260792985103401772604456960
201989802462518266206892863540370349681431833214976
100997755726644545023208548342124073830988682100736
50500305111014978471485332457031486410630723796992
25250866179353842215683195371500467952883553271808
12625789901600097597811862257242596350225872322560
6313073356761637043891063414367479362004984004608
3156625881361612644438097850056830274448515923968
1578357542171203383465331996464960433947269922816
789201071830800222355807533950752865335140941824
394611686287999376489474534834512756848323461120
197311418330299320900522651346824540514538225664
98658496758299476778154017638196351302457360384
49330642175724646553023354801490216173822803968
24666017986149777358484850391941128348208463872
123333574422186157202290116915685743047527628800 6166852945681171380607799093583292217700646912
3083513585126517450550546170691148641512456192 1541800348706224605398596397295325587087294464
770921952424595242760959854622538426709180416 385471865248039091411310755298707029937356800

192741377141890280721070791643072423260061696 96373410829880507868243102818395665775722496
48188066544407937687975404907627559960707072 24094713836937810720914629203028643516776448
12047697200835826298920777976121753526599680 6024018741601373618692120675364592647405568
3012094441392417043961926181334154265755648 1506089755992073639288896012493006103904256
753066145643969378298414467159467537465344 376543706645950968476190464036216011489280
188277170234958623901586847246349127122944 94141243573470881782539231237295124250624
47071951014731225807142519425707842469888 23536640121363505361507711616384061407232
11768652367680698909722081759957100789760 5884492337339822569345153855861085437952
2942329245419647841914633415871810240512 1471206161084692199578344951906538881024
735623849729780239099686597938586320896 367822309458607189205100359961951600640
183916347026162129430178710477305020416 91960769661510332128903620486817120256
45981682904969799771358942867490865152 22991490489592216739133037745786585088
11496069763349766796293302028913868800 5748197140951712611510042592467222528
2874179700114270912436717085238755328 1437130414876342759559206437121949696
718585489847775031450027165812260864 359302886128689341560225556531773440
179656513666745583697718765078708224 89830792134573248307662375945764864
44916663717886852383232684676087808 22458965684243540306317090689646592
11229799754771827210508919520624640 5615058333710942133929646848212992
2807608395017985331302416968056832 1403843811590249797820005256003584
701941712835753464994401013989376 350980759938191015539399699988480
175495331729252649290799446491136 87750141744704895405949521494016
43876308812391733083249659871232 21938773376215509231762279497728
10969696173117575960949864529920 5485002829063698653009294655488
2742578785784304662771828523008 137132807851837999519504859136 685683382072303833826547728384
342851362442708833946671513600 171430516924632875490034581504 85717676313955667003366703104
42860047082797448130858057728 21430628004308531380016381952 10715616233609169347301867520
5357959232532036502297772032 2679055174129744165472305152 1339565365996735039897862144
669801572464298998529785856 334910230965115238555320320 167459837849040488922873856
83732280107761679284043776 41867320645501557053325312 20934250618561137232314368
10467420457185747968983040 5233857802545463660904448 2617002688249026668658688
1308538237612660753432576 654287565550404086267904 327153006147238897909760
163581114759637876342784 81792863222828151865344 40897584532918682779648
20449368727211644813312 10224972593981974118400 5112630412179062915072 2556387263683569385472
1278229660638803656704 639132844717911310336 319575429558210396160 159792218378732568576
79898360989179969536 39950306394496827392 19975716147201835008 9988139548577628160
4994210511777169408 2497175624632762368 1248622996688470016 624329090530279424
312173341358161920 156091068725592064 78047733386051584 39024966204653568 19513032858140672
9756791306977280 4878533092442112 2439335265697792 1219701992587264 609868176162816
304942678016000 152475633975296 76239964471296 38121055977472 19061064859648 9530800865280
4765534650368 2382834434048 1191450771456 595742162944 297879470080 148943929344 74474061824
37238079488 18619564032 9310044160 4655153152 2327642112 1163853824 581943296 290979840 145494016
72749056 36375552 18188288 9094400 4547328 2273728 1136896 568464 284240 142124 71064 35533 17767
35535 71072 142148 284304 568624 1137280 2274624 4549376 9099008 18198528 36398080 72798208
145600512 291209216 582434816 1164902400 2329870336 4659871744 9320005632 18640535552 37282119680
74566336512 149136867328 298282123264 596581023744 1193195601920 2386458312704 4773050843136
9546370121728 19093277114368 38187627970560 76377403424768 152759101816832 305526793568256
611070767005696 1222175893749760 2444420506976256 4888978452905984 9778231783718912
19557013323251712 39115126158131200 78232451339517952 156469300725547008 312947397544116224
625912387274276864 1251859958920642560 2503790286585462784 5007721310659280896
10015724096295272448 20032011142543966208 40065148184994775040 80132548169803235328
160269599939233841152 320548207077722423296 641114428553954328576 1282264885904927621120
2564601829403893170176 5129347773995862196224 10258983778367876104192 20518544017488055631872

410382409564807181110720 820787877559706499153920 1641621871979597272186880
3283335977679563092131840 6566856422799863279779840 13134081780481200750592000
26268901430725349883248640 52539278600976596530626560 105081508681004986589511680
210168920320113560235540480 420349646556434294584115200 840722904945282937394298880
1681493033555394571240734720 3363080514440446535385743360 6726349923540207856580034560
13453077636399045284777164800 26906910851435349712788520960 53815332860145217712045424640
107633688034839471997027614720 215273420698777017139928760320 430558930655750180571604582400
861142039827892653726703288320 1722332436688569892620394823680 3444761587442708955574766141440
6889716603016556251817485271040 13779820062295389184970876518400 27560413837115331732613564989440
55122375099279770190570753884160 110247845048657753831828755578880
220501879797511934565032006778880 441016138995416722932813004800000
882057036791619153471123992084480 1764163591184809722153243949137920
3528426217572762274728479828213760 7057050505551810210300943516303360
14111449715191619174228985475235840 28229786585457526127955644944220160
56461157734165337542663160767447040 112925484594831245658830063292907520
225857307442663632464667610101841920 451727291391329547223350187235737600
903479935794663659034730308535582720 1807010577613336447245520485199380480
3614122567274691152843160706655191040 7228447958645418822390560885823242240
14457301565482910678189600716672204800 28915414427349967423196159323395850240
57832451447468226980026234426894581760 115668148080473038227320300413994926080
231342786532019244989176263948401377280 462698553806184827047423854137625804800
925423069096662328232990360756897710080 1850898061161910004742266026477087621120
3701899968260990706037102662880759644160 7404007628396322805179346545614688092160
14808430640541328396568975530935713792000 29617692048580022365558515941284102799360
59237045632154775875958161641393556029440 118477414334299014041598582800437812920320
236961474808576952662561684636177027563520 473936241897111754483852407342956858572800
947899068354139207285162890827119324037120 1895851305828109811205241933936649861857280
3791808949895882415680316172438122151280640 7583830576271090417900296954005889157693440
15168086505500832008879923126271068025651200 30337023716918966363918504689060715471831040
60675748845672537420154326251158589784719360 121354900515014284224943286248391497251553280
242716606677366987219155839988931629867335680 485446824649410811976850214962160530463129600
970920871888175299030777499892915602383175680 1941896188955057948215709139723020287680184320
3883901268267530596739726559320418741188034560 7768020317249890594096069678389593814031400960
15536476195929439989425372476276700291373465600 31073823514718197581317211191548425909368258560
62149389275155030367567354861086902471979171840 124302263041747331145000574678153906250443653120
248611495066369203109732879268268015113857925120
497236928098487487858929218360456435453657088000
994501732128473138996785356368753681359196651520
1989059216119942604551424552033188983622158254080
3978229935965877862218556782657741209051846410240
7956682879383741030668528922498208901718752624640
15913811773671452673799888559361870770667624857600
31828515577150846572525438547454647475795488931840
63658815213917575594902199952371106820511456296960
127321198547066916089507045619665837378863869460480
254649533332597361978419382669178922233409652654080
509313339142121783555649348198052339418183132774400
1018655223238097686308919862115493668739093920481280
2037367536383903611013082055669765317283643150827520
4074849252583223698816648774217086594178196921384960
8149926864797280351214266874189285107578215082229760
16300310448856226609590472399888794053600072643379200

3260153433623578503350482210279803578408743024459776
6520489554951823369565739881163696692194943040487424
13041344485312979464861030683535572455144800064110592
26083419721444624381181163209487503051799428094492672
52168300944526579665280530103807722386618512121528320
104339524892327821136397467577280877339276336108142592
208684895791204965884467749893892619810631295946457088
417381483595508578992281129266446969885419839353257984
834786351217214452431253517490217400299154173627203584
1669619470486823493755889552895081721654937337095782400
3339332477078436165298544141619457285423132653874315264
6678852026366450686170618354897502255072781267114131456
13358078197152058083488296853112179878598594452959264768
26716904683142429589270713992858710494103252743380533248
534353059439614860231296685558986122462018633161685073920
106873605043276225735435818264509647871661521673218162688
213753196397258958849224598822094101638571554046132355072
427518365415930932455155122230337815067640129491656769536
855060676074687894423722093632974853716274301782097657856
1710169242635027847874267885610548154594536689161763553280
3420434266241359813802183167910293203513049549518663581696
6841060094425327863711661129198980195674051441427600113664
13682503312735872199637911845154747968644007567635746127872
27365772873242177343705002863823071091879824504832584056832
54733078242025220576268364074673292492943267748787351715840
109469221475132172930253444843400885604253772975819070636032
218944572932427809415940323074910372445242020908126875680768
437901405829182545942747512926037947363952991729231220178944
875827331587018946107228759404510299674843883284417377992704
1751703703031345600657924985913889409243563566220744631255040
3503505485777306618202784906037516438274878731745309013049344
7007207130983844070179439680494508116125260662098257527177216
14014806580826149807906619097827966711401527721411794056511488
70075994498423057377271794173334585952762670593135365293334272
3503897804635768285750524642876466917425885128960588015206400
1751997942175191851318729788543102268606818364132203882872832
876023491016249779881098627823985539250347081892056879071232
438024005472451817051416180688209972098642490859005908353024
219018132718389372081141523732213587286055720385991688585216
109512131350276417818287478560161094261395097671240211496960
54757598170679074798002097627107697439881167574742289350656
27379565333109970343430227987067423874532393156932236476416
13690165790440201643929703580290499514562101263246664138752
6845274457162709058072146583523643545929002974013605019648
3422733009552658647089720688451018667288477658201938984960
1711414395261981382571684042570107780806226914698537730048
855731142873816720799253870457353113984107500148052983808
427877544058321375156332859814826168782550771473418551296
213944758339867194956519392200487890286523896436405305344
106975372325286851167436177246781348090886203568050667520
53489182740320052428306329166593755192555229458949341184
26745339658998339636447284884964038496533678566936674304

13373043973918326529370702585799194616719871202199339008
6686709059168741620258881364558184992586451560465170432
3343448065689159987916205718108386338406483759915335680
1671770800896974582851485376968840090259870869799043072
835908784474684585872433947441743505658249929820209152
417966084250440940159562603199533483093282212370448384
208988888131769793691454116339097606678719729915396096
104497367069159558651563465539214235905399176822784000
52250145036217110228699936454439834235719244343934976
26125803268927220565809070069636275259369450138238976
13063267009872943008634085956026316700439639052255232
6531816192641137867181818438617247885597276517695488
3265999440172902115023296949610668710487366754631680
1633045392012617648227842339956356739088047625207808
816545531969392119472018102553689561466205936549888
408284183966237707415057517564600376694194030247936
204147800973889677547052991926177986327642546110464
102076754982330250693288612535027892154094038548480
51039804738857831306525364553483395572183402020864
25520615993275268633203211419726422533659892383744
12760664808560810806571870281355573640614041878528
6380510810241993648271067426423968007199068782592
3190344608101790946628099856085074597045558312960
1595216905541292534560332999479082595245791117312
797630753515844797903308035457813945984401539072
398826527130521664263224785588043297172953759744
199418838751560464787397776723589810676853374976
99712206968930048721591580326578986383614935040
49857497281039932524742136145681533714401591296
24929445538807420344344241064036787118497857536
12465071218547437213158707027616403689897459712 6232709833845582127072646761607206910272995328
3116442029208722823782970004703105987798630400 1558264570747327292014808314301304260230381568
779154063445146586069065813125527763280723968 389587920758314763065363734550201698223128576
194799404897028116548097281268819757402947584 97402424707449425781756347631269332847165440
48702573483192396644732027314064393496428544 24351967306330040199292940406247060284637184
12176323935531941038109933577730961910530048 6088332108949430988286698476169196839370752
3044251125066445728759215081736456361738240 1522168097829087981687540462694157151895552
761105316562476549497736692260043061460992 380563292105204554075851576586503773487104
190286962964585416701417403521493008121856 95146139938284278182454509374867064750080
47574399197137924007100158494493812719616 23787864212566854461486531150777046532096
11894264413282373459711491527153593352192 5947298360140659844339858739459331719168
2973732256820066479411985857670933381120 1486907666784901518327021172806100451328
743474602579884898474024708388367106048 371747685883659518892269415186841993216
185879035238688294273763238089750216704 92942113767773414550695884293039718400
46472354958101340982255074770602164224 23236826516157987344581103697342234624
11618737776632652099017335004691693568 5809531147593155262872059080356134912
2904846703434992238117725329183211520 1452463916536703422399710559094177792
7262522240677955362870279226798374912 363136261543779507270351587024830464
181573201374290666552781780325236736 90789135988345789735193883569029120
45395835644773123096998438487719936 22698551647686675663199967595462656
11349592736493394888950357973532672 5674954824571725973150366074667008
2837556640448377250912776581283840 1418817934305445757625185062617088

709428774193351444896990917296128 354724290616990005490694651641856
177367097068652144266446922317824 88686024414404642893773259407360
44344250147241606827161528827904 22172744043640446103718213976064
11086681506830044396927831769088 5543495495919932870998278275072
2771825119212421771766320332800 1385951245232438554016750764032 692994965429333111075170680832
346507154121223472570982989824 173258412763890194802190319616 86631624233584326659444572160
43317021042611777958896992256 21659114984215696294035849216 10829585972356275180431160152
5415080977508827730802638848 2707616046618139779724738560 1353845802240932847024078848
676941790586397902092894208 338480340026164690336874496 169244892379565214813650944
84624807373024042229432320 42313584278132738526019584 21157382434876727968661504
10578986365343543337156608 5289640756624361344991232 2644894165288475510702080
1322483976132385174454272 661260434810266296778752 330634407771700003165184
165324332074603428970496 82664471880310928179200 41333388861660070936576
20667270891582338891776 10333923676167321157632 5167105955271736434688 2583625034229906145280
1291845459119720366608 645942287354495500288 322980150876502491136 161494579037878616064
80749541318752993280 40375896559283339264 20188511229595090944 10094537089774256128
5047409282375483392 2523775009931919360 1261922689338048512 630978936855068672
315498264520556544 157753530306789376 78878964176650240 39440581599952896 19720840555790336
9860695155802112 4930485016854528 2465311227904000 1232689973690368 616362166714368
308189673291776 154099131613184 77051713290240 38526930386944 19264002064384 9632269467648
4816268951552 2408201584640 1204134346752 602083950592 301050363904 150529376256 75266785280
37634441216 18817744896 9409134592 4704698368 2352414720 1176240128 588136448 294076416 147042304
73523200 36762624 18381824 9191168 4595712 2297920 1148992 574512 287264 143636 71820 35911 17956
35913 71828 143660 287328 574672 1149376 2298816 4597760 9195776 18392064 36785152 73572352
147148800 294305792 588627968 1177288704 2354642944 4709416960 9419096064 18838716416 37678481408
75359059968 150722314240 301453017088 602922811392 1205879177216 2411825463296 4823785144320
9647838724096 19296214319104 38593502380032 77189152243712 154382599454720 308773788844032
617564757557248 1235163874852864 2470396469182464 4940930377318400 9882135632543744
19764821020901376 39530741553430528 79063682130116608 158131762306744320 316272320706510848
632562233599066112 1265159651570221056 2530389671884619776 5060920081257594880
10122121637491900416 20244806224937222144 40490738349781286912 80983728499376259072
161971960598379888640 323952928396014518272 647923871190538518528 1295883771178096001024
2591839599950229929984 5183823315088535715840 10367934860553223143424 20736446181858749710336
41474045285222106267648 82950396413453426229248 165905404512925279846400
331820032397887414468608 663658511539848538488832 1327353916567844496080896
2654781620111983830368256 5309710814176557337149440 10619716776258294027124736
21240023848326946759901184 42481228288274610931105792 84964817759790656684818432
169934357886064183014850560 339878160505094105320128512 679775210476119689221111808
1359588199884102335603933184 2719251957631930585531285504 5438655030991312999709409280
10877612293437529656712495104 21755829049784866628012343296 43512867025389347885199392768
87028151902417925028748197888 174061139508114308574195220480 348131950422785534181788090368
696283243658684902430371479552 1392605172943597472994333556736 2785287717139650282255848308736
5570730176784211237046059008000 11141769838578243819160842797056
22284158647176130328459135156224 44569555234391546037193169436672
89141586348861662834936137121792 178288124457880467190971870740480
356586152436075217424142934474752 713192111912779000932684254937088
1426423837906815134034165281849344 2852926903976144532405924107649024
5706012264277317593487035303198720 11412341441204692244324444782198784
22825316707709498603349637916000256 45651901066019225436100772535205888
91306337433238907331004538476822528 182617745468878727579615063766466560
365245632142559280994442101158576128 730511546694722213659308149568438272
1461063658208651730659464193639448576 2922208446055718068000624176284041216

5844579151388265349364639930578370560 1168948282133018912545606301717731737 6
2337961467976769510436569234639578726 4 4676052743375002391563851731687387955 2
9352365101592931524509129988191236915 2 1870524943287171653178111302601539584 00
3741153732511514002908793215129663569 92 7482515156897369398922727650112495943 68
1496544569754342158405573773993132949 504 2993172216258420874053204035927533420 544
5986510586016314862590521047737601884 160 1197335347903157595414926804724027385 4464
2394737157206104436623498799801068788 1216 4789607237211787364834287980308165610 7008
9579480320022731712843156722028387290 3168 1915949233124377739203547496688088671 84640
3832004804488418271676927297940999771 25888 7664222285456162129893519205011644397 64992
1532886992387097543286636762828257850 556416 3065859055365925321189139369308373643 165696
6131888251915311116100104259204631704 37120 1226411678619754316168348422644835810 9085696
2452891413712892820029389520211157975 4594304 4905918940372554015444164390265288658 2034432
9812110106638644781659099480216522730 9760512 1962476466506436306485974035980493629 10904320
3925061823370287313280256351835365424 04575232 7850341427455404027177129263419487179 74683648
1570111841634046685558749164633648702 280433664 3140310795554025131364144953166799937 223000064
6280795815679913783221583154132604939 770265600 1256194008050355460742975280386322001 0189062144
2512457705929456329683267859892246028167518617 6
5025054791516403475761170318023696108594449612 8
1005038834234778858431160983252580032170772398 08
2010133420332554043420175805800841685245309747 20
4020378344391100739956059290193046612298149396 48
8040979696234186786143533937568819708211358597 12
1608240540737234418474989858950309238365283680 256
3216570284455262959442545860773709070176591282 176
6433318974872114163870224007293599327245230407 680
1286699476166740481771071258607956102827455650 2016
2573470314718116261536195431514384680411730437 7344
5147083354205503119060496691625714310337099150 1312
1029445215794954743009720504044531851970147484 95872
2058947521497617724414683339527841683745750279 78240
4118009222810651925619851341933239327102411179 29472
8236246805252136804820672009621590573426643598 04928
1647295032976593951680328267075340498529692967 501824
3294681409805521084793044264452725764748114430 787584
6589545507315708532450863989509541064873685853 143040
1317945639004074979063127890022726120050228568 9421824
2635964353090016503272165964287088054251439934 5115136
5272074856343766096836152297057447736804845462 2773248
1054444201301499837425594533108143873021362211 10632448
2108946862668492910967917213609596397363510659 51436800
4218010645467972294169290722005810097368593793 63217408
8436255131197917532805494033584854800020332536 47122432
1687297794291978095454481324631617881060695497 135620096
3374689124688745368695727685092529604234648973 953990656
6749565321587069092964985441843646892695813907 273482240
1349950478759329489707770310270044691538446597 3327966336
2699975786402490321644818234064328904459538330 4017936384
5400101230572643327748460525455527956300289428 2959880192
1080050177668061202441456916556479620736300439 15767775232
2160160218443187478666443456043707300425085985 31231580160
4320440163100505104899946157948910718755142184 61855219712
8641119778629270504934010807620813673320224797 22494558208

1728271846211506160013625859868761181826033045042557353984
3456639473394316438080899116426719257976042261280251183104
6913470508731241112269093026231832304600036864950775316480
13827324141347698696752775639220452186495978414682096533504
27655414530465830337934730451954479527583766198925284868096
55312361556472526564727819250936109364351151136972753338368
110627788104026784907172355195926519347069539752189873881088
221261706190217033369778143779961639930873554460868482170880
442535672344760993850423154336140482335216058834714433159168
885095864618175841922580042224715369617370017495383803953152
1770240769093659392288627551554299549128615834642677483175936
3540579617901934201464190037318336718044983268589174716891136
7081355395233099236702249943056148675665469735785988934860800
14163103109324660140952239622951247830481945868787256871878656
28326990856366243616999958719580396619265904532005071748071424
14163887747041583476047719096629148788783958663217814876192768
7082140032950022571797729416734049633967482530216546939174912
3541168096189626702785799642576762436771492864412093220126720
1770633087952121059836367288393250028279622231857956485332992
885341063904714384139917377749059419086749015754933180301312
442682791916684119180825555650746912016843457790444058968064
221347525940505523145846211213482057245156203851710763892736
110676827961334493350639822300795329240945339404099749150720
55339946476208112564178269497424814929656288441172058177536
27670739485874489226518313922225982619419953590147120889856
13835752866822461085473746547869778887005881479854106345472
6918067995353838778844168067313283232150893082317326123008
3459129778648223507475731430345838510399422712353799536640
1729612779809763812764689413517517702361699441774468005888
864830335147707935895756555931058074761843763686018121728
432427140195266982704584202551678649171418903242401120256
216219556408339998730645063568914130480957962320896589824
108112771359523253054499012930994468188103236510296309760
54057882257438253371837747038765935567863745930072162304
27029689417557440108212993806017318520837936802498084864
13515218853197876765253557046325834628872000319980044288
6757796498808516738200308594821504998662516119355523072
3378991785509047546886919333240046341444516039360512000
1689542660806918362336842184534670091778887009521631232
844794714429656475615112351224658506417757999681503232
422409049227926885030901805090990983473036247301095424
211210370620512766127123717284826356868596747380719616
105608108313531044869398266012078611000337685555445760
52805515658402853337617336690872021783188498710265856
26403488580020092120267770187852369033104077321404416
13202109665419378785863436015134363587306952643837952
6601237520414355755796493468171271329030933313486848
3300710104059511059330634464387680432204195152527360
1650400723955922120381511097344862599946461824155648
825223197941044355548852481247942491895413036023808
412623016952063825453474706911726841908797579984896
206317217466802736566261586599741218934944320978944

10316146322878678020289290987180950845774492598272O
5158215886208609606132751322187420372400884573798 4
257917930548894010106042857539218266095726142423O4
128962533393678769952724074484532756785704028O78O8
64483050756455267426213360099728190261772492472 32
322424174080355749380323414785950010653464854528O
161216547189217580814790014536629534999033623347 2
80610503669128643469709160840142032335667409715 2
403063668718242482660116572059846485859090038784
201537409545420873985843669959491405019921514496
100771492365860253320814526944529783555149004800
50387139979505034824353609454656932300168626176
25194266888039971494149977718524486411381374976
12597481893163712788061575354860253336339218432 62989151711537199145240809252291317334938746 88
31495544697862791717508687086514068399409070080 15748159050743617388776668552067854660356014 08
7874297306086638095004950835782683661833338 88 39372575434007337478107836977657199967443353 6
19686832168790742240595459888200490812860006 4 98436883102889078710685006437861908209991680
49219802680912223109196356717360681177841664 24610581905189953431525105107895204125343744
1230563123496189765422601592855503383088332 8 61529857586644092963447396515812327995473 92
30765779499239348827882356694424743418265 60 15383315102578325587020507565471661419397 12
76918702277684883800499183918654755648307 2 384604145212390698329479150049756020998144
19230738951817848882823119025311913187737 6 96156353215080814245861402740680126627840
48079505835536192038803605177400343658496 24040417531765988477338254492230312001536
12020541072881940467637353197880226086912 60104366899404433483027895748226480865 28
30053014217199582313934512753525915648 00 15026922492348473943177538816469295431 68
75136689380485783646939106280878165196 8 37569383149614598788995259239704926617 6
18785210804493152877260482669485385318 4 93928650170895031800116678595591536640
46965623159662149606965471921878073344 23483460616938391656936302272980189184
11742054827022854255194934292510670848 58711896727882563409608587242656235 52
29356759660325427771621251511379558 40 14678785478354786919219104700715499 52
7339595563273429976313791822870609 92 36698991936847332465090156476917350 4
1835000302866375752430567691974082 56 91752550444519244080331378005114880
45877542872859850269567185705762816 22939405261730039249484341204484096
11470019543515076682092544778043392 57351682280825668697214594769223 68
28676633422037976991983232824115 20 14338712851831559817679584131809 28
7169554496322065569683775925780 48 35848762833641756152638798928281 6
1792487659283659222842935911383 04 8962685884426153190269659381760 0
4481466736217005133162319603302 4 2240795265110466835594904757862 4
1120428581056215552304324857036 8 56022976477859884340559866757 12
2801226195145449553295174533120 1400651783198952444781177864192 7003452344125900564573842309 12
350182288612851945262089764864 175095980009704431147743707136 875504078564914448322212659 20
437764128540653370452853391 36 218888108899424758372300226 56 1094470767642614157590868787 2
5472504953940522616601182208 2736328034833987222624010240 136820179634885656847371468 8
684119787640359762817712128 342069338553145620699283456 171039391643055679994855424
855220570047692748200345 60 427622090940053548213207 04 213816948428130361163120 64
1069114256931169741098188 8 534571885860843838190387 2 267293321628051402915840 0
133650350162840443368243 2 668270197558275926392832 334144322151174817972224
167076772761605836374016 835406922238121318809 60 417714990334106727874 56
20886325977457639817216 10443451219104971620352 52218697247405616660 48 261100691996431876096 0
1305539488779178344448 652787758788098654208 326402886593304068096 16320594689627940454 4
816052252479533875 20 408037385238835363 84 204024322118951895 04 102014975809243054 08
51008895279505080 32 25505151327194316 80 12752927507318046 72 63766396755194675 2

3188407798689955584 1594247879810088896 797145930137600000 398583960185077776 199297477650677776
99651487604408832 49827118191738888 24914246290636880 12457466742702208 6228905170042888
3114538484367736 1557312191856664 778677570764880 389349522800064 194680130109444 97342749409928
48672716881922 24337029529960 12168850309112 6084592926672 3042380349444 1521232117776 760637030440
380329000996 190169743336 95087493122 47545057228 23773184000 11886919668 5943623668 2971893766 1485987844
743014440 371517444 185763844 92884448 4643522 23222400 11611522 5805922 2903044 1451566 725800 362911 181466
362933 725888 1451800 2903688 5807522 11615366 23231366 46464000 92930566 185866244 371742722 743505922
1487052800 2974187522 5948538888 11897405444 23795466244 47592243200 95187107844 190379458566 380769402888
761559777288 1523161497600 3046406881288 6092981534722 12186298613766 24373268316166 48747878809600
97498441973766 195002252656664 390015242731522 780051960299522 1560146870272000 3120379639889922
6240931078471688 12482205754327044 24965098703421444 49931571796377600 99865892371824644
199737282301788166 399485559719854088 798993109672263688 1598030199809638400 3196148360549498888
6392472642959441922 12785297129639772166 25571297946721320966 51144003268326195200
102290821286419496966 204587272072373207044 409185803143814840322 818394124285766533122
1636833284567806771200 3273756641128160952322 6547693426241416724488 13095747140453023088644
26192214856846425456644 52385870865573609472000 104774624034908736061444 209555012677340506357766
419121554569727081185288 838266167569546299310088 1676578451999276872499200
3353249137718922292756488 6706682742878581681029122 13413734420638637553090566
26828206711040223488245766 53657889161606343740620800 107318729802264481009500166
214643362562632549075517444 429298531041472272264069122 858620673915358892754206722
1717288571495546481960550400 3434671590320750356825374722 6869532075300815499459297288
13739441939920260570535690244 27479639458477780284305571844 54960790074230078855079526400
109924602463009194283095818244 219855249555116461712065167366 439722588368429069715877396488
879469355253250432015248916488 1758987067539285449197486080000 3518070849144140068728948654088
7036335126419418478125850296322 14073057109101113637587606568966 28146887930726780637847025090566
56295323286502668001037674086400 112593741423103549452762595983366
225193672546403525806899687587844 450399724493199905416548366417922
900824207787185518438594715320322 1801697933175942452088185395609600
3603494901555027734598362721157122 7207187873516341130040709302190088
14414771887845253581769386324131844 28830336057315649806914708087767044
57662256677881584900581287054540800 115327682482263740374666315867095044
230661703217528621896340115250216966 461336082941059526086695197532487688
922697518894123616761420329129082888 1845445743812256362698900526386380800
3690992899672530983749920789029191688 7382188623441098484204081050571243522
14764782895074270001816641046168207366 29530377086532686070450239982387855366
59062376765833664274534395744878592000 118127998717203912816336623049962946566
236262487805480994167208909220337418244 472537956353108325403489144681497886722
945101874190509324945120941844641873922 1890255671349603998166527188652575948800
3780615188636378692885624987231736299522 7561438069147098778876391194316641402888
15123291522042880343963064828339620413444 30247413811583126260346694536091916042244
60496489158160983665534518831009182515200 120996301386311429620751297179669065891844
241999248912601783820867113394639533506566 484011790105161416800463264859881870458888
968050164770238531918384605860969347809288 1936153498660308460471685364004349909401600
3872413335560279714213203032573522246369288 7745039347599885014966070674276689347870722
15490504048158421203011470566812668406005766 30981858802234144752181599570143916232540166
61965419016302894196680516013324991306137600 123934240856274997779956657727243002943897666
247875287359888414325260599037597235953008644 495764186014453666189059733059491742634475522
991555594618261007455196536087578026725867522 1983165634415229365064547212112345136365568000
3966440159187873430437402704099068438558801922 7933098099090576261491421967946893208772935688
15866631759610811324216077055391299080856535044 31734134642080940250898620349777623488334397444
63470011529880515706730173177545297629911449600 126943507551198301823326211311070696566308208644
253893984085271144466384152534101595745587036166

507801906136291370572231764892123596717115310 08
101563168820408090442339044943208800388611309568
203131912827115813540463473815985762867599114240
406274976026830892392497715491107849915951218688
812572252798860315408136966700488348193408417792
1625189107088117692062557004837521993109828796416
3250467417157029506617680152548134579665681514496
6501113240275647258220492590842450346223410872320
13002583292474471006411249753177263066230917431296
26005880208795294992763028649339250880030026235904
52013187665283295945407115584647951255196435218432
104029222982595200381057634774123480150066563593011 2
208064168642674831460676928626347400981876802846720
416139755266891310600402323540450397924844667666432
832302346496865916558901579656411987771871459278848
1664650364919898423833997024463846359388107166449664
3329392073692130029100381779229737486464942828683264
6658966835088926421065539019063564508307342648934400
13318299045587185567860628959335308087369599281004544
26637328841993036587180359761086974316249026528280576
53276119185623404077278923207006664915517708989104128
106555161374521469960394253783678762397074729843294208
213116168755592263532461322306688389926222808341676032 0
426244029524283174288268274092038510116613414293864448
852511443074763643023227807141400480761541323508416512
1705069654201921874939838132197447882579711636858208256
3410232844508632927666441300224189607272681253399166976
6820652761226844210906412672106966898771878466163834880
13641679666872845132959885487531109165996788851058671616
27284107622584003688213891261696569068899641539579346944
54569711822844634221016023096661839611611410754082701312
109142416801042522131208527339861082170847076858013417472
218290819912791551640770016972796970236942664415722864640
436593612446996118038245958531743552264382350230837788672
873211170136818265589903766235786328109758743260459696128
1746470230759288590206631230816171103381505572118487629824
3493036242489881298466909858321539101086987315432111734784
6986264046922370833041114510021471990821926973254496419840
13972911217729958138296818606799731558939758631289538740224
27946588683230349221022816387113038272471326632140169281536
55894709862001564330903991121253226854126272003402522165248
111792484715084860439524698936560754326619781485049411534848
223591099412333184434482831261230109889974037926587557478400
447194458788993295979832529298677422253417025766152583774208
894413437506640446181398792149789249453771951358260105183232
1788875914870588600806265051404447308801419702368430085636096
3577849909455792618499465037018632237390591004040679921811456
7155895978340816070772799942456739714356685206688999344701440
14312184275540093809093339621752429907864376810593277691559936
28625153188797110953282158717182760774030766415617113387433984
57251875653028068576755276381721323464665558420095342783496192
28626722464230957623473117664538562690634792004478229396062208

14313753550973940479284298569108231824468402399454393700188160
71570729349162010734160191529735911518097043983348363511726O8
35786345471727159535949445106965331956926037984712379261255668
17893663134436656852409397224531354077401776988875288383332416
89470767665048669684220359477900210881702674926971935680 1024
44736609828957027553196866416571825688198232454783714721792 0
22368917912694860132141776547096772967772563723040730801766 4
11184765455455603243842559942953816545723005609344802121318 4
5592535977281888210807115806179623303779864678584619420876 8
2796344613417987399846475820441169167349113276248418890547 2
1398210619097515347144696868896263341404147106602264035328 0
699124465743018497183077913785971049566868787540159312691 2
349571810968639660396903696561905214215832010889593303859 2
174790694532885036101134218115412451824114814004553475686 4
87397741790725121001908293974936148270156811282155149721 6
43700068157503861976624739446083035314128107781016780800 0
21850632709822581726147665952348998246588904960477993369 6
10925615670446616231991481090828239418056878015223798169 6
5462957492990970800454564602740989856409651775104299827 2
2731553575379316742456694330033930001895432271298350284 8
1365814202131574042343053179348682537793019327522275328 0
682925808286744856728879596840200037319161259697687756 8
341472257753851346143116302003029402870906427817118924 8
170740805682165131960896402792979393541116112892696985 6
85372741243702295425117327292222042823389505938417254 4
42687539823161012434893226593977194438110477715243008 0
21344354512235438578613894770921683732263101230638694 4
10672469556445185469890588122427385122735481801827942 4
5336380928386325825237114429696964189669706494168268 8
2668263539275029457764467399090117908985836043711283 2
1334168307178448001455188791665876861568409420169216 0
667102422359690637014071941893347384321950409241395 2
333560345565078636650274743976878168929848054199091 2
166784739975155977396756758503541322849360451888742 4
83394653583886318234188072509321780616898438338969 6
41698468590097323884998882883436449904558325366784 0
20849805194125744326451864756106004750333715782041 6
10425188046601413355202144035246892274194134440345 6
5212736748069977273589177846220391086610705494835 2
2606439736419623934788641837408668018062171884748 8
1303255549402129616391347375853570246409495511040 0
651645615297223632694186916501403241893952539852 8
325831727946691228596350072538010680291578662092 8
162920324122385320422803343412659869818090527129 6
81462392135712513273715825278157199745195861606 4
40732311105116183168014989424992232290673229824 0
20366713071188054849586033105452932354374264422 4
10183635294909009057582285749204874281705956966 4
5091957027111995345185777472841641193112390860 8
2546048203384743080790206035540422622685901619 2
12730589466067442444937616673300123244078039040 636546895760558474296210158444906668736328499 2
31828215910887241317276974161240358763437752 32 15914543516873279459871720200117692045029539 84

795748953915146913055247665980760235417010176 397885365993314926558454660977817934291271680
198948127514528198294642744482627875437019136 994767860161994666550290792381733391864201216
49739754137567417081368393117516423004946432 24870557633517550417611123307973075038896128
12435619099125696147269025028593969287659520 6217979690746308542866244201600700527935488
3109074915964884506048987944452208206020608 1554579993278307370332426894052033074036736
777311264287086243820179907938981022531584 388666265967509401237073184425972754022400
194338449895737840282028207441227498389504 97171883403860489972759911334734309883904
48587270929926029902252859474427435286528 24294300078960907409062881640743857815552
12147482346479399933499666772136998993920 6073907326739173081233946361951034540032
3037036740119323097859029668916784791552 1518559908434529827550543078429026156544
759300723404699053085785661199829958656 379660746296066596198149891592573419520
189835565444891832926703476292615929856 94920378870875183877166003394472574976
47461487509652225645490134321318592512 23731392791933429676198633472700448768
11866020914520373264826099892370800640 5933172716537015845776441524195688448
2966667487906922529569916551102988288 1483374308772668568125806170054066176
741707436795937935733327032278319104 370863859602770793701875489764802560
185437000403786309768543731695222784 92721035503093611343074859254022144
46361785402147033900938926330216448 23181526526373631065170211516710912
11591080175836872589935479934156800 5795698544243464823642927054979072
2897928500284246676159057071439872 1449003864223380470248325307695104
724521739152318801208561039835136 362270773096473683646479712911360
181140338308393983344339452952576 90572645034275562432719524724736
45287560457177066596634661486592 22644399198608175988454780305408
11322509084313909339296114933760 5661409284661865342182419857408
2830782013583388007358391123968 1415429692417921671812786159616 707734189022074669973188378624
353876765917594252019991838720 176943218662075584526694744064 88474027182677021521696784384
44238222517158125390023098368 22119715721488870009598902272 11060160092199338662093127680
5530231161827121159693402112 2765191138777286494170120192 1382633348320506204246769664
691335563626184580704239616 345677226546058029642547200 172843335639511884466487296
86424029002997377055850496 43213195093119405939228672 21607187842370061675266048
10803889069090210190458880 5402092108497694771642368 2701119841225142224027648
1350596814100718531117056 675316853794432975110144 337667650269253342330880
168838436820645098553344 84421524253331762970624 42211915048170488332288
21106533984837547589632 10553555222794925506560 5276921726585538609152 2638532920886807232512
1319302489240422580224 659669259018720772096 329843636708615127040 164926321953934934016
82465412776781152256 41233832288297418752 20617479094102130688 10309021022027776000
5154651248502243328 2577395992995299328 1288733180869738496 644384182620913664
322200887403479040 161104841748250624 80554619897380864 40278409460318208 20139754485972992
10070152120893440 5035213499400192 2517675469176832 1258872094326784 629453227032576
314735203450880 157371896692736 78688095830016 39345121656832 19673097699328 9836817285120
4918542860288 2459338539008 1229702823936 614868189184 307442483200 153725435904 76864815104
38433456128 19217252352 9608888320 4804575232 2402353152 1201209344 600621056 300318720 150163456
75083776 37542912 18771968 9386240 4693248 2346688 1173376 586704 293360 146684 73344 36673 18337
36675 73352 146708 293424 586864 1173760 2347584 4695296 9390848 18782208 37565440 75132928
150269952 300548096 601112576 1202257920 2404581376 4809293824 9618849792 19238223872 38477496320
76957089792 153918373888 307845136384 615707049984 1231447654400 2462962417664 4926059053056
9852386541568 19705309954048 39411693649920 78825534783488 157655364534272 315319319003136
630655817875456 1261345995489280 2522760710455296 5045658859864064 10091592597635072
20183734951084032 40368569413795840 80739337850847232 161483073748205568 322974943589433344
645967479364911104 1291970143101911040 2584010654947999744 5168162047384354816
10336605569745420288 20673774089444261888 41348674078795366400 82699599957404418048
165403703514436206592 330816414228127154176 661650842854763790336 1323337714506546544640

2646747486607131017216 5293639088402337890304 10587566407180827492352 21175709275113958408192
42352571471732523663360 84707448786474261020672 169419509258966949429248
338848241889970753634304 677714930524015216820224 1355466754536177852743680
2711007296048650543693824 5422162166049890763800576 10844619480004960880427008
21698929255820280466505728 43380839103261278344314880 86764039389763991511236608
173532801146010852667686912 347075047024987444625801216 694168983515906367832457216
1388375745963675692826624000 2776827049791077299976667136 5553805215309606428600172544
11107912662074116514494021632 22216429787058040343575396352 44434068499935695316325498880
88870554851510619891000410112 177745945406299698298699644928 355501562219156313630796939264
711022467251426461328389177344 1422083620129080590790368952320 2844244611510616517847919099904
5688643965526143708230200590336 11377597416062108761529125961728
22755813802143860213195701485568 45512865544327005806666302095360
91028206968732582373882402439168 182061365697622306268864401375232
364132634915558895579927995744256 728285076871746357244254377476096
1456609767824749846657305526927360 2913298763812013957652204597805056
5826755983949056443979596283510784 11653828880548169945309566742822912
23308291586396454005319881837248512 46617850823393136240041260377702400
93238236947986728938885514161815552 186481544498374370795377015136452608
372973230201550567425966003898548224 745966742812704786522355955048382464
1491974050444616876385559804599336960 2984029230527648359452815398203817984
5968220720332125932269022374417924096 11936765959217910291264827904856424448
23874180955543137435983222121754001408 47749659985300908578873576867590307840
95501916119031084571561418983345225728 191009024534920703970751368463019671552
382028433663558477596759797918697783296 764077636514551094504033717822712446976
1528196811403970467629095679616058654720 3056476699557677492500247847173384830976
6113119552614828099484608670229304705024 12226571412228602427937443292223679496192
24453807438455097313811338487977499164672 48908944104905979543495580783015278673920
97820546665803528918736969180151118036992 195646410243590197500965553588543357452288
391303454311146674328914337633568957661184 782628176270225907311795136180102400835584
1565298887836316931931523194186133772697600 3130682846264364098478912232024125487448064
6261535833712188666189556151351966859001856 12523411949791298270842575677311365486215168
25047504464316438418612078103837594508853248 50096370058100560591078009706104916090552320
100195462375136488689863726409069286326796288 200396369268143712395142866811857480944975872
400803627572028894821116561611152778472718336 801629033215540729703894779197181190110969856
1603330162257404733953111287034411364655300608 3206690357434026439308872364587729825768144896
6413554939439916399111037976974464716860555264 12827458328023559839208662449546939564369641472
25655613554334573760390497890289899390036344832
51312620905244055684727341762971839302666813440
102628029403637927697347375490727759650521874432
205261633993575488050480134911023681391420243968
410534418359750241412531037681183686963593478144
821091137464699013448203611080640022288692936704
1642226876419795088142690293597825341300397834240
3284542955820384298777946730068741276046819590144
6569264317602356842541025745883663738985687023616
13138885447127890175052316063259689851755469733888
26278484518102133330045161269504104451079130841088
52558396283896972619971380824977658397294644428800
105119647063179357159704878221894215784862054350848
210245003117129538158933989587666229550269639688192
420501424215800723996916445463088055061630341349376
841025684394684743351929823501687302045442806644736

1682097040715536077420053512154396987935249861181440
3364285425283405336272494754610838743559228218146816
6728753538271477035409764969825767022495913427861504
13457872451952286796549080860859713115746740838858752
26916475654723239044557263564135784373003309643988992
53834412811083808992032730813104285029026275220520960
107671748625442279789901868995874002624091862306127872
215349343257433883191476552731078870380262348342427648
430710378527966413606298734940819471024681944145199104
861444141082130121659288728838962402577678383211085824
1722935050216654832211959975592571726211985756263546880
3445963636538098842210684987014437294537229492209844224
6892114345285776039994900045687462273300974943785189376
13784602834990708791136860234692099915054981806301380608
27569953958819731004567840756018550567016027450064764928
55141404495316088853723922085305802607844182575053537280
110285802145985431396624325317149008163312620499955089408
220577590602677370171601612927372822221873751699606208512
441167153826767755099909150440895256234244524798604476416
882358252896361539713230150054089736049483092395993071616
1764764396278375138453283998452777919260954270389554380800
3529624573528054394960215393594752732845884711974245236736
7059440708998717026027725580567899254339721766338763423744
14119264541882650524270040747892586085975348217458072748032
28239295331535733992969260669298747326542505804477237297152
56480123158612333874796879685624644962268630348076658196480
112963311308306399527310476065303590542904498174397683597312
225932752598776262610054385518715782322543471305284101603328
451877765161879452330975637813648767118555892523545672024064
903780050252412758883685009179731939184049684873046281682944
1807609140362133226210837485464332688261975169398002438635520
3615316360438881869308609905138402996311701938099824627810304
7230828880306994572391089678696281232198907074807288756699136
14462050079472450812329919094231512943548820546829856515555328
28924884796661824959755317662140926845399653888090271035424768
57851338868757496589701594271637655607403333365041658079477760
115705816288382686519785106437986915048014717907805548176211968
57854477419625189930083512166349259440611384542763890096734208
28928023347529518300137235556852530678607705065812503052681216
14464403992623220817616357515265215818454858930121530528497664
7232398155740841242582048626052083148802932663668404765327360
3616297157585036038177959247235779194189217931138022133202944
1808197618649825727532447090722758406988484765220920941871104
904123329253566717987957278913813608441180282436415408570368
452073924591110286104845506233124006694059091131185173102592
226043092277718606607856186504670604583764020522081320960000
113024611129941035081644809946389602910249247739285027684352
56513838060511383429680763320221951764308242608764697444352
28257685278026124659269560833624551036745930673943440523264
14129225762898278801849370003569063095668870021752266162176
7064804443391747637031979795162925336482387353266406031360
3532498002667177936569637294270659562565169847828339490816

1766296891819241027311642345479928228444573009511737982976
883172391152446543169233021912263337803280547554653110272
441598168197636286341322435542281280692137295176718614528
220805070409524650549014180064215446241317158288055336960
110405528360115578963683571178645126068282834493875683328
55204260757734416326430026162591264507953544921861849088
27602878667705521585509133367929982990882836298392928256
13801813478271917503901626827282166863894450067927465984
6901093811345537107524343485299671116173740993329233920
3450640441777557731548936778479129400200128476347367424
1725366988941173454667850907154211621156693228015058944
862706878496784021780616712534429271106661108928217088
431365131261490658113653985745876365817487801924452352
215688411637294652668499807612269048040822524692398080
107847128821921988140086311175799956586450574211284992
53925025912598324972961359272732694576244943038185472
26963243707117827937939781478782705429632299485364224
13481987228968246694699441660599531785571063725817856
6741176302188789710214496290903855428162988854476800
3370679494946728036539635875753972481770222923022336
1685385419399530608986011803028008624729475709403136
842715545662848599851102834089515504286919978647552
421369190812965947604599883332513348104551051296768
210690304397253797641824174810134472032821056634880
105348006694012310740674203977006135006683293810688
52675430594698861330218160274472516998478029651968
26338428921195783645049609280220983246807206199296
13169571272521068312495069211602853997187698786304
6584964042222122401232666891547608185485897236480
3292571224091855323108899588646894686188972539904
1646330213536324722800732865759992639817498230784
823187407513360892023507968598268968270255095808
411604854129279711323324752158270808315880538112
205808002250939488317447760008703566248316764160
102906788718619560486616571969135864169346629632
514547881558846884072546319669599972607267438592
25728090976229798285600488974676006564930781184
12864393937258626183786830982936013413113921536 6432371193201176612386708739267011771881226240
3216272708886520066440000993533008418602745856 1608179910586225913343323808716255475632439296
804111733364595896733323560333003370981752832 402066755718039418397492608153939502073643008
201038822376890444214161718070688659328204800 100522133447380589614788566032203783809794048
50262427853157978561248136514531618977742848 251318944913128311575550995006480673025294336
12566287528023336517238960877847768280858624 62833139051951287278512121262227600024535040
3141742023189294598541471906765657954320384 15709135468905124165786668875208757948186624
785478041093188766943300898517343459606528 392749654370560662798633679715153972559872
196380144097263471062808455085818107658240 98192730504623305363150035157029614518272
49097694480307437597447921385575087603712 245495118541516112566660412596317683974144
12275088234074751857298432249923912073216 6137710270536849043133329100844491079680
3068938212018161078808721038363513061376 1534510644383948818025388763152390291456
767276091379408548323208503561512026112 383648430283421343816861312773414453248
191829407438569206736059186883036446720 95917299867713870781843858689682833408
47959948008071569097829061968923721728 23980623041143101402368097296503013376

1199063603912520912791083180427208294 45995480278839433777318807480146329600
2997821269058131495341099529078308864 1498951199348273051011397659041726464
749495882083740177176122776772149248 374758082246671914423273362011717632
187384111725736870129242667818680320 93694591164068891523424327315750912
46848563232634673991113660361080832 23424915441617451110257578532143104
11712774633458782612479163441872896 5856545773054419834914768808837120
2928352114689724181794977948368896 1464215671426119223066285746159616
732127642753688177617541259067392 366073724897158371850969822527488
183041814208736327446584507760640 91523382984446734483842052128768
45762929432262652622195925188608 22882083686150969001235412156416
11441351328085305845686430859264 5720830406547563595377577820160
2860492574526237133955970105344 1430284972889346235111575650304 715161829257786951622583123968
357590586035450392844689211392 178800128721003654939043430400 89402482212141056727871127552
44702450031890142993110269952 22351829478854878811142488064 11176216970882343062864920576
5588259601168623360079298560 2794205358448037594363068416 1397140458155881754343243776
698589118543872355752476672 349304004004901917166665728 174656724368933828228546560
87330723367708348936880128 43666542275474891879743488 21833861433547804645523456
10917225864679081675587584 5458760506292130514206720 2729454040122360095309824
1364763913549327466758144 682400403518737442930688 341209425131405576241152
170609324251721215508480 85306967968869821448192 42654636905939517571072
21327894913722062209024 10664235687237182816256 5332261958806667264000 2666203036997371559936
1333137547295704743936 666586788046361853952 333302401222435667968 166655704210845204480
83330103905236287488 41666177852524986368 20833651876215914496 10417107413084667904
5208694444030689280 2604417590759522304 1302243979751849984 651139582061969408
325578587124006912 162793691608514560 81399044827512832 40700621925384192 20350860718505984
10175705237159936 5087990057533440 2544063748243456 1272066233860096 636050296799232
318033738334208 159021164134400 79512729550848 39757438517248 19879256129536 9939896500224
4970082467840 2485108342784 1242587725824 621310640128 310663708672 155336048640 77670121472
38836109312 19418578944 9709551616 4854906880 2427518976 1213792256 606912512 303464448 151736320
75870208 37936128 18968576 9484544 4742400 2371264 1185664 592848 296432 148220 74112 37057 18529
37059 74120 148244 296496 593008 1186048 2372160 4744448 9489152 18978816 37958656 75919360
151842816 303693824 607404032 1214840832 2429747200 4859625472 9719513088 19439550464 38880149504
77762396160 155528986624 311066361856 622149500928 1244332556288 2488732221440 4977598660608
9955465756672 19911468384256 39824010510336 79650168504320 159304631975936 318617853886464
637252887642112 1274540135022592 2549148989521920 5098435417997312 10197145713901568
20394841183617024 40790781878861824 81583762780979200 163171923608469504 326352643309961216
652722878805966848 1305480941984022528 2611032252712222720 5222205242912800768
10444691960802312192 20889946871558045696 41781019643022934016 83564291085859553280
167133085771346477056 334275178741947695104 668568371882404872192 1337172772561828708352
2674417602717695344640 5348979320623466545152 10698246871623084802048 21397070203998473027584
42795293329501552902144 85592892502012319498240 171190396690043066384384
342390016752122987544576 684798480248319684640768 1369633853984786788384768
2739341494945868414976000 5478830563844326506364928 10957956275593832365555712
21916502846998023436763136 43834186285616764284829696 87670733754474963392266240
175346189875432796429746176 350701824483831332149919744 701422538433594142880694272
1402882855799051242923098112 2805841269461828400169615360 5611833654651108628986068992
11223969540757120915265814528 22448543544424049145118982144 44898296014667712919412670464
89799009880974655097174753280 179602855465227768711048331264 359215382337012454455494311936
718450107487138742977783922688 1436938900600505154089158443008 2873955172453465644445498081280
5748065087411841961425358553088 11496439659833505267919441887232
22993498289686653225976333336576 45988234519412591832227565797376
91978944918903754425004929843200 183962841597964650371109456183296

367935586716243583784418105360384 735890980473115733653234596708352

1471821575027488599475265965391872 2943722378217491463288125474734080

5887603212760011455251438037368832 11775523338170079967853250250539008

23551680501640274050407248852680704 47104628653880776330215994408566784

94211792608962009119234982223544320 188428655820324931156075951259910144

376867452845451688147363876145463296 753755188100507027965151699542212608

1507550941020221359271151293586997248 3015183011678857325223998376179138560

6030528282634543863811388330368565248 12061381083822746154349559816757706752

24123411204752809162152685945556566016 48248120483720252031212504515195437056

96498837115869771476239274278555484160 193002866528598077780107079053440188416

386016117650913225215471219099538817024 772053004489260589741456560184394514432

1544147547353389458103941364339422789632 3088378171456515473449939216620113100800

6176922496412504061383991409122761244672 12354177299823954351736208770010592575488

24709019213645801161408869443551325323264 49419367655287387238690642694162930991104

98841393766566344309127093002446422671360 197688104445115828281745801233133966721024

395386842714197935890474832922750176198656 790794953076328430434916126758464837910528

1581632441448521978177765175342858646847488 3163349953488774190971396194337575235747840

6326870048161008851174524075978866355601408 12654080378688938640812511526565164479414272

25308841322111719158551949802345192495251456 50619043773691122070957753103120112063348736

101240809806317611649623213203099678272389120 202487064130505958314661840399918264836161536

404985017296753386660154508787274346255089664 809991812664989713381970673549424325675712512

1620027181472945306887264659048599917682491392 3240141475231822374021175941996702368027115520

6480457175035508268535645131792409801378496512 12961262799214743578057876759182829733405523968

25923222496716941238088926509561679728108109824

51847838790008790640124199001515399978810343424

103698465173167397608141089967814881002808934400

207402505532634427872067563865197924095994363904

414816161437868121055705895589532172372741718016

829654623620934772734553326897336993106989416448

1659353848732266606715389725231219282936990793728

3318796900445327335923345593335529159320005509120

6637772206852242916831823472417239505532058861568

13275901225627662323633911516326841384848213409792

26552516075101677627208352175638407517264618192896

53106459397896061214297762637246264529665619132416

106215773291177534348357641846431428049604003758080

212437255573125892536239516836740654079753538502656

424885929127793432751527499961236904120598138978304

849794694218670160861151932497985000163378401902592

1699635060363506912438497730146992384171121051697152

3399361464579347006309383190596029536030970599178240

6798905616863360375483541841796148607439398189924352

13598176609136053476696634604800476285633710362984448

27197083969090772404852371052017310712777248692240384

54395629439818875712622945788867337708574153317023744

108794181882912413231082298947400107983187618499133440

217594209772374150073837412634131081098453860728438784

435200111557846947371020454746923892461064968917221376

870423607141891189188732168451171245450444432755130368

1740893982336176967270846854816989411957517855351635968

3481881500777143112328458745463272666028293690386022400

6963950073763864580230447562585133016283103340137545728

13928274291946885871607955268487441401019238599006093312
27857296872732085165510030823609233538944541035474190336
55716090323140797175608302220487168551701209745872388096
111435173801634848040393085587511740051026674841592791040
222876333913976203459139133468098285997301860382881611776
445764640449365421674984191522346183785100742165155282944
891553226141556872863380232216991591151195527129094684672
1783154342768765804753584162778581629464379139855757606912
3566404466508835727560815722246360153252734450906651688960
7133000494960279691228926237871114095153421244203576328192
14266384113805775854672442062499015767602747173187698556928
28533534475381984653774063298511606689797303715936488914944
57068601446304835196406484944050363688778226170995161432064
114140267883691402170529686582155027995923689820234690068480
228286665749546267896492806552418657228581854596958114545664
456585591463419462903852479881054516930632659106893697908736
913195702855492780029438693314543438808203218039742333452288
1826440445568293268502344853733955687510282235731394542174208
3652978970851201953891624641677648994808316070766608834887680
7306154101131634741557119151774773229192135340140857170853888
14612700521121731150661978040388496937535277077496993343864832
29226185679960385636419435554454894833372566949424544692043776
58453940635354617943029830056265591583349159487710205392715776
116911019821576929226441578007242786999906370153142642802688000
233828316744889245133646991803908781666228842661729749639888896
116917296923312315907205413796665994666322472508587106837200896
58460217737090004623793665845688799249765261843154669427228672
29230893506261925646992312396522300583184643716007892717928448
14615839071989424491043895935100100770743328255219225361121280
7308115695423943079295817835969525624947167326217252181639168
3654155927426586956534843852194500432261335262412445841358848
1827127003570601186710889393202119026024543430858132795949056
913588021713954447577178430153493917959209615255021335609344
456806270821304150899456081852964161453073757540488136622080
228409265392815539005161474314590681963271353726732802719744
114207697687489501280297453851349641600002914341610768564224
57105381339285616529007085272701971109185075909927567884288
28553456917413241208932721809864560709184347324524875743232
14277111582591837076680950491689067931888078347042983772160
7138747353238526774447770039222927754591991515911764836352
3569469457590567505277532416300660771619971929151018893312
1784782619280935811665589906494928832971974050173077684224
892415254883293935346206802419763640066981067885322960896
446219600063059982429809325796031431823987555342053539840
223115786342236498593257625191090521807242288370722799616
111560886326471502985805293742082663851245399535209414656
55781939740912378337490887444310033399434827442528714752
27891718159294502591039564008789367436623477558726361088
13946233224066408006666842147711859086764770698094182400
6973303684242782358906951145514517227608901308412592128
3486745378226180357240240608586552455917708633889046528
1743419457165484767513502822207923149015483306785898496

87173311260893967820344267006128503503605614831363686
435878248317568486325066964509304247782185321617162240
217944970165333566774206296993982989023171284538753024
108975408085941445192939555866656927077624954134462464
54489165544608053499387981618161179821832132999774208
27245313523122692201153092651496948052425894466158592
13623022136970678826306097246956653096967861216215040
6811693756190005776017824084082416083861387599675392
3405938221947336069441299772343252809619422295621632
1703014782899834625436843751322648788654075395702784
851530227413000608076518808236835586249219821797376
425776531688041951717307870406173389085700972871680
212893974834791799698178168346964492523396017422336
106449841912781311768851200745421145251970774204416
53226348204083361844306658658680022121121769848832
26613887725888033902093858472324735808129076297728
13307300674867193441017193807654730277848633835520
6653828743395184965493729189573546325816364761088
3327003574678386605239430737659863756354206302208
1663546388829590363865998440266477174900115111936
831795495159993712556140755851511235811563536384
415908897952596121589641145784891942086534758400
207960024162597693450605956822014133133643874304
103982799674448663053195670375791147612010184704
51992793633799239690544181170287614328599216128
25997093715187073927245263576339827425596669952
12998895306737264004609218283767923843446865920 6499621877940495522797902389682966987047698432
3249898051256179521645597818740986026185981952 1624992581771055640946122221320244279424057344
812518068957010760534722766634997772877561856 406269923514246850298192211304936703021547520
203140406274994160164511519646187259802157056 101572925396432447589963466819953084046770176
50787823827683907548835586908406269096230912 25394592478575795651344720203417998084538368
12697636521654818764135823476316430810480640 6348988402010869851299643425461931289346048
3174579271597165160265687556382823586725888 1587332171094447697440776700017340764389376
793687353195156407374354810921634867707904 396854310421544483014160635917299676610560
198432472122755381170571933186890959683584 99218894517369260417031774207566040530944
49610776486680415124388790910843300610048 24806052857338100020130847358951790477312
12403358735667996239033649631240965324800 6201845521333471234000937791503017705472
3101005837416472174242525383692776374272 1550544457083104365742290935817021947904
775292997728986322181659589893827854336 387656883458210230746086855939572367360
193833634025963650200671958466115403776 96919413161411092514150244481222311936
48461004654920179963982254864693460992 24231151364567406835444693744387883008
12115900200837361844449130028214517760 6058112359695510135587956592117547008
3029137309486169674475674085063917568 1514609219562292140578684937034530816
757324892190749721959766415768551424 378672587300176686815095181509918720
189341364252489256325153577567780864 94673217427445084621379782190301184
47337876364322770540091387798355968 23669572007461499384746442250780672
11835102916380806749723595301191680 5917709914515431903536984738496512
2958934185420230216106085913198592 1479506706791372240221839728574464
739773160436314686195318250274816 369896483738471626139858318131200
184953193629392954591028755562496 92479072694775048056064176029696
46240774287426809408306987139072 23121006113733047394290943131648
11560812541876345042214196346880 5780561013443083193641460563968

2890357877973996933087911477248 1445217624613226134677546336256 722628155119726901405568466944
361323748966420367736181882880 180666710186488642384789766144 90335772944883550450744295424
45169095398261389854546853888 22585152162040502241860780032 11292878312475154778224066560
5646590271965029217758871552 2823370693846240523202854912 1411723125854983218763137024
705880452393423087962423296 352949670929677283271639040 176479557831321511281033216
88242140098902190463123456 44122250641071812642865152 22061715616346265027084288
11031152956078311866368000 5515724051991745609596928 2757935812972167643004928
1379004799974231240605696 689520846731189329854464 344769646737631519703040
172389435054834187239424 86197023370426307313664 43099664606717760503808
21550408764111183675392 10775492612431743549440 5387890421403947630592 2694017268296011743232
1347044662945024835584 673540345871021899776 336779180134765690880 168394093667010215936
84199298633318793216 42100775216566239232 21050950558236540928 10525756754094981120
5263019114535845888 2631579926012100608 1315825147378139136 657930165875113984
328973879030579200 164491337561800704 82247867804155904 41125033413705728 20563066462666752
10281808109240320 5141041493573632 2570589466263552 1285329092870144 642681726304256
321349453086720 160679021510656 80341658238976 40171902861312 20086488301568 10043512586240
5021890510848 2511012364288 1255539736576 627786645504 313901711360 156955049984 78479622144
39240859648 19620954112 9810739200 4905500672 2452815872 1226440704 613236736 306626560 153317376
76660736 38331392 19166208 9583360 4791808 2395968 1198016 599024 299520 149764 74884 37443 18722
37445 74892 149788 299584 599184 1198400 2396864 4793856 9587968 19176448 38353920 76709888
153423872 306855936 613728256 1227489280 2455044096 4910219264 9820700672 19641925632 39284899840
78571896832 157147987968 314304364544 628625506304 1257284567040 2514636242944 5029406703616
10059081842688 20118700556288 40238474854400 80479097192448 160962489352192 321933568638976
643884317147136 1287802994032640 2575674707542016 5151486854037504 10303248585981952
20607046927777792 41215193367183360 82432585757622272 164869569561755648 329747935216533504
659513462619111424 1319062109610311680 2638194587964801024 5276529913417957376
10553341301812625408 21107245553578672128 42215617007064186880 84433485813942059008
168871475227511488512 337751957654277718016 675521929707064918016 1351079888211148800000
2702231834016335527936 5404607783220746911744 10809503796817645535232 21619584054387594493952
43240321030279795834880 86482947903568805363712 172970507493156038115328
345950238358348931006464 691918923460771571564544 1383874740409690562232320
2767823267795675962671104 5535794109543941601755136 11071883366993062556336128
22144357029796483818323968 44289894651213685047951360 88582150485668804918509568
177169023337820479482232832 354347491408606698254893056 708713872283144875090640896
1417465523498152707342991360 2835006604860031329009401856 5670164325447514486665641984
11340630882349932630624960512 22681866227609672575837274112 45364941381038959780849254400
90732300613717148820047921152 181469436930712756156794667008 362948545267982429346986983424
725916433349078692760769265664 1451871552324385053655129128960 2903820475901225443577439453184
5807795694307361559689241296896 11615900873624544464447207374848
23232420717268731619031864311808 46466079374576748618338627747840
92934634629232067997227053744128 185874221018621277515553703985152
371758345557556838073306600964096 743536498155742242231011587915776
1487112610392741616630819947806720 2974304448947997497599233439563776
5948767354221023523873653967028224 11897851621092104105097682109857792
23796337067484322324896112571318272 47593941785568872879193721845841920
95190418872338202217190437098094592 190385908347077317351986861009010688
380781957898956460539185695643664384 761584198207516572748795338538614784
1523208961234240448838438571579801600 3046499052106895504358572932164747264
6093160363490620222080537442339782656 12186645245534898870887858040700141568
24373939528177114595229282393441435648 48749177130568862897365697410965176320
97500950409566993208545660070094962688 195007093115992521244719850636519145472
390024570825702112144696762265696731136 780069910838838363599907646516710342656

156018136005254500582084353700405444608 0 312044579685482656888374356194937641369 6
624105774720912625225160009978128787046 4 124824478014171987334714261513276458270 72
249655602168322899248793042061854318264 32 499324496616603647656315122194311439974 40
998675577793122993630088320529828486840 32 199740432470607738389509279334206818746 368
399491498765181756106001789124895880249 344 799004265178296070865970039162756246011 904
159805106565245725903987300015144146305 0240 319618720189664475269561184395474086815 3344
639254454497674997462295537521319762041 2416 127854293723204208877093741250338270090 36288
255715393093746836523456749992825175544 95488 511444397482170510585452034969947621818 36800
102291601755369469624798113990848978509 365248 204588648028609674265011641975416865310 113792
409188185092960818560854111938271547202 994176 818398148257404577183369879851418727571 521536
163683985265777503449006307165258872147 4109440 327376681760148182922677276720467997561 0351616
654770785977482717894683878220836501654 4968704 130957641686933813988802640600147401637 38468352
2619222523567421687973370111122550058877 3998592
5238584426792334192341374820484304170014 2120960
1047744761289965001747201883744701644454 72489472
2095545274442926330052257606785084909813 21474048
4191202052611845313220222892161533061433 95938304
8382627112675675932671861141505792606482 97857024
1676570024025532247780655299737703818019 607674880
3353229251031858618053876742348498229485 239271424
6706636908025305481092885770443177645862 526386176
1341363062797378745215603611237871766550 9148459008
2682797487979392788425260136774216007858 6488291328
5365737700728056172838626102145376965230 9359329280
1073176085099465353765346386148464382948 91484151808
2146409260106638945925935103735706745703 28499290112
4292932700028694368642354870348969451017 48060553216
8586093759688221690865679066453050821256 78245052416
1717264423863810928889329678441632548095 720737996800
3434620191579955039211047087185309863880 169971777536
6869423070864576441286869634974709263137 796935122944
1373921151713848560830329019115759759703 0507853381632
2747915378509563666806568222473155333557 0843673034752
5495976907182860423904956813429582295416 1343278612480
1099224611469318702839355436382570784743 61998422310912
2198507683004130641795439020158450220808 02620574793728
4397132286139247755824334335103517744257 62488609931264
8794498412540468456115581259780270093798 39472140550144
1758946450560488280116498769870700939816 307934122475520
3517986437225765738019762575570695721745 873847927701504
7036159946661109831613055222799979127718 263655220903936
1407269403774137637437317058891713362388 9559229172809728
2814613636432106617104046146446861798468 5182295807623168
5629376930631875918666916350220593744318 2492266539253760
1125905317679907720625148081509492778339 89239882926522368
2251870498466880515033825785949733615632 26990465549074432
4503860723147891177634710817760963349169 51002330490208256
9007960898724042650403540127244918934148 96047459764535296
1801640070230460589107531723793582233991 780180517097308160
3603375921432225296268710844276361362307 536532229331091456
7206943404807058828644716481931116513263 025406848935133184
1441426993349933412950402255061902060382 1955498478416166912

28829306114769101203437224274751616362235720366517924134912
57660144725079068295732806896530383033655059472158031872000
115323354441239868369182330487115066685677356422560430948352
230652838864643200293798094362338734608089187801609596305408
461317937693613327698463055500894671689647325516196661428224
922660395315880509618659844554223748326232550858348260491264
1845369830489068727680787156213316306546340901368606396252160
3690837740692752872248509246636370232880433402041032543043584
7381871640814736578270888361692215705336370002689704587165696
14764135600487934824089516460223381889823746402594688176488448
29529055838692792983274512394124664737949505599619934357291008
59059680952819432636739983735605131392503036788100984723210240
118122500456506558613861885365921866618214124753924201463676928
236251278014748503908487606521266940902844351863292866961866752
472515110232967781178502884621380297138520908437474661992759296
236263832218219277270015278100113356235676556574181795030892544
118135054659977331975389556944768281951046329464813129532702720
59069096605422512657885737419739942892127190321267680774979584
29535332940428179664038348183547872404365607955064398391803904
14768058789072551499566913828612886681333810374747478198059008
7384225553965506583557326782725918580242408385981378600108032
3692210856697368708665598325572696909908955792294509050593280
1846154468205992062776266629891217264848353695799164400566272
923101754031649885609867048498043037371114747725537137917952
461563136980151869915800391025238721159026323775746037776384
230787698472239398513333628900727961816247636844361753296896
115396914227201431034383531144418281526491055900425243852800
57699989609141581406050123919236291072429146689334805528576
28850761052341223647454241133131720690806382714228494565376
14425763650055828295941710153322647922699096041894793183232
7213073386970522384078149870039717749997500363337669541888
3606632474456565310092722331709055769322726352863971246080
1803364127713934714073184864199126331823351262029553860608
901706009099793386550004281271862389492669673813561049088
450864977171309708031708065222080806536831858306172583936
225438474896361361394206994904115209163664439852782321664
112722230603533934386279978598595007529456475276239175680
56362611879443594037728229872566205238540365313043595264
28182054228560110441158235222917453356176246493983801344
14091401258699211931726177754775902046541155165722902528
7045887701559184321436618949046538707497093542226952192
3523037386884381338505074510352563195861804750796226560
1761565461494585258145919773090928518987531365239488512
880806114773489923519651145502787720022080177540431872
440414749399843608983171202230055590275197336230559744
220213220706471128103258415854358660269677291845451776
110109533356510225857465615296844762700877957787811840
55056228179892443831651011333255097633458634826448896
27528844840764887367284607509043906958239145379495936
13764787795791776409371854675730132549874486672883712
6882576585600554567550702798469155810314700328009728
3441379636652610465207739129536622672846078659788800

17207354902524718233200634299193337202674035777786368
860390581089319207018128647535178052055883912839168
430206708526201251188112790055344621989033018392576
215109063253871449433580628171550108975062040182784
107557386122321136636552430657713953477803785584640
53780120308853274278157273614826426234038275538944
26890773778272990119019165950397937864587329142784
13445743701059671549479847546691331306077760258048
6723050256491424019725056059091846839930927972352
3361614331226506132355094172419014013411487907840
1680851767103650127423830157646052303428755914752
8404481842970235943350566145412988000758839377792
4202352425211106247909907512978572421869495910 4
2101231964468551638953349214944610241997239746 56
1050643858165773982755601527120145931450502348 80
5253358670486360730172642233839933709511924121 6
2626749025071925773283638416039568880885668249 6
131340935745033559074047785757958545350768721 92 6567221011823541474195682535696932332862701568
3283697618197702497344487891747968699093483520 1641892365241817128795567257823735615877808128
8209679606923915044594452848867434411044372 48 4104948693819372222605534704308095371349852 16
2052528792088393461456921492091236768588759 04 1026291618633550405805537816014212925751296 00
5131594206114520404413074429914037336041062 4 2565865159530644389899229889878505021662822 4
1282966608002014288795961282399995687652556 8 6415003181193531913211538099303694322368512
3207586661188496191221634893303705103237120 1603835865890113212918750368477781522644992
8019392005929891651133416451518552468357 12 4009802341204608618836540530324098661744 64
2004954339722135706053186417444460544655 36 1002503754420983551344051284863435879219 20
5012651694904496248307546805023207430553 6 2506392308852037369947418592864617732505 6
125322938512591330787053189160881587486 72 626631307912903965383677243392661441740 8
313323961631425638416044270490457473024 0 156666134653199647070124959642292112588 8
78335144245343237466113892019677744332 8 39168610582043325698582652109104716185 6
19584824520707516332054179104185280102 4 97926718751966849074085160769091010560
48964657450198058243949713008627810304 24482977762206345975428422816355057664
12241813399656831414440994564198105088 6121068959105244920583888860109340672
3060615609191037066973640219059814400 1530348369414725836827668004032479232
765194467116966570084257949267525632 382607374763285110877340948259405824
191308757984043468356276460942524416 95656914293222190636941223877672960
47829724797211323547872108642041856 23915496223905775888636802672623616
11958065024602945001668775512113152 5979190968626501029509574843957248
2989674712475764779092380965928960 1494876970319139521714987254939648
747458292200198326941892013457408 373739049620413446513145199722496
186874476570363864777672196358144 93439714165260503149385896427520
46721095022669536954967847337984 23361166481354411167621373231104
11680892725687026928879411396608 5840601105348424136974068088832
292037792392666740475421523968 0 146022764758956137051069821747 2 730133166607894519322144407552
365076254710504176694469853184 182542963058530546863933751296 91273899380904502690316288000
45638158616271865974332850176 22819683771045740301753778176 11410144116977773808170565632
5705223174216338732732121088 2852687144971895280689479680 1426381351417810597506449408
713209565174836777334079488 356614227320384127957467136 178311836026674933623947264
89158279196578901634580480 44580320189910168228593664 22290750390765442819948544
11145670343287900762800128 5572982745596540057812992 2786565159774564867112960
1393319473375429852659712 696678183431788635881472 348348315087931172716544
174178769229984013746176 87091690458001220567040 43546998150505217130496

21774075536004911988736 10887325998378607706112 5443807114377379708928 2721975614782727782400
1361023836188382855168 680529932492700909568 340273973445605195776 170141490322429968384
85072996961028669440 42537624380421177344 21269375140164009984 10634969045058715648
5317625260017713152 2658882998753034240 1329476683748605952 664755934060347392
332386763123195904 166197779608109056 83101088827310080 41551643925282816 20776371718455296
10388460737134592 5194367807520768 2597252623237120 1298660671356928 649347515547648
324682347708416 162345468821504 81174881894400 40588514689024 20294794215424 10147665543168
5073966989312 2537050603520 1268558856192 634296205312 317156491264 158582439936 79293317120
39647707136 19824377856 9912451072 4956356608 2478243840 1239154688 619593728 309805056 154906624
77455360 38728704 19364864 9682688 4841472 2420800 1210432 605232 302624 151316 75660 37831 18916
37833 75668 151340 302688 605392 1210816 2421696 4843520 9687296 19375104 38751232 77504512
155013120 310034432 620085248 1240203264 2480472064 4961075200 9922412544 19845349376 39691747328
79385591808 158775377920 317559144448 635135066112 1270303686656 2540674482176 5081483182080
10163234799616 20327006470144 40655086682112 81312320847872 162628936663040 325266463260672
650550106390528 1301134572519424 2602337864515584 5204813167984640 10409901213876224
20820352183566336 41641803878760448 83285806780776448 166576011608064000 333160819309150208
666339230804344832 1332713645980778496 2665497660705734656 5331136058899824640
10662553592776359936 21325670135506141184 42652466170919124992 85307184141651935232
170618871882931240960 341246750965117222912 682511516328743927808 1365059061454506819584
2730190180503051567104 5460524476194178990080 10921337182764509691904 21843250826281322807296
43687654574067252461568 87377614991143718617088 174759841668305864622080
349528906708648584019968 699076260161370877591552 1398189413810889174286336
2796452614598073186779136 5593052803148736049971200 11186400754202651452768256
22373391804215661611188224 44747964200052040633679872 89498289583345516089966592
179001301533173901825146880 358012047799313542940721152 716042985064558564462297088
1432123749060980086086303744 2864323055985686086496026624 5728797227698824001638891520
11457896686852551660571459584 22916397836614910635730272256 45834004599049435900635250688
91670427049738101059619913728 183345689802754660635938652160 366701051012066238305274953728
733421444837246310677345206272 1466881575300720289488281010176 2933840521853895915243743215616
5867835786212702503021848821760 11735981057435226351112422424576
23472581084890095392362294411264 46946400109819476164999487946752
93895276099717523090548774141952 187795503959592187702197144780800
375600911439498658446593482555392 751221629919625882977585351098368
1502482873920508898123967474171904 3005044976003532060585528492294144
6010248408332092649846244072488960 12020813729314242357042862320779264
24042261283928598828786472993161216 48085790218457425886974442689527808
96174115738115308232751878785466368 192353302078631529383109744383754240
384716745362064884601431462393151488 769453773133733420873286872037588992
1538948111086674145087421638577750016 3077977351811762896856539066160644096
6156116962900355007076469710331576320 12312558444354368440879722576683728896
24625765925816053735213011465408610304 49252829925846741177333155554899525632
98508256000122749768480576357963661312 197021704297104034364589683212256542720
394053793187925138384436427417171525632 788128355563284416079386976819659931648
1576298249501437110779802197609953624064 3152679575752610778801660883161174769664
6305525305004694672087434742204884582400 12611382917008335573143095436174839250944
25222343044801456360422264277587981867417 6 50448190124024912124318189358819917692928
100899038704041394080382186331760396075008 201803394320065927824255987891761913528320
403617422464098134975495206240006069813248 807256112576128828604956873392976625139712
1614554760448122774517846668611882221305856 3229194591487975783651559180875622384664576
6458559324159412036534850049054960653434880 12917458930685745011533163472717353075081216
25835598426105331899993253694649569686585344 51672557981678347553840360887728866446016512
103347838222292062615388428772317187037724672 206701120962454860246192271538353282366832640

413413130960651190523215371064144381316431872 826848039992785321108092398103164395798396928
1653739636128536522339508108156080057927860224 3307566384543004804925662840211662648517853184
6615306993657873130344618928222330362359971840 13230962436459473301675824352042670855368474624
26462621771206400685324821695281361972034011136
52926637338987709534595989372954764466662146048
105856062271125235397084670710693609978512539648
211717699728550103449954725350955382047401574400
423446549829699472211480218561047088275556139008
846915400404597475046101972840366824912618258432
16938754022995920113384870171117278946548248477696
3387840007579978145169540177107648486542520877056
6775858421121544535324212639961478159977089597440
13552073654166265560618689851415318693738274881536
27104860932178884101177908845815362135044741136384
54211149112050474162236875977600173765225865019392
108425152719486360244235868527139246520724495532032
216856014429743544327995970198156291021994522050560
433723446841028736335040406684068178005080106074112
867469729645140768028177745943647547932342336094208
1734985131216448126772549357038317479709048920080384
3470061606285229434977486444378679727106826335944704
6940305900275125232819748349361448989591109663457280
13880977175959583191369047619931077049937133310050304
27762685102737831834197197082278512241384094586372096
55526831707112994571312597849389740765787845105287168
111056586417500650948461603068444914097615002075660288
222119018841550625508596020876220693327308627881492480
444249729696199898240537671231103116918774503223328768
888522843418597090927766601419529694365863501367345152
1777092454889588770748915720753706309788355992576065536
3554278445883966719284596477336706461689969964834881536
7108743963977511794142723026332000607606455889035264000
14217862072374180299432506195981176583665943696801529856
28436472433586674021159132678596703904237951231065063424
56874441444849974886906505930462109282288030137054134272
113751876045053203462989493007461621512200315623956283392
227509738400812914304331948307998048919649141947608596480
455031449423038843365369821202145709629795305294609252352
910086844088903716244151491576590642840584653388002623488
1820221578663459491515126681497779732843157392373573484544
3640538938298223101083900759684756360010290955942283444224
7281269438539054438275096312747906508668534254274839838720
14562922000963325348764782212252600594632973193330225577984
29126610249697083641958743598018776343857755756221542957056
58254752994935033172775845543064702996899130251565269516288
116512570980951798123268407780183706612165497981374906236928
233031271944067059801970248948476014461065470919238546882560
466074803852461046714807364673169231395599891751454562582528
932174127633575947651348462898772867738137683328864062799872
1864397295124459603746164392902414545370151166309638000869376
3728892669963534624379263720014566710528053931923095752278016
7457981499356300082532397308448608660631611062453831005634560

1491635531757106183261253435373616780041422852212294101342617 6
2983349527285904700032054818115023655913046983867644003116646 4
5966855982115194067083205530965627503486496526621399607096115 2
11934025819317157468204602851402415390293798171015022415917875 2
23868679348807853604485589281747151547229206577574491235287040 0
47738614117962784545123945721378944627741633626237875277476659 2
95479739076619723762553425758527172322049708194653536168758476 8
47741124958656939217429480037148227694308074568415660891281817 6
23871190189502008276791123597516434613795647519752276849092198 4
11935908949837773472433753588229377690218628877648361626271744 0
5968111402462271403235972688850269036769716997710292413998694 4
2984134165002828035127534291792924614215059778298202007430758 4
1492106314387260184318541119580357355022630528870628903931084 8
746072773136553175536657546632126201468865584296078402073395 2
373046194539738129457022266737036862713207952078421176090624 0
186528001255599835572857880079005312345991556004401575572275 2
93266452620665303208602313394746096667689567984796281549619 2
46634452306765344315387843374994768581191678983695887656550 4
23317839151599018513237265026308244414269286987496817269145 6
11659226074907682434390304182559552268971367241572845355008 0
5829766287007927806080987925982491165404045494698641037721 6
2914959768281007197483411880342603098161203684305429698969 6
1457518196529025245963164898846980306810192310630769439539 2
728778254458773446592311928761329532269891389554412014796 8
364398705326517135101520704049584455567343311896719654912 0
182204141711823773453442721859252072499870464508116651212 8
91104465380194489678062545846855958608034636533936737484 8
45553429952238546314701865382042940483067020406907574681 6
22777313607189923895186228920328950831058361273423390310 4
11388956119130287316510762574818215710291606171696496640 0
5694627717332806342714205344735978002527015853340649062 4
2847388687550234513586514701031424074954114310416524902 4
1423731758217032927907963364847429574322360347081362636 8
711884586329474299511334689589573555583831769477231411 2
355951646775216067534343848377716162003241682706890752 0
177980500192847492656510175980322773107283740337582899 2
88992588499043475772924213885893732606473319660860211 2
44497463450831602608796669890813039329652384576464486 4
22249316326070733665565616419333960617803405466124933 6
11124950463362833013366448946636346345620958517133312 0
5562621381845149596975044841801444801112444851820953 6
2781383766004441343633432605142358214707205222537625 6
1390728420543153944389671394691997014429094009582387 2
695382479042043608481313243406407460752292703947980 8
347700373906255122383895394733408207145019201552384 0
173854754145744220263567083881806341956946025565388 8
86929660669180439667593235198454290170691225177292 8
43465972132744384601701464228002704681454718785945 6
21733556965449274684803155428389132138781912491622 4
10867063932263178534377789371388455968418232795136 0
5433674690900859863177000514291172933722754672230 4
2716908707835065229582553171444058941618196473446 4

13584900351098502637883030428712657081875078053888
6792628581510839563926647500102509727829586870272
3396403493736213904455889892924345457360817356800
1698246348358504013474228017898718025403420639232
849145474924450537360255544667631661063216300032
424583887834824533991698540192952154712361140224
212297519103711899651634654026044239446557065216
106151547145005766153710018977806200768466780160
53077167369077791240801355471295140906827513856
26539280582826349702373850726843590714710818816
13269988740556901892173511859019805488003940352 6635168594850314466580049177308907809326235648
3317671409711088993536671212553956437325250560 1658879260998510376891658918226729484993691648
829461408570738128507491115088240375662379008 414741593321110534284576385531558004413956096
207376241178426002157703606759497910498361344 103690842848148368586559510376608409394872320
51846782553541868047133608686733931770281984 25924071841504775900493731092581829421563904
12962376203119308888710328920898346478993408 6481358242743114913586896147752889123602432
3240764191963287691409313917528302503854080 1620424631277508963012589880590080222953472
810233583286687040160261401208004596989952 405127425467309799407113931060484541251584
202569029645638039367048580758483392004096 101287173278810589515270097993362256691200
50644915867401079673507952803741408690176 25323122547698432294690428305400844517376
12661893580848162376313440104465492344832 6331112943923554302640833028115281215488
3165639548711513708562473001998908129280 1582861312730625132902264744970087825408
791451425552746705761646494470360793088 395736097370090422536080308227838836736
197873240981903746095668684610248638464 98939216639381140461648607553288929280
49470906393905203937731436400726769664 24736102234059918822319284512404537344
12368375635583617837886425412222844928 6184350077068638132306604284121710592
3092256168172733672834997931065999360 1546168648905574139758346860035571712
773104606862390721549597377269071872 386562444635997186610010662260178944
193286292920399506222611317942910976 96645681761400209570108652377866240
48324108531300333014455822892138496 24162688090950280621928659797671936
12081660958125197368314704074637312 6040988935387627212832539125219328
3020573695856327870753863106560000 1510326462009421067545728325255168
755183038045339099857262548615168 377601422542983832970830467301376
188805663031649058006514830147584 94405307395903099763807213322240
47203891637990835262178505785344 23602564789015060321226702454784
11801591879517351505682076008448 5900950682263586425375400394752
2950552712384248548954881392640 1475315041818351942611031293952 737676863722289805372310945792
368848103267701819719553122304 184428887337129368376475385856 92216861520203913446587105280
46109639685921571352468258816 23055424305870592990821482496 11528014384390200152704417792
5764158307922551904999047168 2882154711825001866822942720 1441115134844363890573180928
720576456888113423867445248 360297673177022451224150016 180153558954994095257288704
90079140660738482451251200 45040750921989958636929024 2252096575756805338024116224
11260778026307848364883968 5630536587106513858854912 2815342080529551767633920
1407707933752923302920192 703872413620535361011712 351945430182304535281664
175977326777170695028736 87990969231594561208320 43996637537301887451136
21998895229403247149056 10999735845077775286272 5500012037726963499008 2750078076457519677440
1375075067025778802688 687555547911398883328 343786781154954182656 171897894177104461824
85951198888365916160 42976725344089800704 21488925621998321664 10744744285975871488
5372512880476291072 2686326808982323200 1343198588863250432 671616886617669632
335817239401857024 167913017747439616 83958707896975360 41980453460115456 20990776485871616
10495663120842752 5247968999374848 2624053219164160 1312060969320448 656047664529408
328032422199296 164020506066944 82012400517120 41007274000384 20504173871104 10252355371008

5126311903232 2563223060480 1281645084672 640839319552 320428048384 160218218496 80111206400
40056651776 20028850176 10014687232 5007474688 2503802880 1251934208 625983488 312999936
156504064 78254080 39128064 19564544 9782528 4891392 2445760 1222912 611472 305744 152876 76440
38221 19111 38223 76448 152900 305808 611632 1223296 2446656 4893440 9787136 19574784 39150592
78303232 156610560 313229312 626475008 1252982784 2506031104 5012193280 10024648704 20049821696
40100691968 80203481088 160411156480 320830701568 641678180352 1283389915136 2566846939136
5133828096000 10267924627456 20536386125824 41073845993472 82149839470592 164303973908480
328616537751552 657250255372288 1314534870482944 2629138460442624 5258414359838720
10517103597584384 21034756950982656 42070613413593088 84143425850441728 168291249747394560
336591295587811328 673200183361667072 1346435551095422976 2692941470935023616
5386023679358402560 10772328833693515776 21545220617340452864 43091567134587748352
86185386068989181952 172375275737605734400 344759558674466209792 689537131747441901568
1379110292291902767104 2758292642177843462144 5516729399543762780160 11033747029463677272064
22068070519679657967616 44137293960863922782208 88276893764737059258368
176558399215492545904640 353126021803021946585088 706270490350117602721792
1412577874188382624546816 2825229535353060087300096 5650606644658709851013120
11301508437222599054852096 22603607170255556815355904 45208394932131831042015232
90419151047505096906637312 180843024461493063458488320 361695493655951866207404032
723409876777835210995662848 1446857532487533379153035264 2893790622838792672629489664
5787732361405037173905817600 11575766954264978005105311744 23152138371439763324797976576
46305485668699141278770659328 92613389189037511815890731008 185231614081353482148480286720
370472899569263881330358222848 740965141951641596727511744512 1481968969529510861588614086656
2964015310311477059444409368576 5928185363127864791423181127680
11856680211265550927915087036416 23713979392550744545967623634944
47429196725140774472210146394112 94860869330360119704970091036672
189726690420877380931039778570240 379463284362069044904278750134272
758946375764766655892955886256128 1517932365610790443954708544487424
3035943959384095152247010632925184 6072046375093218833169208353751040
12144440966283649472368790883303424 24289453150973103562078330118209536
48580173952546435353558156939624448 97162883206293327165919307285659648
194330837014987567249444601384140800 388671815234776960334101176393924608
777363912879157572338626300039135232 1554768390577522448018100494580842496
3109617910793459502717896778166829056 6219398080863748218799185134343946240
12439120680281154864325153424708468736 24878890397669626582103873161458089984
49759078869553886871114878946998484992 99520753887537041156044023142161580032
199046700071932617139716576780652380160 398103784737582303934690214553963200512
796228338662598747179894551093243281408 1592498215700065772980817346157120323584
3185079508149868103203691180255508168704 6370325169799209320891495336393551380480
12740982646597364870751216624552172847104 25482629907192622199438885152634485866496
50966589042381029314750674112329252077568 101935836540753628461247155838779064844288
203876989993490396585985926905799251066880 407764613810947072498955084268080744890368
815550495269826703651876629449125975293952 1631143525835518524611686180724180921614336
3262372122262767283839238205100219785281536 6524914385708995036910208097504155454668800
13050169053784911012283879569615742677549056 26101018672303663901494685888446348891521024
52203398474075011556843225275322424855887872 104409519207085390621394157547504303857467392
208824482932041516258203729088727516006318080 417659854899824502547238286164892848595402752
835341487871131945156138228304661330356338688 1670726531885229770435599768559073927043743744
3341540176056391301117846161017650386749620224 6683254576684646122728985569834305838823505920
13366857602513019286444557635266621808295542784
26734412103313492654862288261729263877888147456
53470218003201893473670922505850568278370418688
106943223599553603275234536976485217601929084928

213892022385406839206254457882538597294234664960
427795195143412943724079683624213518769222320128
855612691032024418071300902966699685899950620672
1711269983554445897388884877369944668522913202176
3422629170089685917270335897612979930491850326016
6845436746140960079525804080972141047875748495360
13691230304205096649021872733436644469535592677376
27383174232256546277984274609858013686639376728064
54767775712205798515849607505685476868415136202752
109538405919797008951461331583309852727103037898752
219082520830364841742446896310497503434751606784000
438176459642271331163942258908750602830594275540992
876375755247625957685981450393012397583370675027968
1752797182421418506088156765937047179011105597947904
3505685708695170193608701262176139125710939691679744
7011554105095006750082177984956367786799336374927360
14023473585599346225893906891120914644353586732990464
28047677922017357903246915624658187430217001432252416
56096817345672046709412034934149091143453658797047808
112196557694618755224660477237963614852946629459181568
224398961395786834060993769215258094837971882648535040
448809614804672315345333167909177919940101012757413888
897642613635541925137357594775679300408516520435515392
1795331995323478439168097707466005521873662030712406016
3590757526751746056122960450761304885860582041107562496
7181702125713070467819450973181197455947680041580625920
14363778395845297646785962089679570280348392001892253696
28728305080528908715866044465993491297602847841246511104
57458106738734444276320329505255684069017823357417029632
114919206632822142241817140157048771085659902064682074112
229844399576350791861987242607172348066568314829060177920
459700771774114598480680409800494307923633651057512415232
919425488791055226474772668773287839428261344913808949248
1838898868067762511976369035891174126018510775425186136064
3677893517106829142006385468471545146360997722045508747264
7355978596156266520120065730321484081369947786481290444800
14712340316197749512454721047399755740035800257743126790144
29425446880165931969338621268313086634663409885047345381376
58852426255872729827535600883653323578510438509216874364928
117707917502827191432787918461360947775388114496678115934208
235421964987817846421009270310830496787510703949844966277120
470856189939962619952885407397878196048490357812667401371648
941736899808579094127504548348190797043918615451289740378112
1883522839474465896698476563801250403981713030554489356025856
3767143758663547210283888061812238427751177660412798462590976
7534483676756325254341645992043952095077858519433236426260480
15069359672371112176231031720926854669306723436081751854678016
30139503982459147687557542915531610296915459666594061713670144
60280577240352142045306044778419022510434944922049239435968512
120564293031571977430994007451549648854077941021820710889193472
241134863164879341542751850692522505374571984399085885812899840
482282280533229456447031372963891426081976173509060699694825472

9645896694734004596171180890854756828296167564398992555277025228
19292295557606840126803468644863370269905623317233542233315082224
9646398862872835530632287754008613441609455752834549678038056696
4823324973471125498931420592792770874133049923526164119709286640
2411725257752916616273348654290617513730685985317526700199772216
1205894014385135241540493506092424795197423504435985670272450566
602962699946906087472156342519770416764752008106603995222507752
301489196350622277087032965996664217965396131997607577654394888
150748521363895755218993880366721613774208129970956578848768000
7537622227624018594723563886755559282859096971554684435169228
3768909193526624714248716877587515583930706447881553972297728
1884503636620620065567825905898626601859229023592686861418496
942276338238963887005646686501747705876552411622298368344064
471150429083808870613690210027091055411745155724126652989440
235581344524067898862278538401654128942607052818552060903424
117793737253115681208855985894881365089670763887520397656064
58898401122098706493286351294467832854019000682882382430208
29449966808819786191072354820747491581601309711002283016192
14725366528295109567750766997130533368096559540281687408640
7362874826090163019982678291943660472696232112531116654592
3681533194016385628044986542661027130672092227460694802432
1840814487493844873049316969675112012498034199327915638784
920431188989748466038070334009855229830011142462741938176
460227567116287247775741091591077226705502592630763028480
230119769868850131266223508088613419247999807015077543936
115062878089778319322288235190844112571624158857386786816
57532935622565786505732358168690757759624207103617400832
28767216100121206675160299370979729616718167389270704128
14383982194479760048727209828807040176812115613366353920
7192178169449458379937134986062107772632573766048677888
3596182620829518367755332528860347728429544862707089408
1798138078467153772771048782344820785271401421194919936
899092423259774180832215650129733853164015205518147584
449557903642985737639453454543528656846164850219417600
224784797828042192431399542011095193555161048839880704
112395321917295758021536178375213029343619836285026304
56199122460285209913686292872439230954829574075056128
28100291980961270408302248278635973618924615003799552
14050511365889967929880675060526165880217221485035520
7025438370649650327805112990867172475486067734085632
3512810529177158345333494422573563100543176236282676752
1756450936514745763383665978018837886560245429305344
878248304220456177049929921584930135202304838598656
439135570091769736204013427080220663562243481272320
219573494036655691941530946683988129761667271622656
109789601513713257890527589913932963871106401304576
54896228004549334905144853242935931430689583398912
27448827626121020432512955764452690462912983072768
13724477062498368670622674245371870760524058722304
6862563718453431598098503512605534989512341454848
3431371062207509921541817899175858088202194649088
1715730132594152022017192021024474340824109285376

857887367042274541631737546230509818773560623104
428954833893736536127439540974391233567533301760
214482992133167900719505154416763778874143145984
107244283659733766687645269173165970482259820544
53623535626441791507768980568975025763724034048
26812464711508349835857663275683533143159078912
13406580804897901958915418133439776702228070400 6703464627020814499951002314518893416438300672
3351819425796339010222147781158949240881283072 1675953269041135385234397202529225886771707904
837998412592050632678860257239488576551387136 419010095331766786370260956607182104858460160
209510492183754128200545892297309960720613376 104757968350812431607980653145514434505998336
52380345304873899557844180071186944325844992 26190853217170791655849016784808335699345408
13095766890952316766387971767011599617884160 6548053586659618852425717570809515693047808
3274111863921539660828724629056615788576768 1637098467256634947722295236354236865314816
818570501276250032515114079090082918170624 409295884462091295584540270001523701841920
204653259143028787455761750229002972299264 102329288027505963559626682728622046838784
51165973241748766695686245171371303763968 25583651234872275805779574489215792054272
12792157924435084131858013196372966113280 6396245115717015180413119574069018099712
3198205634608244147448616274975776571392 1599144355678990352345336381458522046464
799592947026929315483182312714577903616 399806858107181727396848217349947392000
199908621350449398526052639171302916096 99956906823653966676840584833816068096
49979751486041617045327425040990339072 24990524780128125376117278832536322048
12495586908617721114785422572288737280 6247955713585689770756102864154656768
3124058986431259492059747221082472448 1562070058034837049370721505043808256
781055311427022176355784699773190144 390537796918312914013104323512238080
195273969061557369924158148568940544 97639519831979141420882067690881024
48821027566589798939842530548645888 24411147608595013584622013625925632
12205890716947563849661380988764160 6103103814798810453505877582282752
3051631135561919491090532335091712 1525855181862216877714062939521024
762947397971737004941429855748096 381483602506182785512914120867840
190746753013248534277556656930816 95375852386702837899328126713856
47689164133390704329938962481152 23845201036714994855106930802688
11922910003367318772622190182400 5961609744188570058845457481728
2980882243346740365689909936128 1490479807299597850978545565696 745259246462912759556068081664
372639294638013296811431690240 186324483022285106922414669824 93164659362781782719556747264
46583538607210505988953079808 23292373766515060309063892992 11646489114712433811825623040
5823395673083668734559649792 2911773394405560281603244032 1455924476134643097963331584
727981127533253027562520576 364000008499592253071687680 182004726616278996181057536
91004724491380932913135616 45503542837311183867871232 22752361714465950639587328
11376476005138154672619520 5688385576521667012722688 2844266575237128344567808
1422170181106711591387136 711103537297429505245184 355560992020751607398400
177785107696394231087104 88894859691206329237504 44448582767107771465728
22224867844306189156352 11112722152529246289920 5556505191452699000832 2778324653320387428352
1389198355457212678144 694617192127115821056 347317603262812651520 173663305231033696256
86833904415330533376 43418078107572109312 21709602003739475968 10855082476846448640
5427681975911579648 2713911356699967488 1356990862722072576 678513023547080704
339265307866562560 169637051979792384 84820725013151744 42411462018203648 21206280764915712
10603415260364800 5301845069135872 2650991254044672 1325529986760704 662782173249536
331399676559360 165704133246976 82854214107136 41428180795392 20714627268608 10357582069760
5178925252608 2589529735168 1294798422016 647415988224 323716382720 161862385664 80933289984
40467693568 20234371072 10117447680 5058854912 2529492992 1264779264 632406016 316211200
158109696 79056896 39529472 19765248 9882880 4941568 2470848 1235456 617744 308880 154444 77224
38613 19307 38615 77232 154468 308944 617904 1235840 2471744 4943616 9887488 19775488 39552000

79106048 158216192 316440576 632897536 1265827840 2531721216 5063573504 10127409152 20255342592
40511733760 81025564672 162055323648 324119035904 648254849024 1296543252480 2593153613824
5186441445376 10373151326208 20746839523328 41494752788480 82991653060608 165987601088512
331983792111616 663984764092416 1328003887923200 2656076495323136 5312290429599744
10624855737106432 21250261230026752 42501621971681280 85005442966618112 170015283979747328
340039364052516864 680096320291078144 1360227824954245120 2720526018652667904
5441192774793691136 10882667024564092928 21765896999081607168 43532919898070056960
87068091595953799168 174140686791534968832 348290380782324678656 696598775963158839296
1393233580723336642560 2786539219040711213056 5573222553269498281984 11146733336915148275712
22294043134582599974912 44589239190669806796800 89180784224348827287552
178366180134716081963008 356741583641469018701824 713501614027011746955264
1427040121542170913013760 2854154030060636664233984 5708455634073863004880896
11417206416052905362587648 22835003127916169430827008 45671186847453056272957440
91344734878147547368521728 182694192122777964382257152 365397828978521668054941696
730814547422974814690738176 1461666873777812586543185920 2923409305419351087409790976
5846969726566154003466420224 11694241684587211664226516992 23389087832084230643040387072
46779384589988075915255480320 93561187031615381088860372992 187127209766509220694419570688
374264090939575358422236790784 748547524692264550911268880384 1497133735010756769956128358400
2994344484127396887617943791206 5988844425052848424893238214656
11977998335115518194855201210368 23956615640250679079847851982848
47914469220540643539970603089920 95831414321159857840491004428288
191667780402476857202081605353472 383345464325267997446362403700736
766710735691164560977123193389056 1533461085463586254123043158753280
3067001399089686772583679861456896 6134161254504402073842546810814464
12268639421658861205035467797430272 24537912668617836524771683946463232
49077092987835901278944864596131840 98156721276872259016692722598674432
196318513156145430950991432010170368 392647167517092687737194837645983744
785314617443789027144813622543253504 1570669799706785357630475139589079040
3141420729051985321942646068183302144 6283003717380799857248683714376892416
12566331953315258141224150584774361088 25133312943737833135901867481589874688
50267923961690299978710867587262054400 100538444071809867371236000422688718848
201082080440478269570100531341706657792 402174545474673608795458123676071755776
804369860136781356901430369337460391936 1608781258648430992423888982645554544640
3217645594046598542089834453232376610816 6435457341592670198663781882347288264704
12871246990184286626295789716459646615552 25743158594366465710528031336449433403392
51487646416728716336928966479959147151360 102977951289449002505603740574038854991872
205961219490881144674699096376318831362048 411933072805728568676381423209119905480704
823887413259389696006729307331204296474624 1647817361814644509321391536488337563975680
3295719794221019253258648916628533070004224 6591609729625498975749029520560782024114176
13183559741617918889961522415728995816439808 26367800047969679656849971580672855169302528
52736961225407043067553796659775437411450880 105476644709749453642815300316410328968593408
210958733937369642301046014626539566228570112 421928356910480754632922857240516949039906816
843878491892444449327507370455909531245346816 1687800539927854778778338052861570328821760000
3375688192141641317803322729622643190305652736 6751550608855146156099938707044291445935570944
13503449666854019353186463909686593022519672832
27007596231995492788346100810569206306336407552
54016586260565893740638547603530453135266938880
108035960114281603809169787171844987315722125312
216077495414862840274124958273258136721820745728
432166141202324945859820684405652597624394481664
864354583149848422342782904529577843610294943744
1728753767790093905931848880495700983943601848320

3457596738560981934356263903864492561333227618304
6915371883083552113697660093475166309558503079936
13831100578090280717365584758442694992901101846528
27662291478002691441467169865987011473333703950663368
55327256807746534789224455605709678961877172879360
110657368110878481498211027783358256914027111251968
221320445212527786835946288710594311808599753490432
442652308406597221350941043708944219578290568953856
885327452776277738059979019993399631078763261853696
1770700577478722066836151905137821646001890771599360
3541492498809777315104691540577688059692510038982656
7083167685324220993074158541759465654762477069533184
14166700746057777471187786800472711038027986812202112
28333413224293421487521483785187057890206956421067571 2
56669725987505760653347879388573874087158784353894400
113342374978286183112532166146813180740356880572874752
226690595963121689836737147032957226612792384875921408
453392883939342026896819923544576183489742017212186624
906809151904881348240331106046475827507798529345060864
1813665071862157285374044730007598576072226048531496960
3627423679829103748534854495844490994257710076745744384
7255034431867785852643239063347569672741936112856989696
14510443008154728416433538270012314713936904144444981248
29021634305147770255161119682665898016477987212635196208
58044765187972167354910634226586661803371871927627939840
116092523531297588398997749599710726554367999205103894528
232191033373301684176348461492496259003984509109903818752
464394039368016383109402847571142129798466039619199696896
928812023978858795732217544314583483177926122037183512576
1857671938443369650491258786973765413517840329671935262720
3715439657858043419036164970636727721359656830539007000576
7431070877658695074179624734651849231367266003468286951424
14862524879202606620573839056060486040030436691717119803392
29725816006175646185557685728563454723465268275299533140787 2
59453164507892158260012072918296244778488984245112846417920
118909394006866048297740862530646790175345205968470060040192
237824917995895560150915158449402181587424886893428854489088
475662095956118047412697183675021565648318723699835177795584
951348711840889949047128100902477536243575347225625293225984
1902746463539087606537723668909823882381026494103160461721600
3805591006792790629962382272029385384549804587510140673982464
7611378173014812093698634412478246008675112373627920849043456
15223148664888085854945008561795442496501231144447112070024396 8
30447081967493095044985496597268785951304475083372799404802048
60895733210420036760161952141893373819212975755606714818232320
121794604971707766860705822178498351471634002688935661653721088
243595487045150920402175480146419910609684107733315787341955072
487203528293772614165878631871686236552200420177520502752935936
974432164994486775054812606901065303770065249776818861643923456
1948914546802856643555735900117516268871459318397193435563950080
3897929527233479474003693172865803860405576274481498295680106496
1949014980430622830447957272748287591534116956084304860116156416

974532598622252961947033979531836626432722887463930286196129792
487278853514597254335044661344764728549193648442854071167090688
243645703859034013848286166461805571941012926576871499618058240
121825990480384700264525001125614389803714514466157981826285568
60914564515626196802453459510162996818461282821940106921771008
30458066895530021736322209228759399367532654205400611465199616
15229425766623472535708844351218650162917333499915584734756864
7614909042740967101628292044028800321034169948565431868456960
3807552601085098967701080956224137780304836573586535684767744
1903825340399857192294007945216937700046294086445177717653504
951937190128582450368737706160903254970084943048543796461568
475980855028618152295235719856668829958511421437249367048192
237996557496472539703051293316443016215990185675113417932800
119001343739318001629242363352275808726362330315801076170752
59502204365199866703479540023165054672364783897022721687552
29751868430370366296168949185096102490774201318072452644864
14876317339070399620299064179304838822683005343816772222976
7438350231477808046256826883030813199989455014298659061760
3719270896710208141182060838204603494318703678344466006016
1859683338840756129617854117446900194321339924769801240576
929865614663204094322338907895749320741664005183684739072
464944779953015061917875378534024272161329023991234428928
232478376287214038337290651560086941975913022695313244160
116242181298960272857821806926580873935580766697504636928
58122587227156763273499144036559138441602511023676325888
29062041902416695059043692304913919957707319349300166656
14531395095627504240668906295774135347306691593381085184
7265884620023330475907983219545655357879861756056043520
3633035846116454415740756645602121521053188857710772224
1816564691110621796763760840715707681583223418696761344
908305729581508192828571679315177301319926204269068288
454164556803852743637631469136250380924120349594877952
227088124408475695430488549307456055594138798527610880
113546985207512509521080682023393460363108711128891392
56774954105393585663458544696529446464574011496988672
28388207803515458283188374190681081373796833714765824
14194469277167061867323738016548719757653330840518656
7097417326288197296526644468878449414204122411827200
3548800006996431829695709964741269474790789701697536
1774445675424382505564048847521657121239759098740736
887245673675274548140121356336339752542061673316352
443634254819178921749109144455925472232121898631168
221822836400360284714078805371840534096606480302080
110914272695565554276801519257859166038576005644288
55458563595475483098281817914899032514424385568768
27729995421584094529081438100434241004780384157696
13865354522715223754510983621709482876174287765504
6932855667319200122240624096600922624979191726080
3466517036640394183612878191173551905935619784704
1733303119810594153052722167023321249690821853184
866673860650495607149502619229933273206916907008
433348080697847068886322077474102960784211443712

216679615535223167098946422666619642482482216960
108342595360761399877365903298093902286429356032
54172691476955608102629297631438991665808801792
27087042636765258133287821806915516094201462784
13543869767526356107630497399055768177749262336 6772109108335041574308541947326889154198896640
3386141666453452547400917597562947109761581056 1693114389369692153823782110731224821211856896
846578972756329016973552711340488043771461632 423300375413905978517607183657681838468497408
211655632224823724274219005822559827525632000 105830538371347229644817209908139367908507648
52916630315141298576262458452499411027099648 26458995722304491165058155975464569049972736
13229838143519166520992541362339716293197824 6615089212943043729728002368473574030704640
3307629677063252099479867027888644957405184 1653857373827491167047866435770251449729024
826949954561678142177899678798090210377728 413485611104805350415933069855527347945472
206748122464385814871458150156004795351040 103376719688184477267474882692122958364672
51689689072088023549610345153121759526912 25845509150041904232741624480091019935744
12923086882019898345339038191810580054016 6461709594509422287153632071787825070080
3230937874004447700818872523835180056576 1615510475377092129030464505888223789056
807776006875980203825746374929428774912 403898388031707171568130248457372827648
201954386312712120611693654725015633920 100979789304785327719661092610672427008
50491192726607297566737678929418518528 25246245400410965636822405776750411776
12623447218759141245137986044395782144 6311885868656399835932384600208179200
3156024063966614524647888089109233664 1578052596802514565664791939057188864
789046580810860934502819916779880448 394533431610232293086621932015583232
197271786407517059460916952820613120 98638428504958986189261469816717312
49320481903079721324032231611564032 24660874776839974776716864157384704
12330754301070044445708806254493696 6165535606860050751529590215147520
3082847031592539640102388651524096 1541463129877526952219991097737216
770751371979392042194393934856192 385385589510010304139396160421888
192697746515162293590797676707840 96351349137659717555948636602368
48176912508869144158249217425408 24089075224454214769262058274816
12044847097236928729699753918464 6022578291123375037384239349760
3011366516814142854959300870144 1505721944033299095613241032704 752880314829763381873415815168
376449828821438607970105556992 188229750113997762501751603200 94117292908638110509225213952
47059855380138669883787313152 23530532152979142256481009664 11765568307944474785534181376
5882935269699689221413928960 2941543192713570525030383616 1470809375288648219676901376
735423577110255588419305472 367721233288093533500080128 1838655339010529636395253760
91935030688506253020233728 45968695935873843921420288 22984938263747280666361856
11492764279778819686006784 5746529713841999519416320 2873338643897294597914624
1436706215436794718060544 718371554462471068581888 359195000603272389066752
179602111987654621921280 89803361836836524654592 44902833839922869174272
22451993380713738010624 11226284920733020717056 5613286575554586214400 2806715345371331035136
1403393701482684481536 701714865139851722752 350866439769180602368 175437723484217671680
87721113541922521088 43861682670868103168 21931404285387472896 10965983617670447104
5483132546323578880 2741636641905967104 1370853505325072384 685444344848580608
342730968517312512 171369882305167360 85687140175839232 42844669599547392 21422884555587584
10711717155700736 5355996016803840 2678066727878656 1339067723677696 669551041708032
334784110788608 167396350361600 83700322664448 41851235074048 20926154407936 10463345639424
5231807037440 2615970627584 1308018868224 654026211328 327021494272 163514941440 81759567872
40880832512 20440940544 10220732416 5110497280 2555314176 1277689856 638861312 319438848
159723520 79863808 39932928 19966976 9983744 4992000 2496064 1248064 624048 312032 156020 78012
39007 19504 39009 78020 156044 312096 624208 1248448 2496960 4994048 9988352 19977216 39955456
79912960 159830016 319668224 639352832 1278738432 2557542400 5115215872 10230693888 20461912064
40924872704 81851842560 163707879424 327424147456 654865072128 1309763698688 2619594506240

5239323230208 10478914895872 20958366662656 41917807067136 83837761617920 167679818203136
335368226340864 670753632550912 1341541624840192 2683151969157120 5366441377267712
10733157632442368 21466865020698624 42934829553025024 85871858129305600 171748114305122304
343505024703266816 687027641592578048 1374090467557244928 2748251303858667520
5496643345205690368 10993568165388091392 21987699280729604096 43976524461366050816
87955300722545786880 175915105044718944256 351839217288692629504 703696448975894740992
1407428926748808445952 2814929911091654819840 5630003937371385495552 11260296105118922702848
22521168670990148829184 45043490263484904505344 90089286369979022704640
180183184425976472797184 360375592223989800370176 720769631192053310291968
1441576155872254039687168 2883226098720802917580800 5766599771394195511574528
11533494690693570375974912 23067579677197499457601536 46136339946015716326506496
92275041075272867475619840 184554804517028604596453376 369119053767022948483334144
738256996999977375547523072 1476551772931817708256755712 2953179103727361330836930560
5906509323182174490320699392 11813320877819252637935075328 23627246218548312590457503744
47255701362916239810089713664 94513820577471708878528839680 189032476858221876273756504064
378074625123000669580910657536 756168593059115173228616613888 1512375871744458014590823825408
3024829114741371365448828846080 6049812971987653403432020082688
12099935428985128151932764946432 24200489827989898994002979454976
48402217596019083368280858034176 96806911072116737497111514316800
193618773904390616515322625130496 387247451329095516072844443254784
774514709698819598230087272497152 1549069033478896328628971316969472
3098217295120306921595536177889280 6196593046565642371866259443679232
12393503005781341801082893063159808 24787639836862797716866534477922304
49576547324325823663134565659049984 99155629949852103785072124724510720
198316330502105120487750236261842944 396642802209012066810712446149328896
793305886827627785291848839549943808 1586652338474462873924545573602459648
3173385806587340354530786936210063360 6346933872451509922424965450430414848
12694192263456678271576714056881405952 25389033564020673396606994425803964416
50779365202255980500121121475690233856 101561326552941228414056508199545077760
203127845402740991655741546895419375616 406266075399199052966740154783497191424
812552919985832245243994431552311263232 1625147378346532769109017107075256287232
3250377833442802095460090702091780096000 6500921820385077305404294380066095235072
13002175947769100839776814711897260556288 26005016509536094137490081327324661284864
52011362247067973190853066461709602914304 104025382950127516213451940537539766517760
208056082812238172090395496303320654413824 416122799448442623507774223063123551584256
832266866544817805669514907039211588681728 1664576268385500728646962735904352148389888
3329237607362731691909791315460562238832640 6658645355908923853051314318224840361771008
13317630994184768644566092011057112491753472 26635942553103379166059110771329088519929856
53273246235674442085972075041087904112705536 106549214730284251679651857079035262371102720
213103873978439238374719128151789433033588736 426218636992619946780269084291016682649944064
852459052056722833622199824556908998465421312 1704961660256411547367722961063569263261908992
3410010432798754854982092546026641059185950720 6820195090169373230457478339852287183696166912
13640738629482473501901543175302584498040864768
27282174157252401085776259341801189257378791424
54565742111079710335498864665994419037351706624
109134271815309236998890421296772919119891660800
218274118816918106653566226523114000330159816704
436559388006435478618703220905364324841072623616
873141076758069487860547977529001298043651227648
1746326755006536036967379026494547892810314416128
3492742712993866196427324195862186379066652753920
6985663831949320637839780677470553945025353351168

139716844758218177656498259264334702638348023889 92

279440825754899885112401809958516652752377961512 96

558895923986726829823614202776727800456119750492 16

1117820392927307778844849571272844590814967155916 80

2235697875762323796084941473984467161435389621698 56

4471509931340064068960367610846490282481689863127 04

8943248222310961091501704547448092484185200965713 92

17886953163883588090165347746406408806814044410347 52

35774819766290507994654572795833265290515373778534 40

71551466409627679617956900197707425934805317472747 52

143106586573348686493209309607496642577159774776852 48

286220480654884027501009637639156866569417829216419 84

572455576326141364031201312126640895969032217758269 44

1144940382685029346120766697949936117598457554167398 40

2289939225435551928358261543293180886517701345636515 84

4579995371002090328949979381372979075676975165876469 76

9160224582266153602366871352319192756637095280959815 68

18320916845056253093667567883784854723840480460333383 68

36642769051160397965202786125862647868813540717494272 00

73287408824416579486140872968311172579892241028643553 28

146578559093024726083752347369794098844314801244597125 12

293164601074432586390445897605931705057690240863814287 36

586344167925631441226774200944550424853501758476868648 96

1172718267404795419345313213354474879183246070452217446 40

2345496397916655912474156049639697817318977247901395189 76

4691112522047441972515371345140891752542924709796710973 44

9382464496523144240164861182004775740895789847581263134 72

18765407897902809070597959347455535953411460551138208645 12

37531773605518659321732392661803040850062682814227782041 60

75065462830463401004537733257390019586604889052358293585 92

150134756899778966731221362382347914946168824952522046177 28

300277176277262262906734516499831581438255743600655010365 44

600569677509933184702052616469934665968347674592531856752 64

1201170004930683687181272399880412338120367723967507385548 80

2402401309683002009916879133641910688608080197499902115184 64

4804925219009273290942426935045993401950849894129578918543 36

9610095637305085124102191205616330853371078786518707213434 88

19220681673183247332639057082281349805680915569556513179566 08

38442344143512648834147463506660075809239347132151223864524 80

76886649881317606006033625697514904014233726250378842739834 88

153777222951219828687544648763419312819977516472910475501240 32

307562292279608890726044092263617635222975160890126531045621 76

615140277313556248153997774000793289611990577668864222177525 76

1230311940135789429711814726948702617556061667114950764527616 00

2460686651288932726231267811791637311776284357784346169400360 96

4921498844612573186077812339371738776880890762677581619490979 84

9843248773294561839386178110320405860418425619572941800362475 52

19686999714727954613233463083794668334150139427581440723485982 72

39375003765733571095389139893897049894926855232033995692494028 80

78752016204022465928622707240409526243106863217810219876032184 32

39377012438288894833233567346512476348180007985776224183538155 52

19689008387283278351077890536410094787403292181323669214530109 44

984475527771105464276949869978197570035829018487961316864557056
492250318089023505500002606567945200350746713954869586501304320
246131436146247139430765139073395807841789459332879257285165056
123068856623991263055764487431409507754102780844161860659838976
61535997587429478198073202663060555793655416010942046338547712
30768783431431662434132080805208178855129720799901581173587968
15384784034574292884613780139443039906715866797166069588951040
7692588176716377276080759938140995192933436597190674295554048
3846392168072804054927314903280235216254469897899156898316288
1923245123893709735907124918744986418021110748601488324427776
961647081875508722175296192924927613957493274126699099848704
480835800902081288198514963238681009452215586976327018741760
240424030433204107654690915007449105962842268444652243779584
120215080207683785605062174197778853599788371700570489094144
60109072599382758691389445445916577109077804589407428149248
30055302547461812290123901896471863709130711664264805875712
15028034397616122617276540534992719431861260516912948838400
7514208760750669544745565060874753504578582600846747369472
3757200161346638890426429927126573646613267471618510159872
1878647971158971504240038661907885270468621821406823317504
939347930822311781633431180126241858815304953502195777536
469685938032568905573421514649270541198149498150489948160
234848955326990960165063719617710076494323259774941003776
117427470818848733771708340955392441194785885237318516736
58715231987100993730442411050964922071205070293583265792
29358364282388810287515325812116811772508598984253636608
14679556285613561854904723049375581254707331410857820160
7339965215016359283025891596346378311580181664794411008
3670076143612968819299710834002482997903348812079955968
1835084839858878998543237934915888420008303395881353216
917565803955636793718310226415267670532466192861364224
458794593990917044082500742686295565530390343891025920
229403143002007845652923186082478647897273795675684864
114704494504278584632298000410904756514676209702928384
57353708753776623219067203890285094540357760784007168
28677585127706977060992703787558905411688708358275072
14339157939262821256225902814987631776599268162273280
7169761657336076990977726868097905423677091072704512
3584972172520371676921251164350997479527274032136192
1792531758186352429176819447326521123608001263960064
896288715056259509946506656238771753726182755926016
448155775509671402652301794407141472824182439936000
224083596745606525165675130347448534392636750954496
112044652868188674502599681745663166186591140970496
56023753681787043211180899158801032588431953231872
28012590464739874585530978722385241041784167989248
14006652044293113782735753932684982894676179681280
7003504428108145136353009252088672634230137683968
3501841417034866690669070768917426910561092763648
1750965310007830406580818455895258752003558342656
875504955749113733913550763665902024363285151744
437763628247156132268346149692087336362395566080

218887389309877698789958458775611830271574278144
109446482248088665722871921352589996180975386624
54724634920619241025382306658687038613081817088
27363014358597074594664326320539539567837970432
13681855628442264338318749655867779914567516160 6841102038792995689652668075732895022608023552
3420638131682429605072980661765950043966144512 1710362621984180682659813642832726288314138624
855203089063573281391568477391238777322602496 427612433567528110726615066683057205244067840
213811661301634790378722947335247510913417216 106908552909752762697069180664483209602400256
53455637584344065102388443830671331874045952 26728499356905874428121148664550529473445888
13364589960819858152524037706882696504934400 6682465121593389545493750540745064136572928
3341317631388425007362741114024390010339328 1670701350990077620989303478838123976196096
835371943142971369148618200332026473611264 417696605395451963901292330622495479562240
208853619609709121614137780539488861159424 104429468260846130638814697883864991268864
52216063358418850235280252748992775979008 26108696293207317575576578278026528161792
13054680453602605016756515090778334167040 6527506380300775622862370521271702126592
3263836266900124368673241748577118584832 1631959671824930462957649118259193053184
816000605099899370789338681114913406976 408010687143666755049926401550115143680
204010535868691912352591731271386791936 102007864082775223590110130883858006016
51005230115602245501962198066011308032 25503264094908439604434665345046806528
12751956566007878228944115828543979520 6376140542280768327835449492282277888
3188151400778798770599420535146283008 1594116265208606688640558162075713536
797078415013906995990703028289142784 398549348711755323830563487770214400
199279744958278574832887730697928704 99642407780339743875246858755375104
49822471540770100167024926080892928 24911869595685164198213211392049152
12456251710492639156456979871825920 6228284311571348106903677023813632
3114221383948188317789432055857152 1557150306055351291063512799903744
778594960068304211616154785939456 389307383554466388850276585963520
194658643537390335946237889478656 97331797648773738733668742987776
48667136764426154747109270618112 24334187352232720063692084871168
12167403161126181376914767216640 6083856323068001360991745998848
3042005532786456016763054194688 1521041452019455676515117694976 760540068822841672324354146304
380279705817977753195574722560 190144688612267335114486185984 95074762157772896815592505344
47538590004706063036970958848 23769899465262838833072832512 11885251964086323073830092800
5942777097770613365561884672 2971464106749032597104361472 1485769832306379255713890304
742903805619121106437799936 371461347542526292509327360 185735396137746015899877376
92870059252114442772545536 46436210217677938797576192 23218695404649328104439808
11609642850229843405045760 5804968999067511378935808 2902558286510050527674368
1451316036743172682940416 725676465115660051021824 362847455929866880286720
181428339650951867531264 90716475668485147459584 45359390755747180576768
22680271838625893711872 11340424149689098567680 5670356190032625139712 2835250152610350497792
1417661105102194212864 708848566949606588416 354433290674058035200 177221148936656388096
88612826268141879296 44307539033977782272 22154332466942312448 11077447708447866880
5538864591712288768 2769502664600322048 1384786516672249856 692410850522169344
346214221354106880 173111508723564544 86557953385037824 43280076204146688 21640587857887232
10820568806850560 5410421842378752 2705279640666112 1352674180071424 676354269904896
338185724887040 169097157410816 84550726189056 42276436836352 21138755289088 10569646080000
5284957257728 2642545737728 1321306423296 660669988864 330343383040 165175885824 82590040064
41296068608 20648558592 10324541440 5162401792 2581266432 1290665984 645349376 322682880
161345536 80674816 40338432 20169728 10085120 5042688 2521408 1260736 630384 315200 157604 78804
39403 19702 39405 78812 157628 315264 630544 1261120 2522304 5044736 10089728 20179968 40360960
80723968 161452032 322912256 645840896 1291714560 2583494656 5167120384 10334502912 20669530112
41340108800 82682314752 165368823808 330746036224 661508849664 1323051253760 2646169616384

5292473450496 10585215336448 21170967543808 42343008829440 84688165142528 169380625252352
338769840439296 677556860747776 1355148081233920 2710364881944576 5420867202842624
10842009283592192 21684568322998272 43370236157624320 86742671338504192 173489740723519488
346988277540061184 693994147266166784 1388023478904422400 2776117326553022464
5552375390594400256 11105032256165511168 22210627462284443648 44422380824475729920
88847013448765145088 177698530497157660672 355406068193570062336 710830150785649606656
1421696330368318177280 2843464718330674282496 5687073551849424420864 11374435334075000553472
22749447128902304530432 45500047179309215907840 91002400201627645509632
182009412089273718407168 364028047550584291590144 728074541845242292731904
1456185977178632004567040 2912445741333558847340544 5825039056619707371094016
11650373261144594095013888 23301336818099546895679488 46603854227819811202662400
93210069638881057227931648 186424861644244984101076992 372859168021455707492581376
745737225508842893566017536 1491512229949548744293744640 2983100017762823402910908416
5966351151253098634468655104 11933004533961100926230986752 23866613530832009167049326592
47734435987483632963273359360 95471289826606495184896131072 190947415356491448886491086848
381904502119539814806379823104 763828347052193463679554945024 1527695379730614595492700487680
3055468130713684527252582170624 6111091003932279727039526731776
12222491492874380799147778244608 24445601955768404288433006051328
48892441851576093957140911226880 97787359583230758674831620702208
195579670926618658870762837901312 391169245373551600783724868796416
782358297787731767651848123580416 1564756209656720667472493019136000
3129591647475955599282579582222336 6259341751276939727240346252345344
12519000415203936511831066680492032 25038634655707987138362881712586752
50078536962016202506127260128378880 100159609225232861471057513663168512
200324289052866635859721014139158528 400658719310535097554654001903960064
801337721030673846779731951059206144 1602716006880554996900311796620984320
3205513143399524600482319382247112704 6411188546075878414328030342504513536
12822701610705415255382843841029603328 25646052258518147364219253994100359168
51293402591250928435345640612283023360 102589401330931124284505546472730656768
205183994958720783396639623441790533632 410378374511158636448536307876239507456
820777518209751412207586737737795895296 1641596574794371103036201719446225551360
3283276226338478763314459926833718624256 6566718606176430641113032829549972291584
13133769519351807511194291610865014669312 26268203652701507480325035125260169510912
52537736533398799876522974057580619366400 105078131522789169584791755729281799421952
210161579957561478833075126686804720222208 420333793739089236993133483830091683201024
840688855126111032640233428573147851915264 1681420245548087182588399778972224674856960
3362925561687904599792665401596307291766784 6726021264559269668817062490496330467639296
13452382811485460276097588355600092703490048 26905446187704762429122103460415048943403008
53812253504877208612098060419259824959651840 107627229268689784731903827835379104064995328
215259903055250304479223069664477116421373952 430530695146242078989276967316392049425514496
861083168363967098040215590607659732016562176 1722209892870900076203754493165070730364190720
3444506898027731912654155610229643993390514176 6889188020627327345801604468258293052105293824
13778724490398381732589795432114596234859118592
27558145879084217547152763855425212731015299072
55117685554743343258251873693242465984624721920
110238158702636502844396439351269013014437691392
220481892591572638344578262632106188119251877888
440974935555744542000727293123348700419256745984
881972171856687614624596121964970049200019472384
1763988945203772290495475315366485395123050905600
3528067093388338703483516773606061383692125732864
7056312592738265651952165832958303954276299309056

14112981997399707793874596237408970282336694304768
28226677618645768567689721617802665312241579982848
56454782484984243095260501521574780119619542712320
112912419465353898110283119615088459229511850917888
225830547921478620060090472374054716439569232822272
451672513824498887799229411035865028840229527617536
903367863612081070956555754647241249602641179181056
1806781399150328732629305374445504883049646606254080
3613654142152990646690998479193054533788021708292096
7227490972010647656246772418990198602953500408152064
14455347319430628038223095759188576276661914799439872
28911425389679921527905293360793510694833657565151232
57824312280997173958728790406419737672686971062845440
115651547565269009723293988182504907911413253990776832
231308941137087343058260791104340680954905131711725568
462629574287273333339867211687343092173967510883794944
925282532600743961126425682332009644876249516688277504
1850611833253882511146233882578666210809128023217930240
3701317202612554200079232800986626263731514026118610944
7402821477434686755731995673631840211689544011602722816
14806017099288530222611051490580855791832119941936447488
29612782487415373867516223267796062320570303721334898688
59227061552507374579620687108860826114952735117593804800
118457116260368002848417855364259055177529725585035624448
236920218831442513075188673021592916250307961869767278592
473852410284298040907083270629335444291112945138926616576
947728765811422111327578390430970112163219933076637351936
1895505422108496281681980479206538671488427951750842941440
3791106625188296681417608355102274237300832074696822358016
7582404812319201598942511503582942263249616491783917666304
15165192748523619670099612593922672103795137668348381233152
30331151744817672284628404361358919362182084706257854267392
60663835985176210458115167069744989033547788151637892136960
121330736961434152693947050833544278685462813781520151478272
242667603905031768943327535055197158607660102519529037365248
485347467774390464997521936886611519688789154952035543547904
970719455477434784216777607325657444324516209730026024730624
1941487950812177276877022681756183698542908219111961924730880
3883073981338969970640980297722105016873568037527743600001024
7766344122107170775055830463863685273322639273663126701080576
15533080563072803217659400664566321025796284944541532404318208
31066945763862529770414280802810543009894582683513622812950528
62135460803158906211019520552976887936393190955888361634529280
124274060157185505762420959000665379705994433089498955286315008
248554397416106398205605753790753967078404968534442374607142912
497121349035683569772739179160354349489642141779773677283311616
994267806478308686268533701478401529644948692981325210704674816
1988585829770500465983178089272188720621226204806206133685452800
3977272093168767118858577551175148763905110047299523691923111936
7954745053593066611501597847611840173135535369973270232950636544
15909891841697197970572081185746765636921701290694986164110098432
7955146788104131359070483338134925463786165920721715931159461888

3977673827679831866427463041698234054555740598047969390131937280
1988887130653799026659842207164502688609199117867540407342071808
994468673733841060052976446739944174970263968355548059809087488
497246891070391303388015894948818502817964188888662957973569536
248629722636931038374771783263832459075398196799775943021297664
124317999869333212527767809526627833370907149577610203527905280
62160569210100452934074863710669718602057600377666217772580864
31081069242767149802132911329012760259330812983263666890604544
15540926940242036568614195401345330608816412888847112447459328
7770659629550249118080967569092140543983709643031195724808192
3885427894489739975927418718755807891779606420819417612943360
1942762987102177696407176826482772755783679010061618681741312
971406013479742702425322146793820782838777404856764278505472
485715266704198278323527940173127593892857652341359608070144
242863763334262602717197403474672398183163301127168538443776
121434946658213033136315418431390499709948888041828636426240
60719005824647382457016067562722400164158062760036501815296
30360269160094124172937212954874775236670840749579342708736
15180517703932278558683196064194175195631325059570217254912
7590450413908747515448892825475481386463614872175381577728
3795320987925677875778093809426937587555783607282827264000
1897708384448490996915870603058067240939879889238981869568
948878137467071527971347150701332844050933987418275053568
4744451041354948778742379499936816033815964015108529586176
237231506988180896749542712261482822803230518253960822784
118618746649443702063947837277278814349239514476828426240
59310869902398477876562159211908108648431884913338220544
29656183240037552360575199892588405061122006294131113984
14828465764437932891434660089611377899014035065796558848
7414419954428544801290860116464276633733533492263780352
3707303513319061578432195094061432158980024725814640640
1853698524711925378109480064945363000546641352748695552
926872646382159983501431291430004960801635171295035392
463448015204178638974061275193664210664974833107861504
231729853608638643098703452336162970464566040284102656
115867849807593983355188133537746917798322332007137280
57935386405434322580512270453706175182180821936111616
28968423953535826741715237069269445732600238934327296
14484577352177246096587169455842901937055033450299392
72422471363793289411158360188525540503904973716717568
3621327025748977887011567824564815019641215354142720
1810709184800655534221977777433429893664971924963328
905377428363411062469085821292226138754668086427648
452700132163247178913591376933868665338425105186816
226355775072394413296319921610812130649758083579904
113180742031582618567922077377344964315151807283200
56591798263484015243842096974641931652712286388224
28296612755588360601861577630305690573924334567424
14148663189717356790901053386645207660746262970368
7074510000820266640435658979068785017265179328512
3537344203390927442710395632407483102078613585920
1768716703185860782601480887640286847762318753792

884380652338128921923881979538416072242665357312
442201476541663726273511757628344360302085668864
221106313457131495792541262743740342241419329536
110555944321715564224163323336654252165897912320
55279365957432690276028007650719166605543079936
27640379877003799219987176816555603564068601856
13820538387645626650980174903875811912682831872 6910443418394676845983380699736911021665681408
3455308821483270183238336973767958043494973440 1727697966884600971742491798833730288078553088
863870761513783425932907555391740777204809728 431946269792633182997284605683308205185171456
215978579414187326514057716835373010883969024 107992011966029030764736565414545959587676160
53997367112482199136222136205702706866683904 26999364120974941445037994852066216969764864
13500022342854391660982460800640540253093888 6750181312610656299722962087623986010652672
3375175726897058384477346887463850947379200 1687630398744394309546606365557854444716032
843836467020129713427269643691891707871232 421928867334031136040618052302428096692224
210969750578998707683800641379455169724416 105487533745490923673646128303848145551360
52745096100741246752695967958984353120256 26373212664368515834284435883022316732416
13186938639183204146110443893276228452352 6593635473091075187539334922520649269248
3296900813295274151011723949201592156160 1648491945022505354126890218571429838848
824266741698686816373959231271031799808 412143755443060477842236676628174340096
206077070018388773748746868810416390144 103041131157623654288187699653372805120
51521863653026460851000982450768707584 25761580863620547278954057537425506304
12881114950363932066203811924733329408 6440719734458795246465297540376952832
3220440996867812229914344559193620480 1610261063253113418298020174099382272
805150814036160360819434034300977152 402585548222882006244928990776131584
201297844713841916040070482200887296 100651457658121414478838234506854400
50326996479660935468820613956632576 25164132065130581849111055329918976
12582382945215347981905901840760832 6291349928932702519628138008281088
3145754192628865524151662548090880 1572916710395689894244628046020608
786478162238473513206712408997888 393248984639551039645555397492736
196629444079932661343877295243264 98317197920044901432488445870080
49159836900061736096519122059264 24580537420050510738397010591744
12290578195035076714267230076928 6145443840022449029667977428992
3072799291263679851101169909760 1536438331258067593684175552512 768238508442147630908883075072
384128925627630732487839186944 192069298517093824760618418176 96037067110186141638658621440
48019742480912685448504016896 24010475703366150038839361536 12005540083137978676713357312
6002921157296441167003516928 3001536136511946497825177600 1500805847187836206074298368
750421813059849581618003968 375220351262890530099429376 187614897997928134694928384
93809810182205502170071040 46906085682723468496338944 23453633137172092953821184
11727111716491225829736448 5863703432198202591281152 2931925503075396133847040
1465999645025845486026752 733018269256996452564992 366518358000535081058304
183263790686285967917056 91634201186152197652480 45818253514580705673216
22909703218042656260096 11455139839397479841792 5727714034886815776768 2863929075037445816320
1432000566315741872128 716018297556380418048 358018155977444950016 179013581588349845504
89509042593988608000 44755647196901146624 22378386548403994624 11189474749178707968
5594878112077709312 2797509424783032320 1398789896763604992 699412540567846912
349715066376945664 174861931234983936 87433164640747520 43717681832001536 21859390671814656
10929970213814272 5465122545860608 2732629992407040 1366349355941888 683191857840128
341604518854656 170806554394624 85405424680960 42703786082304 21352429912064 10676483391488
5338375913472 2669255065600 1334661087232 667347320832 333682049024 166845218816 83424706560
41713401856 20857225216 10428874752 5214568448 2607349760 1303707648 651870208 325943296
162975744 81489920 40745984 20373504 10187008 5093632 2546880 1273472 636752 318384 159196 79600
39801 19901 39803 79608 159220 318448 636912 1273856 2547776 5095680 10191616 20383744 40768512

81539072 163082240 326172672 652361728 1304756224 2609577984 5219287040 10438836224 20878196736
41757442048 83516981248 167038156800 334084702208 668186181632 1336405917696 2672878944256
5345892106240 10692052647936 21384642166784 42770358075392 85542863634432 171090022236160
342188634406912 684394448683008 1368823257104384 2737715233685504 5475567906324480
10951410690555904 21903371136925696 43807841785479168 87617882594213888 175240163234938880
350489122562899968 700995837311844352 1402026858995777536 2804124086735732736
5608388910959820800 11217059296896352256 22434681543746125824 44870488987399094272
89743229774611873792 179490963148851118080 358990933496956977152 717999881392423436288
1436035791581865836544 2872143640757769601024 5744431396703615057920 11489151023783381827584
22978878508319067078656 45958909938142741004288 91920125719294695702528
183844863124607818792960 367698949621252492361728 735416345986578694275072
1470869585461304807653376 2941812957898904453513216 5883773489750398583439360
11767842127405976519704576 23536274550622311745060864 47073729692865340901425152
94149820568972116625457152 188304363504427102896128000 376618171741819945082683392
753255232949571368746221568 1506548244831005694654152704 3013172047525737303631724544
6026495210778926435910287360 12053292653012756529114251264 24107189768935320372815855616
48215588463690255374806417408 96433594779019740007962247168 192872025261317938532623319040
385753721929192794098644287488 771526786671499422264083873792 1543092258969226512661758345216
3086261889190908361590697885696 6172678520886727395715758161920
12345666526783276136500241104896 24691952023586194963137931771904
49385141987211675306550762668032 98772759854501921373651323584512
197550471469160984268402243665920 395110846458636251579003680325632
790241499957901069242405746638848 1580522613997059270653608265252864
3161124456156632805644810074456064 6322407368638294139964807236812800
12645131649926645337279988649426944 25290897125153404789260725650456576
50583061900907037807922948004118528 101168659103014532074648889414647808
202342388808429977066903765642117120 404694918821661779969019504909877248
809410120052927211608462957071040512 1618860804925061726557773808644653056
3237802739488538059797243406294450176 6475767738253905332957878390599188480
12951859995061469092642539937218953216 25904369027230255038738646186479058944
51810036128675143784384424997040422912 103622668405779554982583115242245455872
207250529108417644792794760980820131840 414511442810552359240846582954298703872
829043654808538857792207287893914288128 1658128847991945994205442819758462337024
3316340772733628545652942127458192195584 6632847698966730205789997230798919434240
13266027704932406640548220413362908954624 26532720023862705739032892730255958081536
53066769275721196393938689267572196507648 106136197007433962619623186149264953704448
212277710926851064902737987526771028787200 424566055677668409132459205510024300331008
849153379003269376918884871933013086175232 1698349293302403871145702665691955143376896
3396783657196537976907271175035768228806656 6793737455576536423046274037375252341719040
13587815193519993784556011449357936451649536 27176310951773829446038949647930736439721984
54353983033015342645931752794291199952289792 108710688324966052799571212585441854050271232
217426821167802840614557839164602616391925760 434864531371347151259946506316643049366618112
869750840814177242581554668608161731898769408 1739545237771320365286432649166074730128605184
3479177587828572490819511922231651992919343104 6958529400229008502132317092262309051162951680
13917407249601744045251220680122628232974434304
27835511397490942172475614351441276727245930496
55672416591556792508897574685274593977085984768
111347620776263401345687841335333268999360217088
222700816738826435347161066600234700089096929280
445412783850252136005892901059605724358946848768
890847868445702802634927337837484097079399677952
1781740338381802666516137747111513490881811316736

3563569879744399455524841637096117575209646555136
7127318165450387156034815559938416337311340953600
14254993142823950802039895691369195048406777593856
28510699909494254584020320525723114844381746561024
57022827066681215127921699337415679183899875868672
114048508628747842175605515246770257358072517230592
228102726248266508190735263637418312696690565447680
456216870478074664060518993562592221354472192868352
912456576919232623479134919700695634631126509682688
1824958825764631837674463704552413653106617267257344
3650008995381596856781315139406872073901963030298624
7300200678467860076427405739417833683181383052165120
14600766732345052878584362400043846437117680087465984
29202264215508771208627826642504051015745188141203456
58405989932654873320173856969840818314510032214949888
116814902868584408446184121309347069195059376294985728
233635651743718140504041057358025003522197376320143360
467282995500534928231427744194711737308552000100630528
934589375027267150909546747346746935145418495121948672
1869225518106928890712476012608140791347465980085272576
3738544572318646959211717061045575424808189939853295616
7477276216846872273996964193749738533842895839072092160
14954926578112901259140988530816652436138823596875186176
29910601445064115940576097348267655609183711031212376064
59822699467804858725740435269804012692179549737348759552
119648392090962971140657351686145428331983354824545533952
239302770492632449659667665665365662559215220348787097600
478617513606677914076041255916880936908927462096966254592
957258972456181857665494361006061097398848966992716627968
1914565835398015774357812420356720641959686019583001493504
3829227451767335666769272237402638178243348210361139462144
7658646465477279569645839268183670145134648763112551874560
15317676054839775611506268123124127867565202211005649649664
30636118357449984167441715419761830889722213791572391100416
61273769210440834223741789186550812088628046322266965803008
122550603411963400225200295067155924795623330122778298810368
245107336806090264005834023522420450827981135202045332029440
490226933576507455122534913821058104129431220317068132876288
980478387081668764466803561194550613205800340460091203387392
1961005814020645237377074589493970036305476480572092282044416
3922109707755905891641084113197677692398704560448004314628096
7844415574941042617056038094814830624372912319503648130334720
15689223468740546901659815926468611727896831036222575262826496
31379231575198017138415111326615124414095674866875708529967104
62760032425829880947021181600586050744795375322612533068562432
125523203402527455234424281095883705322798801822947298154381312
251052683906790297149612397981190618312013706001339060343275520
502117922017051367660752467541227651956859616713567048755576832
1004260952441044282044560278240148134579383642848911953649205248
2008572121695971657535231242795681930490096104541379619574513664
4017244677019709501962683858222135183642849846769870663701233664
8034690221294951377709810461705813012611014968913964176506880000

16069782177100967502988506413934711315872660488576374051222585344
32140367823224064501114783808915593213046582078649639498862821376
16070585646123096998126277394980881897173921590073265447640236032
8035493690317080872847581442751983593912276070410855572924530688
4017847278786306623316012094006763119618795672892539211014471680
2008973856207036405104116733318767221140726655289825317783339008
1004512036510459749275113709817076441236027737066690515029721088
502268572458700647999084526487384635950846073244234185583886336
251140563331085710680306099033115525641839138977561556826456064
125573420216410548680534967411269366654127620666503010430484480
62788279383639121010458442652990485243667835922112621223870464
31394924329536483840324700800173143580135930755486868616249344
15697854483626703587710090136925522269218971774958713310281728
7849123401242582627628914936882236374184989086086996156219392
3924659780335906730701392402650855806880246142347317828648960
1962378930025261073794163668430296713333998870825568789594112
981213984941284391118815567767582761613937335238739332431872
490619252434969122670274650660008583280437617532347135033344
245315756199648024890570758718112892876953283722662301925376
122660943090905744223002096053110747056843879339575518167040
61332004040993738000359406373582523837605558408909942685696
30666768268267301944608882360304837073394588574016063143936
15333767258018867444519030766909206113993198971788577472512
7667075190952041958366810176832996845644551828284561686528
3833633376447325097237052485105695317146252085337417318400
1916864578709314607645349940897446105735114128266276896768
958456234597483333336086819621022276448551106931922567168
479240089920154681424749334396660750014772574865353342976
239626031270783848090727629491405180902634798132372701184
119816008790745177734540295892239993398941654416034365440
59909500973049215711858388519388698173282954882941190144
29955498775362921278223314546328699823547541278932598784
14978123532100617350258717416481525280226802558197301248
7489248838259887030702888779899350324339917238464151552
3744717955234732693138209425778969004283216598914826240
1872405745669760935462487230804131423198237289298788352
936226256861077762177934874359389172127433139570081792
468124820443637528312313066658356316327873817245384704
2340682562228368087767829348068509023296015532352864256
1170370511117458705689751081403919944214047078041518080
58519987060366683747793744386792688390043194953302016
29260724281002007325355974035812702336531425442922496
14630727515910336388407537939114530239020626704596992
7315546445659834557068544430161354654887770343866368
3657864566682250459966659945382722095132613667717120
1828977955267291820699523837842383431410671081750528
914511813596729205707858851496702907627517664821248
457267324779906250532977892036107049774849894383616
228639371380723949106013179161931322867970478178304
114322540185747386472768706152904560424258004582400
57162697340566399196265411362421729707265385037824
28582062294129552578073234824195589601200883892224

14291387958987952779006881983590157174384537632768
7145872385455564634488573277541259774084316659712
3573025395708576439736852781643720480488182251520
1786557299344685281114709462258405536967103086592
893300950417541171180496266847475416845057523712
446661625581369850901818901282874032603281752064
223336387976984558106694834571005178392017371136
111670981581642095381240109250286670241196933120
55836884587395955854566400607535375643192590336
27919139191985432009256373294963708082893357056
13959918045136443045614773143079864172095209472 6980133247140085043300679819338937151371870208
3490153735855974281896986533568971108348067840 1745120424070953021071816578734236820505100288
872581990106959450597569945341994043418083328 436301884089221195329615800658434838291808256
218156386562481332680223314322936327437287424 109080915540176033847819364158327617864335360
54541818899555700677763535577593536005013504 27271590014511692215808694538011631538929664
13636135289622767046367810643613247537676288 6818237785994843992415637009110339652943872
3409203963589152230823684348207027768524800 1704644517090441232719775095929442855288832
852343526193153175013854008877685913157632 426182396920542866833910234895325199335424
213096515372254573080446732675903721046016 106550916142118856371969173952072421212160
53276787299055213101857490783096490950656 26639058263525499008865197295078385647616
13319861438761695733400824599304262909952 6660096872880320981184525275534666498048
3330131513189897047834319125708600770560 1665107294969816802538187806824934146048
832574416672342540579608025397783953408 416297592929888339945061073691550416896
208153988761802704800159067342104428544 104079590529330619813893798919216824320
52041093338879943613854032083690717184 26021195706547288660380582353886511104
13010922371827302756917074332963831808 6505623445190480591821928744492204032
3252892852233654902592660161251246080 1626486990936034754637177975128195072
813263777877621028989012934815383552 406664203014361234032971844103334784
203326085674207083082465207329488896 101665578138303998000035597071155200
50834056719752227229419295238782976 25417662185176227729410395970994176
12709148005238170922055572161298432 6354732458944113989702973168549888
3177445457634571259189080128225280 1588762342898542761763336836087808
794400978489899946966066804031488 397210392765264256525232595009536
198610148142789269783715894001664 99307549951473205652407745249280
49655012915775888206478771748864 24828125427907586793376835436544
12414372198963614741757142499328 6207340841986718043412933640192
3103747792245814357973648015360 1551912581749134847120414605312 775975633687681257627002601472
387997488250397545846898950144 194003579828477231440148299776 97004207765877844978423562240
48503312808758537118386487296 24252260867289075873780596736 12126432665099441594183974912
6063367448277172625738825728 3031759282002312227192832000 1515917419933019070758125568
757977599432441013959917568 378998244449186246270386176 189503844591075992780406784
94754283478779431212810240 47378322331010433017708544 23689751461315575214505984
11845170878562966960078848 5922733013234073156452352 2961440293593331416432640
1480757040284813127319552 740396966886480273211392 370207706815276991381504
185108465093656923078656 92556538389837675233280 46279422116423444463616
23140287518964025655296 11570431989858164539392 5785360110117158125568 2892752112652616990720
1446412085123327459328 723224056960173211648 361621035679341346816 180815021439298043904
90409762519462707200 45206007159638196224 22603566529772519424 11302064739862970368
5651173107419840512 2825656922454097920 1412863645599137792 706449414985613312
353233503585828864 176621149839425536 88312773942968320 44157486483111936 22079292997369856
11039921376591872 5520098127249408 2760117783101440 1380093251289088 690063805513728
345040492691456 172524541313024 86264418140160 43133282811904 21567178276864 10783857573888

5392063004672 2696098611200 1348082860032 674058207232 337037492224 168522940416 84263567360
42132832256 21066940416 10533732352 5266997248 2633564160 1316814848 658423808 329220096
164614144 82309120 41155584 20578304 10289408 5144832 2572480 1286272 643152 321584 160796 80400
40201 20101 40203 80408 160820 321648 643312 1286656 2573376 5146880 10294016 20588544 41178112
82358272 164720640 329449472 658915328 1317863424 2635792384 5271715840 10543693824 21087911936
42176872448 84355842048 168715878400 337440145408 674897068032 1349827690496 2699722489856
5399579197440 10799426830336 21599390531584 43199854804992 86401857093632 172808009154560
345624608243712 691266396356608 1382567152451584 2765203024379904 5530543487713280
11061361853333504 22123273462480896 44247646436589568 88497491896434688 176999381839380480
354007559771783168 708032711729610752 1416100607831310336 2832271584406798336
5664683906301952000 11329649287580614656 22659861525114650624 45320848950136143872
90643949700085972992 181292402999799316480 362593813198853373952 725205640796216229888
1450447310389451423744 2900966678372940775424 5802077471933957406720 11604443174244066525184
23209462809240436473856 46420078539985479794688 92842462922980173283328
185689537531978773954560 371388298435994402684928 742795043616062514921472
1485626980720272448946176 2971327748416839736098816 5942803070786269148610560
11885901289477717650046976 23772392874765794005745664 47545966341152305422794752
95094293865546045668196352 190193310097574960981606400 380396064928115661253640192
760811019322162801088135168 1521659817576188559337979904 3043395193016103032999378944
6086941501759657894645596160 12174185234974219446584868864 24348974932858246207757090816
48699158791536107044688887808 97400735434711443347727187968 194806306572701345212153200640
389622284551959607457704050688 779263911917033048982203400192 1558566509460293766097997398016
3117210390173042868463175991296 6234575522850996409460714373120
12469460530711814163990153527296 24939540031443271018117756616704
49880318002925827416510412357632 99763111885930225593570622963712
199531175532017592708240842424320 399072254584349468458680877842432
798164316209327503001760141672448 1596368246499121381723170553200644
3192815721162338540682227654590464 6385789898649705610039642397081600
12771896709949468277429658969964544 25544427245199050669560066291531776
51090122140998329568521629286268928 102182779583197115595846251978948608
204370629768795144109298490770718720 408751400742392114053808955167080448
817523083894387879778041857585446912 1635086732607983062896931609673465856
3270254594854380732475559008352075776 6540671448985590678314509594714439680
13081667416524839783355802345449455616 26163983870156996420165171002940063744
52329265814528626547237474629962432512 104661127777486520508289214508089475072
209327447851831575844206959512508170240 418665280297380221343670980017674780672
837351329782194581997856082020666441728 1674744197939257442616740408011966644224
3349571472628251442475537303965200809984 6699309098755975999435187583812936663040
13398950504510898227838601119390943412224 26798565623019688913613654142312026996736
53598460474035162743100212091684334338048 107199579404061895317946231797489229365248
214404475720106930299384078823219580108800 428819585264180139925751388102921402974208
857660438176292838505469237118807291461632 1715363411648450794318871396063543553949696
3430811893888631823253608635778945049952256 6861793928960724115738948958861605984010240
13723928140288369169941361292330643736231936 27448536845310580216809649333876151008886784
54898434820088844187473152166182029090619392 109799591899113055882654011329223512326930432
219604628316096846780723436652165932945244160 439220145667935163592277701291769682473254912
878462069407353267246217058558414998112043008 1756967694957672414615757429066581262555152384
3514022502201276589478161482032665057772437504 7028219228974416699449616211864335180869140480
14056786907092560439885818919326680492386811904
28114270712472574961744810829849381246070685696
56229935221520058087435967642090803014735495168
112462658036189932502764627248965687074659237888

224930891258679497661314638427499536239694970880
449872932889958260634200044714135396660142931968
899768166525115051891541625146543441681791844352
1799580934540627165029366321729632180086595649536
3599251072062048452551298786332354953619215220736
7198680550085685150087729858410891094130478284800
14397717912094546790145724288314144562045052256256
28796149448035446560231977719613013871658295885824
57593726143763599080345013725195477238452974518272
115190306782912610080452144022329853467178714529792
230386322556596044000428521188537504914902960046080
460784063094733735679905508664830605790896982065152
921590962152550766717907949905172403503976088076288
1843227596231268124152009764961367190852316424044544
3686546536314869429736407260224779149393361343873024
7373275760334405222337589981053647834164179679313920
14746916896078143170404730883315474739083273341763584
29494564542974951792268563609047307619676374649798656
58990590587587234487455330902927331522372405232140288
117984104178449130780747069175520095610784122329366528
235974054363447585173166953090371056353646868388904960
471959800739993817569679535659403842971450984238153728
943942985506184929586050330276131146471216463396995072
1887932739064764448065483178466909213999061916635365376
3775959014234318073917731392763112270111381812953481216
7552105100678214503408992857184812224449279585272463360
15104584345775585717965045857686799817351591089275928576
30209916980389484858224212002007950371609246016013860864
60421330538455596561036664577284602217030619706951729152
120845654232264446811249810301106607381685494763751473152
241697294775235401000852582895288020658619500227198976000
483406562171883816758411090376725653107736021853790011392
966837069586593663030234029925750529796466086506364141568
1933722029658839385087291758196099506754920258610296520704
3867539840288982888228230913081395907833816688415729516544
7735271242520574012563756619541185604315585719221731983360
15470925608926364497342102825839158785927076123224009867264
30942617465623161939113384825191892726445961616009111535616
61886767426787189767085127997410935762075541971140406673408
123776599844656111311886972688876172142518321420525180551168
247559329671475686179207378765860945521771117797539095511040
495130919307278299469281624307939093517011185508055659839488
990286358543210453160296982168312591980960270842066257313792
1980621756943728614764061431441493993855796341336042389897216
3961341593602072646415057797092725607499344281975904530333696
7922879346633376126603985462604926454574191762559448561745920
15846151012125213920755710662048803388299389922334176125648896
31693086661967351176606900797775507734900792639098910255611904
63387742599368549023404760542906817386405610867058936519852032
126778623749604791387191438980525238606019272911840105056960512
253563524600944969455146713750473684878454648179124674148433920
507139603405360712271821099079793785089741501069138276365893632

1014304315217662971266697541317280400845147411560054408869838848
2028658847249209035979505768949946463021623641963664530015780864
4057418128126184258851232910530664248705904921614440484583768064
8115037123507900891486908566322871142737125118603103818271948800
16230475981526866530542702623168827576124880787954653334752722944
32461755432075862556223176227383825733551022677406198065923096576
64925117802195984102721894416859992629704567557806178924681494528
32463362370120121546498718189476166896153544880399980858758397952
16232082919571125520818244585261168738727402990948436127588024320
8116242327041095134193565037892127014689016770848440912898424832
4058221597148313753989003891576834830007166023111331881001418752
2029161015388039970440612632103803076334911830399221652776812544
1014605616100961531943361659209594368833120324621388682526457856
507315362253951539333208501183643599749392367021583269332254720
253663958228711156347368086381245007541112285866236098700640256
126835117665223271514065961085334107603764194110840281367576576
63419128108045482427223939490022855718486122644281256692416512
31710348691739664548707449218689328817545074116571186350522368
15855566664728293941901464346183614887923543455500872177418240
7927979491793377804724602041511282683537274926358075589787648
3964087825611304319249235954965378961556389062482857545433088
1982092952662959868068085444587558290672070330893338647986176
991070996260133788255776455846213550282973065272624261627904
495547758094393821238755094699323977614955482549289599631360
247780009029360374174810980737770590044212216231133534224384
123893069505761918865122207062939595640473345593811134316544
61948067248421825321419461878496948129420291536027750760448
30974799871981345605138910112762049219301955137574967181312
15487783059875889274784044643137812186946882253568029491200
7744083091880552873499317114947299882121393469174287695872
3872137326911580554803305954162846835384672905782280323072
1936116553941442336428476675426021864854324538488708399104
968082222213547197727650186885310156008156312043138318336
484053083728186613620531018028804689794575177420961218560
242032528174799814188618471307477150792536099410176638976
121019257242753160783485716800275978343892305054936334336
60511125199053207236331098973406690645758280202392174592
30256310888364917040459669773337696059785203938658091008
15128529588601615231376895029986023398345633888060047360
7564451866510385971261977586651599383399332903395524608
3782319469359982163417753829155093533812924431380512768
1891206502732385670602259432492193687963091205531631616
945626635392390129747820975203420304509860097686503424
472825009709293712097256117080371882519087296303595520
236418350861196179660300873279516806391622271881969664
118212098433872751635986844009423835761850447806070784
59107510718573706720911625689544634163944879835578368
29555448611010551881191491468718867522348226788406067 2
14777608430462092131687008264802516682496047925166080
7388986902935712428708279593005347876625480954150912
3694584795320189395786527526804718706001468972859392
1847338069586261288609457628553381736845098734321664

9236918707562139396628257468522020603447314911 06816
4618573533596486175104613397138566261334568075 26400
2309343856705951325947549030008061110472739347 49696
1154700473306829782171395680723419545139097328 68096
577364509130341950684508423221404267520912491 80672
288689390803634505141659503040549381236138159 63648
144348263521049017470532397235198314355910036 68480
721759158201403911851175214750609690468754967 7568
360888499398781368174844221662613904578979876 0448
180448709848430390212050417974961481961791134 1056
902265849987350481683393625593080058170461650 944
451144075366274506153267580655676353265983815 680
225577612869436885732419174257406338723368402 944
112791594027868259194102279093487250406872449 024
56397190810509037760997485529135665726030348 288
28199292303541972962471915755763853124312236 032
1409999460091471352222254437347993669280464896 0 7050171525029220281604565434538973411726589952
352517287480054190104892934116898923852542771 2 1762629993543236830647787982534245885593780224
881336774843101355385555647241998575962423296 4406792764572921477236086516084371045639782 40
220345082746516808877219739797937460573372416 1101752636321937719463175768958281844323778 56
550889929455645697270126419463438192890347 52 2754517703751612674043324772238677318094028 8
1377292880112498430868008723580081835868160 0 688663454174595262357177530520412506344652 8
344340234146470654640175349625392047377612 8 172174370602821839050880966995288920791449 6
86089312066204175390837129588940908947046 4 43045719415498715628116887840118678749184 0
21523391398947671780407605442883451512422 4 10761961545072992873378383482853781825126 4
5381113695336074928276482122132918947020 8 2690623309067826709931886251419473490739 2
1345344885233807977862765720886243753984 0 672689057966851300379794158031375381299 2
336352836658399305914102727809814442803 2 168180572166686480819154188301970597478 4
84092363002086654340628506349516986777 6 42047219960415034135839959274024337408 0
21024129209893370550682832686645090713 6 10512324219789612016722842868139006361 6
5256291917316269379052134696477733683 2 2628210862368866374871423979442982092 8
1314137883039799030108390305323548672 0 657085167447582436390534310462803148 8
328550696687632678863436734131915980 8 164279404825737069765803156516215193 6
82141730653828900049943972983236198 4 41071879447394632608493183854182400 0
20536446783937407596007190608373350 4 10268476922088749443883894644827750 4
5134365226104397544882096992734412 8 256724599558221018391123331527475 2
1283654689056110797690599083343872 0 64184319016055825171281825046200 32
320929517896530552290168479626035 2 16046872035639098936196391701053 44
80236340882258351289421797104025 6 4011916079316060394893081785139 20
20060075572596016126575368575385 6 1003028537430586513934266411253 76
5015266481156861107698821968691 2 250769513758039482286315594055 68
1253878517291179545938450448384 0 62695473289608084022266146324 48
313485103573285953738048851148 8 1567464203492657436823834853376 7837514445594425524787127255 04
391885393686278193272754012160 1959475325464175551530758307 84 9797618412484800683488732774 4
489893009882436180466183700 48 2449525495703161633789653811 2 1224792970997071182624194560 0
612411597071280774176781107 2 3062133543220129785207324672 1531104550541927849765371904
765571164736895403463540736 382795027101413441022197760 191402235917189590156312576
95703479141836229900763136 47852920162538832361684992 23927050377079774886494208
11963820336445066796072960 5982057742175123074449408 2991102658063856375431168
1495588222520075606818816 747812558004111512961024 373915502374092611256320
186962362873064733016064 93483487279541580201984 46742896561275396947968
23372024741390001897472 11686300601071152660480 5843294415723652186112 2921719265455864020992

1460895661524950974464 730465845160984969216 365241929779747225600 182625468489500983296
91314986044564176896 45658618922188931072 22829872411047886848 11415217680500654080
5707749577738682368 2853945157613518848 1427007763178848256 713521473775468544
356769532980756480 178389164536889344 89196781291700224 44599490157477888 22300294834552832
11150422295183360 5575348586545152 2787743012749312 1393905866113024 696970112925696
348493646397440 174251118166016 87127706566656 43564927025152 21783000383488 10891768627200
5446018531328 2723076374528 1361571741696 680802648064 340409712640 170209050624 85106622464
42554359808 21277704192 10639114240 5319688192 2659909632 1329987584 665010176 332513280
166260736 83132416 41567232 20784128 10392320 5196288 2598208 1299136 649584 324800 162404 81204
40603 20302 40605 81212 162428 324864 649744 1299520 2599104 5198336 10396928 20794368 41589760
83181568 166367232 332742656 665501696 1331036160 2662137856 5324406784 10649075712 21298675712
42598400000 85198897152 170401988608 340812365824 681641508864 1363316572160 2726700253184
5453534724096 10907337883648 21815212638208 43631499018240 87265145520128 174534586007552
349077761949696 698172703768576 1396379767275520 2792828254027776 5585793947009024
11171862771924992 22344275299663872 44689650110955520 89381499245166592 178767396536844288
357543589166710784 715104770519465984 1430244725411020800 2860559819566219264
5721260376620793856 11442802228218298368 22886167406390018048 45773460712686878720
91549173225187442688 183102850050002255872 366214707299259252736 732447428997027987456
1464930886791074938880 2929933831176187805696 5860011777540451467264 11720311785457054646272
23441200031666412716032 46883552984837432279040 93769411812684078252032
187543435311386583891968 375096093994810022559744 750210634733693754671104
1500458162955534928445440 3000990112887364695097344 6002127799727319066607616
12004550747359817486041088 24009691790529993677733888 48020564172680704766771200
96043489528602844356149248 192091701423688558357512192 384192847580342856005451776
768404584626617190591758336 1536846948185097338345226240 3073769454233920591013871616
6147690024195293010674581504 12295682279845489678642839552 24591969022600786671873032192
49185146971021187972920770560 98372711793681605204190953472 196750259290641668925080731648
393510189987840254883559112704 787039722788794343833913524224 1574118131203816355801417646080
3148313633660088047870016487424 6296782009825086768274395365376
12593873504659994881617515511808 25188365979339632453372480585728
50377969898718550287019860295680 100758415677515671334589518839808
201521783115188484190278634176512 403053469750691251422756461346816
806126746542011068929911308681216 1612293107165279270028619389337600
3224665442493072804394832322625536 6449489341311174137464851733151744
12899295595272405332280077642104832 25799225015844924779260903635812352
51599717682290077787923303974830080 103201970665780612034649601356070912
206409011933962136986905189524963328 412828165072726099809022352675569664
825676612555055851288468652602425344 1651393789929319005917785199707422720
3302868709497052618517266188419989504 6605899678270934450397923954850267136
13212123875095527327522631065721110528 26424896787298371508498828443483373568
52851091648811376723904789511049052160 105704779446052020861623844270262714368
211414751188962576550876219036854648832 422839886971642227570094990663677737856
845700543130718584824533120118052356096 1691442624636305448270094484206738472960
3382968326022347453782245456354744467456 6766102805544168022048603888592023977984
13532537918087282273065433728949118042112 27065740450172457004067319361428376256512
54132810128340698924007542529917032857600 108268278712672967679760892673954626404352
216541874337329075023013400576150374187008 433094382498624429373010031608782991130624
866210032645181417399986524130530467774464 1732462600586227952107905970086989906575360
3465010271764186138831677783825837755203584 6930190684711832746895087254955391394512896
13860721651790586432253637884518214557237248 27722123868315014741434202518251292650897408
55445608866097713236722258534932312374640640 110893939991130793981152224066724078894972928
221793324500132322977719862127167066081329152 443597538036006115986270552241771948745424896

887216854143495172034202760458419530656382976 1774477264429956224191728832866590327643832320
3549041641145844208630104289632683187949797376 7098257506863551937753501827064371441223860224
14196863462870830916493590149726753013096251392
28394423824029115914960353290649526287489564672
56790241444633139993867052563691093097573253120
113583270482416096315626797092166267240334753792
227172116151131825287038978113900696571046002688
454355382674862915885648724086937717322844995584
908733066094924362394438983892148083007195971584
1817510733680245786035161039220841462737403904000
3635110670341285694562888221314773518920831729664
7270399746664415963411090872837572822473371130265 6
14541156305211495758192082028243818823251518291968
29083026234269344496324693199472362394071227957248
58167479716231394952530444684914174283278838661120
116337813927848201824823005941767247556830442815488
232681336846467227489170245027412293094206416617472
465374091674476102657388956342580182149503895207936
930771019312035500672874845260671556221189914361856
1861587710550237592061943555672365496286744076615680
3723266764952808365556274841646775760262216649015296
7446716217610283093977325143897641055901890289598464
14893979781062989891368420120900346118255869456233267 2
29788326372078463278827504260423280506627217090936832
59578114245794257460573212205679277296274090114416640
119159151494863176726982831781023987158587492093919232
238324148996275677065638478301378839449253607918010368
476659990005650001354622586081419409162664463296364544
953343364037497297155936431120162278853643421513416704
1906733496127389183205255380154971478763915832868208640
3813560528359567544197275796139236799641089645419167744
7627308128928713443968081663937061283508695250203836416
15254990402276583599083223471191297935470422419138674688
30510729093391480620460567229016946607846908675739353088
61022954764459588085509375031302594689505945026402713600
122048902684272429860195231209142592326636145402653442048
244103791679251367098743424711359990548520801505002913792
488219555979915748954192774008869592887538624409397886976
976463057202657527421797397190038409356071291617579892736
1952974004890967113870418492724675265874130668832728023040
3906043790753238345794484382138547426072237508860592521216
7812279143449084927696263557655488640792427360111457992704
15624941410783386327607116702067764858880759405003461885952
31250649069337205599643412577649104872353328179568015572992
62502830634215277088145183502325360053890275098258214748160
125008726259512285954007083698705020726147787674760796700672
250023582501188035463447600785518642689030050306010327810048
500059424966702998037762068347254487851529050524998124437504
1000143369862059850297257870246943380649996000875951186509824
2000335779581427409037983207598755571193867801403812248289280
4000769638877470234962901349407248762175487202111444247117824
8001735437184171303699672567233972763926477602830527995314176

16003863193226804274947084871306896007003961602876334992785408
3200851102417053188498964921629169297230993600018322798988492 8
6401859132377491044017025737993918786122389758922757198839808 0
1280403211984175142207224326545899795556558463561773759940526 08
2560869194985704151222087010986031667777277950677992160226181 12
5121863932006116036059450737760527488882877948464873601142620 16
1024397894808164753934945490709798328442239999114752576366575 616
2048846006430212601316001667734982318215808817073060865009254 400
4097792446488191389524224708100735959094275271833233154570715 136
8195785760231915152832892161463014563513865819040689158245842 944
1639197325497489505323466981344911441767836218882982401470051 1232
3278474997897191960160711060794439941665798547915653942581867 3152
6557110689598809819348976317798113999591849316130686164447264 7680
1311454276680647143675306102801469623170420307286012888746158 98112
6557432077207661617404084710216582232112353756729442723014325 0432
3278796385506043758215819453212908174186302988514410501148927 5904
1639438366204128353864798275658762616158214549332049820395346 3296
8197392698276174143108434123555355726116388022034471951081144 320
4098796782765853258446438434408449185720851648704347400092778 496
2049448608196809722669329903519610254191754643195729412322492 416
1024749412505346408057720294917497957761541731019642562299297 792
5123872604561439773903878190375953942136030702207102092186746 88
2561999073298073753759577453082209047732176374657995686438502 40
1281030922157713810283607905488220562198168699106220163391815 68
6405311538331953718437135422176683002651246054417212417821900 8
3202734232937669192728115658456131597155824306651662009342361 6
1601406348354680763118831802911960846493012793047358904886886 4
8007227901202634649368028882979279472040567163844434025512960
4003712030315932741570949375699377355808035181226036763295744
2001905055015274079228942154954557487797893390264928256917504
1000977047436290893836204811029713148845884594958419066093568
5005007836824723740289692722910737768964112473921870018641 92
2502565218233996505699180695336454896849400986525822353408 00
1251313259027815570626757514608770454608372868045354848747 52
6256719544693164442019623407746567303960226214138992603955 2
3128436397123625515452729621224641167439294044025605482086 4
1564256510950334404947823769287999341449237490490857331097 6
7821474116694280260846413639818390495894139794844559605760
3910832839318444248476854216598392142271046068617416278016
1955464310144874183265250806643794518297511119906276376576
9777561003152631211460372524941964827297496027519223070 72
4888900227790445753297245508332478531553718227753532129 28
2444509977002287950432152377096987324729344220873726361 60
1222284920054676512107841000013867691840914663935343329 28
6111574258041045244998029057396208606585786087169117388 8
3055861957904353964728426557361539376983499427330759065 6
1527968393339409265347891929301248722533705290553847971 84
7640029039180041622968126536721023810911780487057899520
3820108055694809989270828304189805747569148223211700224
1910100795899799583528796670009549794841203101447225344
9550737819760970862110895939620983579489160456443002 88
4775485830011471903288904264597109092386152702824939 52

2387801375071229187761180279691863197513862588714188880
1193929917568361211938954213542585924417324413007953 92
596979573800553914998659143619620125038858765829406 72
298497094408463612013920590233973643934527662577418 24
149252200958325133264255804329068612674812971120066 56
746279273562092302607756567705752016911810554757 1200
373148771166279469447117056883080485224778126956 9536
186578952775756393795177914956642480996825488267 6736
932917599841865264333986507358723596906309565284352
466470217902474279846041719967117394414245844615168
233240817942007963762545093127436495187668453294080
116623263466389393801034663135657146584106992140288
583130589808874028603983898537980227871898788167 68
291572431142900544101397240698837361411631307816 96
145789783690682036950401266064342304443656610775 04
728966759049569009250519558896329640907487838208 0
364492299822863916874516393735473879798346311270 4
182250610060471664561886504011391469571474351718 4
911275351047556853432574055775229996218877739008
455648825896377692027857795746751322290191859712
227829988134488478669714281802943823235472424960
113917781660394055662749832866255992662924460032
569602846267719359953212624155200368540563537 92
284808392116734220796338041989560386883252387 84
1424076805498043808080348859507602947481115033 6 712055825206208256089503754533701980272984064 0
356036623831697304069416539656801243402705305 6 178022667530145240047040601023375748334459289 6
890135115722209140296864661091754374837829632 445078446896846040179263158533315004001681408
222544667966293755105046993260376410292224000 111275056242082245060231203627047659291803648
556388892505088062839694553119535567187476 48 278201251899882450189116544051916418957967 36
139104028773610434479192905772032527161098 24 695537157986398219319137697590534224216064 0
347777086052372133121155433160452906313318 4 173892796555772578291371008762819350259302 4
869485250426795450110821504727061236809728 434753259037364004382393982820012861161472
217381946430665141854688606638247551959040 108693631671324140759090110933244336668672
543481450636578552954179592736824486789 12 271747371458268201056454315403713645117 44
135877008799123562817909417219507523420 16 679401659345565125537958383685791121408 0
339709137347756218493184840637022312857 6 169858722511364937108695244715574532505 6
849314381744258824853990345563189542912 424667575465846482082252233774253211648
212338980029781775868754647383455825920 106172086163320155348191588939892523008
530873411558747113810029270940285665 28 265443196150446725439550298590554357 76
132724843260759946987042980855482941 44 663640442231482656271554062078443520 0
331828334079582788803946609939736166 4 165918223521712124736058094420125286 4
829611400018164275350714419351912448 414815841213883963510569183301599232
207412991209342894672890578463621120 103709030905871903795248282638221312
518557831035361801270256380223160 32 259285253770682041782135673627607 04
129645796011841591464571578571816 96 648244825691710810190376601649152 0
324130335662106831528947655219609 6 162069129239179128981353504807321 6
810365453236524210991165910024192 405192630138576388537782148005888
202601266829445335789990670499840 101303109294801238655545133498368
506527925874399047080474658734 08 253270152637395950441611824988 16
126638171168796188671493160304 64 633206330094472010610902040576 0
316610902172481538932169139814 4 158309319648863536279443629670 4 791565941057431515464013447168
395792641935272674765404372992 197901156670914795899401011200 989529961870966272080499917952

49477707019367928233199665152 24739457972593771431187185664 12370031217751789372887269376
6185166724603346515090472960 3092658920165399171868655616 1546367239014562543096037376
773202508973212750128873472 386610699219572114354864128 193310071976268926822645760
96657397171375898233929728 48329879177308666528268288 24165529884464691969785856
12083060090137525337718784 6041677619021352345272320 3020912596486971010842624
1510493191731632924524544 755265042609890171813888 377641744676981940682752
188825484024509397729280 94415047855263912558592 47208676849136563126272
23604914885320584986624 11802745673036444205056 5901516951706297958400 2950830533447186907136
1475451295520612417536 7377435662158815690752 368880838278662586368 184444922738958663680
92224713169293017088 46113482484553351168 23057304192230096896 11528933571091759104
5764607523034234880 2882374130261295104 1441222249502736384 720628716937412608
360323154561728512 180165975327375360 90085186686943232 45043692855099392 22522396183363584
11261472969588736 5630873923747840 2815505681350656 1407787200413696 703910780076032
351963979972608 175986284953600 87995289960448 43998718722048 21999896231936 11000216551424
5500242493440 2750188355584 1375127732224 687580643328 343798710272 171903549440 85953871872
42977984512 21489516544 10745020416 5372641280 2686386176 1343225856 671629312 335822848
167915520 83959808 41980928 20990976 10495744 5248000 2624064 1312064 656048 328032 164020 82012
41007 20504 41009 82020 164044 328096 656208 1312448 2624960 5250048 10500352 21001216 42003456
84008960 168022016 336052224 672120832 1344274432 2688614400 5377359872 10754981888 21510488064
43022024704 86046146560 172096487424 344201363456 688419504128 1376872562688 2753812234240
5507758686208 11015785807872 22032108486656 44065290715136 88132728913920 176269752795136
352548095524864 705113370918912 1410261101576192 2820590922629120 5641319284211712
11282913446330368 22566376648474624 45133852808577024 90269904640409600 180544207327330304
361097210747682816 722212013681410048 1444459211734908928 2888988792213995520
5778118321916346368 11556518118809403392 23113599187572228096 46228324275051298816
92458900349916282880 184922304299459936256 369853615798174613504 739725245994858708992
1479486520786736381952 2959045099167510691840 5918234313523097239552 11836756857422346190848
23674090175596995805184 47349333272698598457344 94700972388406410608640
189406556462831248605184 378822336297699351986176 757663119339472413523968
1515363132167092246151168 3030800051310479330508800 6061747676573548337430528
12123790501052276027686912 24248171297914910761025536 48497523187450538933354496
96997407558142512689315840 193999537482767895023845376 388008519698501529338118144
776035928862934537257091072 1552109636657732031675891712 3104294831179189977675202560
6208740778085831783997243392 12417783787626567225288163328 24836172038162941765163679744
49673553002145498159502065664 99349523855930225577353543680 198703883415138909671405912064
397417438236834736376209473536 794854219286783306819214245888 1589747124199794281772019089408
3179571619652043899811219374080 6359297981808998472156801138688
12718905448627818289382327058432 25438429867275279268902103678976
50878097674589843918079106482176 101758671229258258596708011212800
203522294218673658714515618922496 407054491957661600471230430838784
814128790955951767026859247665152 1628297195993160666222515267305472
3256673620148835596782624078561280 6513505696622699722240435245023232
13027328305895456501831244665847808 26055290437091027118363237683298304
52111848524782282466127972069801984 104226232350765021391058937546014720
208457535303930955699723861904850944 416925211812663737234659697435344896
833870706034931126139743342121975808 1667781976889069555620334578746523648
3335645083416553717922364946498191360 6671452426109936649208121471006670848
13343229370773531725143026098033917952 26687107778654380303739618508108988416
53375513631523394314386369640300281856 106753623411476056042587004528765173760
213512439119810646912802539553859567616 427035262833338363480862140100377575424
854091294854110866272238402186072031232 1708224128083090011165505048342777823232
3416531332915916579573066584626823168000 6833228819331306273630246145136181379072

13666789945661558776228718242037432844288 27334244505321010010393888387605005860864
54669818238637804936660680582270292066304 109342294933267179705067168778661144821760
218689906778517499073625952785563411021824 4373790447381001277474235136027609064800256
874802162409935113602436732968182615113728 1749646860115735344512806387762294201253888
3499378790823200923641478619176446344560640 6998927722829862316514688925656608573227008
13998195728026645571492841225920648914665472 27997072020787133019912609201056161365753856
55995505171041949793679071900542049804353536 111993732601019267095065850797943553754398720
223992909719909269205547115589606015800180736 447996708475560008441925059166649848183128064
896015195022602956945511774308175329531789312 1792073946188171794014346860566101925394644992
3584235004662275348275340345031706383451422720 7168644233896414217043973937862417832227110912
14337636916936555475074534371322845795102752768
28675970732160565032122241733841711851502567424
57353335260896038228190829450075464225599258624
114709458114941892784274350864935009496386764800
229424491416183418224334085659438181083150024704
458860133204966101760238939178012686347053039616
917742567155130734143619414074298021055612059648
1835529735800658529533521899585141338834236080128
36711148674582111181559609942043373271114496081920
7342475755125810608104352169832927729121040007168
14685308322174797706178968911158217832026175700992
29371330268195948392298466965301160411620542775296
58744087784084602744477992216571770318377468297216
117491030063554617408718101005082439627027702087680
234987769117880058656960435154042677234600935161856
469986956217301764992969336595840950430292932296704
939996748397686825344035605767193092782767988539392
1880039168721540241404265076685408569409900224970752
3760169681295413664240917883672861906508528945725440
7520522050295493691346611227949813348394514883018752
15041409476000320108422773377107805767543943749173248
30083549702819305668304648596631969676597715464617984
60168560907275942239527500878096655636215086861778944
120340044817826546284891409125858743838469485588643840
240685935642202416181455632991048352809017594907459584
481383563297503479586256895460758435882192437275262976
962790510621204253619205049878840332292699369471213568
1925627789294803096131792617672327585642027728783802368
3851349114694395370050350271173949013397313437250355200
7702885301598369095674230614006485711021142833866211328
15406144747615894902495521371330146790495317586463424512
30813037784070103227285163029294644317896699010388852736
61627572145816833299158566631857990109605525695701712896
123258137446986920287493614410253383166835306741251440640
246522261204680347953340191113581572228919124182198910976
493056495030773710663386306813312756248335269763789881344
986136935304373450840184462798924736077664582326363881472
1972321761094398960707192623942447919317317250250296000512
3944739303160102039468032644574092732958610671695728476160
7889670168262812315043360082526579254565173685781729902592
15779723460410841102301309751809946086426252056344005705728
31560213168592115149031798677133467327444313482249103212544

63121958832725096186921955701294084964072245703620390027264
126246982656531924151560628096642470546511728885485147258880
252500095295227311858554689581393542329757932727459028926464
505012450554781550827976245939004287132984815367895526670336
1010049421038216955877686225430442979212907530561745990975488
2020147881933741620198839917965754768319690860775401857220608
4040393843582098657284614770141247156427133320854623464980480
8080983846593428148343099408701969552429769840316886431039488
16162360012045317964233938554242889584010546077849051864236032
32325504661807559263563356582163680126323104950128661732786176
64652578599048965197317672111683162169250235489118439474200576
129308295748965623735017262118077928171708522155959110965657600
258622868599666634150798360025579064009833146667362685965828096
517258291402804041663124391630004543352498498045614300000681984
1034541691212549630049304126417701917370661405513006456139415552
2069133599238982353544718939150789496072651629869568624554934272
4138367632105730893981659250932350314807960897426248673662074880
8276936131466994161747761247126243274941237070226720196428562432
16554273997445053071064407984775571840533104691201886091065950208
33109351463912235637266586950597314262367470483900663578549551104
66220309865868730264808715863286969687337463170795109949934403584
132443833607825978510168515650758621699879970747577785485539409920
264894094967828992981439199149886608050170030307130702142420025344
132450261360003014471270683499127986350290059559552916656880615424
66226737618045766225910883711656334337747551982770241121275609088
33114172278045012608093212836874337750175037092882011957055455232
16557487873533571051615491908960254165738149097189451676736552960
8278944804022317899592188699741669728194389823968948687472689152
4139572835638925136688315722501606186759852549671585768288550912
2069836634633345661790268547566188754711255093679348596420378624
1034943425723614377618189616940787208021291956261452154348240896
517484267065277962170622480049240019343478182841615005243146240
258748410634374367766075075814043217338155193776251966656086016
129377343868054877223419455801733212502285648065848215345299456
64690241209461285281900686848222408167746849621785223681277952
32345905242447565976045822897789105042175437605323169844953088
16173344940082244655570651185733503000238725199876863924633600
8086868629470353161559195461286226739694865798546071463395328
4043532394449791997666532664852850989635184498576855482236928
2021815237082203707276733799531294304711468048940337616388096
1010932138469755707860100633318081557302671924296123745828864
505478329199204781040917183435257981124804912061039341731840
252745294581765854075892025105737591799136930987008405274624
126375712281964658815662729246923096517935702971748569841664
63189388636523195296689722970488698568151470224996468523008
31595460566032030592774040658757924438667544482059326062592
15798113406901231768601609916135749796629676925810208931840
7899248265393224120408099751446268686962790805295377416192
3949719913667916178257697272412331237805371573842825183232
1974907847319610148155672334550764066064673872518980829184
9874778689026311035912480164476812566133309790582745333760
4937509070727285665523299328099902400971625109285293260800

246881439847070790654517928698069925943829766163960692736
123443713078888649016435445495572365919539138431828361216
61723353117120951352805963321054884433581696890838188032
30862424847398789098697101947161792953696912282881097728
15431586568118551260495611116898071845301488060171550720
7715980356268853985821335630107623606877259989451276288
3858083714239216170697432850883105645551887974408388608
1929088625172002674242098943356199743832572977045569536
964567696612198631567740730635423332444600983443472384
482295540319197963007215994796373396486457739182080000
241153616166148305115280812137517563375307493321211904
120579731086348814363476813438424214253693058525691904
60291327044811738084656610404044823409866185195388928
30146394273224534493787407044438769846442920563965952
15073562512021599972623254443427563993976374265118720
7536963943715466349176402682317871532365644124127232
3768573315710066356020589071460980533871550557847552
1884332329781199768726488400881512650780139526815744
942189000853683179721341133016267517312251887353856
471105918408383237539719032795889354617217005649920
235558668194962442609383749541822475289154033811456
117782188592866633224453991342850136634849782398976
58892521544126022572108053957394517812561273946112
29446974395909364265994556121681983653848828346368
14723844009877858622967542632333354200708509859840
7362100410900517556468903601912858287246302773248
3681139408431052900727017943829519737069175308288
1840614305705923511609792043351305165257599614976
920329453598160286428037557393925230990305787904
460175877171679408525589546556098939675905884160
230093513772139336918580157207617631928329437184
115049544479219484787182770568592897009352966144
57526166036184650557537731266688489027270606848
28763779916379779360742038624540264774932365312
14382238407333616721357605807868142518114713600 7191293428238671881172096151733076324381622272
3595733826405267700832694699766040694852943872 1797910469345599730539670661832771613757538304
898977012744282805331496986891261440044302336 449499395407882872696579321433068536604917760
224755142221812171363705074710253176593842176 112380293369841453189560244351986042442612736
56191507814388410348633975674422748294152192 28096434471928047051243914586426237683499008
14048557518330944464085420667820550609960960 7024448900348932701274442021213991189086208
3512309520766196585253086854258853536595968 1756197295678963409934476348955355739324416
878119915487414263621204635390642355175424 439070591567673411137585548151803420344320
219540612695819845232284389304142831550464 109772964803901492447888002266191976464384
54887811629946531139816904940156268576768 27444570428971158027844904373608274460672
13722617521484525242890678138569207316480 6861474914241735735929452045167138701312
3430820533870604425206782510524836872192 1715451805310170491224419499233052196864
857746671842519384922723871601842978816 428883720514976762116618996793579929600
214447052554346915885938028893119184896 107226122425602725356783279694724202496
53614359287015996385298772471444406272 26807828680615315046102952547763355648
13404238858861315949778259429902254080 6702281688707487188252521292961415168
3351221973992158200807956435485851648 1675651551815286403744826112245497856
837846058317246853542837003374034944 418933170363425252606630475312660480

209471655784113539220921224469151744 104738363193257226069263605640986624
52370449247228841264033299523698688 26185858448914534746717398113452032
13093246137107324430709073232527360 6546781524878690744029723704164352
3273469990601859636352455396032512 1636774609382186950345024469991424
818407111731722041256910620983296 409213459386175303670654503485440
204611681453244793356426848239616 102308316606700967438763222368256
51155396243389769099656510308352 25578317091714527239965704716288
12789468030867084965051577139200 6394888757938453155060150960128
3197521750221681913797256675328 1598799560737068625032218935296 799419123181648146582904766464
399719232997380990324850032640 199864452201968953679123841024 99934643952623706097911332864
49968530902131467678130372608 24984869913975541153652539392 12492737188442674234119946240
6246519709948788945706811392 3123335412838120387176824832 1561705485350923150750121984
780871632141393053955915776 390445260803662266268385280 195227352768314002779406336
97616037567398436212310016 48809199375319935517458432 24405189983470326464380928
12202890139640342585016320 61016092643772760968921088 3050870108862675322667008
1525471947919485080436736 762754420703816249769984 381386433723944979660800
190697828547990917218304 95351220117004672303104 47676762980006942998528
23838957950755774922752 11919767205754039173120 5960027718065095442432 2980085916626585649152
1490078987110311788544 745057507953665376256 372537761176087429120 186273384187671085056
93138943893649227776 46570597846731456512 23285861873319149568 11643212411636285440
5821746943306498048 2910963840397426688 1455507104570802176 727771144471445504
363894368328744960 181951582210883584 90977990128697344 45490094575976448 22745597043802112
11373073399808000 5686674138857472 2843405788905472 1421737254191104 710885806964736
355451493416960 177730041675776 88867168321536 44434657902592 22217865822208 11109201346560
5554734891008 2777434554368 1388750831616 694392193024 347204485120 173606436864 86805315584
43403706368 21702377472 10851450880 5425856512 2712993792 1356529664 678281216 339148800
169578496 84791296 42396672 21198848 10599680 5299968 2650048 1325056 662544 331280 165644 82824
41413 20707 41415 82832 165668 331344 662704 1325440 2650944 5302016 10604288 21209088 42419200
84840448 169684992 339378176 678772736 1357578240 2715222016 5430575104 10861412352 21723348992
43447746560 86897590272 173799374848 347607138304 695231053824 1390495662080 2781058433024
5562251083776 11124770603008 22250078076928 44501229895680 89004607275008 178013509517312
356035608969216 712088397807616 1424211155353600 2848491030183936 5697119499321344
11394513876549632 22789577508913152 45580254529454080 91162708082163712 182329814210838528
364668424514699264 729354441215442944 1458744066802974720 2917558502350127104
5835257742188609536 11670796959353929728 23342156868661280768 46685439637229404160
93373131074272493568 186750765748172357632 373510538695599456256 747039091789708394496
1494114212376435752960 2988300482346909433856 5976745079881894723584 11953778390139941158912
23908133241032185741312 47817419403568978329600 95637144650147170353152
191278900986312768094208 382567025344662390964224 765152497433398491480064
1530341888354944402063360 3060757563686183642333184 6121662701324956961079296
12243620550555093274984448 24487831396920545255620608 48976843385461807922544640
97956047954165050667696128 195916818274812970980605952 391843081282591681251639296
783705052031114841084133376 1567447882994092639329976320 3134971323851911192983371776
6270093763431274214613581824 12540489758317452086520840192 25081583979544711487629033472
50164376884909037604432773120 100331171621457304467214958592 200667178946193067451128741888
401344029298943051935655133184 802707401410999937938105565184 1605453488448227544009801728000
3210984348148910424286784651264 6422123438802731521107931693056
12844556362615284387284588167168 25689731695250211464706625896448
51380701330539708309688150917120 102763878541157987379926100082688
205532708842473116280951796662272 411075321205260515604102786318336
822170449451149597292603958624256 1644380512983556326754004689223680
3288840254129626917845602922397696 6577838964584282364366392932696064

13155994841818621786083160041193472 26312623508937357686867068433989632
52626514668474943603135633571184640 105255564638150343665074260548780032
210516199878701600247754507910381568 421042540962205026330720989446406144
842105364334013704331865926144098304 1684251293487234712004579746790768640
3368583716612884030690855282586681344 6737329692502597274745102143183650816
13474983903558852976216987442387877888 26950616844225022805887541196816908288
53902531762666467931868221501777161216000 107807659673758626051178695283596853248

215620511644375786929985921063522926592 431251407882468643515228903119704293376
862523584952371426340971928224725467136 1725088708279611131302972100420084695040
3450260493308958819848000688781436911616 6900068714011739075418011435344540886630 4
13801706587233727737328454658655887818752 27604077788465347932593361220841915809792
55209484804926480781059626248744111964160 110421628065844531393865060111608784617472
220848573043672202451221735451458690613248 441707777991131068422942670135939962398310 4
883436827470553927112819863631763733479424 1766916190236972971533572649089456437985280
3533917451065676177683011141830770818023424 7068005043314812824597753970965257520152576
14136350368996546587658971316537946808516608 28273381302726935052244869382290757153456128
56548123734921553858343592263011241379758080 113098969728778475224394891522881936905207808
226203383975427685464205197039482782101798912 452417656986596840959241222066403380786364416
904857092044676621980144100107682394738262016 1809757740232319124083611512165116055807590400
3619602592750570008413869648229734644277313536 7239379410073003537321032544258474353878892544
14479107269289734115628651584114958838406316032
28958911436866922313230476159425937938109693952
57919216670308752790407298301243916398813511680
115841220933767321908707288567271913842815270912
231688017053834276473199961064111989776007036928
463387184480267818257970689987360303732767064064
926796669705734167139082915692993255827040108544
1853637940901865395524448902822531808377092177920
3707365084784524913541463948518154210200208277504
7414908575530638072068060182782489607292464398336
14830173962984452634106384937057341588369024483328
29661061549815258248153299017099407924306240339968
59323555034732322245618765632016826534374886342656 0
118649955190031856832137429212275429677770492346368
237305619370834537503799091568428657336086515679232
474622656723210722686646649424612910633264093331456
949268149409504740731390231424737013188710310608896
1898581970745176072178974328000496410221784869109760
3797255528534268532579033638630303758813229823400345 6
7594693258390037014445448233210164711642053459574784
15189751892189406754620447387628508494039020902285312
30380234535197478960699996617673375129587869770842112
60761930572032288824318196920179466542195395474227200
121526784147339239454472801210024365650430102813540352
243059414301227802520618417159379596432938829357252608
486130520615554252264582463797420923130034906174849024
972284425257305798975856186552165306788384307270385664
1944615618567006186845094891018977534633397604382146560
3889324773238801551476954817867248911380053188447043584
7777883661868718145852743970739308550698662233625958809 6
15558047381793519628201939558103346382426276591250178048
31116843052425352678697999402841043501758617019962359808

6223518268252733220198423937895078847732936171484872704O
1244733585204079180931449599044389799022829787795454689 28
24895270335152234356464288210195276569981446825878696755 2
497917379324457701885991688790055143190125957916965994496
995858703891741433285395226752409509961245958632716107776
1991765298269134925597614151849417467084480002863000453120
3983626377509573969248875700388031828492936176921137381376
7967444316961756174605046194154457445633824696232547713024
15935271757808728821424681975065702468563554077245641326592
31871309763387890587278543123644980091718917524052374454272
63744152022316647063415444594317110492621453787226932510720
127491369035715025904547605882688521603610145052698232225792
254988868053593515364528645153485644443954765061885198860288
509989996071513957839924157083188491361378480036747866537984
1020004512071681769901582047718811387669694859899450670710784
2040058064000671248246631562542491585233265519450811216691200
4080214207715957913380198059294720790254282638205442183921664
8160624574861146660534265987008916820084068475018523868921856
16321641468580754988616271710856784119319143347252326740000768
32644067574878433312328022895391469196940299488935211484315648
65289704425190713294847004738138740310484624566731538977259520
130582547401249119930075927370989084454177300311185309971775488
261171371904233626540915690531401376574770702977815083978063872
522355298011938026443359052641649168482373610666519096025153536
1044735570443081759960977344844099116763041163075481604818835865 6
2089521625675518292665657583197367996592152080353187808652820480
4179143684978802772223536539025507315846961798393487041857847296
8358488237213137918231515823312557277019238872161196932820107264
16717378208937340584031917137148199844689108295070839563849039872
33435559886896810663201605255342570270679477691638570524115730432
66872726711837880316678752472777481703961477586270923841066762240
133748667299764278613908588869739645733127999578529413267804127232
267503762351705593188919345587848656116666087969033957706949459968
535020380207765258300043026872436041534152353562018177756581330944
267516617856059665111123681284587385417486265592984220049631870976
133761522804118350536112924566478375033948177202479675610486538240
66882368340103434258332004245331528679576610804233620598078570496
33441987639073846624303773103711934921089566503613701695456935936
16721395554047988059720772042379052751195413802555296545937293312
8360898644279526403644828766451069020923022176651871122073059328
4180549755767529388714635755856305833124168726013046985588736000
2090325094697647787803428564243538577893413181850079205070471168
1045187655755765440624769625279462119612371000346817458673287168
522606382081353493673912484218577475139017704884297657405669376
261309468142412133517720077898711945235924954797593292737347584
130657872622073760099241956844067576451170528576518878385930240
65330505586470726719811937369389590142189289877120555201593344
32666037430952286695001448158372696029396657732990835605110784
16333411034334605015048463816025298493849335263710696804712448
8166901676596533341298101776432124486500170830462987903434752
4083548918012882087535985822425799863037837014535313702256640
2041823498863748752211460378317768741412794306919566726397952

1020936269360528230327463922711318775653335053285738300833792
5104803946445910422745988281318765903001364765558466192343 04
2552463273044589846927328474540468963868027132344120440258 56
1276262286433112241240831404210777488117685940954503892172 80
6381464681719647795089992855756602471506791578684737821081 6
3190808965636867191987914345229658751212576726298478090649 6
1595442795206955243215416131290508133335878831627293635379 2
7977405537977384452184375449830934455327346500526741127168
3988798549959996344145835121604664121987649421458507038720
1994447165465650231099741259146930508155812796326821756928
997247527975651145063282478745764477658900440962194997248
498635736609238587288347163959031850619947241880489558016
249323854615325801022526544272590731205222131639940808704
124664920463016154200439753282832768550235321169818419200
62333956809184703944808117214685085748929788259833217024
31166772669343066539469817889397689361137095796737861222 4
15584237491134489408496149590305622174138510902420307968
7792305817776823059821604866811398771295771410575654912
3896246444993200707697567469234993227761143684970577920
1948169990548994942742166252532143534937200832326664192
974108379300694765817774385223395227996914911084019712
487065881663446030132232822090359344262614703002353664
243538786838272338677789225784510537263385975231348736
121772316422410831144731020261920701197732299480760320
60887619712842746475283713815793066881885805672923136
30444540607240038689100958750312891582452730802733056
15222635679029352070280030296364624861981279384502272
7611500527219342398004790608786401966368096683819008
3805841607462004380434783034695245750872776837693440
1902966475657168780933585382498645259280752666738688
951506073791667685824889623824833821562558457315328
475764454877375490591493278200172506742370290630656
237887936429458569135270872243964051351730676301824
118946822710114696487397552693920924666138103644160
59474838602750054203579834632929911828205434568704
29738132925221380081730446459449680661670908657664
14869423274533866530835487801217202704619550015488
7434890043228521510402876186354782539201822851072
3717534224595054877694004236050481863046935347200
1858811713787924500093285189461786228246479634432
929428157639160780669784130449165762484745797632
464725229192179655646462833083719205423125889024
232368189782389460479016800471427764801939439616
116186882484344546567401092200497963446157967360
58094835038747181447646892082641022245673107456
29048114417661044805796619032516531384133615616
14524405657974249443884896011856275822715338752 7262377053558988242435741253727142976681934848
3631275639065425881464517250763074021003100160 1815681375675678820855581937331288276832616448
907862465909322350489452624640519771581841408 453942121990402645275557140307697702373687296
226976505513072057653193984147567759478226944 113490975015471396334304699070643333884805120
56746848637203381921006203033751394015248384 28374104883335532837430028266090560544047104
14187392724034687357178477507652712040235008 7093866503200804147820970441130071904223232

3547018322192132308526351064216893894164480 1773551696391931271571108453934375918108672
886797115843898194439520687880152444567552 443409191745915376546743574396558465040384
221709912784940827936863402426520353898496 110857614848461983800177508827380737638400
55430136652226776815961658220750649163776 27715732940111280865917281013905464754176
13858198777054586661926866458717802463232 6929265542026766445447546205241436274688
3464715847763119779965829590561985658880 1732399462256428168603943039251626590208
866220500315648223612485641611130175488 433120634751541181461499881798223527936
216565509672629125558378471395440984064 108285350984743830193003500945885102080
54143973566586548803408883097024856064 27072635820400591255158007860553580544
13536642428753954054305787086297366528 6768483473653806240516285121158971392
3384322866465317726939838349584629760 1692201998051866166810767069294886912
846121281435536735075807481898729472 423070781922570193373115714575007744
211540461563686009604163844100325376 105772766083043461260884915456573440
52887650692121958859843954431492096 26444459171361093544622725567348736
13222546498330603829661736959475712 6611431705490330443506055567638528
3305795080907679486090621327769600 1652937154535096875214107435859968
826488384308177003691452103917568 413254095674402784887925244952576
206631999597358533965062218973184 103318475678757837743080907735040
51660475779418204251815352991744 25830856859728744816045126057984
12915737914874193753091287810048 6458023699942007549080006295552
3229089221223459110807184343040 1614583296237957223537182769152 807310990932092445835386683392
403665166872603139951090991104 201837419139580028492244320256 100921127421429243504471572480
50461772636534236381410492416 25231490781176925505292599296 12616047622043366409939976192
6308174926749135033616826368 3154163021238293431131832320 1577119289551009672727625728
788578534241436314944667648 394298711853683896762761216 197154078293324818026594304
98579400329903843835904000 49290880756572639329255424 24646030674096678370279424
12323310484953518537965568 6161802816429348945395712 3080975195190969310904320
1540524491083632074555392 770280692285889746829312 385149569514981728190464
192579396443509291483136 96292004064763859435520 48147154953886536564736
24074153937695571705856 12037365199223937564672 6018826714800044638208 3009485414994060247040
1504778736294049087488 752407382545534025728 376212698472021753856 188110852835638247424
94057678217632808960 47029965008723247104 23515545454315044864 11758054202134233088
5879167838555471872 2939654288021913600 1469862328383045632 734948756377567232
367483174281805824 183745985187414016 91875191616962560 45938695320109056 22969897415868416
11485223585841152 5742749231874048 2871443335413760 1435756027445248 717895193591808
358956186730496 179482388332544 89743341649920 44872744566784 22436909154304 11218723012608
5609495724032 2804814970880 1402441039872 701237297152 350627037184 175317712896 87660953600
43831525376 21916286976 10958405632 5479333888 2739732480 1369899008 684965888 342491136
171249664 85626880 42814464 21407744 10704128 5352192 2676160 1338112 669072 334544 167276 83640
41821 20911 41823 83648 167300 334608 669232 1338496 2677056 5354240 10708736 21417984 42836992
85676032 171356160 342720512 685457408 1370947584 2741960704 5484052480 10968367104 21937258496
43875565568 87753228288 175510650880 351029690368 702076157952 1404185870336 2808438849536
5617011916800 11234292269056 22469121409024 44939316559872 89880780603392 179765856174080
359540302282752 719097784434688 1438229928607744 2876528576692224 5753194592337920
11506664062582784 23013877880979456 46028855273586688 92059909570428928 184124217187368960
368257230467760128 736532053121564672 1473099290615218176 2946268949974614016
5892678637437583360 11785638749851877376 23571840449657176064 47144806799221194752
94291865398256074752 188588234396139520000 377185475991533780992 754388966381577043968
1508813961560173051904 3017699980714384031744 6035544076616843919360 12071376383609839550464
24143329227971982524416 48287811377448571895808 96577928597906357485568
193160468881831142359040 386330161135699139493888 772678769015471988539392
1545394431519091396182016 3090862650014477630570496 6181872873981544937553920

12364040895868269227933696 24728672087546897161519104 49458524766714511734341632
98919410716670458291290112 197843543799823786227793920 395696532332613311746015232
791411954131158102072885248 1582861687194179161307480064 3165798932252084236938379264
6331748980231620302523596800 12663800191918144262340870144 25328204846746095839269093376
50657618619311806307712892928 101317655090262841873775198208 202640145883804142264249221120
405289963174165201561896091648 810599269161444237190587482112 162123722394911614251476561856
3242551819150687621296712318976 6485258380806285915127787028480
12970826246622393175324298838016 25942271463264429040786047238144
51885780866568143461846993600512 103774037613214857684243785449472
207553026986586856889587167395840 415115957493487996821373527785472
830251722027604559727145441558528 1660543058136466251623087655092224
3321165344435446767583768854134784 6642489145195922063842724796170240
13285295203041901185035823768141824 26571224231383916484772395887886336
53143716113368061198946288478978048 106289967527936578856695570364366848
212585005658274070630997127541555200 425180152521349967097206228708753408
850380587452303585864836404668792832 1700801739723814475070520703840157696
3401684609086043556822737196685459456 6803531477448916327008865971381207040
13607387473451491080744515098782990336 27215423984010299014942596509607133184
54432146042235231736792325643296571392 108866888232899730887398916534757752832
217738968762657996602426363565844725760 435488322119033062860109788124347891712
870997413425500265030733698234012663808 1742036365225868808682495640438659088384
3484155807201474174607047768818585698304 6968477767902421463698208513519706439680
13937287842803789156364642978804482965504 27875240299605470770665737861139106103296
55751809827206726457204379529338492551168 111506278110405022746154566672797545791488
223017873132793185155800748573836212961280 446046380089552649638584727604154668679168
892114027827037857931135916121273822871552 1784270590949940833170204754068476616769536
3568626252491611900956275351788811175591936 7137422646166684271144282390881338235289600
14275185574700289480752028156370108238790656 28551051714134420838430983061955080014004224
57103464557736525430715819622339887100854272 114209651374408418369139346241539228347400192
228424747266687571753694106476797364986183680 456860383569116613538219040941032546555133952
913742545209716167138099737856940726275801088 1827528646562398214399522787663632718882668544
3655144405410728189045692199226767970427469824 7310463035393319898584677646252541006179205120
14621274519930366838155941788103092143006941184
29243245938148187758285056567402204547310944256
58487885672871283680516459117196449617216012288
116978558938892383688925610199176980279620272128
233962693064084400033636604327922122649617039360
467936536500768065378843976514980569479987068928
935895373746734661380829488748233787321480118272
1871835348983866384007942048933012871365972197376
3743759900948526890508450240739116336177968316416
7487698207858642026002032767224413859247984476160
14975753227640460541974330105941190092280064638976
29952220079127274063889189354867104932128320651264
59905867405947254087659436995703659359393024049152
119814589307279920095080990563346217709058813591552
239634887605330664029686214270570233398663158169600
479281193192202975738420894828896062758417378312192
958585222347489246834938722233303317439016880570368
1917216116621145084386071309617629018722398009032704
3834523577094623350204530349537302805133524513849344
7669229841893913063273836159678695145644506019266560

153388250591971588522772232405655693620439260216688 64
306783808692129831560135483235474968655976800096092 16
613582232400632972149453003319277100142150159517614 08
122719369483401256235727008033520852594469343768608768
245444584973351836083126830806372570321017311267389440
490900861959802319389599291091406870906191869995122688
981825107945801933225889841140137202340698234910932992
1963696983943998455345162200194921325738025459663241216
3927487503992786088477089436219136493589308899009232896
7855162080195150532527708944096860671405133757383966720
15710698304809457776202478031510896711263299433498935296
31422144898457228974699076349656144159432662704459874304
62845786374591084793986393272580989792677453083843756032
125694565904535423277149267691699382532979161517535526912
251395118119777353932651497676473570961206833734767083520
502802208860967722622008919939096753712910688868926226432
1005628362964761474757429689050492731006815420536636571648
2011304616415175008541683076445583909175618926670841380864
4022705013801654135137013549580364712675214024536819236864
8045601589545916506381321892539123213998380391463911424000
16091586302977049484977233371835034005292665467708368748544
32183938853724531914383645917183643165177140304977829298176
64369410202989929717625650181394436639537899349077842198528
128741885397061591212968017056843173897443036176400051601408
257489900776286645981369467501794949031620547309288837611520
514992061516900219073605801779807100536710044531555144040448
1030008642962454292368945337112048606020357988889065225715712
2060066325782216293181358141328966021934591777430040326701056
4120230731279048003249651216867669663656935154163900403941376
8240657621987326840273172302154814566889373506935440308961280
16481707562833115348094084341148579612929753411086159620079616
32964199763383154031283648155975060184161519616602877244473344
65929968802200154732758255259305922284927064822066870497574912
131863076155268002805898428413323448403062180821855973012406272
263732429412271392292560692616070104472540463999156410059325440
527477413028013557946649056810986624277913132709201748187676672
1054979934462968662616353456779666079221490674840181352513404928
2110010085739820418678817599874717819774310168523918417302913024
4220120605107407024249856572380206962211277974734948259158032384
8440442077470346422284155890021956569747871224844119367420477440
16881285889451757592137197270566998430146373000436684433049780224
33763375247925644679412165522180167441594007102370260262517211136
67528357433895548349099873006452676045790536407734303317869723648
135059928743879614678750829937090034416786117221456172221410050048
270126285239936265318603827722549433483982323254887475614161305600
540265425984226602559411991141837596268784824133725213571005022208
1080556562977161348963232653677152651139210003515350951827374866432
540291136992934746403820662535315054870425179381625738256369844224
270151996248644409163012499116026892085622678502788000299526127616
135079212000410722562057333482198128368016383657381565735433666560
67541212938249620271304208703191405346610714031684565660552134656
33771409938146939630789875332641873254606618117339174226693718016

16886106703584534562963823156844021917953939609418032811555684352
8443254219047799655266354323683553604302285080083239254882254848
4221727543151666014525398534472548124813800177728731051993333760
2110913988389716100708809953551659723738228907707921238272770048
1055482102601799597077460319933522692534778863275738475274436608
527753605504370571900257831545607761600221636348758165706244096
263883079853920672630892751562227088466526920529823546887634944
131944678477828029655828293675825148066471511442634005461073920
65973908514347861498105105785268375949839781310178118739165184
32987738894890854084148032366312088933221903449519617373896704
16494261766303888709621755919994994945761958121975087689105408
8247327042581175188584747828416972712456482259595183345631232
4123761601005203011179308848418223976015992729101411423354880
2061929840359909214033121891313980797901872164202615586947072
1030989440108608461238294679209424803897873981927262731108352
515506980018631157730014206380929604422405940876608834371584
257759619991479042420440536578573403447937445394793151594496
128882874986821252987936984983341002342335960175640943001600
64442969988951492382826850838697651480351598826942655102976
32222251242246179135842604592862400894767608783032419352576
16111508745008306040135891883187988024679709076296755576832
8055945934446761256175240734972387800987806880538650738688
4028068748194684746141267764175390794817879611464461844480
2014082264582994432097457580432293844570927891329799159808
1007065077534323245562140639388446145866457988463683698688
503544511388574637537776244280372684723726015631233908736
251778242004993826147241084433261148257111518515312984064
125892114157850166762797023363167977076180014607504506880
62947553656601710225986752254852690011902134978676260864
31474525117139168535287496414060695742857131326800134144
15737636702988740978790808350347523239881597582131068928
7869005423703948844968934246832349304167314750431035392
3934596247956763600271232159245468494196915354898268160
1967344892030776389028998597537381168155086667290509312
983695830041585488961190557726014044605857828565942272
491859607033891391703940908341668752567086161743314944
245935649523495019463643268910165241415621704601829376
122970747765022171537658041824748053273850164166000640
61486835384148416671747224597206742919944738015543296
30744148442892873787332714141019729601482196974043136
15372439596855769619395907991718043871496012470157312
7686402486132551172562729456463111471125463226646528
3843292586918608767713752458533600503251460109107200
1921691965385470974573070094417822635470094302445568
960868818655818782644631979784422509657229275168768
480445827309451039001364456179966850789705699557376
240228622645496343340206461233861223375398380765184
120117165818133583589865347188869510677971955875840
60060010156759497754813731880404204834122360684544
30030718702226101857347395083186827164629371715584
15015716163036227418643962113085775956098781544448
7508036487479701954307113342289069164941438615552

3754107446720645099646122814017625175916743229440
1877098324850719611069344478445357884681383575552
938571463170558336157813774940951590702197768192
469296881957878433390477655329612119531851874304
234654016165238849351024211594374221856302432256
117329795675769241003404797761971191973339463680
58666291634459528665648744863377636509263855616
29333842715517218414797545422884838515928989696
14667269806902336248385359207040429388613025792 7333809128023031644685972851319219759630778368
3666991676297447582589633049559112412477521920 1833539394291689671418139836729307472569827328
916791475217327775770731574339529369450446848 458406626644405357916196615157202501307990016
229208757840073413973513721572320158945378304 114607101178972074494464567783019533618380800
57304911718953721001086137389939493882036224 28653136424210702377469995444184610477441024
14326908494472272127198461096699737006931968 7163624388419596532830962235653584387571712
3581897264801528501031346961478650135838720 1790991167696629367823606402565254038945792
895516851496247242565769662195591504986112 447769059572089900609868061554277995249664
223889846698028089968425646005380119003136 111947581805005614815958630616810620190720
55975120130498592323852219115465590439936 27988224679247188619862561461262935392256
13994444646622540538899506682396537782272 6997388476810743383933866317080803934208
3498777315155108249208989646481669488640 1749430195952422403225523067211468505088
874735867163645340923275655591051132928 437378318175539740116894888788184006656
218694351384628404886075974890421223424 109349771840743469856852252693375221760
54676183994586368635333258970769915904 27338741034400501171120195797426110464
13669695035753909012286881054733631488 6835009777153783719506832105377103872
3417586018215306466435111841693696000 1708833573926860536558403815349420032
854437069373033919949625854925996032 427228675891318785810024901088641024
213619408548060305822618437357142016 106812239575230609370112212084981760
53407387438215532914457602745696256 26704327544407880571929549724450816
13352480684853997343315149038026752 6676398798752027200332761606914048
3338278627538527864503974347407360 1669178927850521064420783945678848
834609270965889098294790358827008 417314539003258832189594372407296
208662221261786557615896782700544 104333586510971849568498189598720
52168031195525210164523993923584 26084634567782247772399446523904
13042626768900945231268448043008 6521468126955383288168586412032
3260811434730146980351474401280 1630444402991301158309327798272 815241544308764413221459197952
407630443560939123644127248384 203820057483748020338762448896 101912446593513239427730636800
50957432222576234343040024576 25479320574197924486107365376 12739962518553865900347359232
6370132375004384778820517888 3185141745365918303733678080 1592608651614822109028548608
796323215273342533095129088 398171052369637005837991936 199090248551301372564209664
99547485458892121104711680 49774923321066777963659264 24888051956343747687481344
12444321126077053196566528 6222308136991116274696192 3111227855471852975554560
1555650821224073906880512 777843857356110662991872 388931152050092186271744
194470187711064520523776 97237399698541473955840 48619852770775343824896
24310502846139975335936 12155539653446139379712 6077913941911145545728 3039029028549610700800
1519550543071824314368 759793285934421639168 379905650166465560576 189957328682860150784
94980916141243760640 47491583970528722944 23746354935217782784 11873458942585602048
5936870208781156352 2968505473134755840 1484287920939466752 742161552655777792
371089572420911104 185549184256966656 92776791151738880 46389495087497216 23195297299562496
11597923527688192 5799099202797568 2899618320875520 1449843520176128 724938939957248
362478059913216 181243324923904 90623809945600 45312978714624 22657026228224 11328781549568
5664524992512 2832329605120 1416198356992 708115955712 354066366464 177037377536 88520785920
44261441536 22131245056 11065884672 5533073408 2766602240 1383333888 691683328 345849856

172929024 86466560 43234304 21617664 10809088 5404672 2702400 1351232 675632 337824 168916 84460 42231 21116 42233 84468 168940 337888 675792 1351616 2703296 5406720 10813696 21627904 43256832 86515712 173035520 346079232 692174848 1384382464 2768830464 5537792000 11075846144 22152216576 44305481728 88613060608 177230315520 354469019648 708954816512 1417943187456 2835953483776 5672041185280 11344350806016 22689238482944 45379550707712 90761248899072 181526792765440 363062175465472 726141530800128 1452317421338624 2904703562153984 5809544563261440 11619364004429824 23239277764673536 46479655040974848 92961509105205248 185927416256921600 371863628606865408 743744849399775232 1487524883171639296 2975120135087456256 5950381007663267840 11901043490303246336 23802649930559913984 47606425761026670592 95215103321867026432 190434710243361423360 380878427685977587712 761774869770464657408 1523585768337948278784 3047243594269934485504 6094631303727944826880 12189550837832041365504 24379678136416386154496 48760509194337379155968 97523324231683972005888 195051260149386371399680 390111743670809597575168 780241934085692904701952 1560520761659533228507136 3121115310295361295220736 6242378194543312266854400 12485051536991803886534656 24970693369793966478721024 49942567331208650368745472 99887495845658735560097792 199779714057800340765409280 399568872848566420821245952 799156635163064320223346688 1598351049257991597608402944 3196777656379709109540225024 6393706428486870047727288320 12787715088428643752748253184 25576034639767094820083859456 51153278205353804269342425088 102308974262346837797034262528 204622784227972134110767349760 409255239862501185254932348928 818529822538116204576659996672 1637098330702460077286910590976 3274274032657375490841002377216 6548702807819661654216367144960 13097715100649144653501459070976 26196049171317931997140367704064 52393336282675149374555634532352 104789148445428869509661067313152 209583248651014880540421731123200 419176400822344044123042655240192 838372608685316654330483696467968 1676784831451890440829764164911104 3353648891066295145997121873772544 6707456238457618820669430835445760 13415229389565294698689235846692864 26831092604430703512079220044988416 53663452859461635253559936793182208 107329441020123726965922866992775168 214663952642648366849451720798371840 429338046490098559534115415222386688 858696375389800770738654777696059392 1717433315598808844818157449894690816 3434947760836032296318010688794525696 6870057780948893805999412955599339520 13740440080451446038725609067219255296 27481529198010208930904784446479663104 54964356470235051568716701517041631232 109931309088899370551247668282247872512 219867810474657275930123867060824965120 439746005543031621515504795114308370432 879512780273497382341523712213933621248 1759067098921863043304075668398501003264 3518217274593462643850207824738269528064 7036600702686398402184528625359074099200 14073533712371743033337283202483218284544 28147732038741378524611018308496576741376 56296793305478541965094940424053433827328 112596245066948653761935688462227428343808 225197807045880447187362992152695978065920 450406247915727173701709214761874198888448 900833763479386906057384890436712883290112 1801710062254638929422702702699354737606656 3603505195101008093461271249050567417266176 7207180531385476656154274185404850718638080 14414701345137874250772011745417133205487616 28830083255009590378470950240049129947398144 57661527639486864510795753978527986967642112 115325777537909096529299214953915428080975872 230656999593688928074013843901549764453335040 461324888223119326178858515790537345489436672 922671554517721592419378687555950324144406528 1845386665178409064962080687061651914619879424 3690860442642749890170807998022806361901891584 7381895109857363300834909243844617789128048640 14764138668858453642656404983287245708904628224 29528974236004361367285982957770511679106318336 59059342268583630898518311897933063880806760448 118121472130317078124929315760650208806801768448 236248519446933788905644015450868579703980032000 472508189266466843122858798760873483588713054208

9450386792781322168688591332400196155389320888832
18901219600466614949840013379165845278008761384 96
37803331230741171124605688187062596490477761986 56
75608446521098224699062699231587004849876002406 40
15122046116142821429782804417809763343759296167936
30244805856131995839506137978604251435086783709184
60491038959956697638893334243177952365309950164992
120984932415298807197548785058294803720892665823232
241975573821368438234621803260467405422330862632960
483962565624278524148292072808690406805752787238912
967947967211640343654681078192892005533687698423808
1935941606349447278025556021536806394911739644739584
3871974556551227737483499773375657557512207785263104
7744131800807121837831775007355404650401872562094080
15488628977023576401393100935918988371558659107323904
30977988704865818254245303714254334884627146180919296
61957438911368967411408811113341386052273948294381568
123917800826012596628654029596348204670587208453849088
247841447658574516868980873932027274473253040637870080
495694587330247680961307377342716279210663328736083968
991412558686692656369306013642756018949641152392855552
1982871885425779901631994545200158958955911294627086336
3965837306956348981050754126229611760025080568936923136
7931861686122276317675038324117811204276677097239347200
15864097516663709346497136791552797777006386113209696256
31728943322165732115288393869739946290918836063881396224
63459383222008091075165028312748594055649799802686799872
126921759599369435839506537772034591058923854955221614592
253849505509445379057366037837143988013096220610139258880
507710983640303772871438000260437587816689462619670577152
1015445912523433575256287849693174399214372968038125273088
2030939715532519209539399397730947245590734021673818783744
4061975212036342537132446192151091385505444214542774042624
8124141986015293310372187177680576559658840771475821035520
16248766709591580309295896394211794069661358622773218797 1584
32498100439602039130347107057749456547818981825025467744256
64997733374744944149552572462526063404821582389173119090688
129998531740571620076821861619106427428010402256590605385728
260003193463306703709077156626321456092755279469669945180160
520018646890940334529021180028860114658979508852317359177728
1040061813710534523279776093610154634264896917530589655990272
2080172667278376755003019654325178078423669634713089187250176
4160443414271368926892974242860093776635090868729998125039616
8321082987971968687559818354139662792845684936067635751157760
16642558294802399042667376445118276064842376269350550504472576
33285901227321721420430232363914453087986765333131659013259264
66573371730077289511051423675184708092577556255124434035146752
133149882011022272362484765245081020018363163687971100087549952
266306041123779931405733366279585247703142429731386664209612800
532624636451030636172994404138016910739117064173662256488251392
1065274381309002819069044151433726652143898537769102369114554368
2130598979431888731584198989182838965619125894381760450505211904

4261298392491543650060619350996449253900909426450632325562630144
8522797652238619673905681447254441153127134128275487500229672960
17045997038988304095380248385031967596904898807299420698668171264
34092797546998737685898267751110105775111058716095732793753993216
68187202032041734362072077464312552712824639635185248380343287808
136377617940171986704695238852809787750854323676358062346357178368
272761663632521009370492645553988940152118736164691255864055562240
545536182769396090663189626804716609605057649953332774070793535488
1091098076547500325170787925002910677811755655154566072826951892992
2182247575112416938030393192792776272826792020804933195024633430016
1091149498564916612859605267789865595015036365650367122197681537024
545587604786812378352006969591671526808338360449133823441523179520
272800230145583225137105652644205128054579269036542042892102795264
136403328948880130549103910246287246352494678924258587031722000384
68203271412484324264827497085235964338849861665123076308696301568
34102439175264291627551519523664152750726191934058429550765801472
17051621322143210561344645252355161666013726517777660473591726080
8526011528327137654456765371439123478332178534263053085900275712
4263106197791335014120604058350333061828746904818637967502344192
2131603315709550600506412715490552192245702271252874696027275264
1065826766261716846976261700902968926788515545048215204151689216
532925937334329196849658522030330878727089977234996530144870400
266469245768899985105593096804588647029961090972942729106948096
133237761435317685893178466297005927348188596664193596570730496
66620449993092689616780192095858765590698323920957914293993472
33311009634263268143485575521607283753651174754909515151310848
16655897135990095739290527497642592355976593774670036577812480
8328144727424278703419133617240771417563800085942657789984768
4164170443426754768596501742830123328569651642275148645531648
2082134261570685092741718338519930474178701620789484198035456
1041091650713996400592592902812399642036288710220697036652544
520558085321325127407163318182417023491613305023325987143680
260285172642826027259015092479317112982541127468151727980544
130145651312494745407224262933712857109637801212320231194624
65074358151788238592470489813883578864002519345282299199488
32537945323664552240664424080455364586593069042202241400832
16269355785717492592546801626984469870592439205881666600960
8134869454801354532380695606870628723944171945331106250752
4067530508371981384243995200124511256296062143860689600512
2033813144671642751148821298406854075310019157527913037824
1016930517578647405087822498375726261236003621562740637696
508477231410736717300617173774012742408498832180762378240
254244602016074866028661549180081177099497926790077218816
127125294163390686703507255736577991497373218744886624256
63564143659371970196341868441557697222498737047367319552
31782820118524298520465054507413199348155432361145663488
15891784203681305971379587397023775042530748099303833600
7946079174050231341263323770170475205491890009017417728
3973133123129904848418426920914531444859202984191459328
1986613329617347013102595978371912643486230481937104896
9933300488348708009979892481432797822714297358892400064
496676716430534047722340253550301621399872115404963840

248344204221816347472842941514481675832014681432653824
124175025114182835542257878126906270482046652581412864
620889740587287486740471427482858515240429822232494O8
31045217780183039788482673216559283903531319077896192
15522974265500852619970887529487821022520573522083840
7761669820455092672850219225348000046637743752609792
3880926254079879517857497342976044791007600372088832
1940508798966106349644942536639044779348164433936384
970277235446136470180568200895033581596264340914176
485150035704609882769332566735272386759223232430080
242580726843075765224190516511513991360157147201536
121293217916923294531857374827695894670351339094016
60648036206154353225809745699817396830312052293632
30324731726923529592845401992893423162724217520128
15162722675384941286392965567939073955146204446720
7581539743654058888181615069715718164465150066688
3790859074807823566583373677730949675678598955008
1895474138894308844537969910302020134562311438336
947759370192352952892126490869282715642661699584
473890835468775741757634013293777682002083840000
236950992920687503534602390576457003091418415104
118478284053493568095193887253012582590897455104
59240535823321692211543289608898331818042851328
29620964809948300187744817795645186170318487552
14810830854117877134858995393420603215807774720 7405589651630802087922790944509306673228152832
3702881938101332804208042096154155869276209152 1851484525193632282227344360026829200969170944
925764040668299081175333835988290233650118656 462892909369891010618497745981582933407825920
231451899202816240324664286984510374995296256 115728671860343487670039850489114641643339776
57865697059639427588873778742987047894515712 28933529094553555671363816120708387483680768
14467104829643698774145371434961625510051840 7233722556005309856304417404784528639131648
3616946348594385162768074546044122261618688 1808515709593057698691970194847990101835776
904279122444461407999951558336959536431104 452150195046196983326959009624962010972160
226080414435081631326971120040722126864384 113042865673532385495231367634481624121344
56522762064761977663488587624301092405248 28262045646378881289680745715680686374912
14131355130188386873808598809605413273600 7065843718593666551388412380685241679872
3533004936046569832936262678283888361472 1766544006398153195089159583112577941504
883292772386510736855093913541605851136 441656770786972438082804017763461365760
220833577690344753869030539378059902976 110419384993601644348329534937194561536
55210990571015455881071900092679585792 27606144322615044793989516358380945408
13803396679861180823721541335211048960 6901860599207419625224162245615812608
3451011429242124419293776911813050368 1725546279440269512987736350409097216
862793422129738408164292122455834624 431406852269671029917358034853560320
215708496737236427876285004239601664 107856783669818670396945495526211584
53929659485509563427874244466311168 26965463568054895828637870584758272
13483048696677504971669309468180480 6741682804663781014509841821990912
3370920630494404771592514454945792 1685499929328459517965053999448064
842769771704858325066925385711616 421394789372743445575661885849600
210702346446528864308930539421696 105353649103343002915015067959296
52678062491710786837782433103872 26339650215875036109028666114048
13170134592947339399583057838080 6585222038978580372325891309568
3292688390741745522430126850048 1646382880997100429348654022656 823210783311664048741122310144
411615063062388941403958804480 205812367234472929218678226944 102908601468875693867688525824

51455509660257461563018969088 25728359293038538096096837632 12864481877974172705342095360
6432392054714538181317885952 3216271585220995004982362112 1608173571542360459652890624
804105675237111708407300096 402062282351521593494077440 201035863542243666392252416
100520292954363268018733056 50261327068802351420669952 25131253830211534415986688
12565922063010946560819200 6283108605458062956822528 3141628089705326316617728
1570850938340810577412096 785443915914478998257664 392731181329276353904640
196370202350656604340224 98187407018337515864064 49094856430673364779008
24548004676088985812992 12274290568420644618240 6137289399398398164992 3068716757293237010432
1534394407443637469184 767215218120328216576 383616616259418849280 191812811729336795136
95908657664482082816 47955454732147884032 23978290316027363328 11989426632990392320
5994854053983551488 2997497395735953408 1498783882240065536 749409533306077184
374713562746060800 187361179419541504 93682788733026304 46842493878140928 23421796694884352
11711173225349120 5855724051628032 2927930745290752 1463999732383744 732017046061056
366017112965120 183012851449856 91508573208576 45755360346112 22878217043968 11439376957440
5719822696448 2859978457088 1430022782976 715028168704 357522472960 178765430784 89384812544
44693454848 22347251712 11173888000 5587075072 2793603072 1396834304 698433536 349224960
174616576 87310336 43656192 21828608 10914560 5457408 2728768 1364416 682224 341120 170564 85284
42643 21322 42645 85292 170588 341184 682384 1364800 2729664 5459456 10919168 21838848 43678720
87359488 174723072 349454336 698925056 1397882880 2795831296 5591793664 11183849472 22368223232
44737495040 89477087232 178958368768 357925126144 715867029504 1431767613440 2863602335744
5727338889216 11454946213888 22910429298688 45821932339200 91646012162048 183296319291392
366601228517376 733219636903936 1466473633546240 2933015986569216 5866169412091904
11732613702090752 23465777159995392 46932653831618560 93867506686492672 187739411419496448
375487618932015104 750992830050074624 1502020844472238080 3004112057688653824
6008364852865662976 12017011180708036608 24034585311369494528 48070296522645831680
96142844845105348608 192290193289838067712 384589393778930876416 769196801956371234816
1538429632709761433600 3076931323013560795136 6154006761215197446144 12308301752806546604032
24617179966365396631552 49235512854235400110080 98473331551480013914112
196951274788978455216128 393911772949993765208064 787841992644061239967744
1575720878776269899038720 3151515544528834636283904 6303178663010258948980736
12606652473925697250787328 25213895243661753207226368 50428971078944223825756160
100860303341129882474119168 201725329048742634593452032 403460102830451008477331456
806939095126833495535517696 1613915969185529948232744960 3227907496234785810788909056
6455966108197023450224656384 12912234447848950557742989312 25825073358607708430073331712
51651355643035031489321369600 103305129137709292236992151552 206615093978697042990683127808
413239859363951003014763905024 826499061541015840096323108864 1653036808708259348326236815360
3306150988668974032919654825984 6612456719842858738373672042496
13225222924695538821816068866048 26451064819410720333769587294208
52903367578860726047814073712640 105809211037800022856177945673728
211623373835757187233455487844352 423256651191828657509110168682496
846533109424285881102618723352576 1693105832929828894374034218680320
3386290894022172053085661981310976 6772740244369372634846511050522624
13545797401388802327043396276846592 27092228628077718768787540905295872
54185724906755665766976578513797120 108373985114711787992756150434004992
216753040831824488903118287680831488 433516222868450803641448548987305984
867052728146505258953321045225897984 1734146021112217821247489984954368000
3468373171862850249176675758913880064 6936908603002529711716743095838048256
13874141724558717850160269347696672768 27748932486224752553774105007434498048
55499163046664138814455342638951301120 111000922241757545042724950526067212288
222007036780373624913078431548463644672 444024458154464319481413924089585729536
888069685496362778273341970164488339456 1776180909367593835167712184299610439680
3552444895484924227577480856540488400896 7105055944469321569639074688963511844864

14210444195937589368246375329692093775872 28421553005873071194429202562914327724032

56844435239741927304731308932888935792640 113691528935475424441208425479898432274432

227388374782933988545908466188037985927168 454787383389834256418800162832558214610944

909596034427601071491566786578080914735104 1819234604151067260291066494982090800496640

3638554278893864755197998833616039543046144 7277278698971189979627729354535794970198016

14554897680309300897718922083679021708607488 29110475925352443672364770916572906953637888

58222312980172571098583395331575540980121600 116447348219280509704874497660010536105934848

232900140956431754425164409313739980503252992 465811170948604978881159646614917777589272576

9316444119968692897823980949204711188344078336 1863331796080351675771285210359173643019223040

3726750704446635111789217044617849818700578816 7453675633465133744071727337034704702725423104

14907699716073994529130041169667419536099377152

29816096330435443140233255330530859333495816192

59633586457445794444412856643453759189585756160

119269960508041405216718405251691599424359759872

238545496202382443089222194432951360939096014848

477102142777364151490015156725039046058945019904

954226586299926833603171849168350740479396020224

1908497774090250728452626769773246777681804001280

3817084751161295579397819682419584148809631924224

7634347908284179403780771650585349484511311691776

15269052628491535297531807872663061342806719070208

30538818880829423575004144888310847433181629513728

61079065009351553109889348062591144361499641774080

122160984514088518139540812697121187713272049041408

244327678018947860118605858538120173407089629069312

488666774019437367916260183363995942775270320111616

977356384001958031190617299303503077472722764169216

1954758439930082653097428463758028538789809776230400

3909608223712498487627244657818101845268348048244736

7819369135129663338119264776240293225914153088057344

15639163645668659401968080473688765522583220159250432

31279058042155984255395262789793889186676268284772352

62559577585949299413708729264420494656372192502087680

125122078175173260633253865898506421878783696869261312

250250002356895844878180546536343708889646017468694528

500511696726890336979706722551349148043449282397732864

1001046777479977968406104704060021756615213059716153344

2002140323012350525705591926034690434287055109273681920

4004374182129490229197948887898674710687368198230114304

8008935436468558813969427847455937105601252355825729536

16018245017356274339085915838229049579655536630382460928

32037238323550862100465951963092449896217137098226925568

64075973224778351045520144499453601266246401871377858560

128154939604909955780216770145444605480117059092603731968

256315865520526418938786502583964016855482628884903493632

512643703662465852634278929754077645501462279169199046656

1025311352567757734781969708680454514583918601137182212096

2050670595621167528590763115705507476329825287871932661760

4101436972213639175235173628100211846983626746939001798656

8203065506369886586577642049578817482615205836268276547584

16406514136624989645369873685914422542526316357317098995712

32813794521020412235168926545342420239644442084195289792512

6562912153758169035919621143771199078847250290751276318720
1312613080662451124961091395694782821953122432932698935787520
2625287461146536885476517125270651656273589615430285215662080
5250697521936343042061702918303475337281868729990345119498240
10501640243159224626340743172131294724033116458240239615344640
21003770884891526337116161015311277547004990912999577983385600
42008522566929206843101671372719931291887497819037353472163840
84019006728150722023942041429634614979530027624151101955112960
168041936644886060723361480227658734750570119220454993931796480
336091719666941354797677755192096479084160366385215567906734080
672199132088221176297265099857750977334360988659042295899750400
1344429649685119285998349378662617993000802489095306911972065280
2688922070387592438804337115219468062665766001745058464289259520
5377969682809892611223950946227400278659854050599006209268776960
10756190449689200689678455324031728863976352195415790979918069760
21512883067517232313818017511217314341265992579267139082597171200
43026770471312126496558248748742341909158561535405392410716405760
86055549615179576730960924950100110271570275824553013312476938240
172115116575469800937610704805431073449646857156590483607042129920
344238267841160896826599119421323852712306325328149881178260766720
688492605062764383555953658463571117050637872686237590284874547200
1377017348886413946917418156168989057353326189432350836426455121920
2754098975294598253445857990821671761210753266984452984566322298880
5508326505632737226113759338610730815429708310208408592559468707840
11016910121352555890671605391156236216875820172895822431972585635840
22034334462879274658231384210182021605784447450749655357652467712000
44069697366106875070239115276103141555634509111415331702719528304640
22035362903227600412007731065921119949850061660665676345067060592640
11017938561700881644447952246895334560941434382811843419387178516480
5509097835893981541446019480415054573478918967645424333120413368320
2754613195468761130334031418691220933243560371942463478273618739200
1377338736495265744972526548587457289873830630031107394993515397120
688685437628075462389018693914652056562940537045491525425110712320
344350753504259026145887056767787734094482879537714676676731863040
172179394097240160548632383289124719953747745276341795320454184960
86091705721175404012160619097177786430127025391913126151271219200
43046857196865363875002523274896606441690089072827677321157672960
21523930766571512871962368500602159834158332724849395783339868160
10762216467355171903211737681878008223735810456642476453050449920
5381233775712293685221145556727468265196227275430127507215482880
2690679658873500709418211136257966209262274661269508393952870400
1345371214945427288112924747076099142963217842411976517148999680
672701300227052110758371963011607590647649177094599418660782080
336358496490695288730140776242582804906844716491605289373532160
168183171433932261040547785489680907244932422217955434708336640
84093547311258438858012591429035206018221243095054112364953600
42047754452775373597875645056614979206988137540565253687869440
21024367624960763883372497199356177702432826766801725596631040
10512429011766920483903585935202432900685792381660412174663680
5256337105526729513060461635363388475077585689959980775505920
2628229852584999392084565151562780249906137594544877731840000
1314145576203317013819449742721933131136741172054882537963520

65708811305706716579830845483123806866020677341866310500352
32855171900623791234344601915075478587602148040494247051264
16427969074197112089386890544294526871096978705027669426176
8214176099041164280800740065525657224196441694904107663360
4107183830491886258454017429452025506422197018647190306816
2053639805731595188253832413070611200373086594921163390976
1026843848108623623640328055707604823767537340259365814272
513433896675724826576869952439952023674265691529074966528
256722934648568920666787938513050817732381356464233512960
128364460479637714022570450403062811813814933581964771328
64183726817495483855873465774800107380719594465906393088
32092611697586055350230853174034404427265861070415200256
16046679993212184386262486730334377582085962453938601984
8023527068815670548704773436825776475269497186334801920
4011857070512624452139151754242182079748006572850151424
2005975303308706814962958395035737960930632276266450944
1003011035680550701928170456475192440993630633053913088
501517209853373998187430857716257950760972563987300352
250764450933236322705388243597459840512564905723822080
125385148469892823158530529168395352822321764726996992
62694035736583742482183468269030392694180538296041472
31347748619110536692550835976931554488600097114292224
15674239684964601072004968909673956315054962540281856
78373025301869666898867259915441067692904938261708800
3918742608945816630866017688022578614141197626638336
1959416976399074906149202709162311690914963061211136
979731324162620748432698287156667037379663654551552
489877080062852021895397609866089114650922889248768
244944249022196834787223038076922355306006975610880
122474979006483829313373635610400076643276253298688
61238916750934620616567876091169487816774509395968
30620171999313663288224467188569468655955446071296
15310442811580008134082498165777096701761818722304
7655399811751592312026381368634729537772957204480
3827789108856590278505756827190455362332502523904
1913939155918692200499161485031772977889263222784
956991878704544630872722278234159137306137591808
478507089724871580747931906976215892833821786112
239259120048735423029751337417676108507287388160
119632347617517527842768360673622135298831941632
59817567605333672085330526319203108172010094592
29909480700954290124638436150797574347302109184
14955088799620872103305804570996797304299585536 7477718624382299572146195533297403717474058240
3738946424477081546319744390548204391399161856 1869516768381506653283195507223853462030647296
934780162262236266703259409586802364180856832 467400970166859603382460532780838998673195008
233705929601300536706645680384138407627980800 116855687059585635861030547188928657959682048
58429204659260501684369127092894056052686848 29215282894364092719111490295661891562766336
14607981729548967298019208522438377549594624 7304161005957944118241335948522904658903040
3652165573570702293736533817913310271504384 1826125322081216264176199830782584106778624
913083928688540690742066376304256538902528 456552598168236624698016418608610512207872
228281615996101452012499824532546377482240 114143466454042295837995719880393749430272
57073062455016932834870763747257155059712 28537195841506358875371833777158717702144

14268930227752125666654142840344428937216 7134631267375535947811184396054749511680
3567398710437504531147648685968642277376 1783740893593620544194852586954954899456
891891215984244411407940415462794330112 445955992585839275359227268724055605248
222983188589778172507242164858357022720 111494190443318353667435347677343121408
55748393295873810540624806462753865728 27874845685044222123765969543418085376
13937747361075769488609767927729618944 6969035939814713957668275541875097600
3484599099545771585515833559942692864 1742340114592093096098764674473918464
871190339705650199719806284488245248 435605311057626925695115115869765632
217807726131214375765163544747704320 108906398366807644341384765780262912
54454466834004050400093879593336832 27227867242302139314747688148271104
13614250533801126714724218249936896 6807283723225591886037296212869120
3403721089775310207356241650384896 1701900158968912235846917597167616
850969886525084684007857184571392 425494846782856625046127785279488
212752375151585454044163489136640 106378663455871297782631542816768
53190569667974934271590670532608 26595903804007109825932784828416
13298261387013376258035117195264 6649285436011598801551920988160
3324720089258254737043141689344 1662398730255355036655161442304 831218707940791352394376019968
415619025376952593230585659392 207814348391754755131991654400 103909592047516606824345239552
51956004949577918041347325952 25978606937698766335261016064 12989605700304286824924184576
6494953965879595241108930560 3247552540803523534877884416 1623814049333624724600651776
811925914132743840881180672 405972401799337659731017728 202990923266151699510722560
101497822816317284577968128 50750091999779359700287488 25375636295700038555795456
12688113295755198630723584 6344204221830188991774720 3172175897891389334093824
1586124842433842086150144 793080867960994752626688 396549657352534231089152
198279440362285542932480 99142026024151985160192 49572165933580599427072
24786659427542603137024 12393617944147453280256 6196953087261802496000 3098548601224939175936
1549310329409488551936 774673179103253757952 387345596750881619968 193677301975068180480
96840902787347775488 48421577293580730368 24211351596743786496 12105957273348603904
6053119374162657280 3026630055825506304 1513350212284841984 756692698328465408
378355145257254912 189181970675138560 94593184360824832 47297691692040192 23649395601833984
11824972678823936 5912623778365440 2956380608659456 1478224664068096 739129511903232
369573345886208 184790967910400 92397631438848 46199889461248 23100481601536 11550509236224
5775388835840 2887761526784 1443914317824 721973936128 360995356672 180501872640 90253033472
45127565312 22564306944 11282415616 5641338880 2820734976 1410400256 705216512 352616448
176312320 88158208 44080128 22040576 11020544 5510400 2755264 1377664 688848 344432 172220 86112
43057 21529 43059 86120 172244 344496 689008 1378048 2756160 5512448 11025152 22050816 44102656
88207360 176418816 352845824 705708032 1411448832 2822963200 5646057472 11292377088 22585278464
45171605504 90345308160 180694810624 361398009856 722812796928 1445659148288 2891385405440
5782905028608 11566078492672 23132693856256 46266461454336 92535070392320 185074435751936
370157461438464 740332102746112 1480698565230592 2961465849937920 5923069138829312
11846413155565568 23693376066945024 47387851645517824 94777902314291200 189560202675093504
379129201443209216 758275995072462848 1516587174517014528 3033244717778206720
6066630173044768768 12133541821066248192 24267646592085917696 48536419084078678016
97075089967971041280 194154683535569453056 388318374270393647104 776654762939296776192
1553345554675612516352 3106763166945262960640 6213670449078601777152 12427629128533355266048
24855834717819013955584 49712822357142634758144 99427950557294483210240
198860512800607393808384 397730248973251642392576 795478944690576994336768
1590994782869301407776768 3182063352714897653760000 6364274279382384983932928
12728843706669949320691712 25458277709150257347035136 50917736009921232105373696
101837833203083899033354240 203680388772650667711922176 407370222278267074714271744
814759334022465628009398272 1629556446976794213180506112 3259188451817314340684431360
6518528019362080510015700992 13037358270179064677325078528 26075321003267936669237510144

52151850932355487967649726464 104306119716350205193648865280 208617075135978868903996555264
417243821678514654841390759936 834506986170143143749576818688 1669052657966513955632744235008
3338182687185483247532669665280 6676520116875877167599701721088
13353349718761575680268128223232 26707318407542794050673706008576
53415874755124873481622311141376 106834225390328317723794420531200
213673402540813776968688437559296 427356708601941836979576068112384
854733224244512240043550522212352 1709506062570281612255897816399872
3419091353303077488849389176750080 6838341162931183506373965441400832
13676999238512424070098305058603008 27354632302324962254897358468808704
54710532255250152739196213640822784 109423599811700761937195420688056320
218852270225802436791996828188934144 437714681656406699419205630003511296
875449645722417050508835207258308608 1750939856264041404358518309019189248
3501960842166497415398732407043522560 7004083943609824044160856392097333248
14008492405773306515048495940215242752 28017633848653929883550558192471638016
56036565771522493474008249009025581056 112075727691474254361830763266215772160
224156647679807043551290057028760764416 448323679953331156757837175050179969024
896668129094096452826188472085676818432 1793337796563061184273405188141987397632
3586838669875858925788866864225242316800 7173843493251190966061846704333019676672
14348019293501328161091919360431109439488 28696703201000548780120290624392359051264
57394735629996882476113485055844998447104 114792129715985334783972777725810557583360
229589576343953809231437170679862236545024 459189786511873897789857571816206715846656
918400840671680354233681604545377917206528 1836844216639225825775296130916684805439488
3673773503870181886166458105485227552931840 7347717148923824241564647898274170989969408
14695774580214569421592759171155773748150272 29392229725162980720112445091526411032723456
58785820579793645194078743681482549138292736 117574363418522657895865194359824552422277120
235154171354916050807145802713368013135937536 470319231745573571645122433414173842854641664
940660241562630083351906522803223318874816512 1881364039268226046827136357556197904080699392
3762815190822383853900919339011898340823531520 7525804606216631228295131925822801746971328512
15051957661576989497576850347243613624591187968
30104612221441433077126873685683247510479437824
60210618239457774318200093353758535543552999424
120424024072065364964292878672301152132294246400
240853623330430362584371141274170466354964987904
481718397033459990480313050407477256890682966016
963459094812118511583767636533227162142871912448
1926962791114634084413818344502999621008755785728
3854014785210062291320202831879089835463535493120
7708207976381712827625537949504360857819118829568
15416772764686602145221340470501084089422333345792
30834259153219557270383210083986892926412858064896
61669945554131820500647478453943235347962098876416
123342745603649052921057073479825369686196963246080
246691200198068929681638380103528537352939457478656
493393818377679507042325226494812670666969976930304
986810472718442309442747385565136533256122077806592
1973666617363051209601688636281295450356608403505152
3947424578578435600635765002864635668401945302794240
7895031844861537564136305466333360872181347597156352
15790429065132407854002161853874900815117609177448448
31581588881083481159463425550166159771745046321168384
63164639263804293221845054785165035826509748574879744
126332201530883248249526516939995504219058809014845440

252670249068315820110725848619321873570196241759862784
505352190149730287444797326717305477404549730980069376
1010727764325657869336285912391934415337413956880826368
2021502296703710327565954342698515751731456903603027968
4043098129512209832918673721226325345576171786888806400
8086383331233998021410877514111238375378859533143113728
16173140806887152753968815171539652119210750985017229312
32347029902612618930231750629713654975327565807496462336
64695556382901864705051741832696011424467259289916932096
129394105921156983099279964811929425796558773929681879040
258794198153020473576912891916933657488366058559059787776
517600368927453961910531708420016926767229138517511634944
1035224683097733953334475266012333077115452319833807388672
2070497256681119965695774230369264601392892725265183014912
4141090294333544049445195857427726097109761621725502504960
8282372150609696334997686508233845982867475585841277960192
16565127425104609142209962603224479543030855856463101820928
33131021097979651228849104379962534240653521082487295442944
66263574691500168346556567106952218790490660904096774488064
132530214374082068470829850907958738199348559286437916180480
265066558730327600497093135204026077635431593529364566769664
530145377424982128105053137184269357744332136971706602356736
1060315274778618110431840007920973120435602173769368142348288
2120679589414543929307147482946815050765080147190646159966208
4241457258543703275501229900103367721317911893685112070471680
8483110676516637384776329668626210682211326985977863642021888
16966613671891736437100399074091371843573660369171006286200832
33934011981500396209296277621860644645449333532772570576715776
67869593238434639088783514191077091207502692654406257162059776
135742325027736971517948946276865786248213436486534746341376000
271490927157209329716661728343154780162842975328513956717264896
542994408517889432794851128265155975658518155367916841503555584
1086013925442720412312757599688004781982700720157611539145162752
2172078067699323918071625885691395225296730259158778790566428672
4344256569026414023035473144013561773256118156004669005685063680
8688714005308360419855389033288666191837551587383560860474540032
17377829745127785587279663557100417674325733725515567419157905408
347564629592777006696970980952470059299527285552528026234733461504
69514532856599660329669738152586353022507979308049835262302224384
139032279589287838639890560229357388370221003022087236110275051520
278070986930752713240883288307084141390852094856149603391891308544
556154829365859498403970912310907012082524367336249469126465028096
1112335369740427140652350496015291482766689089920399462938294878208
2224722161498270568993518334817537882736658890336599975247319400448
4449547167031373713364671355208985599879879201664802049236098088960
8899300022132412577484612081565790868572881245312808295955114754048
4449752855101038864119940726356805268693002043648006246719016665088
2224927849567935719748787705965357551549781732319804172730237976576
1112489635792676003718802524376156234376531221407802611050483810304
556257673400692073781605597884816846489085788327851567867924316160
278135264452523072851904966790777787894952982975900915105303363584
139070846102350054406503567319573576272681535893938023138322284544

69537029989219286193527325621879129298943290149962794361996443648
34769318463631772591901433791985735230772906176478288577415872512
17385060966326951043519602386515952906037083638987589986916761600
8692731350419007895544243938519519098343857094868017842562793472
4346466108837270134664343341890530871834586185121120345833603072
2173283271232518160778282357260651097248621911404115885192904704
1086666744023200627112196521788018379289975365123835798734503936
543345926215071086917625932472855604977819887272806827436277760
271679240209270930139576802025851010155326045991847877752651776
135842758655503158410170318907637108910871074173646170893582336
67922948603185425875276118401174356372039562675684201455419392
33962258939309636272733538674265079144321794132272658732023808
16981521788513279803914509073971490051311903463351608368168960
8490957053685870735731124405405220265231454930283443685163008
4245576606557550784752497136912347752403479064445541593120768
2122837343136083100819716035561042686095615331874680671830016
1061443191496695404631591751332955747994745565763295273549824
530733855712674629426662742442695076470841732794625105592320
265373057838500778268764804609456139472155341353801287204864
132689593910332120912099118998782370354444908155145010806784
66346329450706926344907917846418335486406072816694689005568
33173930973123896116883138096722742897794845777908436303872
16587348610447164530656158635118159026193327573734764052480
8293865867166190501435374110937473301744616129257654976512
4147028714554399368771334452157933545196284235823963963392
2073562247762851743412490924423565219760130203509550219264
1036805069124251901219657311384081833461059144553559228416
518414507183538965366534580278190528521026593676171673600
259213239902475990061620252432170070155761807537781866496
129609613106591248719986607362622438025505159118738948096
64806303130972251204581544254579920486564707234293481472
32403899854324439024584892413924310980188417454608744448
16202324071581376223439506350279330858547240646035374080
8101349108000266467293283246798253113500136282383187968
4050768090104922411433406659228420398863326120874344448
2025430813104855794610085847528857120488292050278547456
1012738790578625191751734182721752020772460520059961344
506381087302411243099212720839537740650387507490324480
253196389657754945161279175159099735457272377475334144
126601117832152134386475994949215300294675500602753024
63302020417713398096156201159440366430357406233919488
31651740959675364499537202422136541356688531083231232
15826235855247014975498152132276449749099179524751360
7913300615328173850613851526742314409927046753943552
3956741651516420106739313493673201972652251872755712
1978416497684376644085850611987623370170490184269824
989231084805271617401022238569322877007427216080896
494626960384177456379559585572417034464804670013440
247319189182859552029304025930086315212947865993216
123662449086815187934414129536982056596746698489856
61832651791100299927088123054460477793509731991552
30917039519396502943484590670214963644323057369088

15458876571621427961712559906599844195945624371200
7729616691772302225841412239046103284864860028928
3864897548866945235413272262396142235878453936128
1932493375923869678952919202634616414662238928896
966268988707133370099601137035580855692625444864
483145644726165950361371336376926752027065712640
241578397549382607836471052118031538103909351424
120791986367841120246128218023799850097142923264
60397386980495468287010454994291965571165585408
30199390388535188225478400488342003046879854592
15100043643411321153725786739769011654088458240 7550196046277524097356186617683510892368494592
3775185135424693808924739932741257978846380032 1887636123855312784585693278320380255754256384
943839839999139332354508295135065761042661376 471930809035311136208084975554970697104097280
235970849035526303119457901771204256843431936 117988146776698519067436657882461582567407616
58995434517816943287572182439660518356549632 29498397823642313520713017969045122714697728
14749539194188077698819972359129993125560320 7374939738277499318641717866868712446885888
3687554939730479893936724777086214165495808 1843820005161105064276295310369036053774336
921931270228485090792114116097482512400384 460976268938208824723040288505223498956800
230493451381087552025011759480852870856704 115249384146535345844251687354546996117504
57626021301263457837998747484333778403328 28813675264629621376935825645697029373952
14407169939313756917436138774613584773120 7203751123156351573202182363189327429632
3601958638327912343843147669535931236352 1801020857538824450542602078738599378944
900531197956846364581815161354616569856 450275983572140251946164641669966725120
225143184082928660800710851331312582656 112574188189893597814169690913820901376
56288392169161432613991978080992755712 28144845121688033160449555352537530368
14072747079397675006951560832289341440 7036535798975666716839171994154958848
3518349029126247965101281786082623488 1759215079382331285891488787543883776
879627822100769294616168341023227904 439824052255186473143296144137256960
219917096729994149489254058881449984 109961083666197531203430022847135744
54981809483698993831116508126773248 27491538567149611030259002414989312
13746086196224862572479875383296000 6873201554437459814915124779548672
3436680005381244171795155933724672 1718379616771879218066374738837504
859209615426568175117585755406336 429614711233598370600992070696960
214812307376956326821595631845376 107408629568556734171347614171136
53705552724317652465948706209792 26853395332178468923111802667008
13427007151099055806624626114560 6713658318054438575846675447808
3356906530279674624190518919168 1678491950766064980228850057216 839265318196146324181220327424
419642330504630079124007813120 209826000955593498078702731264 104915418329435978297700777984
52458918090537603778025095168 26230063508178609203599900672 13115333985544208259093626880
6557818108499555958193651712 3278984612113503893420244992 1639530084988614903871832064
819783931960238930516770816 409901410713085204548812800 204955427723025471919620096
102480075044754170782416896 51241218113997802802511872 25621199352809260106907648
12810894824309809406279680 6405594986107494379552768 3202871280030042027982848
1601472533503168433094656 800754713495657926098944 400386580119865817825280
200197901745951336300544 100101256715984881844224 50051781279497047769088
25026467100500827308032 12513521780626565365760 6256905005501358538752 3128524560344717197312
1564298308969377562624 782167168883198263296 391092591640853872640 195550799420054306816
97777651509840838656 48889951654827261952 24445538777367052288 12223050863660236800
6111666169318473728 3055903453403414528 1527986911073796096 764011047722942464
382014319954493440 191011558023757824 95507978035134464 47755088529195008 23878094020411392
11939321888112640 5969798383009792 2984967910981632 1492518315229184 746276337483776
373146758676480 186577674305536 93290984636416 46646566060032 23323819900928 11662178385920

5831223410688 2915678814208 1457872961536 728953257984 364485017600 182246703104 91125448704
45563772928 22782410752 11391467520 5695864832 2847997952 1424031744 712032256 356024320
178016256 89010176 44506112 22253568 11127040 5563648 2781888 1390976 695504 347760 173884 86944
43473 21737 43475 86952 173908 347824 695664 1391360 2782784 5565696 11131648 22263808 44528640
89059328 178122752 356253696 712523776 1425080320 2850226176 5700583424 11401428992 22803382272
45607813120 91217723392 182439641088 364887670784 729792118784 1459617792000 2919302692864
5838739603456 11677747642368 23356032155648 46713138053120 93428423589888 186861142147072
373730874228736 747478928326656 1494992216391680 2990053152260096 5980243743473664
11960762364854272 23922074485522432 47845248482672640 95692695988600832 191389790023712768
382788376140447744 765594344466939904 1531223873305968640 3062518115356114944
6125176968200585216 12250635411377881088 24501833772709183488 49004793445325209600
98011838690464104448 196028180980555579392 392065369160365899776 784148752719241281536
1568333534235501527040 3136739126065040982016 6273622367318157819904 12547532965012467351552
25095642390777238126592 50192437703059083100160 100387181249127379894272
200778974184273187176448 401567171740583229128704 803152790225240167809024
1606342473938627754721280 3212758734853550347649024 6425665043659690371710976
12851625235224560096247808 25703840766259478898147328 51408862124139675207598080
102820085431520785237803008 205644893229524440120819712 411299231192014619532066816
822617351849960717644988416 1645272482631784392451686400 3290620523127294699226791936
6581392161982041227100422144 13163086555418986111494520832 26326777573747779537576394752
52654764073315173704327495680 105311945998269576667004403712 210628727699817611850707632128
421267126806192140734812913664 842553596425498115536421126144 1685145878477223899206432849920
3370369128206903134680046895104 6740892998918716941894456180736
13482095482847255228857637142528 26964809935714153147852723847168
53930857811467591675980346818560 107864191503013754112510491885568
215733334766184649746120580268032 431476573052683582534440353529856
862972953145995731153279093047296 1725985520373248594475354958069760
3452050268909011453288303460089856 6904258994143051435251794008080384
13808834900936159927853962191962112 27618303627172433970408672735526912
55237874904945096170218842174259200 110478285111090648799240677754929152
220961640824582210516087342322679808 441933422853966246867386658271002624
883887128117536145405197263793291264 1767814821054279594151242422089154560
3535710771746973794984180633183453184 7071583802770776803331752844377194496
14143492124095212033390288844774965248 28287633285297740920234144001591083008
56576564644810115547375420627264471040 113155725438049498508565106502693552128
226316643172957531844758743501716324352 452643670939632133344774547996091088896
905308111066698406000063217977499058176 1810657760508265090621154679925631877120
3621398597766266738484365847792531275776 7242963349032006591452844671467597594624
14486259005062959411873915294700265275392 28973182624123811281684282492930670723072
57947694476243407479241468792921621790720 115898047408478384790228745199963804270592
231801411728939909243949105628168729919488 463613457281846097814881441712819702595584
927248182211624754283729344338603890704384 1854538899719114625875391610503136752435200
3709162870029959486366649064658131446923264 7418495881243379441965029816619978777952256
14837332044853679822393523007847389324115968 29675344654441201521713972764909642184654848
59352050438350086797281799028249011442155520 118706823135635541102271305053357477030002688
237419090789141817219958024100433862351388672 474849070614025104470746876188305541285543936
949719919299533149003155408351486715736621056 1899483394742032178129634128652724697804308480
3799053901769996116505914881204951928270749696 7598282028111855753505123010208908921865764864
15196912505367438547996832516015827974380060672
30394521909022331177966838023227676210057183232
60790437614619570519880022028847392942708490240
121583662822388957367652736022478866930605228032

243172900831077547391090855974525895951586951168
486356952034754360093752479808188116083926892544
972736204814707250810646495334648880529359765504
1945517011119811562867576062105843057781731491840
3891123225220417248227718267084776709009486905344
7782424856402422741440568819915734604911021654016
15565206524728021972851402211323831583606138994688
31131126673302396925643333565632387914780469362688
62263680594297499811167725417234225324697321472000
124530215683980411542097567406407349639667408437248
249066140358731646923719367956692497259880347860992
498143698699004941526487202201140590480851757694976
996310233361093178411071336977792372883885639335936
1992666138648352947538336539106607129612135526563840
3985423621149039076509060808515259026912999548911616
7971029930002744515882897077634607589203456089391104
15942425235414821757495345076477394249161826161917952
31885581221648308966449791995371146639833480290107392
63772623944933948835817787675575009562686616512757760
127548170893142559477471982720815451691412544890601472
255102187792834442566616780180961768514903713511374848
510216067598767532356579189840585267293964674483093504
1020455519223732359159849638638493995116243843886874624
2040957806499859307213081795191634911289116677615124480
4082009149104507792212928626212563664691491334912999424
8164205370418593939999387324083715013609498629191499776
16328784885256344591145834791484605395672029177114001408
32658318059351002604585789869603561528250122191690006528
65318132696378632053759820312475824530312372058304020480
130639258548110517796696121771489052008248999466456055808
261284503406927542971745205836052909911746509632608141312
522580979435268100700196336258255431613990040664608342016
1045185904113362230913804521688810086808974124128000802816
2090419698712376520854432741722218620779936333853569843200
4180935178396057159762512880133634135883848838902276161536
8362061918734722555632320553645662060415650020194825273344
16724506961354661583479230694048111698127204725170196447232
33449780170479756111387640561609798550846218819901484695552
66901092836500378111633639470246747410876056378925152993280
133805250664082488000983995634547795440119350236094673190912
267616631310328439557401424657204192116973175428678080790528
535245522584983806225669716090625586707415300770333630398464
1070515565098621466673073165733685578361768501366622198431744
2141080170054550641789613798572239966617412802385154272133120
4282258419823716700466162531354217553022577204074128294805504
8564712999076664234706194931127910345620657606755896090689536
17129818317011790136960129599094771170392321610727071183536128
34260421271740503609015738671867443299086656015884700371386368
68522411818914853888222436291090688514777337620630516751400960
137047962188697401116826790476892980862762726418983265520058368
274102201479130188914417416743209169391941555193410995074629632
548216957161731151190362505065264754116715315097710918218285056

1096459022730403849103780353288222338899095039617199692574621696
2192968262274690791653671392891830339129518898077955097425346560
4386036958177147770199564158414432000921695433843021619402899456
8772274783609827914183571062090406647168706143060266087910211584
17544951301730720575936027614703898584988042836868977874029248512
35090706072483570647009826210453967751277346775234847144476147712
70183019083011400284295194383000276665157215753463477081787596800
140369252042111318549141472690185235655519475912914519749245796352
280744931836399673059385113228739835961449040637804170669832798208
561502719177153418040974562154218401223718258899558603682348007424
1123031149363014979926357795701914261049076873047017732050060836864
2246113720743446247541532934190783439301434456589836513470851317760
4492330285521725070460700553955476713009430334171275125683161923584
8984866259113115291676670479058773094831983510325754448849242423296
17970143894365560884863879700413185527290212704617917292664321998848
8985277635252445593187209221354412432458229194292162843815079575552
4492741661661055371971239296251116050635676018137683520648999075840
2246422252847943973674436990912512942521118719564642809695229181952
1123236837432680130681627166849733929862199715030221929532979412992
561631274220694137263017919121605694231920035139061227109172117504
280822064862524104592611127409172211766370106381505744725927264256
140414246307350570276856647628770788208390097596740437948634234880
70208730091719544128703865776477735266797571001364001767152418816
35105168514881901559489703869285038214700046602178892279993860096
17552985991952015527313737425165604398000653851837891838205755392
8776693863231540137441311457844344844325642201293168768207290368
4388447365243536255612877101552943744825478738333695808655851520
2194273899435651221252549237091857533744068188010403616604028928
1097162058124767157349329961703621597537698503426979664440066048
548593583265854352036192652430657214101681456424378760289058816
274303068734662562698860162004751814717256830567633844179042304
137154672918198974689811998897087511191836466461539154106777600
68578905734533334015096958395899557512522258819630693062017024
34290237504983590342643958671627679714563142204245904535322624
17145511071350256838869719072652790336432577499338231269818368
8572951695104359253208729404745870407791791948276755135987712
4286573927266795043491299636582672823683647573442197318533120
2143336003490705230189117285396205221735699586373008534536192
1071692521674006469316292376250537015814787693012459204902912
535858520801330161769013054901485710380862796419207071268864
267935390382828544439939960838851456427165873166092270043136
133970760182496003997686697113480028831950174061290502225920
66986912586788867887701706903767164725158705769767434715136
33494222541164866888280032625397157517171162254444809158656
16747494394467649916354605899455366335881485812002950479872
8373938759176433194284597743106076956588695248391748190208
4187065160559520715195946268242235372618323795391010570240
2093580470765412416624796832465716133471149983293073522688
1046814180625532237825810265405157290316569034445320880128
523419062934179133669611057288728256948781538622052499456
261715517777796074213158490937438934369639280010722279424
130860752044251290795755726615256870132443895355209154560

65431872599802272242466103880897136540034075352528584704
32716684588739449543527172227082919006923101513726296064
16358716438788881482910646256858634871914582675594149888
8179545291604019097028853200087905120183807297162575872
4089866181906798726301191635873246402205161628264038400
2044979859005793952043978335851270122159209803973394432
1022513313529094270468680426882958521607919396907384832
511268348777645782457685842920140991068116945914036224
255640020395372214840515736199401360666137096687190016
127822933200960769226094275469366112899107860208680960
63912928102117715515965341419515772732573586036883456
31957194801877523209441772552174244507796620984713216
15978962776348094330450437197295301324653224475492352
7989664075878713528089994059251740197704069229314048
3994923381791689945477384759927914866540763110440960
1997507362822011563454886245114979817114745803112448
998776517374089077085540055133001100479555025502208
499399676668586186221818493854256146200868574724096
249705547325063916950433480071005871080979818348544
124855628157917370394978856607441834530762674667520
62429241326651391157370486589690366760517720080384
31215334287172048558625772437829908127827051413504
15608023955509200769283150790407316437697621393408
7804190383716188629626707680949839405740858540032
3902184394838888437305919983348010296316453191680
1951136798909841279899243063110550444881238556672
975590700200119170572763067273547870802125258752
487806500472658850597952301495910259581815619584
243908825422629057954761534677523291881284304896
121957200304464345305273459303545726985830400000
609799993948807080816583075634164904015509323776
30490693872690994490264710808278472269051723776
15245695385489224286118941899737246265174392832 7623021917316475663552764197667628197911461888
3811598070944169592023028722733316631617863680 1905842591615050676134837673316409582139998208
952943073879008278129080492633080424235532288 476482425975245609095371074303978028700532736
238246657505493539563100951145707922641649664 119126051011682137289258182569713415466516480
59564386635308752398482944783286434806104064 29782873882388218076168399140858080939474944
14891777223561029976547662945036472237948928 7446058752963975457505563159821952003080192
3723114447073717963368647423562833943592960 1861599758832724098992256633607345942822912
930821147064294608150094777716637456924672 465421207356113583402030619314800971218944
232715920590039931364506924885641606987776 1163606187510115355513999270056941364183040
58181638603501552672872538835530962436096 29091483915748668794372721321295621390336
14546074264873280626154586612412880781312 7273203285936113427561406282088975433728
3636684719717793271022759628985755238400 1818383898233764914132408058463511379968
909212718304316596376718151217072570368 454616743745875367843616136601194725376
227313564169796218749436598796926582784 113659378233327376788532564646627901440
56830987190878322101173414947396255744 28416142632546477904040273785739280384
14208395834826897378746920048890216448 7104360176690277902736851602455396352
3552261217983553558050121590232842240 17761711173810984082365908689618993152
888105869315095692853378292060782592 444063075862349672261901119656034304
222036608533575749048556546640838656 111020839567988330983081266726830080
55511687434594393720942130066620416 27756477542597310975171813384912896

13878555683948712544936280868257792 6939436298299384801143327522029568
3469797377312206664909257304965120 1734938302737360464623425424457728
867488958409308798396111098216448 433754382724968682240254742102016
216882143122641482641226967547904 108443547441399312081163282022400
54223011660738941420856540135424 27112124800389113400565719629824
13556371885204378045351584595968 6778340685107099695210154688512
3389247713806005183872258539520 1694662542529230260069719867392 847350614077728964101655232512
423684978445421399084225265664 211847324925989158058811457536 105926080314633808287755141120
52964249083136518773052276736 26482729004478066701113491456 13241666733693937007850422272
6620984482574420332572049408 3310567799150936080609443840 1655321678507330997466431488
827679728719596977314070528 413849309092764227947462656 206929376912864983618945024
103467049639673926632079360 51734705411457680727343104 25867943001539199069323264
12934266648674778887487488 6467280898289979120156672 3233714236121284398284800
1616894011548789618245632 808465452518468518674432 404241949631271114113024
202125586501653984444416 101065099093836205916160 50533702468422709805056
25267427694963658326016 12634002077857980874752 6317145154117066293248 3158644634652571074560
1579358346123304501248 789697187460161732608 394857600929335607296 197433304064295174144
98718903831961272320 49360577815887478784 24680851857897160704 12340707403925291008
6170494439451000832 3085317588469678080 1542693978606927872 771364581489508352
385691086837776384 192849941465399296 96427169755955200 48214684389605376 24107891950616576
12054220853215232 6027247865561088 3013692652257280 1506880685867008 753457522802688
376737351335936 188372970635264 94188632801280 47095390142464 23548231942144 11774384406528
5887326420992 2943730319360 1471898714112 735966134272 367991455744 183999922176 92002058240
46002077696 23001563136 11501043712 5750652928 2875392000 1437728768 718880768 359448576
179728384 89866240 44934144 22467584 11234048 5617152 2808640 1404352 702192 351104 175556 87780
43891 21946 43893 87788 175580 351168 702352 1404736 2809536 5619200 11238656 22477824 44956672
89915392 179834880 359677952 719372288 1438777344 2877620224 5755371520 11511005184 23022534656
46046117888 92094332928 184192860160 368394108928 736804995072 1473643544576 2947354198016
5894842613760 11789953662976 23580444196864 47161962135552 94326071754752 188656438476800
377321466888192 754660113645568 1509354587029504 3018777893535744 6037693226024960
12075661329956864 24151872415727616 48304844343083008 96611887709421568 193228173465354240
386465143023730688 772947878233505792 1545930940839100416 3091932250422378496
6184005238333112320 12368291951642935296 24737146853239291904 49475419606385426432
98953091012584538112 197910685624796446720 395830378448847634432 791678771296204750848
1583393571389428465664 3166859200372894859264 6333862515933865574400 12668013262243882860544
25336602985240069144576 50674358891984745136128 101351023626978703966208
202706658939975835320320 405422541251988525416448 810863529248050760384512
1621763951984248939872256 3243601690944792717950976 6487350955842175112314880
12974997059589529577455616 25950584414989417860562944 51902349421599553132429312
103807060026440541087465472 207618842419363951820144640 415247129571693642930716672
830513148609318764442288128 1661064076150500486046285824 3322203710164726886415990784
6644558536056905601478819840 13289419303568714860251316224 26579443070047237035089985536
53160095065914088699354677248 106322607983467406657058766848 212650051670213271830816358400
425309774746983460695030366208 850638892307080755456856031232 1701316470240389179047302660096
3402710311733233694361786515456 6805575365971378061257935421440
13611460216952577467584595623936 27223539403924797625306640809984
54448316747888880630888180744192 108899109375856332022326159736832
217803170511869805565751915970560 435616244544053894173703024934912
871252296128736354431804435857408 1742544206338729841032405643689984
3485167640839973946402404831330304 6970493738004976421479996750561280
13941304388660009900310367676923904 27883242602620133915321483705450496
55767752855840496060044464114106368 111538041012881448578891921634623488

223081152628163810075389830082068480 446172244646112944598599163378977 9968
8923651753318625436424072148308 45952 1784770915482932390625662324164263936
3569622960604279387933020437333671936 7139408180485387989229432452677632000
14279140879524434405185648061375840256 28558930796156185663824862434792833024
57119159666527005034556857493667971072 114240915481483277482927980235500552192
228487023259825089793484490967330324480 456984431113367249242226042927319089152
913989631414168637794966207839955058688 1828020801203205554210960659650543878144
3656124679156147665663977807242355277824 7312415511811768445812068590367245598720
14625163330622483120592363132499561283584 29250991275242858699121178168529262739456
58503311778481502314115260144118805823488 117009282012954574459976327902358172336128
234023880937892288583444271032957466050560 468058395699750856493871772522397174857728
936138059047434271641710005957758835228672 1872318653390733660591352933741446641483776
3744722377373197555798571711134751225020416 7489654895929855580828875109573218334146560
14979570074226632100121213593753868436504576 29959820713187106077169353936722600409432064
59921002555841895908192561371874927891709952 119844727370619159324092829740609309929111552
239694899259109053663601073474937528149606400 479400687553959577358032974937312872881979392
95882315317940209477727605849501378929491968 1917689862501770069678778523648754024190050304
3835466837289471899604203671197010581042233344 7671107899150807319701700590193026227408732160
15342564247445341680389987675984062585465995264
30685825393178137442753148343164145432229052416
61373044582931183049452642668720331387052228608
122748876759012182426797977302224743819292704768
245503328704323997509381338534017649728961904640
491017807781247260330333444927171623638676799488
982057916307693051283808425572615895638859579392
1964160434105783163813899922581777088000731119616
3928410071192360450120365988036644769447486160896
7856998548346309145225864261819470725787020165120
15714353908615794780421993095131303825358136016896
31429421441077942540784515333247332398284463407104
62860270129848591041450088952464114291705309560832
125723394755082594002662294476867127573683384614912
251452498500936011844848822097612053127912300216320
502916414983413671368746110482979702216915662405632
1005855665929910638095589153541470596356013448757248
2011757003785987866907372172233963576556391145406464
4023605351424308915247132074769971920801510786596864
8047393390553284193359039610144033376980478564761600
16095152156515901112447630141496245824715871112658944
32191035063850467676354362125408849790941570191589376
64383531629338266255626927935650415864902796315721728
128769986261951194317090263240966264295844904496529408
257545818530451712245853341221263393723768432723230720
515103329074002071715052311921188517711694112906805248
1030230042174201437876795882799700495951702720734298112
2060506852400797464646974283514047912960034431309971456
4121107240906384107080713602857389668033326842302693376
8242401554022346569734957277373367020293169643970887680
16485177252463849850616974698063909409039371206672777216
32971102793766013123528069682762169554984806250807558144
65943702165208653091644379938793040583781740176539123712
131890397485770559872465241024123484115187735702926262272

263786781282247627123283444341321774125623982105548554240
527585535185908269003272813268793160041744985610489167872
1055195015614642567519957475709885543664484014019762454528
2110437921714937194066738649764369534490956113637093146624
4220971624401178506187124696217935963305888398469322768384
8442134810744965248481544185814265715259729139328918487040
16884652745375146969177677958385319007815362963438382874624
33770071738520726882784535090284213170222535296437857550336
67541675972582319654427428527595576649628689331997898702848
135086416936246371086571573749245453917624616142240164610048
270178963854656205728576580886599509071983707240969063628800
540370187673639338568020028549416220617436364394915596075008
1080764895275932531357773790651266846181810628615786129784832
2161578830409172771159015048407402502257497056883482134839296
4323255740532960959204965031024542624302745713070784020217856
8646707640495152752183799930468560488180994624749207541514240
17293807599848767171915339597776071455512995646713694085185536
34588399837414457678926158669230043869328004087857946174685184
69178368950262762028043276285815889655260033764577008357998592
138359876451393217396468470466343383143728118706876248733253632
276726030004521821473700776722109973953872339769196961501020160
553464614212514416308929225023066363240576884249282851071066112
1106954336831970379340913793203825557146818177920343558280183808
2213958890477823852127938272723036775624965174684242828836470784
4428018214583413891148097918076844873912587987055597082225147904
8856237296422360156080638581415232393150491249485417013554708480
17712876327355785059730162653353550076951613049719279725318242304
35426556123733699614598096287753270735204487200935450847054135296
70854719185511658219471734537598882633011496604864684486943571968
141712652247111834419494552999382447591228037615716934559557746688
283431732246400704800091273847134259832866164043409000290456698880
566876319997155481522386883391007248966552505710768262923595808768
1133778351003019106889182438175491956534745366669437050532556439552
2267608124023454501467182219137938830272771443834675150435842523136
4535319092081741578311999123849787494952104308660952399613144334336
9070843872233148307379267618847394658717331459305108996709207244800
18142099120605626916269073979990428655060908602576626388384251641856
36285020993489914435559225444572135985374308573086069566700177588224
18142921872884287519290151464581707330313399970509443178315925946368
9071666624511808910400345103438673333969822827237925786640881549312
4535936156290737030577807237293246501391472834610564992061900062720
2268019500162784802977720961433578167899017127801083545401679675392
1134035461090100545333269152110266542551148919148442297386204659712
567030586049404344588838911751872000576394637198171411035784740864
283521720776879208255521623724305364938607407411060836689233575936
141764074264528122108311895786337364794508748111517983930287390720
70883644070308320044431489855261023559856896258752774757978996736
35442625504176289517353515908676682361229709230873278775407149056
17721714486599209506245643444861426471265485166185085085912399872
8861058110555137126907264467692255880958057858466765392060612608
4430629488905334750345853606476899263141686566920494120582512640
2215364961266550468619037489553835292902172102303802772567359488

1107707589040216781032574087934610477116750460573679242421731328

5538663487235791638778147155461516538912074349977285492798914 56

2769394514635249686196711935624990346120198198543087386744586 24

1384728642826301776502175146759611211392179611048766013544857 60

692380014167489354952997162853363624862130061412994166858711 04

346197853460913910827453376163460822014085158650802663472496 64

173102849919041572089204085450119915798552643297554121757818 88

865533865538130943823407414092547102950313536348534558896947 2

432776740740527013600397200467247313453931928104649254502400 0

216393274355994277644545346944110537716353544017515614778163 2

108199089170862524244446046827298709352870561991353301152563 2

541007705818639548333097100912710749237821759869743974580224

270509982891483237721981983844463975855645354891360721698816

135258056436823350638707708616286288546189914923924728053760

676305607139525412082122126551702945822785762010845476290 56

338160466047467035485352855010987224457310974701033656156 16

169084064262585682464822323373061488001614534198322287083 52

845439477507189235934841096203146818872867905230638730444 8

422729316850725029772785287770493098868831569734833012736 0

211369447473927720789075013719706394150614593427173330124 8

105687118261246463345878691777083119433406700993465076940 8

528447563927645331486099383471565208957530526366717444096

264229768274529173121402654028857410374013773883054751744

132117877292617840249877808160966108134631142291375390720

660604352239855469695271446537517555411276988206117027 84

330309659008310869070576926135102285074699132477678551 04

165158570948347001646759064500722896221879885426149294 08

825811561962692843791148329669473249532051023067296563 2

412915134591825339674250668417666008977351309501923328 0

206462244101152128726463586000297696594338553735099187 2

103233460453195793807900918895881194350000726359618355 2

516178994279077616262850223958067702014160879258435584

258095343146088131743097926718364716139159063359389696

129050594576318727677385370728847790635618843544780800

645267587897966947416108890492566116008290777049333 76

322641101457170128222645463670446639419243668187381 76

161324204482678391368618241047305110417170973925048 32

806639291183858593129568751296934505623600568782028 8

403328779977162614708023148678671729580673133969408 0

201668957181197966425630960854438103174772991773900 8

100836762186907312748625173684770170779604708281548 8

504195228916078211422174334711606449859114603380736

252103323448809929550611400499681022910102832676864

126054516219790376695067816821779410445324181831680

630286853575878943074149666968591547177984736624 64

315150563026403001336480124914143021064674282045 44

157578849632433265567942708171995134270178097889 28

787912088758325152338226769434593790040095273779 2

393964964677241988418369999004605954364650029056 0

196986942487660700333813306645957506854626210611 2

984957013183502032292208068948060182634637033472

492489656964350281457674802333166415498071506944

246250403668474773384622785096151369839412248576
123127989427387203020204084512859765964894371840
61565388510268509674048388238821923505041309696
30783391153421708918997367110606982013817716736
15392044025854581500485270050901501137557389312 7696196237499154270735928273249755634102960128
3848185231035508895614610760524380349713612800 1924136171660720327930628692211941441187872768
962089863901843104026976002080846353759469568 481055820986663022044318829027860993462501376
240533355011202246037574828507649405022633984 120269399764536490526495121250684156657008640
60136061011735929017101414123771805401350144 30068711070601806385477633811100766237097984
15034695817667824131202280280157814886760448 7517518050017372534832871827382623327485952
3758844095600416502032301757343169605795840 1879464583096073368324083800497513773924352
939753559195969242816008361161721372475392 469887413421950900734987411037342928994304
234949023622958590030985320746912585875456 117477170267470864847238467987576853626880
58739914361731217339492137800848707158016 29370621794863501127682520803954493751296
14685643204430696792809486353742316961792 7342987755714821510888856152753693523968
3671576954607147312686484564318114283520 1835830015678441934964270526129690902528
917935777026655106792649385050162331648 458978273107044623051581753517739606016
229494328850380846353419407255199023104 114749760573619690590523968875764121600
57376178361024479002169117061964365824 28688738217619556354538124843023335424
14344693627363436603995845577532243968 7172509072958547515361314366776410112
3586335666117688364362352972393349120 17932083978780514855522024380699246592
896624481348629394431436137600909312 448322381879116523050930042426097664
224166261541959174443071008025870336 112085666072180043680338497419345920
56044100686690250069570745412878336 28022684168645239149486121058041856
14011658996972676632093434704822272 7005987954811366844721904440311808
3503073205568197686698545764106240 1751576216865355975518069654028288
875807915473306553843433213001728 437913861256967559963915799494656
218961882388640921503057496244224 109483417074399031512078546370560
54742946477238801136314172309504 27372092208639043258294535716864
13686355589329342974215992639488 6843332537169582159642358710272
3421743639837246416088360550400 1710910505544850876177770872832 855474595585539272155680735232
427746969199326553111238017024 213878320302941735072317833216 106941578003110096794508328960
53471997927374663026428870656 26736603426597138827801788416 13368603944753473071194570752
6684453088104188364244123648 3342302101915820096445480960 1671188829889773005384450048
835613304410817981273079808 417816096938374729926967296 208912770835670234608697344
104458746601076552126955520 52230553892158993474781184 26115867241889855443042304
13058228768850107074347008 6529261958377643213586432 3264704766165116444999680
1632389276570705641603072 816213085029426530353152 408115765886750119952384
204062494629393487364096 102033553157705957376000 51017929500357585534976
25509541210931096190976 12755058835841699807232 6377673533108925759488 3188908824148500807680
1594490440871269367808 797263234834144165888 398640624616326823936 199324815907790782464
99664659753709076480 49833455776761380864 24917290838334111744 12458926894143766528
6229604184560238592 3114872461024296960 1557471414884237312 778753299628163072
389385445907103744 194697121000062976 97350759523287040 48676479273271296 24338789392449536
12169669574131712 6084972226019328 3042554832486400 1521311775981568 760673067859968
380345123864576 190176856899584 95090575933440 47546361708544 23773717725184 11887127298048
5943697866752 2971916042240 1485991575552 743012564992 371514671104 185761529856 92882862080
46442479616 23221764096 11611144192 5805703168 2902917120 1451491328 725762048 362889216
181448704 90726400 45364224 22682624 11341568 5670912 2835520 1417792 708912 354464 177236 88620
44311 22156 44313 88628 177260 354528 709072 1418176 2836416 5672960 11346176 22692864 45386752
90775552 181555200 363118592 726253568 1452539904 2905145344 5810421760 11621105664 23242735616
46486519808 92975136768 185954467840 371917324288 743851425792 1487736406016 2975539920896

5951214059520 11902696554496 23805929979904 47612933701632 95228014886912 190460324741120
380929239416832 761875658702848 1523785677144064 3047640073764864 6095417586483200
12191110050873344 24382769857560576 48766639226748928 97535477476753408 195075353000017920
390159502093058048 780336596372160512 1560708377116409856 3121487122976997376
6243114983442350080 12486511441861410816 24973585833676242944 49948297567259328512
99898846934332342272 199802197468292055040 399613402135838851072 799244818670187184128
1598525666137393332224 3197123389868824592384 6394390894925725040640 12789070020227601793024
25578716501207507009536 51158585923919620866048 102319477690848455426048
204643567067715338240000 409296357507467531255808 818611161759008772063232
1637259217006164963229696 3274592220988624764665856 6549332015929839205744640
13098959179764857764315136 26198508655340074234281984 52398197902300865879867392
104798756987843166582341632 209602236342169202809896960 419213917417304144910221312
838446724300539768401297408 1676931227532942493964304384 3353938012929610902252027904
6708027141586673633150894080 13416356514628250923595464704 26833317492166309161778282496
53667843910152232952731271168 107338105671943695163811954688 214681047047165848844322734080
429371765500888614722043117568 858762873814891063510881533952 1717564433256009795155353665536
3435206237764474926577888526336 6870567218033860525690139443200
13741443921077542396449003667456 27483506812174727483035456897024
54968251564388740346345812918272 109938979008856051453241424084992
219882909777869244427582444666880 439775723076052771897364082327552
879571253192734109879126550642688 1759182120466725351927049873260544
3518443469095964968191693290471424 7037045394516958465058573668843520
14074407701683973987467521513488384 28149449228668062089635791378579456
56300166107936352408673079460364288 112602867517073161276149152327139328
225210805636547235469904291467100160 450431752477896296775020556559843328
900883787365396245220465060370972672 1801808139549999793781778015244517376
3603697408738414194245251819494178816 7207557076753657601853895216998645760
14415438672060973630434573590017867776 28831526381229264114322713492076888064
57664350836673161935552559608236081152 115331297821775591284919384464636772352
230667787940409717397467299425602764800 461345960474536504450191659843863969792
922712690136507148210897441673044819968 1845466918647882575042823127316723400704
3691016914045501707327702742574714322944 7382199981590476529139518461031963688960
14764732270179899287247262873828997464064 29530129154357691032430977651188135100416
59061587536711166980734859109436550545408 118125833529413903793215525832993661779968
236256983970810947249922666894228444938240 472524601765588173826828564244939132633088
945070471179108906307623589402842750779392 1890183477654082929923180100631614472585216
3780452025899896094462226044915086887223296 7561074192983252658156183777133889658552320
15122488668333426254775830928875211085316096 30245657901400694386478588606965285707055104
60492676932269072526811030712360298486956032 120988076123473512561329768421580051119603712
241981596764817760138074950836879010530590720 483974082565376990306980729661195837643948032
967969943202236920675623115297267308453429248 1935983442547439721474569542544285883237924864
3872053997380811203195785708988074299137982464 7744282219333485926884864665775153663600230400
15488912887810698894756315827148317457848991744
30978522673908851871485804645492655176995045376
61958439144392611906917955273377350876584214528
123919665881935040141728602511538782798356676608
247844906950169712939242588952645727687089848320
495700964272938691190055945764427779554932686848
991424229291075913003253427247128207471371354112
1982893060072548887252789925930801711665754669056
3965875323125891896998145994734694016777533259776
7931929052213372038981424275215569220447114362880

15864214916349920567933113121923500814678324412416

31729143456546194115806755386831726376924840198144

63459714160785094191494569059632902248986063142912

126922282816955600302751254691204703488244891779072

253850274624682024445026742526287204957035314544640

507711967230905696569101951340330005875161691062272

1015446770424894688496300835256171203672505506070528

2030939212775955967708795535663364791189375260033024

4061969769404245116849978801628774350067479015849984

8124122226513156596564733063861638235512415023267840

16248609828435645918859017048931455541779744029671424

32497950407689957289177135940279269225069316025614336

64997362317017245481272475565391254733158287983771648

129997647637309152768381358500447942032355887832629248

260001141281167629148435531740226749196790399395430400

520013974575433905520216692959115228657738046251204608

1040051333177065105487124644875553917843790587423096832

2080149434406524799867631807665754756744210164687568896

4160392404917838777522028651160803355601678309057888256

8320971882045255910617587373980194395429872577481277440

16642317908509668532382234891277564159312777073693556736

33285384105857650487058590069189479055531617984849117184

66572264789391927818705420711647659584875363644622241792

133147522734137109326587322569832722117374982639092498432

266301031778980726031527607432740250129998475977881026560

532614036179374466819761139451630112050493973355154112512

1065252017601574963152934128075559447681981989509092343808

2130551925688801985332691954495717342525952064615752925184

4261199632348908088719031305680631579375880300426642325504

8522590826640424413545357404739656947399712943243557601280

17045564777166065299305304396236101472095330571267661103104

34091895802102563543039787965985778098782470512096414007296

68185324099745992974937934278998706506748559763315011616768

136373713190573717727592585252051713631864357004874390437888

272753556363310899010618603892212028500463188966237515284480

545519372690948725132104074560641259474395327845452499386368

1091063265310551304485941882673716923895728555516859936407552

2182175570478410317415351232452302657685332910685629748084736

4364449220671436051717637399114342935158417420675079246708736

8729094600772102937209144666648161109892338039957797994496000

17458581520402667541966029070135272698935682477130874991149056

34917947678522258419027537613948446356173377748692307986612224

69837464632478363508246034175252694628950781086245731981852672

139678067815824420356873986245216993091109613350213695980961792

279362412733384227394511808279857193848635329055871855996436480

558737379670239228150551288138560803030102862822632640061898752

1117499867747420003024157919434814436725870135067043136261849088

2235049952308723099494426525185014534783069088977641984799801344

4470200338245212385881074423000800392228795815642395394151809024

8940601543745957145546591591263143429782906906659013637408030720

17881604822002979038662068673049372150216444364066472973024886784

35764013113028087572461908327144914881734149829629837342467424256

715296331641004341351993586163821709260708218622534574777701498888
143062480204289386250949801156949024177346688130494480541210902528
286131388160755808463001770162267413005103465072964092253763010560
572275631825865688848207876021273555311027107769878446850208432128
114457697466043952154082442343602456922369457078765741838578168627 2
228920537133829533077046618965900405565066985207111588614229301657 6
457851358671142323691856706489191794570790112513383387102604532121 6
915723286149251162459240350093165556022892509225087193953500921856 0
1831487709912435355069534574415895045808409586846815227403585558937 6
3663057695052736770441176897290917959142068310486912133800338548326 4
7326279940561205661486569291500091653334634894560387625587011957555 2
3663222245508468891045392394209173694192566584073475491786673409228 8
1831652260368167475673750071334150780858907860433378585389920419840 0
915846698991050252912401972781857357310766214415009712443252067532 8
457933633899008383993964454948319662096039249306665066095771962572 8
228971959151245820765863961752855322768347695702912637984958945689 6
114488550676493724767372848015775407244337883376246371461015955046 4
57245560888682269575906857577561576552250959450518211964776218624 0
28623423219558838384063645573617724741166488606456619099522229862 4
14312032997388270990086931179227330603103748743827066108328175206 4
7156177192498561394071019785822899417812126592212911333447617740 8
3578168943151493646549286991016066767036189406256144806365573939 2
1789124645026853298031532044560341912583157758202916973003669504 0
894582409238979886394210296806325220824110406638880771412276019 2
447301247982266561886327285666239742678320967088151528161358643 2
223655645672521590287774711464658437472293365428431335308289638 4
111830333676954949816192890048098501802713123656393453267949977 6
55916422258824552244249212181933892434639782299285619440877568 0
27958838839585814790200989669909266983961501385187256123890073 6
13979733274879676729138686624425793875301555810365851263670681 6
6990023564983223031588439206948477129311180464069037232698163 2
3495090246263303849303767550842028660485791511477574416780492 8
1747584355017498091406657749104909378157996395460315108605952 0
873811793451672129080715861394402213036548517590921504410828 8
436915704697297606229051424118174868497049418725842727259340 8
218462756334379573958872458769574315237912289328112351156633 6
109233830160055172401609602740030598113649934646651669341798 4
54618141076460278911891488047637019304171862314623581552640 0
27309683536446485811489087362629369775759378652960664217190 4
13655148267331416083516215350720114949716413074304768829030 4
6827727383219794630643943510062772505776568411064602774732 8
3413940316386940609764889672382743768347465142488410567475 2
1707008470581991952103903794867050641903323039722259873792 0
853523391485256799662681376771364699816456754100157231923 2
426771273839758811636705428054602039340625994169592263475 2
213390425968444611721035083861760864386511805644552955494 4
106697607508504908811858726848110354551355307102154889625 6
53350001016393755881599955882670138454727355691016650752 0
26675599139267528678635274170642549816888528915477928345 6
13338098885169089708235285199975015203206689992723765657 6
6669199100352207538576466657314377748984557763854283571 2
3334674379059935111517645357320623948182885265673342156 8

16673746039718832268735286929920295109367458247097712640
8337060092068994489941173536618735238910245082914357248
4168623582139286422757351804138661461568380521139929088
2084358559122037800272058419983977651840819250411339776
1042202663587216194582720468949312286448724120126357504
521113023806706744514705863953317873488519307523522560
260562357909902695869025746715989801876338277491933184
130284101958226009740349280727660333504208450611052544
651435124807503357730928440486662883035123881238069248
32572486991193833338005523866747799659071768585306112
16286608871006249394732312854582078900290798275788800
8143487123207791060230931887895128985522856129462272
4071834905456228711547853674249609260450156560515072
2035963124654280946490120702275827014069442528149504
1018004398290223768603157283713424698956903388020736
509013617126653531980627108144467945439542755983360
254512517554097589829837787216111770700316908978176
127259113272434206834681010179994784340431219982336
63630983883909809377221563375966841665351992737792
31816205565801257668551310830968145580244187742208
15908459594823805324245919986976435163906189557760
7954408203373490907108092279234398768845142622208
3977293304667539576046612282490289977868595232768
1988691253824166849269589212681690285657309577216
994367927657281955257936142059117791190160769024
497195114201240242940538838888695219775833374720
248603132286919754126054803373915771978293182464
124304353736609693390920093651741967034334838784
62153570664879754859406392808263024039761543168
31077482230727331511676369395327532281177833472
15539089564507392796824771193261776271237447680 7769719006825559918905678844429893200942989312
3884946615698711719699486046114449133133627392 1942516863992321739973066335006975832897880064
971280210067643810048194823478363549614473216 485650994069563375054928239726619591390003200
242830941552652422542879533857028703986384896 121418193035261578779147473925373806138884096
60710457647098473143427590461116630142287872 30355909388283078448640721979773178607566848
15178294976508460162783824364494021071994880 7589317629437690550623643869550726420103168
3794743885310575509927687778427221152104448 1897414477951152872271776811039539547078656
948728506623508994789854866432734259052544 474374887135720776721910663672849372282880
237192760479843528024446947064665807519744 118599038695913333843969281146453464449024
59300848575952451837857544380287012569088 29651088901974118376865224093673646456832
14825876757986005417400837998601893314560 7413104532492475823184531975183481700352
3706635342995974468834322475533008371712 18533359209872855513038189481737137946624
926700374123861895829608862853885853696 463360571655648017570061492419601367040
231685478124682543612659276706129903616 115845335210770539220143903601229561856
57923965679599903316979084424697085952 28962631876907268511943108524389695488
14481640457007292682698337418215424000 7240982487780475554712560287118000128
3620572373528652384037975932564144128 1810326751583533495359835860784644096
905183658201370399350341877643608064 452601970305487025510382912447447040
226306055755144425672797443036545024 113155563178772669295201714924683264
56579049239986562877002354165547008 28290158445293395553201925434376192
14145396135296754833951336892989440 7072856523973405945650855534395392
3536507490149217237163021311148032 1768293359155865750750307427549184

884166486618561441459552099762176 442093146829595003771975242874880
221051525174954643407087217934336 110528238467555892464093407215616
55265357173817231612321602732032 27633297556928258496298250928128
13816958263473950593217850245120 6908633874241885969143287513088
3454394308373398320838824951808 1727235839812926828553003073536 863637262719577248343296835584
431828302766345541205046067200 215918987086451229119221858304 107961911394864843817960341504
53982164623252036538154876928 26991686774535825583664791552 13496145618722816449126072320
6748223925088860053209874432 3374187520408155940928356352 1687131539135940927625887744
843584659033901942393798656 421801774249916710487326720 210905609491441224888877056
105455165928962047267045376 52728763556101741044826112 26364972073861229228064768
13182781184835793966858240 6591538166370486659842048 3295842870161538168127488
1647958328568916503166976 823997611028531961135104 412008028886302835343360
206008626129169845059584 103006618907594136223744 51504462375301674958848
25752807648403140902912 12876692054577722163200 6438490142476936937472 3219317128832506396672
1609694593213272162304 804865311005145563136 402441662701827522560 201225334950541131776
100614919275084251136 50308585537448968192 25154855718677905408 12577709334315663360
6288995404646187008 3144568071067271168 1572319219905724416 786177202138906624
393097397162475520 196553096627748864 98278747337129984 49140473180192768 24570786345910272
12285668050862080 6142971464384512 3071554451668992 1535811585572864 767922972655616
383970076262400 191989333098496 95996814032896 47999480758272 24000277250048 12000407060480
6000337747968 3000235982848 1500151545856 750092550144 375054663680 187531526144 93767860224
46884978688 23443013632 11721768960 5861015552 2930573312 1465319424 732676096 366346240
183177216 91590656 45796352 22898688 11449600 5724928 2862528 1431296 715664 357840 178924 89464
44733 22367 44735 89472 178948 357904 715824 1431680 2863424 5726976 11454208 22908928 45818880
91639808 183283712 366575616 733167616 1466368000 2932801536 5865734144 11731730432 23463985152
46929018880 93860134912 187724464128 375457316864 750931410944 1501896376320 3003859861504
6007853940736 12015976316928 24032489504768 48066052751360 96134252986368 192272800940032
384554191814656 769125563498496 1538285486735360 3076639692947456 6153416824848384
12307108527603712 24614766811021312 49230633133670400 98463465290596352 196931328627703808
393871453348429824 787760498882904064 1575556182137896960 3151182733019971584
6302506203528298496 12605293882033307648 25211150714020036608 50423427327946915840
100849106455707516928 201702716511042404352 403414440221339549696 806846894841188581376
1613729818479396126720 3227531694552830181376 6455207504293736218624 12910703238963624148992
25821982938679551721472 51645118798863710289920 103292543440736634273792
206589698567491695935488 413188620507020246646784 826395687758114202845184
1652828269004375824793600 3305730324985046487793664 6611608223922682652000256
13223511595750544656826368 26447613487311448019304448 52896407566243613449912320
105795176315728661722431488 211595074997940193090076672 423199594728846125470580736
846418078923623729522016256 1692873936779110416205742080 3385823431421946746734903296
6771797978571345322116644864 13543898188597594301526966272 27088400840104995917641285632
54178010606029606464457277440 108358439063698442187263967232 216721713830675342891226759168
433453099067907602815851167744 866925540948929039698497634304 1733889767524085747530585866240
3467856906300626831328352927744 6935868555106164335191068246016
13872046595222150015450861273088 27744712160463942721039172108288
55490662260967170822353243340800 110983800402012912405256284930048
221972552564182966331612166356992 443955008648680215705423525707776
887929824337988997495245437403136 1775899262757235127159287646781440
3551877753676984518656168837513216 7103913963678997565987524762927104
14208144840008052189325423701655552 28416923505316218493351595754913792
56835114661232665216104688213032960 113672764623665786891012369832476672
227350599849732486699630726477774848 454711340904266799234473426581192704
909442964218137250139370800413671424 1818926493255481803619589495329914880

36379341161493782139208747796649738244 7276030491575585641205141137340235776
145523855017048297091370654307010478088 2910542004051697627172769717344324812 8
5821213815524858625036252697096880128 0 1164268724589264399145393191901022126 08
2328589372147114146567071688765336453 12 4657282590231398989686713987457257308 16
9314772872337139372478569194767683420 16 1862996112842296153116742082924170444800
3726075302434328863475540653789608411136 7452316758368130841435194283461751865344
14904965823735207911838614518688573816832 29810596261468308281613680940907287805952
59622521750932401479100265688874855956480 119247701957856372789946338991870272602112
238500720827695885243384293211981666582528 477012075479358049813751816880445575921664
954045418606648658281470094673855637356544 1908133372509162433870873111173640245739520
3816351815610055102357612065999138433531904 7632873772403570673946955819301992751169536
15266087827174062286357375013211417270550528 30532856219081966449641676775637698077523968
61067073567631616653137207049705123227893760 122213686939419860081398212109626970060147916 8
244279183306267936643379656186258309494341632 488569255648277343317590140359954435571449856
977160289368037626696841936694784504308432896 1954364134879041133517007185339320274947932160
3908815382044014027280660994578143082557997056 7817804988659891575054615236955291230440259584
15635958426463510191095816969508592591529050112
31272613751214474464164806930213205444355162112
62546621299003857092275959842818451411304448000
125096030191157530512444611650420983867797143552
250197635568614693680674607230410129825970782208
500406421509828652672919982319956583832694554624
1000835143764855835968981500358185816026895089664
2001714889020108733184246072152916928776802140160
4003518981021011588861058287178924450999628201984
8007216368003611422707248860104030088891304247296
16014789547930399335384762291700422551566704181248
32030292719707151650710053726385569850701599735808
64062012687107009261301165738740589196539582218240
128126879869599430442364448049420077383351929929728
256259468729969684724253129242717952747249390845952
512530355441481017127554724773191501455589843664896
1025083546846045329613206382121894194833361811275776
2050212765618257249942606629394810773511087870443520
4100516875088847681317600989091666314710904236670976
8201216437882361725499977438787422164799265464909824
16402798251174056176729505798783023400353444912955392
32806327253166777804918113439982404942216717792182272
65614116007970886512754430564797526167453091516907520
131231155019216434831345268499260484900945494898900992
262468156044982193274363351737851834933969613527973888
524948004103063033772072332954365400132096474516291584
1049919392232323361990835924866054260792507443953270784
2099885552517041312875054367646755442641643877747916800
4199864641138871803536873771122804727396545735178584064
8399916354487321962647277613904197139019607429722669056
16800206853393800636441615371125569646492246778176339968
33601161995625914695177351028885490029890557393814683648
67203820568928456234942942631039681533593242462553374720
134410634293210166159062366408616766014810740274954764288
268827254897126839696477695110308337924869991249605558272
537666482415666694149661314806766287640237003898603175936

1075356910074159417812734478785831798861468050595990470656

2150761710633970894652292655916262044884924186789549178880

4301619202239245907358232708521720984093824544774234832896

8603429966421100050823760210421835756835601431938742616064

17207243056727416573862110007600459090967107548658031132672

34415252361225266092153399188714493336526024466877154066432

68832037217991398073165156724456136982235667672876491735040

137667139427064527924047030142966574582838572823997350674432

275334040883629251940352749367404175040241162060448343575756 8

550693077636911965917921854124300703278292191121944340332544

1101410675202477786057577441801035811503522282069843618299904

2202870390262263280558622350706940432900920363791597111869440

4405838860239141978004179635623618485589592326887013974278144

8811873879907514789782229139666712210754687852381667449634816

17624140078673491247112198016172374900660382101978613901426688

35249064795063905829319875506022650759622776998387785807167488

70499698865561658328830709959401103435849579585636687622963200

141002536281991009998043337813513810704907210348995607263182848

282011349665717406676850511416450829076230523053435678560878592

564035253534905586715228694411748073485293250817760285190782976

1128095615476752720153512731981188977636250911057298426519617536

2256241447767388533753136150277763616603830640958152565315338240

4512583329162543254398493673186298555870318919603416555182882816

9025367525580618882581430091634139757065953114581055959470178304

18051136785672302512731745673791364804782536779910557617149181952

36103077040366734520601262328628900190866334661318006630716014592

72207761018777728031478066619350141544335191525629796054267330560

144418735913643974043507217162884965413875427457247157694205263872

288843899579464984048116602174139295478160943726469446559751733248

577700654663284040018437540045017320257142065076889155462185877504

1155427020335276223881283751483512099115924485401678835609736577024

2310905462687968735451384845753979115435129681299158720590202798080

4621913769410770046280404377081868065276820783589919539921864884224

9244033226891205243316078125311555799366764409163043277326648344576

18488477829921740788142694992918750936359774502292494949619133841408

36977778412122142179306467470428780547972040372517806689169941987328

73957202328801605556465508991004011844644906348090124695820323258368 0

147917695666717853541394489758445351593908092433533761076133162385408

73960493337916247976739399848405233147459028952632514117929929801728

36981069421236784591390777408793895248982005844249073848896639205376

18490946086757722597205927446692586962117248606090945319414156754944

9245678731448526449358233094494113149871747145028676857189996953600

4622942209759095800056751232820966409342434993505940527336457764864

2311522526896964187717192959197438121874498207248771313038958526464

1155786974457190237703005150992196519538889458872286181204844085248

577906342732949190773706911192836989070264907060093352945104453632

288959599118651631347955623444787859185542542342021807643893432320

144483013435414333654528895646578611917976315576998469407617318912

72243113655575142581753998978538164712159067999149301749664396083 2

36122360296897842403907765873736994142096601097243400144739631104

18061581882959985949522768427391582361698931099370145770578640896

9030991808735525348545826958957333826174780825059295734393733120

4515596337995528861165134852109438235750048050216759291749072896
2257848385811647524028678112370104779206352843951935358150639616
1128949301312765308737394399342745220268840831397745535213371392
564487204859853427730224871250219025467252620409761695675711488
282249879531662100545876271414532720400042412560325311872368640
141128078316698743613320053601977964033229257457884887953440768
70565608433783218476850985748344783933218654317803559985348608
35283588854608532573520972347850292924911339953332337996988416
17642186746162727954308225910764096941606676373881448000651264
8821289532510594810927982823801523710378841385548363501404160
4410742845969912822350926346110499474977172292078001501241344
2205420462842264119618930640160118547382461945690910625890304
1102734751349785914031199053632493678638168872671410250579968
551379635639219884126466393592464041792553386248682594107392
275695947801773405618666630184340622133011168080830031462400
137851038891968434587050031786224611684872821518659382935552
68927051941525083182383374240139456151620029498451875069952
34464292218532974535620866293583303230401824118787029336064
17232529233151703740025022733548439192496816744174060568576
8616456178518460106119806160152613384896360714477303234560
4308323870230534171113550476765503586772156528433788092416
2154209825600919144583598936727350240548066349814462283776
1077128858043285601805211317535974343855027217706015260672
538576401643055815659311583354136783718010630252399689728
269294187132234415208008753970143197754253825825895874560
134650086721470461293180858131609001824751168262795952128
67326539938411857491178669639073202386187711806321983488
33664018258044242167883455106170951929999919740622995456
16832383273441277795088787696402651333452991789042499584
8416378708930217253117923919859913350953011853886750720
4208282890569897804345726995759250517589763906626125824
2104188213337343491066246015794272179851510943154438144
1052117490694869039979814266854459550454069966497906688
526070437360533167213252762905891505491192230709297152
263041064686815907218299196192276617877674739084820480
131523455346682615414986005465803741504876681407496192
65763189174978638610411206417734587035457996636291072
32882325338307984756664705051283651659238826284417024
16441528044563325104061903446850004900374327125344256
8220946709986328914895727184029091985564620554240000
4110564698845497638880251322316590760471038772903936
2055328021348915410156319526309317764079883634343936
10276868466637541000436256695730170073962123941117952
513854841300312147897176814152840632942153032531968
256933129640926897788112640220298114451622047252480
1284694193158488860813818436682087956216083789119488
64236136905617136366790276627013427603178277306368
32118782076654921163335667456491438549157330026496
16059747850250637071638098299738081648362760699904
8030052331086906780804181435615222011073428193280
4015115368524247512894656860680701598982738018304
2007602285752520817693611501776896096214380969984

1003823443621458939469947286606720696468696465408
501922872183328735046544411162496672415101222912
250967011277964000179057589510816498297927106560
125486293232131816417421486720192330194151800832
62744540412640816372657089342488205619670024192
31372967104607862268301717662440123071132073984
15686832001447658175137445326818071666214567936 7843590225295692608062015911208040898431549440
3921882224933778064277654579503522981877907456 1960984668609854912262150601701512757270020096
980514112376410396192736956825632011800543232 490267945223946668127199306400253822483038208
245139417129844069079015067193845819532902400 122572430823857402047215240593782363912142848
61287576541396384777461473795320909028917248 30644468835432034265657663646875318050881536
15322574700082938071292295198045090793652224 7661457491224929504877879286326261280931840
3830813816204194987054805486814988582518784 1915449443397962610835335665233423262285824
957745989346913864071634293529676116656128 478883628497423211362800377221320301084672
239447131160694745344891803838901271920640 119726224036338942504191709533571196649472
59864441246165256167968758573845878669312 29932885237080520541920831190453079506944
14966774925539206499928641546991609839616 7483553616269076364448433749378339962880
3741859884884274739466273362630437502976 1870971480817005648354164925285852512256
935506509595936963487596584628243136512 467763639391685551399055353306780008448
233887011992701310527156207149719224320 116946102144779922677392368823024222208
58474349146604595045603317035594416128 29237823610409614376255224829838360576
14619236323758465614854395570939756544 7309780421156062020790589363480166400
3654971340216445617076990470745227264 1827526234927430111879343129875185664
913783399873318707610095512188878848 456901841141461179640259729720082432
228455991173131502737735851672862720 114230530887766207827670919242842112
57116533094483332143236956324626432 28558900372541780186319226513915904
14279767098920947150509987432759296 7140042005785502103930180804280320
3570100231055265316302683946090496 1785089729608889790320138745020416
892564671845073461244467758497792 446292239442851013664433072242688
223151071481582648353316132618240 111578011620869894937207864557568
55790243750474232848878831403008 27895740845256759114576865263616
13948179907638200902357157412864 6974244696324011123712941096960
3487199719414460898123651743744 1743638545333458117195416469504 871838615479842892664503533568
435928979146478363365649416192 217969325276517640199523532800 108987080489898049358111178752
54494749170768639308230295552 27247979048294126968702500864 13624291755601967141644926976
6812296993528435399469301760 3406224054627943614058070016 1703149806245834764190744576
851593792588848860676227072 425806341027390169628540928 212907892880177954459484160
106456307623330412052348928 53229334403285923437477888 26615257497453320424390656
13307923896631839565021184 6654109522268509458923520 3327128548110549567668224
1663601167543422202937344 831819030515784811020288 415918738629929260285952
207963981000983057530880 103984296343500742459392 51993301093254978076672
25997227007379792461824 12998901734066047942656 6499594982221099827200 3249869548704587841536
1624970803149312884736 812503415973165924352 406260715185837703168 203134861192546222080
101569682396086796288 50785967097950240768 25393546498928541696 12697054724440981504
6348668099708846080 3174404418598600704 1587237393671389184 793636289021739008
396826940603891712 198417868348456960 99211133197484032 49606666110369792 24803882810998784
12402216283406336 6201245580656640 3100691509805056 1550380114640896 775207237189632
387612208529408 193810399232000 96907347099648 48454747291648 24227910516736 12114223693824
6057246064640 3028690141184 1514378625024 757206089728 378611433472 189309911040 94657052672
47329574912 23665311744 11832918016 5916590080 2958360576 1479213056 739622912 369819648
184913920 92459008 46230528 23115776 11558144 5779200 2889664 1444864 722448 361232 180620 90312
45157 22579 45159 90320 180644 361296 722608 1445248 2890560 5781248 11562752 23126016 46253056

92508160 185020416 370049024 740114432 1480261632 2960588800 5921308672 11842879488 23686283264
47373615104 94749327360 189502849024 379014086656 758044950528 1516123455488 3032314019840
6064762257408 12129792950272 24260122771456 48521319284736 97044786053120 194093867073536
388196324081664 776409828032512 1552854015803392 3105776751083520 6211690941120512
12423656760147968 24847863276109824 49696826063847424 99395851150950400 198796100348411904
397600996789846016 795219585765736448 1590474355903561728 3181019080551301120
6362178898590957568 12724639272158625792 25449841494270672896 50900808888448188416
101803869576710062080 203612242753047494656 407233492705349730304 814484999809208942592
1629006028415436849152 3258084114424911626240 6516312344037899108352 13032912918451949928448
26066402297656203280384 52133957516817013407744 104270220876643240509440
208545053439304908406784 417099330250646671589376 834217107245367052730368
1668471107978881524563968 3337016002934057887334400 6674179579820705451081728
13348654307546590254989312 26697898910903539215630336 53396978413427795842564096
106796318010097026507735040 213597358386676922660683776 427204161506319584611794944
854427212478570647804444672 1708892203889004252770598912 3417859965641734419864616960
6835871047010920668376072192 13672044325476744994045820928 27344693113863297302678994944
54690595153546209234532696064 109383608158731647727414804480 218772052020741753971528433664
437553775448040424976454516736 875126893709194684019704332288 1750292473044617036172999262208
3500662317341689408613179719680 7001479377188289489760721829888
14003268239386400324590168440832 28007155448792443339317786443776
56015548837624172058910472011776 112033573555326914878370742272000
224072098870810971277841081040896 448154101261936225597881355075584
896328009564501017280161096138752 1792695633210259166729118964252672
3585470494583032597795831472455680 7171099445491093724266850032812032
14342515803632244505884074241425408 28685665432564603126468896834453504
57372598515729434482339290372112384 114747732332659325423481574150635520
229500535267719563764569135114092544 459011211740240953364350243853828096
918042705890085558399124434958942208 1836125976599378420139096764420456448
3672333082837171446959889317846056960 7344828424951172107283170213702402048
14689981368456002641293123583425380352 29380611774019322136039813478891913216
58762521622253277978986759581866131456 117527639392935823371787784411896872960
235060471082730181571204099320122966016 470131326759177432797665259632904372224
940283422705789004905844641251125624832 1880608383786446288432717526472885010432
3761299844322629134107491540887037542400 7522765842144731382699096057656610127872
15045863991288408994366418067078290341888 30092392596574710446669288037686720856064
60186114421145205809211479882433722056704 120374887298281981450168767378988004802560
240755091508547102563829149986217130983424 481520816841060484454641530428916504723456
963062901330053527563249521770797494960128 1926168337955972172434431965367523960946688
3852421746503674579484729774386905863946240 7705013634190809628201191236077527611998208
15410367550748540194865845846762486992207872 30821415666230922266658618442739837520838656
61644192461929528287171090383909402114523136 123291107182794424082049887764678258374737920
246587658883459583179515189523075425040859136 493186206802660636389861207033588666664484864
986394191676804212841384070042052966494502912 1972831939496574305806091452033857199320072192
3945750991279080371858829527967216931302277120 7891676207130024264210952303733438927928819712
15783700863403775569408491103064887986506170368
31568098625095005220790155197325796234309402624
63137591046764918605526656377043632991212929024
126277969686679653538946004718871347027614105600
252561514559658939733677393367310856145604706304
505134179491917144778925554593758036471962402816
1010290659729032820180992644905788721305430786048
2020625920948462701608268361248122739333873532928

4041341044877719525709102865369336072113770987520
8082860495717027296403338016484853331119589818368
16166077803357231082776940604462069036023275323392
32332869230560815145494410351908862819614742020096
64667165708814336250869878989787175134365866786816
129337185913014084421501874551513249259004499066880
258680080816798992682527982246904296498554529120256
517371579615139633044104430781564188958200120213504
1034765995193362561446305794138639569838582364372992
2069577662312891713608805453428301523521528976637952
4139246668478116608649998637158647814731786449059840
8278676024660899580164772734921385164841029889687552
16557717424731131886059096391050949400436973762510848
33116165600280929223577294624518256942383775491293184
66233792702199189350072792933869230167787206915129344
132470508407673040505981993237403892901613725695344640
264946862821895404623636801214138650935306075120861184
529905417656889456470619231906939032134769397702066176
1059834219339976207387929722771201524797853290324819968
2119715206732347003669241963457049970652335570491015168
4239523949569483185125248962743393783417929120664780800
8479234971348544725824027997145375251062374200695062528
16958875140871162461627951161376079258705777803201211126912
33918436463070805747884352561850202478061624477704257536
67838369503818238340356945696969106429935376630332522496
135679732162989730369890372540475615807495008610513059840
271365450636685968118133707374026037510238527920722149376
542742873894784950992973339334201686810974077240836358144
1085509693032395931499358527840702597202942197280456835072
2171067276550443922025540754026003641567872480158481907712
4342230334072191962104728904741204177459721131512100290560
8684652230086992160316752602860802143567394605414473531392
17369687584059200792848094792478391864430693895609492963328
34740141415888834530125368758470358883453197160780077727744
69481815327318534949109095863967868076090013060682339057664
138966695645718801675934908421990036770547263599609045319680
277939521273601066907303250232088674777829002155706825048064
555891302511529060925473367240394552029126954224391118913536
1111807124951711976072680468033223509005191808274737175461888
2223663289760731660588828403171315827904259416201384226193408
4447424659236078738064591740552369275596270431706588202926080
8895045477901388309903053349524213790768044062020815906930688
17790483274661238287353846435887378060687094521256910816018432
35581751187039399909803172345452657079676201836944379636350976
71165071649512646489797303638261116075956429262749875281330176
142333281849892986319976525171233835985120909703221982579916800
284672840801521359320716886131890879636657921761888429194346496
569358235806513492002961443842628174606148048234665786457718784
1138741580019968530728978230842949179877960505891109429053489152
2277533376853820154904067148001284021087249830625774570383081472
4555167187335406496700355668633339364837157298938660565318369280
9110535241926345367185154082528221374999629873251543979741151232

1822147221836375548193919365557952804064989029725153365769112780
8364437479057496404590161582922052266626010416959999587117999063
04728891027495435399083078585465027944878046055949937002164351139
84145781419375175597797166801017190271300814255595974966018540830
720291569266502528231555435769882749907252038600003925063208422866
944583151388509410535033075875462238543804897377631800388759528144
896116632848802752921391056042231795454621143511051150130220442111
1808233270839807247471550993818742286400962615093151880365377957186
7648466551964017978200639751106041963785365886328402920940630060302
3360933124496842922916355029149198709537613084941004162301008412462
2848186629013129977886286111217262698300898879445040496544151340863
9795237326625378274237860243320937130938855028380376032125620199847
100416746548962611057969265287968434444350605617434879298848202630
428098561493130835313262362651419036252539848221334524475910368002
527828377602986327490808817573544524271272381990462868358386446079
199589601116161493196655495555210893105235019842142241533833910535
711197061772738567466147827933508175069741672017466446258167443139
24191397164372459523733156189194620147836978584467861090654332858
36290263597898929274881866619232211243104069543166463494479089790
99757809215748607836160009333301849125880671102985203465292064262
07782987366498491331038412846667537685977729109291272873065558665
37600335928434591198114480128233342830631630274315338098644023285
04720808784600183449697868840961166739864166859515421099164613593
88383768079447790969717025826406458338278758778382963275391800353
56712196605748629051109278115430402916978215460689507774791268501
37200260240376243427686635246977024145852124649122993369290647349
25328245532523252770140890329409126472927669262605755674920865636
71898239026513846684448724448234700836464638100325007332598203799
40566177643383033491913501865882419218232720784673568413867987390
22591617886754571820801320753823744091165612595923165807184364403
74500734759088134478229452873531392455838106342392447725143959281
80216900422017049262261509889720322279240748525845332071830482724
39650635242967130666878777058918411396454826698642127589705845198
91083841879245075112250023346176569835295538402879741012963838791
95725377182724844505308069888028492392487093682655127031770881918
62933020159796669905748623361424651009863361066160170767491211969
79859059167555727304687616712341197686018999781994973219164004065
33555172638979660972032356178445220178733241952281346361011615687
90380750047834800128178093145798673983296453538041570010599354015
87590302919557120890485344936292999859654677049797576954320399240
27909608570884452524804396080416185208319458725504559353595505215
23096780822263114420553479165360716268342315621735525974045174907
53536111318024095632781248976954696955018603371419852821368301158
45566023804424908333557516402609922954903260491770843103232002783
07320203408880233309253518604748981897499545030889570304139156725
09278617189437134345335667510931598725075981198950469579895041933
95183604403007370548786384161236450208959692834790713768737408862
45119421036631908651261555181213659955217395740008253920903440186
69193994712055221246068661420032086980615660695686878273881393483
67348924058572733580050432434912656400608846196734146636338056878
60054575619265003522174611172488696290010494431526288731554990814
378531487744

10873295314871741745186590649354435893584894499880498626556
5436767383650001020160354570538714064697417463934169907200
2718443554932065583863706908200105091301193838964045250566
1359251709019565328823618265565426575126839472980502773766
6796408202865489328576915385154003023015410132394914611200
3398278930316576006517869721240436585198311449943657676800
1699176879600203674373640874951935829444458916844929024000
8496071470210596727441734446418266831448810543590146048
4248129271210087541507632259038427257837663251477823488
2124111403657438359647198647433860549975460615580286976
1062079085854916474270290582674253735516044802710831104
5310512349405568843584909208157885980221796488157593600
2655314634768277657909182751472251641431684481380515840
1327686547416885447012955449432780146376235359341117440
6638578887248160325356597615647172360183142389959884800
3319362518705946707824208992065221994242553991607091200
1659717796893906626485059588153428904196768394117120000
8298771672174199495290073401371234056361298962153472
4149477179939432929077424430987661795869377976860672
2074784261895883055254906080644853281779053236322304
1037414966911024822985549972897937832811708742107136
5187189014370540591718234527367245123669454330265600
2593651597092978534254359595122400541640182474997760
1296854343500343386324800963280589260722818892431360
6484414442270987527612110644999891253127732736819200
3242278583520129061800108236798418101320685505740800
1621174972952382179897080575548445288038752321536000
8106053270723499144470535163488407627085809451008
4053115838342543694727833724617294406988928647168
2026602520661668908610199933745192500217476284416
1013323561076032984928241502590868898470244122624
5066729309106157577756915191545707734158750515200
2533420406416075115436311435068535487983140208640
1266738079139535720997082637182108554443452579840
6333829775355169421380047784149746824476675276800
3166984577506330118887341191194475438368043827200
1583527133667537763542329245157038732248875008000 7917809892909552338204939473584198726568640512
3958992058740707929349116360691601895946452992 1979539585513319844797881492295552214304292864
9897915708281428624606024021226517403176796160 4949066744498129012611320290487636867416064000
2474587817427771856459814285181007516621864960 1237321131303239603306984212559098299767848960
6186741769462966391920306412638464206123827200 3093438941204867383652845881240718456704204800
1546753498839125785672769278081102405173248000 7733937635379089397595578077709227909971968
3867053888281274933413654882506471897038848 1933569479436502584014760363079164919545856
9668060073661838506613466424525469452861440 4834136375070582046576565516827557153996800
2417121356655122419923198910696189790781440 1208587262887476908279057531489300502282240
6043069237236963032982578038152530545868800 3021601080018270762284934209429279290163200
1510833770709030004039289699891146653696000 7554335007044623134680561475338268311552
3777250580272048124582337225610401677312 1888666828510892340912196856775834599424
9443541834428803097666125503732341800960 4721874763151572245385633361792755302400
2360989304544371470969101985859669852160 1180520613756478409622693645411481026560
5902732876203855418804181489465635635200 2951431341812659394747447375936933068800
1475748122761695540046402003570524160000 7378902873085306913595401595862908928

3689532566181068063479396586936598528 1844806847909741335080546187970871296
9224237063644743192106970412367216664 4612219943870389854405604942440003840
2306160677959204056378862339348234224 1153105691991606592777461103738222464
5765655225018055786827455189011660808 2882890995039039304883802429666099992
1441477188784525358176938632413184400 7207544400247655319559880249966592
360385142828634192411753366893363632 180196532822442809422756360644191984
90100247115284261319818018920857576 4505111390967355896412892875980080
22526052130852493634174424029593536 11263273653434103893142191839648416
563176062072098048459858583224320 28159422073624545113130378723328
1408002052182209390163391414272720 704016500341595762335131946188888
352015987296043414794284092620808 17601186221064447421050110607336 880078653866336205119300829184
4400489983397250195930480640000 2200293348731409683132228567044 1100170852882097134149608407044
550097515699244713366551265288 2750548024787204298291491635252 1375304235539092514875113472200
68766722934229144030224056322 3438411704575183115834621952 1719243631219454515079020544
85964070507565873612036505606 4298297972707951073506099200 2149196210018804233205186566
10746217168418164648286617676 537322664337115406527365112 268667235126661290320199688
13433656904238243868835840 671697602607171161083084848 3358561800012150643621888
1679317793494222740914176 839677343491185080008704 419847895117629394780160
2099285592448331247779848 1049665854654257760829444 524844456542174948884488
26242799287861050867712 1312168787430667714556000 656098805234141442867272 3280566083764745142272
1640319070679391535104 8201775497382052495366 4100977820683573657600 2050533946338060533766
10252894911671671193636 512656004582651985922 2563336317908602060808 1281696306451972096060
640862226974821580808 320438150361828556868 160222593618123161616 801130560276660224
400574076231352320 200291436162187264 100147917104349184 50075058063802368 25038078787715072
12519314271764480 6259794574835712 3129966006894592 1565017363185664 782525861462016
391271520665600 195640055300096 97822175133696 48912161308672 24456617525248 12228577198080
6114422816768 3057278517248 1528672813056 764353183744 382184980480 191096684544 95550439424
47776268288 23888658432 11944591360 5972426752 2986278912 1493172224 746602496 373309440
186658816 93331456 46666752 23333888 11667200 5833728 2916928 1458496 729264 364640 182324 91164
45583 22792 45585 91172 182348 364704 729424 1458880 2917824 5835776 11671808 23344128 46689280
93380608 186765312 373538816 747094016 1494220800 2988507136 5977145344 11954552832 23909629952
47820308480 95642714112 191289622528 382587633664 765192044544 1530417643520 3060902395904
6121939009536 12244146454528 24488829779968 48978733301760 97959614087168 195923523141632
391855636217856 783728452304896 1567491264348160 3135051248173056 6270239935299584
12540754748506112 25082059252826112 50165218017280000 100332635057815552 200669668162142208
401348132417306624 802713857020657664 1605462898413404160 3210996165570985984
6422133068630327296 12844547612237365248 25689658174428151808 51380442248763146240
102763136297339977728 205530776194307325952 411070559587869392896 822159133574248267776
1644354295945515499520 3288780649485068926976 6577705414158213709824 13155699058692579131392
26311974578137461686272 52625102077779530219520 105252509998568274132992
210509631683154975653888 421028486738346806083584 842075420220767321718784
1684187733929682062540800 3368449254835658963288064 6737046083623907602989056
13474387315152994558803968 26949364926116347823259648 53899910443853413057822720
107802182070948260938252288 215609086508379391521718272 431227617749724522333863936
862474124965380523248582656 1724986028862624003658874880 3450047615588973921641168896
6900246346905399671929176064 13800794925265703001152028672 27602194313441213316891410432
55205597552702041262957527040 110413612957043311784264466432 220832061617365082085227757568
441673794641287081203853164544 883366932095687996474501627904 1766772549817603661082593853440
3533622470887662658432368902144 7067399684280235989399100194816
14135108853570293323866925170688 28270836677160229337871299903488
56542911294359744056017498931200 113088298468798058872584796110848
226181548697753259266269188718592 452373000915820801574737570430976

9047658088722701692338735268495361 8095712318257974706365438256742400

3619221691814109205610681195298816 7238601839953246939896549478498304
14477520592556550937143473132797952 28955675010413215988987694617198592
57912617671426660207376885937602560 115827770644053776873556765281615872
231660611890508466664719517376053248 463331364985818759164651008377749504
926683012381241169999725964006785024 1853340658958168964334029982251614208 0
3706894308801793893362295434037428224 7413950876880417000087982446085144576
14828226272314492426902748048190865408 29657101581736301707259062408422883328
59315501237687237121425257440928071680 118633598623803741656664780130020753408
237272389544466018140958090756370726912 474555163682649105937173242505399894016
949131096552732351184860606996116668416 1898303731480332980990749457962867097600
3796690539710402519223555403867001716736 7593547232920278152931223783616538476544
15187426772839502534830673518998147039232 30375518159676897527597798941526434250752
60752365547349579971068501690113148846080 121507389550690729773882810994346858381312
243020096013364599211257237216934838140928 486050825850695477749497704890351919038464
972122919349323514152961870693668323590144 1944288373994512145613856663213265618206720
3888661818580754525843579170078389178466304 7777493778344969520918890027460494241038336
15555327839056859980301243429528420250288128 31111336242847561837529413608271704036999168
62224033615162807428912680714973135146844160 124445078948926098236553306842680572443937996 8
248907023496392699746481550847330357170143232 497824936028526869523793929682098530923053056
995671650128536679109249515339072695011639296 1991386856400039238341822342627896656354344960
3982860825086010236930291309155295845370822656 7965895874743883994353875866109596756065910784
15932140198631495029694338227817203642780352512
31864977295550444141361849446830427546857766912
63731348387675796446670044876052895616309657600
127465484368501409221232781716889872277807562752
254936543923302451098250947363347906645991620608
509884238219204167508072662585832137472736231424
1019790777183606865639286860889936923306978443264
2039626155857610792524856793216419143336968847360
4079341514696015707542279729305928880119961616384
8158861435353619660069691744358038947131971076096
16318079682630415810109648060208440268048037838848
32636872989107184600159825263401605283664267051008
65275173225907075160200708812772660062464916848640
130553200947199562240163534197484219115202599190528
261112110885169948319851301538846236210950729367552
522223563975188154431875106936544806838299252070809 6
1044494115466846383995599071306407328688167165362176
2089033902859859358707392007763837041220698578616320
4178159149572051898847171745829718850130125653016576
8356500986848770160559118952263527235637708297601024
16713367349106873046847788825735233542030330578337792
33427465449032411545154679493886825225570489122947072
66856392399702153993227562672606366734160634178437120
133715707802678969792291532714878166034360580221960192
267437261611907263196255880169087197200799784174092288
534886215236913173615857389816836124665756815808528384
1069795814500023641678406038590995709859828126537744384
2139638397052441872250194595096638340776285242916864000
4279370330209672922287154226022570523665828465516478464
8558927732628924200147838523703728731558172890398457856

17118229609677005111442737190724632831569377699527917568
34237207508192323645179594668083616400044819236517838848
68475911594061274134947429909435934273901766147959685120
136954816343475801959071340965409271495427787645767385088
273915618997658111296495644223893348886104085991230799872
547843210616729237349697213033936309562705193381853659136
1095710366476284504212806275240171842706404429562491437056
2191468623438221067452436248824942132574796944722551111680
4383033027847746252958519894339081159473570060640238698496
8766257617638100742024334582056556107595092463670750347264
17532898359161417956263258750869899792486089612122046595072
35066562966093268856955696675253374739563988593805184991232
70134658427727403602769751697533899788311595926732553584640
140272381846536538983256220089122100194990429331709474373632
280550893675236541521945873566352801626715333619907683155968
561114047314800010154758613908922805726899617152792835129344
11222252614558253874531250961370280016400737134131540607893504
2244554268973815457505969389845428842695350067914991091056640
4489206617662246331898873713900595305178451735133801932652544
89786093947537234975716172962206658499324066668875243366383616
17957611108365908662690974329280282179015819734965765734924288
35916006854448740660477428132238465316333652264362089474162688
71833582984331327991145815211832732549271330117585294956953600
143670304519530349322673548318377068931750711412892821931163648
287346886140796085326110932426177345529917525181230107896840192
574706326485062944013749536431201106392667255073349143862706176
1149437761377067434750554416020095043450998919568476143863463936
2298925739568017962947219518355575748233326657980508000003031040
4597951912763802112786660409341922819129310953648127424558268416
9196104692783136599357763563945388283583937182670477698220949504
18392611120077337946284412618413861857818504916089401094650724352
36786025709176805387706596217873894296938270933675693585719099392
73573658356397869765688734397840129756479064070345169964273500160
147150530588884257511928552719864941838163172546677905514217603072
294307488929945550984959273288099248326736433905330942199776411648
588627833364245173892122882272937225954293045434612146742235234304
1177281377737198491628654435939351910510226446117124818169835290624
2354614177491813270946126214665658738223733602730050685710400225280
4709331199018459117269887114905227310854028626451703470162259738624
9418868086106583385295043600958274290521180094886611137807438053376
18838147548352497072100625944212187918668605873739630670580713259008
37677117848983654747222329373015654512589703115412078131093100822528
75355881202524630700486813715213866375684388966689789842049550254080
150715053414163903813057937368792847452378743405110846843825797726208
301436688846557092450284494614315924306777417753684228007104989888512
602886541729572754548906228982092307417594697394293524653116768649216
301449852883015662098621734367776383110817279640609296646011778760704
150728217450622473461395177122253306256418605292035915482732586598400
75365754229868557936739743530309210478714285381883591321229641908224
37683699867212939571390949249745883914609634058874612450546495258624
18842261309745800087206013367168581294931062713403714620239084781568
9421336342942565194358276054732110316278654198685061507602460966912

4710771015506115172556772712939964992545888520334132852542689771520
2355436929770473873967203699256937413476224970662867475642074529792
1177744175893945080828010521021946165339752840579334262506402086912
588884943451326612336209596207711811970696597913617393595883454464
294448899477840342129206965952225270635758387768783827969282932736
147227663615008689045154566900297317643084238290379479570312069120
73615438745548603512852825412240999984144641348183522577991335936
36808522841796431251564183687166670573373581775588652685413318656
18404663155409280373350977334106420577337421438542772040915484672
9202532444960172560459931412314752933994025994645608869562155008
4601366656107852467122187078788147789659670635009915859333283840
2300733544867809327007204225709459556161164136348513641942745088
1150391880840846210226657456012422608746246477596034677109424128
575208494623893878474856399585057719705955443508906266623737856
287610524413682325918192035581952067519393824109897597346381824
1438084007577708856299477935685687637592904963232671030690447360
71905769654288274819929926790199620713056507205196631353851904
35953669464861060745060442868777711314830266397028873681240064
17977227051288992040077961171227806136566139595729715842777088
8988809685073726853812850454033378307858572996472497422467072
4494502922251478843793360161226426773717038097540068461772800
2247300500983047130340147547718082196752394848421944106156032
1123674770420177419391807507411475503323135324036926990712832
561849645174415636806770620481954954135036611931440964173824
280930952569371281958818743629086078304252780922209216495616
140468541275767372757126088508597339770493627939348975452160
70235803133424552267421402601325820194430432708796671328256
35118667814482709078139880474176485251807025723959427465216
17559717031126571011284529823845030203199417546760259633152
8780050077505893741749559705300908890247661115770402766848
4390120819724250988928427249339651339447806729080337858560
2195108300347777553491037323014424116885891450137737166848
1097578095416714806258930510679511282023939767867652702208
548801020329770417886171179925905252802466905333218410496
274406496475591716321438552256027432296481963366305234944
137206241393149111849895757274551119095865237033000632320
68604617274251182769536119210544261021744746191424323584
34303056925963904807062179891906481247778436933174165504
17151902607401109114678150089270415992342250385318084608
8576138375910132912912605116293795680397641152024543232
4288162724059855634243067593976191682312078555695022080
2144128130082322406014916314902742762212668267688886272
1072087449067358497454149416408694841634648628765130752
536055416546777895950420337683009151081481561842909184
268033554279938271586882983580835440672819404651626496
134019700143243797599277899160083152902449014190899200
67011311573259229702557153264874292734244163027992576
33506386537448280302737678474853504508631909480267776
16753558644133472877098390158634931325070868723269632
8376962009771402801413970539921555197912891353202688
4188572348738034582139373000262822366645174172385280
2094331846295183881785880365282433567166951334084608

1047188759110675236251037115216727975505657790988288
5236057975368792658045670238961195837139199574671366
2618086077592104567418077450919375898375055097200064
1309071583749906402906659891179076939090255203532800
6545500643518802610521405284492329644964914292326400
3272821684144036603254755556544637297239276283494400
1636446523264335950624404235421554885998047710412800
818241102228326799810715346285395561688228639539200
409129471412242812154614287430006840188716711936000
204569195855161112201935450858657949766659552051200
102286828002100409163281879001156239719480374067200
511445290383101311127980162864917522778154860544
255728220377850288219775465362026923479453925376
127866897782074960437780424645797542784915210240
63934842687612388382836558305290811915051728896
31968118242093648273391452143841426218822926336
15984407570190551177682312567518723240059994112 7992378009667139109334449531558366685354262528
3996276117119501314913871389678685875339264000 1998181614702716537580259006789094204000698368
999112585422841208851791159369422735165882368 499567181747162074456726407672149184165707776
249789035391451772243778617829793500374237184 124897239954661253629597015911756204332810240
62449981106798310568652361454307829239250944 31225671118132997161253107476368778156048384
15613175841433419519090017112791820846235648 7806758061900170228776740243699626307223552
3903464101541815349004235965501671095664640 1951774586066772791810050904576764518858752
975908560681318954558991913201346744942592 487964914164625756606479187057155615227904
243987773994296017966731208756818928992256 121996545453139578815111411992530025185280
60999601954565574323428609803325292937216 30500465591280679619650756805192786640896
15250565102639286038793604354361463406592 7625448704819116133880915153063266746368
3812807429159294624182514064472900894720 1906445252954515590712285276207084208128
953243395664691934666656760088858984448 476632082426063036988585441037087932416
238321233509890053321921251014873186304 119163212903374294074774890755601203200
59582904525901780744294578001882906624 29792101300058207225600855312982605824
14896375168582762039527210812511879168 7448349843568210233126996984266227712
3724256051422519723245194281138257920 1862168590530467164963445035071700992
931104577674837234152146464787136512 465562430042220442911285206019211264
232786285623511134373248589822427136 116395678112956023645427288317624320
58199106707078240052115140862017536 29100187178839234140758318782611456
14550410502069674127729533567107072 7275363707359865592539953871454208
3637761081842447060607570479677440 1818920155002480662472582011813888
909479884541868897320689391894528 454749845791248731702543888941056
227379874655781507372371540967424 113692413207969324446735568732160
56847444544023947603642683490304 28424341242031616491958791307264
14212480106025629591048120434688 7106394795517725468058422607872
3553274769011318070296392499200 1776676070131886703281786847232 888357377879057185707688722432
444188360346085550988724201062 4 222099015876321213460319830016 111051925789799835988509327360
55527171820719532623429369856 27764190373269573626302038016 13882397418089690470444695552
6941349824772297063869186048 3470750470249874446258012160 1735413014056800180290715648
867725396494331568726212608 433872142980131523653533696 216940793856548631471980544
108472758111515750558597120 54237559647378592690601984 27119370119499655050952704
13559980207655006878302208 6780137677780093115564032 3390142625866341395988480
1695108206421318117097472 847572549954732768100352 423795498349403238825984
211902360860720046800896 105953486273369237094400 52977896058189225394176
26489524489846916120576 13245050475299609772032 6622669352837880741888 3311406734012978298880

1655739395803508113408 827887712300263538688 413952863349386510336 206980935274320625664
103492719436973998080 51747485618393841664 25874305759150342144 12937434354551881728
6468857914764296192 3234499326126325760 1617284847435251712 808660015903670272
404338804044857344 202173800068939776 101089099057725440 50545649040490496 25273374276059136
12636962015936512 6318618446921728 3159377942937600 1579723331207168 789878845472768
394948012670976 197478301302784 98741298135040 49371722809344 24686398275584 12343467573248
6171868004352 3086001111040 1543034109952 771533832192 385775304704 192891846656 96448020480
48225058816 24113053696 12056788992 6028525568 3014328320 1507196928 753614848 376815616
188411904 94208000 47105024 23553024 11776768 5888512 2944320 1472192 736112 368064 184036 92020
46011 23006 46013 92028 184060 368128 736272 1472576 2945216 5890560 11781376 23563264 47127552
94257152 188518400 377044992 754106368 1508245504 3016556544 6033244160 12066750464 24134025216
48269099008 96540295168 193084784640 386177957888 772372692992 1544778940416 3089624989696
6179384197120 12359036829696 24718610530304 49438294802432 98878737088512 197761769144320
395532128223232 791081436315648 1582197232369664 3164463184216064 6329063807385600
12658402492678144 25317354741170176 50635808993968128 101273817011191808 202552032068894720
405112860230811648 810243312647667712 1620521809667424256 3241113988079026176
6482368713646407680 12965018902269526016 25930600754492473344 51862327408891789312
103726906617597263872 207458316834821898240 414925640868898537472 829869296136306556928
1659774621069632077824 3319621299733302083584 6639386714654680023040 13279061659685511757824
26558699780123326939136 53118552481751260725248 106239410806511735144448
212483433299041897676800 424976089970120650129408 849970626684315009810432
1699978146856777438724096 3400030080689849715654656 6800207735332289107722240
13600710618569757568270336 27202011532949873842192384 54405203657520465095688192
108812768498282365013983232 217630259363047599673180160 435269963459060938636787712
870558816384053355854430208 1741155411699969668870569984 3482386381263665252064559104
6964923878254782332775956480 13930149987964468322845589504 27860904438838743960278532096
55723017803497102549731770368 111448453458633434357812953088 222901742620545327232324730880
445813156647647571498047111168 891645656108408977062889521152 1783329997843045622259369639936
3566737366938546580785920475136 7133629476382003834106203340800
14267568437773829013281131462656 28535755845567300716699712487424
57072749631173886813674324099072 114147975142426344387898446446592
228300902045009830296896489390080 456611707610333943635992171773952
913243222261296453356382729535488 1826526058603850038881562231046144
3653131345370214342100718006042624 7306421147065457212876623099985920
14613159206780971483103620375773184 29226952238862057080907989103149056
58455172128324342391217474909503488 116912879557849141241237943225417728
233830829718099195400081873263656960 467671800641000216635375720152956928
935363883691604084941175387557199872 1870768332202415473223198669616971776
3741617794043245553128093128239087616 7483397847363320319619577834488463360
14967120213280299065965938824997502976 29934889463667914985385443962036158464
59871077001550463677678020548154621952 119744750151530194769170306344473853952
239494692599918924365969143185276928000 478999769793554918387195347363212296192
958020308774543976084904816711741472768 1916082155923956230790837877394116706304
3832247388597649018823732242729500934144 7664660930694771152131577461341536911360
15329654168388488533231380874448143908864 30659972950774869524399213652426427990016
61321275129545523964671331111913136324608 122645208715082617761088469837946833338368
245295734342148375185668554904134788055040 490602102508263029698320340264751818866688
981225472664458618050607141442468123246592 1962493480624782353409147204710865217519616
3925072031841294941434160253073588377092096 7850314204866050352100052193450892638289920
15700968692099021642663567761509217044791296 31402617948931885162254062272233297626005504
62806597027331454078361978042896322324856832 125615916313598275664431663082652098795405312
251237277145067286344278740159023105882193920 502485544332587604271938830830548402834715443

1004992664723235025500438272585843689859842048 2010028885589435931124199857121438646050750464
4020144883464803622495046338142379824763633664 8040463991501470765483385924083764714851532800
16081276432146668571953358343765539560351596544
32163249762580791225879889678727099382000254976
64327893321736490615706125339846239286594633728
128658574236622797559304942644476559618377515008
257322723659545227774395269218521281327131525120
514656597691689720860361306296178886835016040448
1029335496128577972343864148310630422031538061312
2058715593747553005934011368057806140786088083456
4117520390475900134360588878988702875018200088576
8235219186913388513706310043723586936928448020480
16470795185749953517382884658939536247640991727616
32942303995346260014706298460863797242850174828544
65886035238385225989293655207697043980836732403712
131774924972155863898349426987332986951946230300672
263555558935082551636223087118543771884437991587840
5271225355851706750951494640524843139729967045148672
1054267907666496797261086213625197471382116214243328
2108581487259160185238366292401417326608596676378624
4217254318370653551909120315104879420905921848541184
8434691324445973466683016090813848377189300688650240
16869748024301279659095583102835875825133515360436224
33740226799421224769650268048088109791776858687143936
67481915100479780442218739781008935866573373306830848
134966753204234222690273886931683304299186058478747648
269939352415017768992220588602697473730450740687667200
539890396843134185207786806684056677725058728835678208
1079804177712465664862264872325436815978431952592044032
2159655123477325918617912262565520553013492895025463296
4319403783059441015022589560960334948140243769733677056
8638994638328460385618709193579257580507003498832855040
17278363421076077482384478530475690529467038916396711936
34557475130990468387063077347585731795840141670255427584
69116446839657563618714395268440165065492411015434862592
138235886834668380926605271683417733078609077380717740032
276477759980043269231563505659910272052466665461131509760
552967492581499553219832935905970155895430352321655078912
1105958930405825135953077720984239535371854747442094276608
2211965751297302330932979140313077517905697580481756790784
4424027283565908779919605677315351930135371332158650056704
8848246129074425795946506148009097648918695006707573063680
17696875382034068064107601882774982875133294698195692027904
35394517011838569072644382939063540904858398765952475856896
70790566519218004034147124225154232118900416271027135315968
141584198029517739846010965144362764856168070020298637836288
283174526041198943247455363676834130949070614997086010081280
566361312046724813605777594129885464371610179907149488979968
1132747144022103481433288921812205333690158259640253915594752
2265543327901514671310045310729279477274192318932417706459136
4531184735517644759507025555668296574336136237168655163457536
9062565630464520352787920979756068388247775672944949827993600

18125523579787502373123581696351087255646557743105178658144256
36251831797291928081342642866380075469595128280640915320602624
72505232870017702832876244680115952855794282150142946649833472
145013604290903099006134407254943509544796615478008125316923392
290033485683541584693032650299310226756009333311460714668359680
580079525570553942747592972177466868844850871333810357405745152
1160184159548049432218241287512626568355366152089398570949541888
2320418535909981957882593261340638798042061123022352854175186944
4640937505447730102657407895312048918746779883731817132902580224
9282075878150992579099258535885640482818875042837857114909573120
18564553490813049905767402562294366256288380636424159928027971584
37129910450648229306672576105634903093878022374345211252473593856
74261427839340717603620694173362147350358566951684205297782489088
148526069554769953187792472270908977025922178309355976181235580928
297058566861716942336687112390187318702254445430687083533812367360
594129989227787956595578560477113366705329068485324429410307145728
1188285689464284057035565792347704192012298492218549383505979113472
2376622800945984401759948927482363301227877694932899816382687870976
4753348445926801378897532540538636436862316810857401731506835030016
9506902579923267908550334452225092542537756463698007660496588636160
19014216535985866118611207646745824422701758611362423715959014424576
38029255824250392840243492778082927520656008590657664221849703153664
76060157153058106886529140525348412391816999917180962023562754916352
152123605315230856185142590989061939484643965306093191206852207050752
304253792648690997194453801854854108371307861555648916733157808537600
608520749333840564037244843463168675546655584998222902105222405947392
1217067826740598267371164166433258268701390893770295941488258389639168
608547077406757703333919322970089593154735308772073039383035983691776
304280120721608136491128281361775025979387585329499054010971386281984
152143351369918710657648450619252627690703758136480794165212390359040
76073321189516676534866380278808871195856861804106030662469543788544
38037483347036998870454267623995714273180922269985832121166446198784
19019153049657829736737672554293496474216706818959324455549060251648
9509782212898580019124105648294567905921476251462866425257448701952
4754993950484122584939687509721193787367299546723035311370183639040
2377548397259477580158661097647551810886930483857318705055821463552
1188799909638446933923739220217253364045105597176559877213275553792
594412810323577538884073945805365411323372976212230200949320187904
297212832913965805403139140751052070312096576918090231646001299456
148609630333071420682120654299710717481253332865032681408671252480
74306422104579969331335869111947699903229188635510123497170927616
37154014521312114160805705537020020532915855419251953145003114496
18577408995167121827971738259033095557108558260374422270710382592
9288905364839093287770311874778090423879594405561433984459603968
46444553116047312830777377310019816534602454840467828416782008320
2322326774837539508834799341325293928632556239077469920667107328
1161188495825711301140455013820339794981942528960512816471605248
580606802116326423931755178489016312823803469191145336304828416
2903096781598985986466641425033931364078317836951017132186927104
145157977630816992663702630411677285872366969653230798110720000
72580558090842343002042274153194444852787510415476515063988224
36291063683138094836116616550275123384695768002168815536308224

18145924160427509085606048011976512171498890398299686770311168
9073158239642985376576893874407731325324948397757482886234112
4536677199536108105175381871413603282450225798182561193656320
2268387639625361761031158402811670451118988698743190472097792
1134218339741334734737312934958269630506432249197550173683712
567121429834994294479523334255352017726685074511752555659264
283566844899660610795195100515784610100077012212365012238336
141786487440912037175314266951946605668405743584426873323520
70894776215996884476515491823000453143386490531335620263936
35448154355768875182686925085013801726285054635228901933056
17724460301769654063558052129263688440438432002394996867072
8862421712827435267886320858010238008867168343587771383808
4431306637385021751996807825694315898757560342989022167040
2215701209178162935025227611191756396540768257092079321088
1107874549831907497026025654768177421851378171344823779328
553949247537366763269718751970238322716186107071803949056
276980610079389889013212338278193967253341564235598004224
138493298195048198195782650285634386574295037467647016960
69248145675200725942479565716085894760959646408747515904
34624821126438676393533903144677298117385887041835761664
17312784707638494907914011715655824427145975439648882688
8656579426028825809530535929486499897799503679189942272
4328383249119202082552033000572543791013009819277721600
2164238392611995630169399018200918816563133899480236032
1082142580332195109531390768057782868809881444660805632
541082982179196201989041013507553164669097969790746624
270547337096147424606193321493107447466627608625545216
135276591551348374108933068116219156299353116177858560
67639757277311517957384737742942294432696214021472256
33820609389474424430151470713887505357857934977007616
16910670070146544940805286278151931749683881471639552
8455517722777938833267418599680055410219397727387648
4227850205241302598066097030142072472798427359477760
2113970774546817889749242380222058620243577927630848
1057008223236492240232718122686540502043971087761408
528515529599787767795407527631025846983076605853696
264263473790664707737227996959390721472083833913344
132134591390717765788376115051634259726314682449920
66068722943051588854069115811786579358293723971584
33035075095372147406975087048878014426715053359104
16517894359609250193457808095931369587141622366208
8259125585766213341714036333711865980462859026432
4129651995863900793349584309729023583677453434880
2064870599422347457921075226301057088561738678272
1032457600456372259583679148868801192642375319552
516239950600785395103410342293536920501940649984
258125550486692330207490555076336622341346820096
1290655562836495981431637969502952392215861657600
64534175214822898879765330733868236630524952576
32267784505698903521855838358130138576559538176
16134240701993178801914505674663079418928300032 8067294575568452921450546085130544774788415488
4033734400070158220971919666464774920056340480 2016910756178044990609283145182138726359236608

100847715616050543536630322856594499634515148 8 504249467115994187713982442270410314755342336

252130178075867828872406635128924065669054464 126067811296869281943911024561321486980218880
630352667779023247258093657790904705629552640 315183139536850042398316096387600988179005440
1575949725920942305837926819398748117716172 8 787991877078817199842136578429745647268659 2
394004445598581623382654873580058617839616 0 19700647632887732342212072897262220602245 12
98505364929231917576457010577607551562547 2 49253745847012586720926828334452000056934 4
24627404614704607326812575690050112166297 6 12313968152951460646580868606437112152064 0
61571169992753088148777246839245841104896 30786249610374436532325075323153060724736
15393457112186164495130763613341600448512 7696894709592555362049494782553335267328
3848530431546014238266803879217935155200 1924306754147875397754430183579601338368
962174146261371838187729213775117549568 481097457724402988749121667880217214976
240553921159060029202189364436437827584 120279556727959282014908947466383523840
60141076438194274714361606357274066944 30071187256204454210634369490678185984
15035918146655885532043967901359669248 7518121332604771979385375552869012275 2
3759141795940800596374383553350205440 1879611462789607601528039671177674752
939826013804407452434443782840123392 469923148107005552052433865045704704
234966644655903688943822919335673856 117485857629152300930714453074247680
58744196465176378694758723240329216 29372732057888303462080109971767296
14686682941594208788390429161684992 7343499927122132922870401668743168
3671829191723580725772794378321920 1835954209943047495055193961136128
917996912012152313611995366555648 459008359526390439848196876271616
229509131523352361445198034632704 114757041641754751483148815564800
57379758760916661121849306906624 28690498350477973251062103015424
14345558660248807970599776288768 7172934072629314657834250534912
3586544407567112665184306462720 1793310889409784000725743828992 896674787518005834429667213312
448347065165559834248231256064 224173682860583756408144527 36 112091601994668417078756638720
56047009923153823168553025536 28024109424486718898863865856 14012356943698263106725609472
7006329587576583382009643008 3503240351652017605328240640 1751657954757871759825829888
875847866844867358493769728 437933378155399418537312256 218971411444182578913869824
109488066905332724279541760 54745214044287079551074304 27373197317953898481188864
13686893806882128593420288 6843594477393653973123072 3421871025673121824768000
1710972406324708331487232 855504649906427875295232 427761548325250792423424
213885385848643823599616 106944998767331125493760 53473652305170169593856
26737402613337388220416 13368989537044845821952 6684638883710498766848 3342391499449287311360
1671231778521662619648 835633903659340791808 417825959028925136896 208917483114089938944
104460993356858654720 52231622578336169984 26116374239121506304 13058468594537463808
6529375034757087232 3264757886122721280 1632415127433449472 816224655902769152
408121124044406784 204064960068714496 102034679057612800 51018439040434176 25509769276030976
12755159515922432 6377717196914688 3188927317934080 1594498018705408 797266189221888
398641684545536 199325137240064 99664716103680 49833431793664 24917252767744 12458894819328
6229581627392 3114857922560 1557462515712 778748035072 389382406144 194695397376 97349795840
48675946496 24338497536 12169510912 6084886528 3042508800 1521287168 760659968 380338176
190173184 95088640 47545344 23773184 11886848 5943552 2971840 1485952 742992 371504 185756 92880
46441 23221 46443 92888 185780 371568 743152 1486336 2972736 5945600 11891456 23783424 47567872
95137792 190279680 380567552 761151488 1522335744 3044737024 6089605120 12179472384 24359469056
48719986688 97442070528 194888335360 389785059328 779586895872 1559207346176 3118481801216
6237097820160 12474464075776 24949465022464 49900003786752 99802155057152 199608605081600
399225800097792 798468780064768 1596971919867904 3194012559212544 6388162557378560
12776599992664064 25553749741142016 51108598993911808 102219397011079168 204443192068669440
408895180230361088 817807952646766592 1635651089665622016 3271372548075421696
6542885833639198720 13086053142255108096 26172669234463637504 52346464368834117632
104695180537481920512 209394864674591211520 418798736548437164032 837615487495383810048

1675267003787786584064 3350606065169611096064 6701356245527298048000 13403000721430747807744
26806577903613799038976 53614308728732204924928 107230923300473623543808
214466458286965674475520 428942139945968203726848 857902726636010117005312
1715842346760167653113856 3431758480496630144434176 6863664534945849965281280
13727624217796879283388416 27455838731404117272428544 54912858054428951956160512
109828077292099338734927872 219660876950681547115069440 439331198634328833520566272
878681286734589145621987328 1757400352401041248405684224 3514876262665808411134787584
7029903641059068650916413440 14060109513573040959126503424 28120823490055889232840359936
56242855905931393094855426048 112488129663502015448060264448 224981095030282489412819353600
449971861467121895859036356608 899963065747357625784868012032 1799964817120942919703326621696
3600007005494341175673834438656 7200168753493593023882031267840
14400646991997007392832787316736 28801912954013657475803024195584
57605063848066600331880947515392 115212603576211771424311693279232
230430158912580684369722983055360 460870221345475651781645159104512
921760249731579869647688704196608 1843560113544416871464174180368384
3687199455251348007265941904687104 7374557366827724543207070897274880
14749431646305506143764515970351104 29499497117911126402229780292304896
59000261886422481033861057287815168 118003059074045418526525107982041088
236011188750491749970656202776903680 472032518705785325776524379179450368
944085319821174303223472705610186752 1888211204461555909787793305722945536
3776503538561526426257282400451035136 7553169336399882065877956378912358400
15106663191353422558482695913845293056 30213975419814161970418958139731738624
60429248913842957647745048903545782272 120861093976115182709304363055256174592
241727380249088900246237256606841569280 483465145091894870147731574206341578752
966951059371223879605977270398000037888 1933943657117316037832982784766633836544
3867970390984368632908022057474535194624 7736106935468210380300157090831605432320
15472546177935366989568540133428280950784 30945756969868626437073532170386702073856
61892843167733037790019968147833684492288 123788344791457645411785743909787929673728
247582006494898430487063103047816980725760 495174646813763140301109436552116204208128
990370561275458839256185334017196893929472 1980783657846782795820303589860322758885376
3961652386285295826256473023372503459823616 7923474913754052121744677734048722803752960
15847290109875025181952818842704877375717376 31695260784483892240832564434624618287857664
63391882698435468235518982367678963648561152 126786487655806303978745671732217381442813952
253578419829483342972906757458153671177011200 507167728694708155976644342903745158936788992
1014357235460899252014950341782365951039111168 2028758027064764384153223995514483168409288704
4057603166415460528553094614928468869480710144 8115380557402784577599482477655942804285685760
16231109563949296196185551450909895739219902464
32462916026186046474344275893015811739736866816
64927225848947001112634897768423664002067857408
129857239291043818553162487501631409049323962368
259720053768387269762110358932830980189024419840
519451257909373804835791485724798284558801829888
1038924816563946140294724507167869217479109640192
2077894234618289341835732085772283731681231241216
4155877672217372806164030314417658056808486404096
8311933750396333857313192914581497300509020651520
16624224312715844204596650400655356974802136989696
33249162249278041389133829944295438697172465352704
66499751746248788738148718174560326889481313452032
133002357987882989396059552921059552769235392397312
266010424966536802631643338985996903519016315781120
532032267914615252942335144259749402999123693535232

1064087371792313801242767221095009997920429511016448
2128220415510794193201728307341042379685223269924864
4256532174873921567835844344984129527059175035633664
8513247037452509498536464150572348589495807062835200
17026859450314351722802479222352876249746528108806144
34054449651447368897064060287122110641002884183883776
68110360804532068697046324259076937565025424300310528
136223644612338799199929055887819307696090160465707008
272453135231226922011530926514969480524258944661585920
544917962475552491246407482508600691312675136783515648
1089876393089773022769395062239745248431536647684877718912
2179765386006999142772394965863696607363958526816813056
4359624308118787463331554967556687056841175033316376576
8719435688447153282236640006771961797908866025998254080
17439245521313463275620340156861098964270763970727510016
34879239331465239973534800600356548665447591778917023744
69759975240607106791657841773981798804707311232758054912
139522943636567467272492164694501000557038877815364124672
279051873583841441923337291682076807009326266330424279040
558115719789095898603380507950303225809149554060240617472
1116255384821017826720172865072905675199293150919265353728
2232558660127687712467169428490409797560574387436098945024
4465213101226679542987986253670016489445124946067334365184
8930617764395967322083267300718426767538202234524941680640
17861361865267715111638112418819364111237230915383042926 1824
35724003553124735177191427549900857379336427677221950324736
71449539601790336243241213446828865067856474093566084251648
142902144194662404264199143587712030754080185665376535707648
285810418371488272083831720563532662744894846287241805824000
571633096707303471278530307903282527963258642487461080465408
1143290713343260796778794349358999460873455184800877098565632
2286630466543829302001056165822867731640786169253664072400896
4573359012802274020889047265855473083069323937811147895341056
9146914185033778875551964400130421405714151074229935291760640
18294220688926019418651668537099793290579308545675149585678336
36589226015568962172398816547877487539460629885780857175670784
73180021306571771014988592043110776995525285360422830359969792
146363181164011235370359101980933157824258621898567892737196032
292732639429757857421482039751289523314933346152580249508904960
585477833062986488204491751081425461962698897016049427086835712
1170980774532914523132038845320543754591062203453876710311723008
2342011765879712139710188376956473170513453225751309132899549184
4684123965387190466312598126543717663689564089189729690351304704
9368448798029913306409638998348977972704443453753682229807022080
18737299330570891360388163487221041236059517458255810157822869504
37475402130163912215914097955488253053420296018008511712063389696
74952411198372083422103737873068847269443114239010806216962080768
149908036272832684824758559670322376864091272884009178019594764288
299822500297842405610619287189014118378592634579993487210530734080
599657856100038883143442910074766966058005446783937236763743879168
1199341423208785910131294491543011390717651248815774998212852580352
2398734268434988107951406325872977698638583208127351045796434804736

4797571380904808791280447337319865231683727837246304190334328897536
9595348449879282733316164045787550132180578516475812578151576371200
1919110827589789576814286683387073960198740271691803355126898989894656
38383039304074452139306811152332757879227296801768883892469654093824
76767724112706225484655777273848073108959576339403401364802656796672
153538739234527093381395864486061260918929118150538069889332010811392
307084060487283471586960348848852751239878167244538674098117416058880
614181285011025512822257937451165961283796196376002416835141620989952
1228388898094968164941190354409252840175672116525854970948096819724288
2456830452335770608475729667832347515567503680599410216451820794937344
1228441554240802443534539313423094675391831564073555245503723975213056
614233941156859791415606896465007796499955643923702691390768776478720
3071235552596659180531972068109234127651997752905313880014837782675456
153565067307444232678070343992982178527008841924388207167145588555776
76784179158279437545077326965673646614009403698059737163436142886912
38392912331418379375559740967428101982257193216962685371649745747968
19196867541848519989290409226009690328754842292447751080790710026240
9598639458993925145400473984152664833190543988207079737878273589248
4799422573531795148077871677650242251001833415095141967680596082688
2399762708783313861727753181612076042704197418043372033211027685376
1199907065400365074708285262199515479953739064269586541290878664704
599966388204536609276346966796496469277689709758743532988121743360
299989621854445340599275651246617599289254943691346897665402077184
149998024803311188280188909547493481969832516251661014418371641344
75000619339699853130369996735839082147518780328824290002021122048
37501113138872056060322769348965711655060651265909036397428211712
18750958303947092777730270165005941118180956183702963896922931200
9375680019229078762649577827764513204415793367225704797565878272
4687940443242305568217010286513027924870554321299963823335145472
2344020438435035877554615829571899623766605979493537623943675904
1172035327624459485500363257943642642548967399168546668109889536
586030218015700516111709300550667736607315904295162262123970560
293021386109585644736618486064757075970074054503025595096498176
146513831605660515708691160927090141818245078429235029565505536
73258485078264104524536539410900872825726564803478630791380992
36630027176848975597363749179128337371165295196169873400004608
18315405907282949466229614326403119164733653995300215702159360
9157899113070705566888677031621034821942330196257747352158208
4579047636249968200331273450020255030758916697432693426618368
2289572857982291808609104192114996325273334148368256588578816
1144810948919799758526285829609932567583604974010083231924224
572417734424226806374009781581183486265271436918019084779520
286214997194276866742438324178700344369370193415498276798464
143110563588220165148935878783404472803052334185993505603584
71556814289650948463326297738729386710709785832118936403968
35779173392595907176092328042878268509946702285620560003072
17889969820183170060260753608195921832269255827590825902080
8945176472034193266237671597476354704782580256185685901312
4472684016988400751172483195427374246715266299287979425792
2236389898979852434613065296058285570519621235241557950464
1118218894732752246819944497201442008840804660419563094016
559121419987789138166678173186870616210899351609173606400

279566696304601076461692048886510114000698186504282832896
139786341307653791920022505589792459947973348601989431296
69894667231503522804599493368164931447798801975918723072
34948081904590074824593866970716816460805464825421365248
17474415096714194123443993628675583598855764331441684480
8737394620566675417295526885996379483654398125086343168
4368790846388126886434528478827483583940457042225922048
2184442191246458032110646757328388713026857510954336256
1092244479649426310502014637621517817041743250397855744
546133931837811802474352948289420638785028872659271680
273072811925455224848849288884041184524593060059807744
136539328966002274230261051811686024828335841894989824
68271125984638468018048729590675728697187576880037888
34136293743137899460483466637754222490103616406290432
17068512246978282455971284240085290315806722186280960
8534438811193807590850417580646734693280818084708352
4267310749449236976857596520625412114329137538138112
2133701046650785079144992125463728441008933016961024
1066873359288475834930592995307375412426648632426496
533448097625779565144344963941443302174415378186240
266729757803660606411696715114599449067753220079616
133367733397215715125610474129238623524149375533056
66685293946300563522686295350588761257211070513152
33343360596996634741283676818279105376173726629888
16672037110421493860612102980631915061870959001600
8336196961172335175291183776062138717827527344128
4168187683566961710138158030904159952359787593728
2084138443273877916315362086888625272902905757696
1042091522382137488780822579162585284812958859264
521056911563668009701982057440428966587232419840
260534030968133637506776412649782645383992705024
130269803077216635081280898289675403737184600064
65136295335183225704586795127229742391186423808
32568844565879066934266570554810891456890273792
16284770732083260508119871773003455859093667840 8142559590613493774553229134300732994871099392
4071366907592678647523261191049869030097682432 2035727009939305203884953907474685781379907584
1017885283041135542004138609712218523855486976 508953530556309241032900132843547078510510080
254482209796025355531865480415492447546638336 127243827156948045273640447204605677919010816
63623274707941706390674077100732566032351232 31812317918704695072263965299581146552598528
15906499241719268474595446024398005044510720 7953419762043094706529454699502718406361088
3976794951613277587880593193403217145233408 1988440011102503911248229518527537543643136
994241273199184514278081220176733257334784 49713127042355853646602384054484871424000
248570952123762407896503535500665557090304 124288134517872773779997575364453339234304
62145396486932171805871691489286949961728 31073362857463978360872297648173615153152
15537013735730935409404374775851877662720 7768673021364940819186300363808473874432
3884419587432206966835206669845504458752 1942251332090971762038631578893385990144
971146435232920020329829911432009875456 485583602210177079820172016708663377920
242796993401947074737714538850660909056 121401092849402804782671534673495064576
60701844498916036098242899960829837312 30351571286565334902575016292456071168
15176110161836325878014291302248611840 7588217340194992152370537229134594048
3794189799735910682866964403572441088 1897135464687162644774330096288792576
948588014753184974057588995395682304 474304148581394312864006471323484160

237157144893098069349609222474563584 118581107747749491133607604643692544
59291821524474973796205299025051648 29646544587537601012803397864128512
14823589206418857563752073107865600 7411953059534457310551223641833472
3706055757929742919613205364867072 1853067493046128591975399454408704
926553553563692862072098113191936 463286680302160714078248249589760
231648291911237498560223721291776 115826621835697320040661658894336
57914548857887945400605728571392 28957893398963615390440313847808
14479256184491629040288881704960 7239782834750725192678803243008
3619968788627817932606582816768 1810023079940136634436882006016 905030882783182151285236301824
452525112798147992676015800320 226267392102352454854706724864 113136113902815456685702774784
56569265877227342297202609356 828285237401523478800600399872 14142920932216643057593876480
7071611581835773357443776512 3535879134878161259304530739 2 1767978453322669253684363264
884008116127266105423036416 442013502796598792001945600 221011473764782265646186496
110508098065632567645700096 55255229624437001234153472 27628205108028859322728448
13814397701919609014190080 6907346424912394183507968 3453746999432491929960448
1726910393204393384083456 863473643346270401593344 431746045045172055572480
215877634208604455174144 107941122947311441281024 53971714395160327487488
26986433658332467167232 13493505059542385295360 6746896644959268503552 3373520380073672179712
1686796218833855053824 843416123815437008896 421717069106973245440 210863038153113993216
105433770876370681856 52718011338092183552 26359568618999513088 13180065784476467200
6590173629726588928 3295157183607472128 1647613776175824896 823824480273956864
411921036230000640 205964916161511424 102984657104011264 51493428063633408 25747263787630592
12873906771722240 6437090824814592 3218614131884032 1609341425680384 804687892709376
402352536289280 201180563111936 100592429039616 50297288261632 25149181001728 12574858936320
6287563685888 3143848951808 1571958030336 785995792384 393006284800 196507336704 98255765504
49128931328 24564989952 12282757120 6141509632 3070820352 1535442944 767737856 383877120
191942656 95973376 47987712 23994368 11997440 5998848 2999488 1499776 749904 374960 187484 93744
46873 23437 46875 93752 187508 375024 750064 1500160 3000384 6000896 12002048 24004608 48010240
96022528 192049152 384106496 768229376 1536491520 3073048576 6146228224 12292718592 24585961472
49172971520 98348040192 196700274688 393408937984 786834653184 1573702860800 3147472830464
6295079878656 12590428192768 25181393256448 50363860254720 100729867993088 201464030953472
402936651841536 805890483552256 1611815326842880 3223699373162496 6447536185278464
12895347248463872 25791244252741632 51583588017111040 103169375057477632 206343148161466368
412695092415954944 825407777017954304 1650850738407997440 3301771845560172544
6603684428608700416 13207650332194111488 26415863614341644288 52832853128590131200
105667958056993947648 211340419713615265792 422689846626485272576 845397707651480027136
1690831444099979018240 3381734945793995964416 6763614006776067784704 13527516243928287281152
27055608948608877985792 54112370818722362818560 108227047480453939331072
216458706646926306050048 432926636665889466875904 865871720075852643303424
1731780333639852705710080 3463634454256000249626624 6927416482464590175666176
13855128112834359704158208 27710846521479078113968128 55422873634578873639239680
110848108452399182101086208 221700939271281233847386112 443411323275528206985199616
886841536016987892551254016 1773720850965838742264217600 3547517259795403398851854336
7095185635318258626350546944 14190673502091420909994770432 28381951467092649134576893952
56765111860004912898328494080 113532641571649055055006400512 227070118846576568626711625728
454149909099710054286820900864 908319161012533942640437100544 1816677007651295553414464798720
3633431386555046443096110792704 7267017515615003558726583975936
14534344516239828462521892732928 29069308002499299615181235027968
58139853945037884610637369180160 116282183770154339981824536608768
232569319300465821484748669714432 465148542121245926011696532422656
930316891283120418107791450832896 1860673396647497968384379673640960
3721426021457510201106352891232256 7443010499240048930887892870365184

14886337911130154919126159916531712 29773309647560423952953068184666112
59547886945721076135307633072537600 119098309192642608729418259551485952
238201688987686130376442505915793408 476413519180174086588096985457229824
952847320769951824846617918165745664 1905735206359110953034083730834063360
3811551542356636512749863250673270784 7623265343990102238863118079356829696
15246855206533862904453019314734235648 30494359450175042662359604941509623808
60990016974564719031626342507101552640 121982630097558705477066950262367715328
243970452491975945781762431021064650752 487951289577668961218781923034787741696
975923348342772061748077968054892363776 1951888235060412402117184180080418488320
3903859546870561361476424848102104498176 7807885247240595837436962672086744039424
15616102801480137903842151295938558164992 31232870216958168265620754495407256502272
62467069661912121447114412797874793349120 124936797779815812725974633209870147387392
249878912471614765115440881647981416153088 499768458767195809557864993752445075062784
999558185182324177769696448417854635638784 1999158905660513472847325818661638242304000
3998402881912757180310517480975134426660864 7996975905008974829852766649253984737427456
15994292092384870598168996673115401243066368 31989264749503583073264920095445666022555648
63979890628474849900383693689321059117957120 127962503515885067308475094375501572381605888
255930451549640869632365602744722053054595072 511871792135023209295562033476881922691956736
1023765362341529358652785722928639478549446656 2047574280826024597428894757807030223429959680
4095235673937980955104436139513562979522052096 8190645572447825430702165526826131024368369664
16381639594039377902390917549250272179385270272
32763976086366209886755008089696564620067602432
65529345969307327937456362161785169762729328640
131061479531764472202805416288354420570646904832
2621228534249828577061396216506277003231670304768
5242682188872256419434363200871690330644093599744
1048558738489711369491867937461653309649693179904
2097162078469819800230018946359851916022398320640
4194413359920433722952604035592794425490820562944
8389005125802455690890340356931770037873688969216
16778367063528087871750945285355902449531473625088
33557447750902528723442419713696529646631138623488
67116322749497763406765897713362508788398659993600
134235499994380938733293911998663916567070085480448
268476708979532701306112057141205631114685701947392
536964835940607050291272580570166858190462465867776
1073952507844297395940642093715844908303107055681536
2147950687614761382597478052582712200450578359255040
4295992719081855946627343835467469168589885214294016
8592168125868378256119463131539027872557227420155904
17184701627146089237968477184286234815869368823447552
34370134005110843927396056210988827773248565613166592
68741729511859018757710316106810371829516787158876160
137486382026992699321257039583286176225072886182838272
274978610060534722254186893905903217582224396095848448
549968912134168091731719417290468165428606039652040704
1099961208294533477910130093538259791385526574224769024
2199969184641461544713642704991166503827682138290913280
4400031905387712267214050445811626849768622256264577024
8800250882985002890001630963281841383763760471894654976
17600875910389162491150322069880858135980552862520311808
35202500109616638404594764426396067008867169562502627328

704064967969099036537777694260608354915464667999292620 80
140815986749173060996732019998659073930717188949706539008
281637959809052629371817002290392953756682888599109107712
563287892239518273500339929166935519303862798597610274816
1126599729721862576514091707506170262188719639994004668416
2253247349929377212055007113356938971539427365585577574400
4506590480830058542163661623403074837402830902366291623936
9013372523602725320434618040184543463453614147122856198144
18027128171090667113083825667125874504203132979026258296832
36055022589951767170596830507765324162998075327613608394752
72111577675444400230052019362557798635179769394349400391680
144226220341970532237820755419169897888726776266943167987712
288458570666104528031074944226448397014188027490375070384128
576929401296535983173016755229113996501845004893727609585664
1153883322521725820567767244010662397950627909613410156806144
2307815684900759349579001955126193605795131618878730188881920
4615729449516134116044938844462124831378014837061280128303104
9231655058461499065863747557343724902331532872730199757684736
18463702435781459799275234851526400283814072142675678517526528
36928189509279842933645949176730701525930157079781915039367168
73857948293993532537482857300817204968464339748424946087362560
147719035138854758415347632496346013770136730674572124191981568
295444347379444903511459100782115235206689563704588712418476032
590901248962360580384445873143076885746211332120066352905977856
1181827606331662707491947089443846602158087073661910561950007296
2363705429477208508430004865203078865647502966167376836176117760
4727511292582183203752231103036929053957663570021865096904441856
9455223452419898781288904951335400753240642415417953042913296384
18910848639350862310146695393193886797131915381584351784035418112
37822500747723854115431161767433944175565091864665594964488486912
75646608433491967221137865496960229513732705932324972721812275200
151296430743072452422826814918105141352670456270637511029295153152
302599289238321940806755797684579647355751001353250153229931511808
605211433980997953535715931065898024012322180330450568802545434624
1210448578970704050915840533525273506626284715908801662290455691264
2420948579958824389520498409837501930455850142313404373951641026560
4842000003952481354418631505248913695318261705618410846644741341184
9684205695974627859592532381645647059449646253220025890772401258496
19368822768088586020695603505586933456525538190406460176510639669248
38738468288455832644412284495765145588303567748745737142952953643008
77478582081468986494866723960712848527112118233357107865769255895040
154960455172052615401817757859790811755234201938445482891265209008128
309927492362334515627804135596311852912488334820353500101983812452352
619868148761127600903945510946084164629016531527632068842874413776896
1239762625595172341104565501399089246866112786829114274963562405298176
2479577907336178960802479961812020328948385021205928824482751966085120
4959261126964026478791657841651724328329088937507258198076758243147776
2479683219627847517989177879839703999380703916301329373594006277062656
1239867937886840898291263419426772917298431681924514824074816716275712
619947132979879018793968949466846917453255702849182480676315147010048
309980148508168794221153094610153688128647782368053774657610967941120
154993365263199039522660857243441958765333856655758154488532181188608

77498328136156840967372583590903536733171911063744710824129439203328
38749986820357081086707369280043047041838446899805172201996393906176
19375404786317870844864223382317162858545469133868994495964034105344
9687908081228600573187381062306401098085857408917701445464935628800
4844056884649132861971325216727110383449490125450452821473927102464
2422079864341982718674479951150510108928025773221027460107693195264
1211065643179699503181648646968732513065653241858414254739211419648
605545677094203823513028659181104985833646798553157389712288120832
302779266299278947717616497438921857567233488088553826027485265920
151392847025727991839359332643645611108821788450264478599413235712
75698030450908254909955208283915146717013416428126022092541919232
37849818694476256950115375123003743939807969315559902442688610304
18925311081749193222626573052024957260554615208528396919553130496
94628564081301289850977292712740212756026222879638421308880977920
4731528637692830679441086008267781960463969077506322078992695296
2365814535660298433166653690449276641563313357596716751772450816
1182932376237090763306382188382331151447321088220136232024276992
591478742322016155014718765770011991056492748820957044081164288
295745648262743464188123218674429203194662476765922986075095040
147875962682239425434443527231926205430539289560683725054803968
73939550616553559387412722563318904631873670369202978536030208
36970559945993703028801840755337353274238847979032047272329216
18485672291855313181948660114507627116270430386731302638321664
9243032305356887424748199925673288797710718391973290820239360
4621614232393059129261034897046382018643110795290465160658944
2310856156053837273073984915628059819215431197297142455599104
1155452597955572490758726191366464314554653498474526165434368
577738558942113172490229962459449359750795699150240551534592
288875409453220049800548414617833281112132324531609010176000
144440769717691756677990924002970941174433399744048872292352
72221917354386744227853820348512620896400318611146619748352
36111724924963805058356089347769885602791968675134401675264
18056245586367119001392634260641730378691889022347746738176
9028314355126167736803611923699258977993896853564146319360
4514252958534387986455453358538826383320924597977209634816
2257174369752846052254550377614011638822450384586173054976
1128611130119249055640687037979305042992219235091870646272
564317537681037542577049443575802133286606638945327382528
282164755151225278666877684080975872538551830172359720960
141085370730965893022615323187025339216900170436027875328
70544181943159573355895902166781371082262212892937945088
35272839260418100100242071370025036278037170283930976256
17636793774628206761268095828329693507471617060696489984
8818583959523681736207577985823434437962324489713745920
4409385515866630045890554028741011061094420224539623424
2204739525985709611838659532285152451603839102111186944
1102393147019052100366021025099899686330234045976281088
551208265522624697406356142028611573429274270448484352
275609978767861672314850885753636651846715758954414080
137807912387205497963261850246483758489397191342292992
68905417695240079884549128808074595527718251603689472
34453439598438705393733666246453655905368953768116224

172270851746286854225963840444350070234393908671938 56

861372527501900907416296748282159304709715242516480 0

430695398136183771851387147171284129123730470836633 6

215352266260708544997312960100744302946301660207513 6

107678416726662602034466173307923270665369042498355 2

538403501614854657851379332827371949287936274464768

269207459798198152765213899557563772624513668218880

134606584394484488302369066350720785302529599602688

673047194449349501110655914613298421464011825479 68

336530733463138280354733248736496458207687826472 96

168268934850800905077069270083171852841684870103 04

841362514850163349883859578990477382897629134848 0

420690177723161087191186403782547750793416959590 4

210349549010620249720221509034928405069009675878 4

105177004579829977922424908089291467370655435980 8

525896173271749154923695308305593661034030170112

262953661822174210117633038082364992607391580160

131479618504236921386709211005966577348884037632

657412030486933688573009514853753291970361425 92

328712984226341385106236487338836848598151331 84

164359976604607962962984108625398525605560975 36 821817305480226166864249867906893134560231424 0

410917363968706259456789596343396820546328985 6 205463037598649717740727129366673536906271129 6

102733696606473152876529730280824331769688883 2 513679372068107234413479479391559475431211008

256845130551924352222155153689498646006988800 128425287534897543618785283841608777149186048

642140048969164555632464954192341156474388 48 321076830131920696585501744588319213601423 36

160541817889629557677385506040233924482826 24 802726103566493835310100698931541210824704 0

401371558842419941116636933830956399617638 4 200690032950796482289111759098071096911462 4

100347143240191497009952525640331997007052 8 501746350024923764376745858658142227791872

250878491924445021851864544557312235274240 125441904418214080757678079892776678326272

627222814371028252947119437534486195077 12 313618053325493051052924237802544499261 44

156812349732735987816144378418922950492 16 784078364013627250529133189682868256768 0

392047489681787280988772243635560880537 6 196027898678380468356488946214843816345 6

980160262579336481092958853059535962112 490090515883385310201736487522426421248

245050450238551189928496774257542430720 122527821267704862378062652376935825408

612652087080670648959384588125502177 28 306332533911408493014227957183162613 76

153169512141240830774381810151787069 44 765863786633887075208248208559964160 0

382940006280784998272293683180496486 4 191474059622313229470231631040505446 4

957390580521169799021582102453813248 478705431465386725346003024852549632

239357786335094275590607499239096320 119681428468747594254106743025958912

598419818849740253564548682161848 32 299216247677871267929281824596951 04

149611292965436204538144654056488 96 748072310459683875558241979072512 0

374044078046093364212880343931289 6 187026000431172395323319849163161 6

935149809196490542700997631803392 467584808118559554392698008895488

233797355819436918717448600944640 116901153789797030119274098720768

584518148349378004399119484846 08 292265263874885429100934238044 16

146135726787540928001154366832 64 730694108188195707259208073216 0

365354791219343387256322156134 4 182681264172294460441520137830 4 913425663674586136274395987968

456722503243849985170595643392 228366087325203451101996646400 114185461514240954809347735552

570939396829400920338485739 52 285475743043798533315116400 64 142740893836448303230494965 76

713719580754986699017158656 0 356867346163865940940921241 6 178437450975119266186631577 6

892206144341527809514012672 446112516903729644047433728 223060980818347691668930560

111532851592415280657072128 557676063878283577398394 88 278843934897245375755714 56

139424918927674481406115 84 697139352033631374671872 0 348577054714445171156582 4

1394249189276744814061158 4 6971393520336313746718720 3485770547144451711565824
1742922167060373274886144 871479530274260346994688 435748988509167028273152
217879105940601941524480 108941858813310184456192 54472082328159699075072
27236617624832152961024 13618597042792228192256 6809442636584189952000 3404793375886132903936
1702432716740085415936 851234372768552189952 425626193583530835968 212817600391392788480
106411051995510079488 53206651897661882368 26603888898784362496 13302225924368891904
6651253699672801280 3325697218580578304 1662883793662377984 831459489017233408
415738540601638912 207873668347330560 103939033196920832 51970616110088192 25985857810857984
12993203783335936 6496739330621440 3248438384787456 1624253552132096 812143955935232
406080567902208 203044578918400 101524436942848 50763292213248 25382182977536 12691359924224
6345814179840 3172974198784 1586520653824 793277104128 396646940672 198327664640 99165929472
49584013312 24792530944 12396527616 6198394880 3099262976 1549664256 774848512 387432448
193720320 96862208 48432128 24216576 12108544 6054400 3027264 1513664 756848 378432 189220 94612
47307 23654 47309 94620 189244 378496 757008 1514048 3028160 6056448 12113152 24226816 48454656
96911360 193826816 387661824 775340032 1550712832 3101491200 6203113472 12406489088 24813502464
49628053504 99258204160 198520602624 397049593856 794115964928 1588265484288 3176598077440
6353330372608 12706929180672 25414395232256 50829864206336 101661875896320 203328046759936
406664683454464 813346546778112 1626727453294592 3253523626065920 6507184691085312
13014644260077568 26029838275969024 52060776063565824 104123751150387200 208251900347285504
416512596787593216 833042785761230848 1666120755894550528 3332311880533278720
6664764498554912768 13329810472086536192 26660183894126493696 53321493688159830016
106645239176133345280 213294981951894061056 426598971103042863104 853215956604595208192
1706467942006209380352 3413007941606456688640 6826159998400989233152 13652608227178130178048
27305792915108563779584 54612738751721734406144 109227783346452682506240
218460178378923792400384 436929580129884439576576 873877607003842588704768
1747792107495832596512768 3495658001967960031232000 6991463577888509738876928
13983222303682198830579712 27967034903174756366811136 55935250397970230144925696
111872861979181895112458240 223750446324846659870130176 447510337382659059030687744
895039564231249596642230272 1790116907394362150446170112 3580309372652450215215759360
7160769861032352259078356992 14321841953519608175450390528 28644288369949023665488134144
57289785665717661960150974464 114581989183074553178651361280 229168814069427564874001547264
458347299545412046781400743936 916713941903937927629596786688 1833466569434103523392784171008
3667010510120662383052749537280 7334175762746235438639861465088
14668661010502292222348447711232 29337940991024227134834344984576
58677119922087739649943589093376 117356715724254050060436976435200
234718383208665241641973549367296 469446669937644766326146291728384
938913146915918098736690969444352 1877865907913093329642178710863872
3755811043988700923621950965678080 7511780544302430375919089019256832
15023878001254917809188552214315008 30048389827809949733077852780232704
60098047306220127695557202263670784 120198629913640711849917397933752320
240402330429682336617440782680326144 480814802064166499070093538986295296
961649886537936649810611025223876608 1923340337895080602962069944950325248
3846761805428575812605835678905794560 7693685870133980838575062935821877248
15387696258821620103876909027664330752 30776041554750557061207384367369814016
61553381183715747829321901358821933056 123109358515860763072458067965808476160
246223909328580060972544666427946172416 492458203250877191600346393848550785024
984937175689188522511206909682418450432 1969915889753245323643442063335470661632
3939914856256227204528940614612208844800 7879995866011927523541994205106952732672
15760324039022801276052214361978975551488 31521312692043495010040880627488091275264
63043954612082774935954665062036462895104 126090567680157119703655137738193486479360
252186452272297379070801890704628094337024 504383538368561037468587011865738431430656
1008788344385054633591140484644441348374528 2017619224065974384490213891114811667775488

4035323518723679003596293625881481277603840 8070817178630818476424318939066678439313408
16141974639628557891312101252740788646838272 32284629843990957659551129254696440830099456
64570620817449599072956112007822608733044736 129143963893834565653619931012504671611781120
258293372305539866322655276018728251514945536 516597633646821202676141380024894319612657664
1033217045365125345413944416024664272390848512 2066477646873216570951212143999079811112763392
4133042406032364902149070911897662154887659520 8266259036636593324791435071594329375099584512
16532866522416913690569456638786668880847699968
33066429943121281463112086268769358022992461824
66134253682817471090170518519930756568579047424
132271294958784758508233729004645594182346342400
264548165103869149672252841938859350455069179904
529107480580337564656076451736855025090891350016
10582737261905873659935294439191982698543288680448
2116519125302144381116871949820510693809589321728
4233127453585082884726310042514111981065202565120
8466433313131754014437752370774405149022452973568
169332223438186684518845769313041172671829001633792
33867160500219722017632067769067070091226194640896
67735748248132149995145193824103589677588772028416
135474350991649711910052504220146078345450309550080
270954410974070247659629241584169954671446150086656
541920239929682142998306949456095505303983362146304
1083863315822447581354710831487702202530148848238592
2167772303571061753425615528126426788904661944369152
4335635950994456688283618786554898345498052384522240
8671454589693579739432013033713886226373561760612352
17343274554796492204593576988635951523502037504360448
34687279860411649860646255819688261188513902974992384
69376021222460630624210715324209238660047461882527744
138754965448195923054257838018083909886134235630141440
277515776902941169720188490775498684904347094990454784
555043245818980986663722611029659100072851437441253376
1110109875664159267774136481016641660674017369803194368
2220266519380713124441655479947930242404663729447763968
4440626574866215426670075995725154326922585438578278400
8881440221942009208913682063108896338071686836522057728
17763254588303175128974424269534968044596405591775117312
35527257465444663680242968825704286826098875021012238336
71056011508565954205074178224677275126009877716948484096
142115016172485162099324837595891953199644010783744983040
284236018655676831577002637484858712294536532267185995776
568484009932766677910711199555867036379570085933764050944
1136991965108359385334834248284033296340134214666312220672
2274031820702370829696492194912665039842256514930192678912
4548159422376045777446631786514526974008489201055521832960
9096510406694699791000558366407447736664930744501316616192
18193403937274616054215706319571683050625766173783179132928
36387574122319665052860591812656941255843341717127450066944
72776680740180195994579541972341032820870302173377083736064
145556426471442123766875800638736366260107841824998534676480
291118982925047711089185034665581333756950158606485803761664
582250225814422349289236936107379869987369267125949076340736

1164524971557498552800207605767194144921676434077853090316288
2329098982972304814043882678639257099737228667807616055902208
4658296045659225044974700291488251819262208934919051862343680
9316788250747680923723270451395978878099921068445743225765888
18633968820353823514994280639630908235350848534106765453688832
37268722278424570365084040752939717429003709862644088911691776
74539013832282987400359040453235236774611445314149293832011776
149081166215433668141099998801182077382430941806020819681280000
298168609532602722962963833391787362431277985967486103397072896
596349773268676219287455338362421140195388176645861134863171584
1192724654944293985297966019882535111056440762713500125864394752
2385499526702471064042042726080455883444210344270555964004892672
4771099487032708314976306824791683089551078326228223352561991680
9542399841320949003737056394844908824427471927830669554228396032
19085201417152962755042998280212902939505574406409784806665617408
38171206303328055005223767541471976460312409914316461009748885504
76344019544700369000723077045036294083227342031626704812333072384
152691252965489255981997238014257270491659728469240975210336747520
305388933683155547925096643876883905633729545750457081592014700544
610790722870665167772397623450506540568279269124864425526711812096
1221607156750038479389203918294490539738198893497629375738788446208
2443265735517493246467225179375935996679678497491059800848306536448
4886634315069819068312085044325781827765918415973721700438072360960
9773474318209303287379439459799383324344959673930647598359063298048
19547360012557936876269417661894405986316165031827703591683963748352
39095542777394534355559912808380090647884821431588223973299601801216
78192731059346389917161980585942738646274625599042081526462552211456
156388753127807422246408271110250591993559216669815430212651801640960
312784088273844129316985162097231413389138364283093394744756997718016
625581340584146828282307563947923285582316590453111858128420784308224
1251189009241210795861289607402767488772712904680073853534655146360832
2502430674628255870315928173819376812761585256907847981624937448210432
5004966661548180297818554265666437295955489408911096512361129207398400
10010143947679697710010504367388241932775616608012994122944767036751872
5005177286131517412191950101721804636820127199101897610583637829353472
2502641299211592984689324009874744153626223047098649079847446070165504
1251346977678713631641336484444292994421191247323174677201536612827136
625686652875815385469005481975606956014635485548512407239675095285760
312849908456136977558671360864533707409337673717718737939290942078976
156428245237183131191419990370631968405678802330590636129372168257536
78215768123148886801752150154498541553344383901160951644549432737792
39108706813853104003897152561840549451924683318513292612206390673408
19554764783065882303459115023215914063588587343223054701069032488960
9777588079602606302484826882755776700607416513594731548017434820608
4888896883836135726620048126951798184710269677788967872750176698368
2444499863935484150998841406262854009558415549390284985745817993216
1222275642976450219343829374524904463380848129943043017558273818624
611150676992579181594119022959190960991244242595471771121819320320
305581766248466626758161679327964845146032210109711016732250865664
152794097000321831359631923588167104898221149460843073951796035584
76398655438205174670091503756175893611713096933415319768733319168
38200131188124716830183522859134117387157809568204551280784310272

19100467328573423162660646920090143984229535334850721338600980480
9550434531542243955114766205306614637440082942799583518404902912
4775317699398888164449604475284078641382699109086903183754657792
2387709066513327175670912923957424982022678373387007304153432064
1193879641663605134558511805136405321677003596115281508214767616
596952375035273340640783574147049076171334002768529682176409600
298482464619372057001155622862947745752083103739709305122717696
149244370860553721840959729326185476709249603047576884578615296
74623754705710707590670823610448540271228827112649558297935872
37312661990572277130430891278902171093916426350755337153282048
18656723314144600232763185376290036026109219572592947578798080
9328557816501530950155462556564493252630112984904113290477568
4664376987965380891964666212491984246102808091755876395778048
2332237533839998154425800573350860932945279845529848073158656
1166143286848652931434634020227864871419577822590878974214144
583083903388653392828183876890149638183257861208416955924480
291548081676490159969525371833183420328363405560697212370944
145777105829326811762479402610646010782548940258592973389824
72890085410204271770098059652350155700458088868418670297088
36445808952872568829478208999688653004820853803770426949632
18223287600321500886953694086601114079706331586665759375360
9111835362103358679584141836678950828501118135723152637952
4556013462022983457845718315028672308574535239056712794112
2278054621497143787949682855858934601449255705125924634624
1139051255991397923488253277101766524305621895361746436096
569537600617111976500832563137032873943307969080265277440
284774786619262495628769243861591242866902495239828668416
142390386464984501503561103077333024381075502969762349056
71196689810168877596368792111935213664349879159805181952
35599093193922752220478516342601957569081003417364594688
17799920741380532821386318314618154152993533627413299200
8900147442899844766266689228967664760723282773072150528
4450167257554711560920109650313126222474899366218825728
2225130396829750369353437343071210032294078672950788096
1112588582441072479123409930492928476675353831396081664
556305983233634886785050594725125968601834163158384640
278158837623366767004198112101893849432995705309364224
139082341814958045307935463420612357282537164519768064
69542632409116353556885935395138894924288238192427008
34772046955376842229902069539985805603653947062484992
17386388853097753840680585691201081872581887514378240
8693377114253543283205068306204630471668400748756992
4346779900979104823034921883404360003522928870162432
2173435622415719002233654806853202385605828682973184
1086740647170942796474924336002112384725096465432576
543381741567013045916510634288811788323639294689280
271696579774277346797779550288283692142365178331136
135851144382524085318651891716080745061455354658816
67926999438954748619207004144009822025864060076032
33964213343323727289544031214989635760500221411328
16982463483585040134742280178987180254034206392320
8491410147754108312356272375239771313909151039488

4245794276857848278670702330492976250400599441408
2122941739919321200581634236683033421923311681536
1061493170704859130913958654059789359323161821184
530757735725028830768550094889031003842333900800
265384443048814048040060431374083664011543445504
132695009117556840347922907651825913050959970304
66348898355353328337907799808304997048074108928
33175146075964118250927072895348518785334116352
16587921487125786166450122943272269523315589120 8294134968134756603718354719435139826982060032
4147154596353310062105823983617072446153162752 2073620854319620911176235303758287489407647744
1036832205231293395649779307854019377869357056 518426991651388167855720481914447505517445120
259218940343564818943275654950942661050105856 129612192430717776979345534472330784670744576
64807457344826572243526620734595119408218112 32404409237147127998690237116512423240531968
16202544900940484937808581932863643388477440 8101442591653702938136022653735537578344448
4050806366418581703683877170519626731225088 2025445718505155969149871507085742336638976
1012744126900510543228902214455835653832704 506382697274221550941434337684400069672960
253196665549093915134208784070441156214784 126600991230538527398850199649341138796544
63301824843265048615298003631730849742848 31651577035630416765585453719395565043712
15826120824814154611760952811462852608000 7913226565906550420364589381613961347072
3956696359703011767424351178748248195072 1978389718226374162333203833344757858304
989215628300621220477116038657695809536 494618198744027679893815080321506344960
247314291668872374774536070657082392576 123659741982865454801082300576705806336
61831169065647361107448282912435208192 30916233569930997407177707768258756608
15458441303519157130315637040149954560 7729382911036407778521210098085265408
3864772585156618495942300838047776768 1932426857397516551311998313526460416
966233711108361927326423104014516224 483126996758982789498423525632901120
241568568981892307666817749629272064 120786819792146610292211868221046784
60394677546673533375507430813728768 30197972598636880802454463758467072
15099303211968497458577606055034880 7549810062309277257963990115418112
3774984259317152893319588601659392 1887531743739833578828591072804864
943785678910545355498693922390016 471902742975586960791546154188800
235956323247950621916872673591296 117980637504053881718986135044096
58991556692066226239767966646272 29496397316052755810021432885248
14748508143036199250079441223680 7374408814023010297574083002368
3687281778263960485054222696448 1843679574758207910660701945856 921859130192217789397146271744
460939236502665811731970785280 230474453954611364382684217344 115239644828944911449691521024
57621031340292070354020466688 28811120133055842491597586432 14405862297982824903092469760
7203082264718864280193073152 3601616690223158054419955712 1800846124043441984371687424
900441951487652470766698496 450230420476791974673776640 225119932604878856982102016
112562327485680863313657856 56282344334461149068132352 28141762463040933239717888
14071176379425645972684800 7035735763665412662755328 3517941668809001169584128
1759007727892648003895296 879522310690397711499264 439770378717235710525440
219889801044636282650624 109947206365327355019264 54974756104168284356608
27487954512836445601792 13744265486794374512640 6872276858585263112192 3436210486886669484032
1718141272240353705984 859088650518686334976 429553332458597908480 214781169828926324736
107392836714276847616 53697544257045266432 26849335078476054528 13424949014214737920
6712615244595724288 3356377991042039808 1678224179893108736 839129682132598784
419573637159321600 209791216626171904 104897807336341504 52450003179798528 26225551345713152
13113050550763520 6556662714335232 3278400076644352 1639234398060544 819634378899456
409825779384320 204917184659456 102460739813376 51231443648512 25616258695168 12808397783040
6404333109248 3202233663488 1601150386176 800591970304 400304373760 200156381184 100080287744
50041192448 25021120512 12510822400 6255542272 3127836672 1563951104 781991936 391004160

195506176 97755136 48878592 24439808 12220160 6110208 3055168 1527616 763824 381920 190964 95484
47743 23872 47745 95492 190988 381984 763984 1528000 3056064 6112256 12224768 24450048 48901120
97804288 195612672 391233536 782483456 1564999680 3130064896 6260260864 12520783872 25042092032
50085232640 100172562432 200349319168 400707026944 801430831104 1602895216640 3205857542144
6411849302016 12823967039488 25648470949888 51298015641600 102598178766848 205200652500992
410409894936576 820836969742336 1641708299223040 3283485317922816 6567108074799104
13134491027505152 26269531810824192 52540163133276160 105082525289807872 210169448626126848
420347693345275904 840712978876596224 1681461142125281280 3362992652994740224
6726126043477835776 13452533561932382208 26905630073818185728 53812386047543214080
107627023894900113408 215258551389427597312 430526109978109935616 861070234354729353216
1722176497506477670400 3444425052606993268736 6888994220402062393344 13778276671180276498432
27557129803112856420352 55115412527730319687680 110233130898469853069312
220470873482958133526528 440950970337953121828864 881920387419979953209344
1763877668328107325521920 3527829123632509489250304 7055805821217608654913536
14111906790340396662652928 28224403876491152030957568 56449988344603021473218560
112902337872447477769043968 225809398111377825183301632 451628240955721389657030656
903275371377374257894916096 1806588521686611472951541760 3613252601236948860226502656
7226656318201349549099843584 14453614867857602755493363712 28907834198625012825574080512
57816877323069640280322867200 115636172497778509818995146752 231277180698835478154689118208
462564032804227873342775885824 925147408421569580752347070464 1850333502469366829638284738560
3700744376191188995543750672384 7401643494887288663621863735296
14803596474784398672312452251648 29607811919588440034762354065408
59216861779216165449799607255040 118436199438510901660149012758528
236877350637178944841397622013952 473764604794672172724994437021696
947549016629972911534387260030976 1895137647341202955237571292037120
3790354522844920174812736128024576 7580867502014868878300659343949824
15162051916679794813951692863700992 30324737658659703742604134079004672
60650742967919635714609764861214720 121304021237039727888022523128840192
242613113076480368693651033070501888 485236367357762563222514039766646784
970493017125128778115452026784579584 1941026599069464859571751948071731200
3882134327777344325825199685148606464 7764430914831517865013790948307501056
15529186348216694156754365052635578368 31059021733540705166962296417312309248
62119341541296044040831725458706923520 124241279231021355495477716165578457088
248487750758901245818583962827486134272 496985886111519561292424986647630708736
993992541410473261895364095280578297856 1988026621195814802411756434531790356480
3976136319141366162065569357004848234496 7952438791782205438615251689892231512064
15905209890563357106198729331549533110272 31811084395124606670333910566629206392832
63623498018244998256540724940318693130240 127249654492481566344827257494757946949632
254504625896946272353146130217757015277568 509019885617858824033275490891996273311744
1018061038883650206720517442696957032136704 2036164613063165530748967807219843035299840
4072414296718061296113801458091544012652544 8144998734619583061459334603486803909410816
16290337751606087061382132581581039587033088 32581356067946015999691191912376942710489088
65164073265359715753236237323183612493824000 130333086878965479901418018164322667913339648
260667182097180333043775777280172266558062592 521345253230102136118382382547782349698891776
1042712284531687212298426421070440332563316736 2085468125206340304720176154090631931457699840
4171023362698612369686998932080766395577532416 8342220949969088259867291111960537856479330304
16684790349081903560721168719519085843607191552
33370277596451261203415510430234191948511444992
66741948989477430570777366842860424419617013760
133486685572104677469447425650504929884422275072
266978946330508987594680235230578021859221045248
533969043033617240500931238320292367899195080704

1067960386812433011625004012358857384159896141824
2135965375115263084496291096154260065042804244480
4272019953211320291485148335181610723531632410624
8544218312384228827955428956109402633955312664576
17088793436691634145881122483711167641694721015808
34178300497229621271702774110407060030957633404928
68358028242151948503286606506783569557051649556480
136718910979689308926335329585506038104376064606208
273443530950149441692194892314889874189297660198912
546898479881840531063438250917535344339686382370816
1093819795726764357484973434410581880601554888687616
2187685263379695305686140733972186145047474025267200
4375461870611723792804669198246417057783676546318336
8751106428928113948474113857096923650944810084204544
175023278233265560622677778635402026372644534151544832
35005887217349786696814659113220410886798896269361152
70013235936336904296547521911273538056617448471265280
140029394875948470398931451192212508679274208807616512
280064635758446264409535717123755882490627041345404928
560140963529991176042417063726173495245411330151153664
1120305311086179646531525386409670451019137155222994944
2240657390224753881956433290733987823094903300287365120
4481408316554296941699631617297269488303064580257480704
8963003705318172238972793306253126660832645119880462336
17926381555055501189092646755823428690118322158491926528
35853511398949315800479413798281208117142708154445856768
71708519375575258445547068169831117708097543983815720960
143420031906503770580270617486199638363819343317479456768
286846050123714048538894197265474082622887197334654943232
573704072868841111834494319117097777036271416068701945856
1147432090980508253182400487406494777653536874936188010496
2294912072446668565391624673157588002469061835469944258560
4589919925864641248836896743004372899262099842135024992256
9180031413671890733781088279387139587172152026660322934784
183604459512289979397776766145531066751640208738101191770112
36721658150228428823982711464575708657872226845763475341312
73444848795997723536823781276178567624928072430649134284800
146892762583077178851364279246411435868223382339542635773952
293791655148317821258161991880931472973181239635574005956608
587595570260962569627190850538080148419831429184125480730624
1175215660450578993476115434628594701786600758194205899096064
2350480360758465695395698336362058213467077316040321673461760
4701058801231546807678331606933854046721906231384463097462784
9402313761892324449130533082287183333019315661376565696004096
18805019842643110565808805901413317145189637719968410394165248
37610824323003144466713091276504535248681288234367378792644608
75223217921440135603617141500364872413966602057595873593917440
150449574393747964547616200895441348661141255292913979205091328
300905425889231315775996237580305904988698612941272422444695552
601823405981933404913520146739458225310229430593433772958416896
1203671920370808356550095636636609281286123270608645402054885376
2407394057555499806546301959588604223903575360060846516385873920

4814888548738765799984825291807979770469808357808804457323954176
9629977964733063973754093328877502186264931990991831763752321024
19260357663977192695077072148278089663180494532732109225713467392
38521518796976514885291915277602349907662250166961109847844585472
77044644531997288760859372517297040977927022536916002488524472320
154092502940083095502269828958778764281059089479819570562719547392
308191433632343226965641825765926893212528267771614272296780300288
616395722769040525853487987228592515725876713167178806936243011584
1232817156546789195551384645850662490053393781582258138557850845184
2465685735110994678791586634488279897310068273660317326486431334400
4931474314256821932960807954550469629026697968312236751714321956864
9863154316583309016676885280248758926866518778607677700911562489856
19726720009305948334864309302793157191359283241181763796788962131968
39454262770890557272749696090177593057971057850296344383509598568448
78910171046338435751541547149537743466447098436458322346882545745920
157823633101791513915167404237440601633904162344647911853491788709888
315653848221812312654503428351611432669828255632758358026436971855872
631320860480083194957344096456683324143696373152441784691780732583936
1262668049033083529211362672420287565895472470078733706661375042912256
2525388754212001337016074303854416967007104387705167687878377241313280
5050882820715671231218846525736517604446527670505735924868008793604096
10101976266014679576811088887528402549757693131202272947958526209163264
20204373781196033382368969447167539781244661842786148092362069662236672
10102397515181353805557880559639137231486968711583875144403543453073408
5051304069882345459965638197847252286175803250887338121313026037514240
2525704691087007008576168057937467978304061072991369335212140174245888
1262878673616420643584758508475654906760110260269534804883883664867328
631452500844668891440716493991287912184094992021692471080848621305856
315732832440563730544526866872374185494067426954308769859877705089024
157869707229396507684347743374552207448043678948885652089665549762560
78936499119255575048216026656458661074526822210308459624696123490304
39469072311906448127129090812820609212515902473087046602279736049664
19734947532092554365075084148705943943884196920509931696105705177088
9867679454115942333292811445500791640755221302238170045535771164672
4933942571092803742024040408324305654784172072110687121509344870400
2467022707563818158700837546949107744595366746551144610125402079232
1233537064790617223194827444868031330899323728523472829748065861632
616781387899662683519618058130754394750482041885686677216715341824
308397121702008377720911196913746562025651109754818469779698876416
154201774727092706841006682381057963338030599283396800475520040960
77102494301590612410778883152621322831617821844692183030595321856
38552050619817435700527212557356831997110172023842982911715311616
19276427044419782597832491769201501289205716562669937154066481152
9638414389465423672700688629862293289928173556709191426137653248
4819307628360478023242565687561917967626744416041707137621032960
2409704030994122105067393530096344645144701026864409281086619648
1204877123904002599256752108205865153238014922853982496681361408
602451116155472072989903725681778991951839666137880176409706496
301231835179471423175715698630312703642335935424384552239366144
150619056140603404928239767209867955654376018889914508136939520
75311097345735549134310842552289779743792035033818370077097984
37656333310584697902250900749822790830198030311339743042863104

188285589741508106186731901117503458942500215528851505235886O8
94144756465046361431104649242946481867005139750502147628728 32
4707335902966933488442167396357061713138008586828927131975680
2353716991340774452664551165283399666462880093066373441257472
1176883015599041080554009316194134238178377946359141658263552
588453767763847467387871524873284321562657923092548297949184
294233013864087197249369195824750762018063436502762883383296
147119571923125330402401314606429681627398955729625808896000
73561318457103531090059015650241991122883096603935088050176
36781425476322198489458686998634570716033357671528635826176
18391095862046315716943933086074072935312583520544863813632
9195739492965766094579261336415430256304244102662704857088
4597965527454187165343278064896912022476098222526488903680
2299030654212745641698462730793054458400037196860812689408
1149539272349198850362643214568826452781012641229190463488
574781608796012439938027531870562838181003342013987291136
287396790708712727347366728228356224985750181706689675264
143701388509709617362859845260715515440499346203192852480
71852190832531435552601816320362645919406180077652043366 4
35926843705104031185303201888447580333936964225722220544
17963795996971172303798661087540965535421514031592112128
8982085070695164507472860615429070451937272975161556992
4491136071452371431523195343543829068081894467263528960
2245614803778580304654980189686561455097576223473139712
1122830785915487446774181353800604188077102606657257472
561427084970842370610436306378963824302708550788972544
280719388491970508916890967928812777283432899124658176
140362617249259916264281891334071821207755761427415040
70182770126267289035059149351868626886897536646250496
35092115813952309968988676518350671584958596289396736
17546423282385487710223889180383514863234212127834112
8773394328897410217976720050795846966994563055484928
4386788508301038290420747755699968251186010023526400
2193439926076685735926567743001006509437369259655168
1096742799001426163321380804076014446640866753773568
548382817482254729339738868325762819281524438859776
274197117731898188509393667306759207621307750416384
137101413361334506174458950225318502800926640701440
68552133928359959047110533398628700895599703097344
34276780588026332503495795842299075195368042921984
17138747105936342741718162492641899971468117147648
8569551958929759615844213532067131172626106417152
4284865182445673930414672908906656179759077130240
2142477192713234026453619525889873386602550525952
1071260897101815543849951298663209341662781243392
535641598923507037236546417190740995012143611904
267826374648053151274058592524938659596448301056
133915974917176391964921988227253410843412398080
66959381255163104146407340096018745944300322816
33480387525869006155176843039205393233447223296
16740542212078230118575008015200706747372142592 8370445330610978579780797255399358439010336768
4185309777591421050137045251599181752167301120 2092698444938676405191845937749342142414716928

1046371000540821142657584624849546704372891648 523196389306152041359623140412211168769212416
261603639170946755695226984199824492675989504 1308045418444087453555321199096771700483686400
65403632051672056431514453046815577314689024 32702496590569870092684153272622652193767424
16351588577651855984805540010918757865095168 8175964430009388461634501692763094816653312
4088067285596424465433116690033405350379520 2044076178094077350024491266842631646216192
1022059356694971233666212094334280308621312 511040312171451896160089277623622397067264
255525472997709087743536254040052319911936 127765394954846113703513934634146720645120
63884026705418841767629871124133640667136 31942677966707313341751387465596960505856
15971671290352602899843919684563550339072 7986001798675774564406072818164310212608
3993083976087623839445092897023422627840 1996583526418680198343574692482345074688
998312532396774238482301468226489417728 499166650792104188896407795105903149056
249588517692910629275832428049280794624 124796854994884582051730479272805007360
62399725571656924732772372260484808704 31200511822935779219839752442283556864
15600580430021548036646659377162354688 7800452474287603231686721266591465472
3900307366782216222525056422300876800 1950194248210315414603376105653010432
975117406514761358972112000077791232 487568844462182505321267973664538624
243789492833492165578239973645090816 121897281717946539247922980228956160
60949908509573497853362986817683456 30475588080086863041382241760444416
15238110952693488578041495056023552 76192139326771772817695934615912448
3809686194498400673185560851906560 1904882711330457468761577197928448
952461162705857300465186984951808 476240484873242933274792685469696
238125194196778608158495939231744 119065072978467874839797767864320
59533774429273222800173783056384 29767506184656254090224341090304
14884062577337948390180895326208 7442186031173884867624810053632
3721170386839397770079586222080 1860623879045926553173383708672 930331282336077110653487153152
465175312574595472360141225984 232592491990576194696769437696 116298663846927326606734131200
58150540849283277932541771776 29075874887551446280858238976 14538239675230626797722796032
7269270953342765227508236288 3634711034535108528077537280 1817393296199417221200478208
908715537565640089181093888 454367213515785783880974336 227188329124375761585700864
113596525745429315615457280 56799443464335375219032064 28400312027978046315167744
14200451161894202510409728 7100373154899690931617792 3550260364426140304015360
1775167075701217571110912 887601984594682495107072 443810215669378102329344
221909719520707478552576 110957165603362952970240 55479735723186083332096
27740444322345345089536 13870510391548824256512 6935399310962487984128 3467771713075281920000
1733921885334659923968 866978957065839443968 433498485732174462976 216753746465714601984
108379125032670986240 54190688416242335744 27095907158074589184 13548235054014005248
6774258264495357952 3387199500991856640 1693634934868017152 846835059620052992
423426325903048704 211717560998035456 105860979522273280 52931589272764416 26466344392196096
13233447074004992 6616860975955968 3308499207454720 1654283963465728 827159161602048
413588170735616 206798380335104 103401337651200 51701742567424 25851408154624 12925972512768
6463120474112 3231627345920 1615847227392 807940390912 403978584064 201993486336 100998840320
50500468736 25250758656 12625641472 6312951808 3156541440 1578303488 789168128 394592256
197300224 98652160 49327104 24664064 12332288 6166272 3083200 1541632 770832 385424 192716 96360
48181 24091 48183 96368 192740 385488 770992 1542016 3084096 6168320 12336896 24674304 49349632
98701312 197406720 394821632 789659648 1579352064 3158769664 6317670400 12635602944 25271730176
50544508928 101091115008 202186424320 404381237248 808779251712 1617592057856 3235251224576
6470636666880 12941541769216 25883620409344 51768314560512 103538776604672 207081848176640
414172286287872 828361752444928 1656757864628224 3313584448733184 6627306336419840
13254887550746624 26510324857307136 53021749226242048 106045697475739648 212095792997990400
424200382089003008 848418356364050432 1696871897100189696 3393814162944557056
6787769063377469440 13575819601731649536 27152202153416720384 54305530206740283392
108613312213294252032 217231128026215874560 434471263251686490112 868960540901882462208

1737957110600783888384 3475986278795605704704 6952116672779287265280 13904521575934726242304
27809619612621755908096 55620392146748118663168 111243090136505451020288
222490791959029329428480 444990807290095513632768 890000061324264736817152
1780037016136676892737536 3560147819249648623681536 7120443212451886923776000
14241181572808953200377856 28482953441428265106407424 56967087474477247624118272
113936536132195930070843392 227877794630874729786900480 455765033994715198864228352
911548957455361876309311488 1823135693842586709780332544 3646346945548899333884084224
7292845006825250496415006720 14585992245105404650123689984 29172588953120616614834733056
58346386832060847858844172288 116695191515760924976037756928 233395218734800308468774338560
466800108876157533970946326528 933619560565428902008687951872 1867277806757085472150966501376
3734632984766626280569114198016 7469420712038163233672590786560
14939150909086147812413906354176 29878920788191938314965262270464
59759079516423162010205423665152 119520634912924894780960645578752
239046221586006931083020887654400 478102346692328145208240968302592
956224500425284856500880322592768 1912488614931826845170557417160704
3825056458026167954678708378271744 7650271372377364438032603844444160
15300859657404785933415581864689664 30602353140109685981531912080982016
61205973930819600192465320865169408 122414483162839656843733635136749568
244834036928080226605073257086320640 489678215060962279045358487798284288
979376712531528209761140922847854592 1958793989882263722863129740198281216
3917669109402942052407955269401706496 7835500478082713318179302116813701120
15671325474719085063085387389647978496 31343299986545486979624341091337109504
62687898047305607666155814806756524032 125378392243040482746125894861677658112
250761976782939500319880320219684536320 501534338159596070295017701432027512832
1003089445506626279900549524849371906048 2006220429388120838422127293669377572864
4012523935525978234086311075280022667264 8025214024551429582656735126442580377600
16050760356101805394281696204650230841344 32102185326201503246499844312830601854976
64205699880398791408872592432721484054528 128414058216789152649490992479563528798208
256833433345561444962473600187368178974720 513677500515089169251930430831218600706048
1027376268678110897157827322575401686925312 2054795072652086911623587566976732344877056
4109675215895904057863040977605322631806976 8219520572975268584957813642514361147719680
16439381428317458108379090659636154063650816 32879443421368758093685108068487171663724544
65760247972205199941224069635404070400294912 131523218203345767390155846267667594946281472
263051880924562269795727106529054098183946240 526114650884866009622285041045546012950659072
1052251079841214959306231738065967659066851328 2104545715825395798735786788081686584464769024
4209178543936723357718220200062875701591670784 8418531312445310235929733647924756468507607040
16837411074034347512846053791447523067663745024
33675519046356149107665280574091066396624551936
67352431889287206379276907130574173315843227648
134707651371724229086446506225932427676874702848
269420877929748090828678396381433017444125900800
538852906232095446968927560622002359069004791808
1077728113209389424560996656962277366499515564032
2155500827909175910368276385361100029722043088896
4311090858799145943229118913595290652890110099456
8622360123559880131443370112936762492672268042240
17245077059042936752857004797365887359128631771136
34490867741932226485654538737716499465825454915584
68983162731557158931190135761402448426787292577792
137969179958499729782142388094743795843847350648832
275944068907770283403809009333365389668240232284160
551899555797082214486666484954486375297571526541312

1103821947557247724331429902484483942517325177028608

4976520001513489948364800235922977099100600362633956000473490128896
9953245691096645047484869842993773867014323567251116198429898833920
19906902758332620396480278428283187071654892818468640791825634820096
39814628268943901395981634341157652818562277004870098373582943944704
79630902042445123998005423651497862987629536745605830327029236498432
159265095094004890408095157241360840676269038962942927813785170214912
318536772206239065640358934359451910754558008869348389947023734865920
637086708448936700929055108472364280313155879625621848532954258604032
1274199744970790541154784696451649478234391483025093834343722094952448
2548452146087415360902918351917140791684942413597887943243071345393664
5097009604466499278992534621861965253802203722291176435597397001764864
10194229833516335672358465079779297848469045234773153969417302625484800
20388880916199345573463721831669330378667366049927910135279622494879744
40777860433073203960442102700756013012079328326061902466344927947757977
6
20389723414532694030957305175890799742125917210691114528169656982700032
10195072331849684129852048424000767211927596395536358362307337113305088
5097641478216510622112722130028067276396117092863579730264922867630080
2548873395254089589649710024027875473414217993979490139688088589303808
1274463025699961934121529491520858654315188720763595207121857872396288
637244676886439536709101985513889785961634222268722672199835725070336
318628920461449053178719612633675122382837042077823870419371256971264
159317751239839169001444116255202675892428486510643202369412325703680
79660521124476905706764213096783895296719225991187234764569511460864
39831083314517113456403184032983226323612104363526434172216430034944
19915953033397887029712130758787252499432297865729625481074052169728
9958182204768608665611334750541445918529271774848016938019944660992
4979193946419136908183302060844632793671197308415610567751431618560
2489648395226984741780468373209271314038879364703606333246445453312
1244849908622200514734642857998113115621080037599703691308587548672
622437809815454329289525764695795287111360196423802107996976185344
311225332659904200605865050196267008206090187023876185169829298176
155615880206040618283483609022318186428250137917925658170585251840
77809547041064568132017346473251434376727591161956611878127927296
38905576989554413561146444217671887769665056682475197335481614336
19453190229288271528142107599359029175483158891986044365949632512
9726795981899668137855496544941057233066894721367245032079228928
4863498424577600255819969645101299939196104998370733940591820800
2431799429102683221356095508866035630929381318028922682572013568
1215924822958283157401103097590710646130355068436239197424058368
607974965682612352062079220374201738398009738929008526781054976
303993759943041562711803445976524076865420971819948727425040384
152000018522388474696283640882973642265918537087696595729776640
76001578536628084018332779388842623049563294132709413873516544
38001573906030965344261869168099212483083659860785264941072384
19001179271873944339678674320888556720692836327607911472693248
9500785795366203003613207028863753599921921362411595237425152
4750490977397716918693538448641614419748712280509617369251840
2375294528556166167790236691425676019768231939906718559895552
1187671784206736938116852079265272414831053869779314217582592
593848152067695396169292906408853409888995884802634577608704
296930206016011161640079886592535306181232417357806023213056
148468167999087312597756659990321953708983446157147378810880

74235616495084522187736688342188127163675341817695873007616
37118574495312694038297523344607638736429480278409028304896
18559670371541563491363351259060606945510644823985060052992
9280026747713389981788970422908697261403274754382802976768
4640109154827999108948132608143545525025613548386537963520
2320102467899651613500890002416371209674794859790837219328
1160075179192651836263856850380484828418391472694202728448
580049562217738932888634349776392025999692757746493423616
290030767419575973822670137181270818895094889572942741504
145018376865141240600511549737172812395171700136319385600
72510685010247247144844015441855107671397977743083700224
36256090793961936994716128007561904572605052709003853824
18128419541400125208505124147098127654755558273232928768
9064396842909640959826092145207651511604295095981965312
4532291957559609657699811108433119597915405527673733120
2266192746832199417743288072131206720014331753678241792
1133119757442297003318335295022926820535480371759808512
566571570734247148882513276990125140531897433340248064
283291631373672898052929453234393435398027340400295936
141648738690111110832301133986862150265052982065233920
70825830846692886319068770678263791415546146965159936
35413646174165108610993487181548253849282901448851456
17707188462491887031226294511982305995396364707561472
8853776918950609878477922716595242533075639345348608
4426979803327638120671349088599666034226548168458240
2213535573589985651051868409450855400957638332121088
1106790622758076120884031137300938892401001290006528
553406729360579708121064034938225042161591706976256
276709073671060677900056250612990319061341384474624
138357391330915750869790241878434058520943457730560
69180122913150581394776179225186478755608111611904
34590775080421643677328618755577964125372247179264
17295744352133998328634573949281344436470219276288
8648050582028587409302419260386853405127157481472
4324114493995087827143775773066517296009602662400
2162101848487940974818170957969803944727813292032
1081073224989169018032227014703174620725412626432
540547762867183774327684275210723634543459303424
270279456619891519819627521534929979362106146816
135142515903095576237706452732249070726241320960
67572651748122696282799572348516575885714784256
33787022772348802223372959165454308204154454016
16893859835318128152673066078325164232725757952 8447104142230927596829826286961587181687144448
4223639183401395558661559767380296123505704960 2111863147843663659454103195639899328083918848
1055953351993314769788713253794825297207492608 527987565032398854925187454884850465186512896
263999227034070162478009141436144140884639744 132002335775970448746712277714931524588011520
66002529017452908127209992355895489366851584 33001945073460295940531922927162608219848704
16501312819097068908729424838188735878135808 8250826550731994923596444106398083823173632
4125498345957727696414087896850899853639680 2062791708274728965514976870251378897846272
1031417121785297041411454896038653934436352 515719194716614800032710678475809209974784
257864914270290539679846954466145726365696 128935115591136839671669284847193423872000
64468887023564204751707546230656992280576 32235108125779994833790225018858636312576

16117886369888943645863338461194388242432 8059109338443944937415782206479729164288
4029637745971709025949947591181132103680 2014860411360722791596002039561199812608
1007450974867795535108515141765916786688 503735872027614837209514631875616833536
251873128310665953432385846434137636864 125939160303762244130007188465233428480
62970878226095755771910726856699019264 31486088150155194739408929740390662144
15743368593631255796431248026215907328 7871846556092457111579015591118241792
3936004407684643162471203584564264960 1968042768661528884576644968678470451 2
984041666740368093958648790643638272 492030974574985872814536368947462144
246020557889893849324874171286552576 123012814246147381121240079049687040
61507674773673918790021536228048896 30754471212137073509711516465627136
15377552518718593812206132408614912 7688934715684325434778253292208128
3844546586004676981726720190054400 1922312907083595623032156867002368
961176260582426377600476819488768 480598033811527471842437602738176
240303968666592087744231839786598 4 120154460213039009481708997181440
60078466804655879012112939771494 4 30039852993299037750702148419584
15020235981659340220419798990848 7510272733345807827442618859 52
3755213737919745727639312138240 1877645554586100531953246666752 938842120106164100043418632192
469430731459638967055106965504 234720201433097942044252307456 117362518568188200280475566080
58682468209913714769412489216 29341838567866664699293597696 14671221515388236006940475392
7335761873421569832117075968 3667956494574510830381957120 1834016026219118372352688128
917026902575490664757198848 458522896020711071669026816 229266170376838405479727104
114635446371660637562470400 57318903777451036192538624 28660042184535876801921024
14330316240173117753786368 7165305694039148553306112 3582726633995869114859520
1791400210486081976532992 895718551987114697818112 447868499365594203684864
223938861368815529230336 111971736527416978309120 55987021185213096001536
27994087053358851424256 13997331757055577423872 6998809993715864567808 3499477054451970211840
1749774556023004069888 874905292410011516928 437461653404260499456 218735330301757620224
109369916950692495360 54686084375253090304 27343605137579966464 13672084043766693888
6836182759371702272 3418161748430028800 1709116058587103232 854575621479596032
427296606832820224 213652701462921216 106828549754716160 53415374388985856 26708236950306816
13354393353060352 6677334115483648 3338735777218560 1669402248347648 834718304043008
417367741956096 208688165945344 104346230456320 52174188969984 26087631355904 13044084113408
6522176274432 3261155246080 1630611177472 815322365952 407669571584 203838980096 101921587200
50961842176 25481445376 12740984832 6370623488 3185377280 1592721408 796377088 398196736
199102464 99553280 49777664 24889344 12444928 6222592 3111360 1555712 777872 388944 194476 97240
48621 24311 48623 97248 194500 389008 778032 1556096 3112256 6224640 12449536 24899584 49800192
99602432 199208960 398426112 796868608 1593769984 3187605504 6375342080 12750946304 25502416896
51005882368 102013861888 204031918080 408072224768 816161226752 1632356007936 3264779124736
6529692467200 13059653369856 26119843610624 52240760963072 104483669409792 208971633786880
417951857508352 835920894885888 1671876149510144 3343821018497024 6687779475947520
13375833829801984 26752217415417856 53505534342463488 107013267708182528 214030933462876160
428070663018774528 856158918223593472 1712353020819275776 3424776410382729216
6849693558253813760 13699668591484338176 27399900132922097664 54800926165751037952
109604104131315761152 219212711862258892800 438434430923772526592 876886876246054535168
1753809781289128034304 3507691620172293996544 7015527355532663848960 14031342941441479409664
28063262343635262242816 56127677608775131332608 112257661060559476359168
224519933807137380106240 449049090986311614988288 898116628716696939528192
1796270150921541298159616 3592614088819377434525696 7185375751591344545464320
14371046651087868443754496 28742683597986095593160704 57486547787592908597624832
114975456758427252017856512 229955635883337373680926720 459920716499640486652280832
919860322465212451885416448 1839758423862287860932542464 3679592405588301636188504064
7359335926904055101023846400 14718974085263013859341369344 29438552633435835033270091776

58878314192691284695714889728 117759046237021798649779191808 235552928177322055816257208320
471055527761201028665912066048 942130398335515891398619430912 1884299482297259450930829459456
3768676335846974238128840114176 7537507414198859148792042618880
15075324313407539642652810018816 30151267596834721975443069599744
60303773133708729331161038323712 120610022147496029422871874895872
241224996055149200366843346288640 482459895630612683775885885571072
964939598301853933636170157129728 1929918810684964999441137086234624
3859916849532444263219867716419584 7719992155389917055114922520739840
15440301223429891167580219217281024 30881236272159896449861186786164736
61763740194920021129123870275534848 123530015691040498717050733957480448
247065101984481910351707454727782400 494140345173765646538626883081207808
988300972757134944747677713413701632 1976642510333477192836203321329975296
3953366150305368992354102431665094656 7906894559887567198071596441340477440
15814113638328792822869976038701531136 31628876313764902499193518389444214784
63259050701744438705294169402970734592 126520697551918144824402604054106079232
253046587400694824476433738604541378560 5061033559395106718608124538201741197312
1012227887977647576526763198388799275008 2024497314330163431674554640748232310784
4049077705410063420591165769437732143104 8098321564319599955666444514757999329280
16196975435638146140301114981281068744704 32394615485274184738538681866092277661696
64790560198544154392950267539244835667968 129583778853079878617646342692610232025088
259172874618142896898784300613461585428480 518356383060252073124551831683405413613568
1036734033768436704903070124279775312740352 2073510602832738527114073170385479596507136
4147106276257207288844012184422817135067136 8294382693697875046919756056149350154240000
16589105669762671032302975486906132076691456 33178891904259183941532877723027127689805824
66359144937986051636919608944483982452457472 132721012134907470781546924885827419050606592
265447468787685676578509263765373746392596480 530905826611112823187849355518185309367959552
1061833431293708586437360367011246251901452288 2123710418730383052998044045972243770133970944
4247507949746697866242734715843990072930074624 8495190124065259252978762679486985211184414720
16990728697274245546944111854571980553017360384
33982154292835945175861396700339981367331782656
67965702382246798515669139383072003257257689088
135934192357643413359230970730928087559703625728
271873959901586459374247325391424337209783746560
543759070175772184060065418641984998600320483328
1087540441096742898743272373002242645562146947072
2175125483683882858732827817441030587847305854976
4350340170348559839958221777755151769140635631616
8700858746658707924901575841256484725173319106560
17402074305240592339773416254005331824130733899776
34804862234327537659487361650995388395829659172864
69611151716347781278855781587960226286795701092352
139225157928080974477473679747859351563864167677952
278456024846932772794471592639596501108273866342400
556923467675407193267991651566948598177638794657792
1113869771313897681894080235709408388277459713261568
2227785214553961954504354336569839160399283674415104
4455661772960257090441096403441723088487295844614144
8911506233625180543746968267487535712352048680796160
17823377842659693813223487456183250495459011344728064
35647486436138053077906076754782859132427850655727616
71296434373913437058730357194398434547875357243998208
142595791751101535923297121758462301661790026353082368

2851974295087523954582670582562554684556586764363366 40
5704065510306034381398797459911726671754746003330170 88
1140836486087404170726450750939668794879263695586721792
2281719740227202930346284019793984510815156381014818816
4563533016559195038479333075417262863743570741712388096
9127253105327968432532196222493113411713657442790277120
18254880355075093576211452588303402191880346804311556096
36510508998988500574717025463241155120666757446085115904
73022514575653627994022291499751011715145642567094239232
146048022306660509677221064146039426377915540484036493312
292102030924027526732795090585153658651079591667769016320
584216034469468068222296105756456929092656204734930092032
1168456014181762165958004060685213081766306452268644302848
2336959918849176390942831819715024610694600990134856843264
4674015618669656899939311036119246115713178151464850161664
9348222799281922035985916865616886020074308645319973273600
18696828722449060544186423317990559617444521975420492447744
37394423692668554032802025809494694389480853320402076696576
74790379880877973954462409966016539088145325379926336995328
149583824752837679686641536626087378794657888238097041195008
299173779487838822928716506640283358826050251432682816798720
598359818940004572968299880056783920125569452778343102414848
1196744157808663000158333493666002245198076805382641142464512
2393537355474633708760134454436873300290029410417192160198656
4787172790663882834407203843083484220367810420138204070936576
9574541740756996502588277554586443680311124038884047642951680
19149475800372454672724294846011837839773254474983374288060416
38299736238461832680544069165701576637848521744397306580434944
76601041752357512031279097278758955192301069077655729169498112
153205222055582717402940112452229514217810189333033690356252672
306416721212900821486644060693882236102036481021511844747018240
612845996629272416334815792966610887536905166753912617563062272
1225717101665486379392686929090914605739474742929603091264176128
2451484420144855852231484544497214872810278304702761894804455424
4903068927391747778913551904616252010682832142470926352141611171 84
9806339415090488156494823668511944781891743769559493277426647040
19613080564692041060558532827546974854434118089867432253062119424
39226964598406211616254836636140120290169497281231755902541889536
78455536134856682222785215234372581742941516765457294597919080448
156914286145801882426121514392929845811088077936902154781508763648
313835000043780800813345196634229056272586244685779440734358732800
627682855591915673548894728965196841845992666995509143811399876608
1255391422192539490942198129323871142293625689238918812308164575232
2510834266402495269573213601434697201790532088973638673987058794496
5021771376839823114524061888443304237987625598938879446715576877056
10043748441749311379803393148034428144788374039860963090914072330240
20087908259637953061117325038364495627202993763688334576793981812736
40176639271554566725255727561320269929658478895309485943519637929984
80354924047666454656553610091823097209821940526484605466902624468992
160713139104447551725191530122011309120653846524700478093531946156032
321432860227124388274551680120752847643327623992863490506517286748160
642878884490707346197440599994966154090695109872652049651941362368512

128578409705433183169155567949685322578946994351915423658169630248140825716208502544979419764603180075482867950993345860087477190197604515845143347012800664441139618554042780244022517564267418044549293831880704
1028690465018466599665263294414092782890967291872563718732109628571648020574230549536006222052057560392590339548621417832876571087209815343104
4114930359740536090159769846500665004255579399642895753506445411850649682300292191477418718182563618456238812028690314384323855908977212653568
4115098859407205781658486515344958876947289631795536632084452309414707220575915546202703137039224248835529066465723739359285356867278790983680
1028816839768468868289300796047313187409749965987044377665614801744691251441895111340128986332018982642496074810687250306224374393283197009922572147411712840727909949908145966638956693810063011493275291315339264
1286100033929337503251649433579904237086426628805355883915459235414016643063181001127321274161956543412577347253176289603010596636406579200
3215381725187929454612495981484365180756465190882640396177715977256961607723772685111151427091090125833737388332250158632869686124960808968038783413881287877739670947547424421992159524379727706416959664947240194739821685099991719432222328400785213288989831455322016472629248
2009778128698188029737025485345983973023289017888213605597407346688010049096331560605299440396797877739533929567931424272225469955309568502465100981513522509783308451277960137134538670373821147643694284825123769269249839002377338850433447178889534038476701551089481154561256214174471200093963275613915149817546117057171735602239838879744
6281199427399541189038421426543136380738787062098180634626018508803140663991221540954130232391755261836873494419168841629026421309441570364134371655656870627035119477741688797653644296470369916682247851981365662704183380689371806622824704240488520860631133113548839260710297335650412041217840079284704822463544101194552083218432196307568831788899535894944105627276430618623227990429742504345609815579308844977350579189950542906466856246436773744336229629952
490789008805025486218181634790222455609078085607398359266702131224539952608390105245370188602664979393767192468805475086096138241227022738826446808991564773290941800354024032862051610442858496
6135239236166941778573100582243173155098442211419147332904550403067682389100824756094188649015818654213382129264018306797404161533872580059089311450913503455025365438771576409231473571266567669519827838831224273663412010707018854260440932268968719155238348383776911079456463796533731436052573314999091902847909888
1917458420731400139577963800370466850543766389676123042611200095874882630862315316636888702718094922943351469882547141345284793842211257731182718779369345642365934919172797947107606528
2396970145486173299802857151777689992861335386050883429072896
1198509592671740504123162309441279401377605592851396652171264
5992670563001971791724480214968569031622717463386757949030402996396581322620531416574441365370528178703481258266318602241498228940572127583485454387623228270273024115411576831344647491297952414724506313107772818856382283482450970102516940837457256009844055475994718037607857066009221624411604385792
1872901112880724421021194860556071611030051549698634809344093646971263462303412132690961587518437982100908834469969924682444344144419288660281944768572816223081216636859973632
2341270062557861703356964670728884855273528693915998224384

11706589765217568811918941845367416512177583897567832309 76
5853414608822914553526530168545204373993762162777836748 80
2926767167518522350546794707203350245949366188385878671 36
1463413515312793712165162165067049152450925647691419484 16
7317217234331631245284634882662115909635841005949498163 2
3658683446049646964871729469994493028508526886720949452 8
1829379137466739153550570749328964051099566635233574912 0
914708275954327412332638381830340793972434913553337548 8
457363491587642623944995694498099781197543254744943820 8
228686422599060770861836099040514582704434526356609433 6
114345549702150114875587175415989637405048712670373478 4
57173944052384922160128150655860991728940081081221120 0
28587556626847393441231356801863582377677902913627750 4
14294070613751162901199319137898334445442882643322470 4
7147181457039314540891479937432438851023406914915532 8
3573663803601523815591650152957855239662686254084915 2
1786868439341695180368780168599745526906834525356032 0
893452488441314226470867630360281716991162961834803 2
446735378605890431378672588210345335264454330495795 2
223372256495561874760955680620274906016663590037094 4
111688411844089266916287533567688572200550007413145 6
55845347720198798226048613412619845696384109903872 0
27923244759176481496976730020697702646246608050585 6
13961907829126781940464576667542741222150580574617 6
6981096639332661566220394162368315560588928561971 2
3490619682050966081104249995482630255051283418316 8
1745345522217800689549151454890551364904051277824 0
872690601705059169273088956019893801141230423244 8
436354221150608996885801092297255959915217603788 8
218181570724344204567528853292282509629909997977 6
109093015436691955346078580217968519651105597030 4
54547622755605904204196366894897892243628097536 0
27274368896432915367676721840405762330851698278 4
13637463207531439316627630116681289269944673894 4
6818870983423210474708449656579848687231749324 8
3409505181540350645551542127409526369745580851 2
1704787435684548026874429713264564197937643520 0 852411140299460365486544181412182605501248307 2
426214281378323358767936753096041556016837427 2 213111496303458267396300770774299590464152535 04
106557925958877427704316519468985515637315993 6 532800518830128608552413425332365394769346560
26640570393293503929162212665990160567605657 6 133205574225402887153518770326810256983719936
6660414824216912733061323866183485556470579 2 333027546858184055422335460801322913187758 08
1665171762527612370958023641467357742759936 0 832602895382152232402184989464050459790540 8
4163099547502491396626790790972110241005568 208159230904711081562132831731198409152921 6
1040817422171487966464630619568956531277824 520419344909710262559298540240960508395520
260214989366838270943140885348721375576064 130110153139410705303316250288481248477184
65056405797701137567531028951300904583168 325288675128484612417019663791805924638 72
16264766063423176849819209141355366318080 813254918521106153939371754656021820211 2
4066357669355267326938915261221376622592 203322037305250194209048587458132207206 4
1016630955713685110355757059275977916416 508325862450559624833135590630647398400
254168123522138347244196325811652919296 127086657909498441035912428153991069696
63544627028963854224863346701077839872 31772962555158924396588523966258007244 8
15886805794348280409669402987310612480 79435651564509694181980930716655943 68

3971863707863899315780742324837941248 1985972418751156961231219056921542656
993006491785182132286033475712057344 496513387097392891978228711481671680
248261764151097358906720342553657344 124133417376749135912163164683239424
62067976338974796185483079044825088 31034621994787512207442287874015232
15517627910043813161071518112808960 7758972411346935109210946144305152
3879565433835981818943066616102912 1939822330999248041640330080026624
969930972540252586904563426000896 484975389790440576494480905994240
242492646655377429768340049494016 121248799207767285644719822995456
60625637543922928202634810621952 30313437741981106791454854873088
15157028356000374740796152217600 7578668920505098042932438499328
3789411831505004357733400444928 1894744601378729847000290820096 947391643502478757566940708864
473705493157796295816868003840 236857582282176606425132826624 118431208992727532470915825664
59216813422183380864632619008 29609011174001497746903662592 14804807818455652530745507840
7402555024955278094019592192 3701353070341364961333215232 1850 7143141025454378283 17184
925376046517204197495013376 462697467991567838037934080 231353456362266788664180736
115679089364374829154697216 57840725273808131988652032 28920952932714424699977728
14460771614262391702814720 7230533381083785527820288 3615340477518187602116608
1807770713224724122016153 6 903872012867694319632384 451945229805884014592000
225977226588960434683904 112990919137489431035904 56496612490249322364928
28248882705876964605952 14124729583314634014720 7062508906845392863232 3531326511016734359552
1765699284305386143744 882867656551202553856 441442835474856017920 220725921337055379456
110365212468341374976 55183732134077530112 27592429016992186368 13796495983472803840
6898388729224757248 3449264733356556288 1724667551050366976 862351367711227904
431184479948636160 215596638020829184 107800518033670144 53901358528462848 26951229020045312
13475889387929600 6738082132918272 3369109785935872 1684589252706304 842311806222336
421164493045760 210586541490176 105295418228736 52648782856192 26324928299008 13162732584960
6581500510208 3290817363968 1645442236416 822737895424 411377336320 205692862464 102848528384
51425312768 25713180672 12856852480 6428557312 3214344192 1607204864 803618816 401817600
200912896 100458496 50230272 25115648 12558080 6279168 3139648 1569856 784944 392480 196244 98124
49063 24532 49065 98132 196268 392544 785104 1570240 3140544 6281216 12562688 25125888 50252800
100507648 201019392 402046976 804110336 1608253440 3216572416 6433275904 12866813952 25734152192
51469352960 102940803072 205885800448 411779989504 823576756224 1647187066880 3294441242624
6589016702976 13178301841408 26357140553728 52715354849280 105432857182208 210870009331712
421748608598016 843514397065216 1687063153868800 3374195027214336 6748527493382144
13497329864671232 26995209485156352 53991518481940480 107985235987136512 215974870020784128
431958536134590464 863934664455225344 1727904513282539520 3455890395309256704
6911899528106868736 13824080531190448128 27648724012334317568 55298573924575477760
110599399648964640768 221203302897556652032 442415612994368045056 884849240387245572096
1769734509571510108160 3539541076737058144256 7079226268662192144384 14158740767700536000512
28318057996153375424512 56637268913811357696000 113276843670631929085952
226558299027282285559808 453125821426601425895424 906270089597276561342464
1812577072682700541788160 3625227932341695921782784 7250603438635981519978496
14501502025177142392782848 29003594346164643491217408 58008369283950004393738240
116019099751141443610083328 232042921868765756865380352 464095288470497253021188096
928209466406925984623230976 1856456711745714926408171520 3712988981355155767139762176
7426129078437763362926362624 14852560388330430383146401792 29705725239570668080880156672
59412659404960950790935019520 118827736661561130840219451392 237660309026400720197137727488
475330289459358357427673104384 950679921731830548922141507584 1901398529089888765977873612800
3802874429432232868222928420864 7605903601369376408980219232256
15212116687748574163029163245568 30424852345516791016195776053248
60850942631072867412666451230720 121704361142224305585882700709888
243413674044605752692864997916672 486837251609525788427929188827136

9736943102596801429402567636418561947428234600617418049310299258880
3894935697363749100436214142468096 7790029851052526729547615372836864
1558037661475511051644560492147507231161387054810335147591958194552832
6232404176022089852458541309231104012465061882164225350797381959103283 2
2493063082456854199335536259948871684986227576961726657023192256154173 44
9972657978019489830750623984821207041994572160423105269490972691466813440
3989225450484625145663641171938770944 7978613160246079504690673921887830016
1595755083904581743610813099979623628831915750715198951725669828311633625088
6383279950461253715824678924734955520012766819515765434173030784374286372044 8
2553415826121672182882442179820566609925106935498180515062317454969567717621 76
1021407868823537151774005115898860404736 2042857276021942582169038475768354570240
4085797628793621721580133439477976662016 8171761411086716557644379854838488367104
1634385512917237934425698566144204682035232688374872342651146450423226414233812992
6537807897268108720877375025988874797056013075881640135374424929330813389805663027 2
2615229497146906281620782314960372346388485230565332533475356511396934485567120343 04
1046134334154627629956245847810077909581824 2092311203605120377220424617446084790190080
4184707477801970989056715078544027522433024 8369585096787402447345161844391770928971776
1673951047594172583315378706339097362615500833479701516617293543234500875996810788732928
6696076416270227084032285525042334865031168013392425058433990918835341749770615144631500 8
2678539456865505533921222489891312111840133125357187804088425768150753259657002389507932 16
1071459338889168093691812307906276111067119616 2142962233921302067506947927762303488465305600
4286011580128535895260542479424109509592743936 8572197384828935311014378206647224084509753344

18420252188342359903212035934494915048120454883414376448
36841252665523033228718192155624180833146973604290756608
73684001908722693302024624884517063140106074883505520640
147370996972798640293225730915571529227836405116859056128
294747980256303787964804424124217864350921320933414141952
589507933134020590686314772834585340492339663266220343296
1179039811510867210886041394841469904565673369331224805376
2358127513507386480798906488027538256293334824260017848320
4716350807986077079651460372744273406910645819715172171776
9432893177914762395410215538866940602469243981820617293824
18866169479714741263035020664490668782234392648421780488192
37733105207199915470499220502494912719060594666404652777472
75467742909940696829856799352016975747304808071931489157120
150938550810963125437430315398088252112976853622107345518592
301883231604089714430294064184285105462688182200703425445888
603778723172506355971454995144787413398845314314384319709184
1207581966273666566164643723842009231744628528454723577053184
2415212972404640840772754914788887273383132856561357029376000
4830524024523897098432444763787512166554017312426533809291264
9661244208477025030638759395994499572683537823460707119661056
19322880735812511728825258528827949624518082044136693241479168
38646546109341946792745996531333800207338176882703944487272448
77294661494117740255682952010023402331280379354269004983173120
154592461539103173851747821914758408495768809886260241983602688
309191200179941734384259479618940024657953722127964948001718272
618394954563354242130046630816726464648739648966818824072462336
1236815017533650030983148604791145759963143707355415504282976256
2473680251881183155412407895897677181257616233554386720842055680
4947460937390132497717037164426125685177890104795884866236317696
9895122742035797369218517074113794015681095484965992581577048064
19790647218582659486005919638750673322012821520680430861362921472
39582097906187448467149610258547517225326904142857753119143493632
79165802750419155924574762479187375613256330488709289031122288640
158334819376926829829700608882559433551717705383406143647915180032
316676066506030695620503385613488231753845499578787418467171565568
633364988516415463163211106923715192808511176781525099277025542144
1266755688041539070170830885240907844218662708810950723239415906304
2533562798100494428030479113268770605640606128117702495849561456640
5067228440235821431438592912111451045687773677227007090440582201344
10134662568541308013632455195370721760188670196437218378364082978816
20269736513221946328775449133037082858003586076840845151694003109888
40540295778722553260571975750665444391259663521614507093319680524288
81082237062002427727186106470513446133024309779094647766502709657600
162167765133119497866456522879392006967058585029920562692732116533248
324342112284468280557081665635514243336137101003303659704917627502592
648697388605395130762500571024488945476314063893532388048742043877376
1297421105283707400821675621555898808560707851560914913375297665499136
2594894866713249080236700202125639452337575150669530101306222486487040
5189895045718166717660098322278962575107469196434460751723699283951616
10380000716019670549693592480613292491079576183059722601669907189858304
20760422681206015328133976633337319663888427946501047399784831623626752
41521687860745379113761536610896108691235407053765299192459697735073792

83045060718157455142510239910235156109387916429057007170699464445788160
166093491429648304114994813197356189672610037501166831912959066842857472
83048430711490848972484573287121033563222121072109824742259602397069312
41525057854078772943735869987781986145069611696818116764019835686354944
20762950176206060700614726666001727754264081428790660578454935087087616
10381685712686367464680759169056231217996678504586131387449976165498880
5190948168634852289527077502555799279430658147388466242836242393726976
2595526740463260423356887710291741474931488521241933395973748352352256
1297789698304547350975118334652791655073823984394816835264687753920512
648908013188732245135896407079856286340951854084333486271250665832448
324460588612595407392116823416658372572495857985629277455078727352320
162233585315412346108142721646694300987257894464545905887266060894208
81118438162263494260113515792529707844133929968138586523496379056128
40560041833410407733077835380856132597319456352002110051679863832576
20280432292844534168049456432723705636285973859967463420805769068544
10140421834491932234779997587509672486956109771966935907885803110400
5070313761280798692767633479328746077884616306975070052684360843264
2535208302657815634072634082451327956145588863983336075712910065664
1267629862337615960880725712619141436674434787239568562541819854848
633827786673162052362567192006309447638037571243734543613592338432
316920321088758062142385763851524088469428874433842402978137374720
158463374420467549051743965849946726559919481622908767074739290112
79233294148278033516147524887065704442562263014448166330204946432
39617450543161146253211533424579022802582392608720974561520123904
19809127006091637874174652202812596691941826855108932978968887296
9904764370301351310871768846667840991296228702928689338588856320
4952482618778441842328105795964691818310771989151456093846634496
2476291526203104014610163584297731570486714813419283759199420416
1238170871508493554028137135306558615909021816131419735737761792
619097989957717550375596239232125723287343112776598795937906688
309555272080594161868561955405486069310087658743743862003466240
154780774591164774274662895597454638488251880549594163018989568
77391956571016233807522406746083121160729965863658197518123008
38696762923225040238856682846719461538666995726259656763375616
19348773780470981786976081160198681248484504260345107383844864
9674583049664721727261910448518815863817755328780193193000960
4837389604546976280517890158469145551696629263693916347039744
2418743842130795848702412546339441585742190431498868048789504
1209396440994051778572940006722155197818033115575388962029568
604710480461352816397336870137294801382485507700671949832192
302361370212839871754101868456756001927977228806824709324800
1511183750097501667654767650922432301582355851881656721866752
75593407544291699716242183808243301100361544679950544535552
37797470019916282802550271077635225704772581709536364068864
18899118133843357873489725125574400429682195539548727934976
9449750628864287172852157356165594003489050112164636917760
4724971095403447704479726074771993896068501227277454934016
23625334381873759112666686735730595395196238699236295704576
1181290664336513985146755217037596921179113392416931971072
590657304789670007330083533104948072380053717607858044928
295334638705541511043394728845548842085275369503625052160
147670312508124009210873845569311823990261940101660540928

7383665283173863145002516335792461346894309772575427788
3691907470470762914730670196559665747137761270033914265
1845991149677297128480041112611550410414183826890057318
9230142820596063997973735634716339736297435093815787520
4615164946402821176773632853187463710261975526590644224
2307629241253805177280198944508378776187616753136697344
1153838004653099883086790731211512848622122871489036288
576930694339648588766740995084418154575218683204861952
288471193176373617995043312281539942419687965332602880
144238519591461470803358063510435403775883294531387392
72120721297368066304597235440050418170961303198236672
36061091399502698603757719562441567226990479565389824
18030911075160682027608410702428962684250153765830656
9015638225285007376668980811818570877502533874483200
4507910456494836869766878136211330206439995433025536
2254000900173585025599632933256687487064361964404736
1127023286049875808157913399203854935454363106148352
563523061006479551758005165889683063688272615047168
281767239494010599718526816088719329824681838510080
140886474242390711779025524616298563902613684748288
70444664368888061849393820594118731446443225120768
35223045808290383904637439440044090470789803933696
176111879716068368442288984291514407609178997653504
8806118263995772466129624431503384991481546670080
4403148334978680355557378358624783089186797256704
2201618768979737239024972250748936841316410589184
1100831685235067150135627661092741069019711275008
550426992990132840379384598405506858690608627712
275219071681366052845477683132321591435680808960
137612323433832842750631533530944876763028652032
68807555513491329539262112747864478904108449792
34404474655033118851604229365128259713351286784
17202585776660286466788701178162139987324174336 8601467112902006753887643836880075058986352640
4300820668736935137190468542339540062155309056 21504538905114334487185557583119521297408720896
1075248723327199664420940447534636281869893632 537635250699341302241301051754755957517713408
268823069867541386136065939871096887050240000 134414257192706060575740676932407897670811648
67208489725820714041724191964633675908251648 33604925427644198897789022731531701490548736
16802802996189020387357974740373282513485824 8401571639277970662910719057490357140848640
4200870890230715566071225372397036512477184 2100477980411222900343545608024447227265024
1050260257853544008825739264925188099145728 525140762750738283739852862919076292329472
262575698287352281533418046687779267543040 131290507599667710598454830958010194460672
65646583027829640215100319286065377574912 32823956127912712565486611546562828959744
16412310370955302511711531725046484566016 8206321338977124370339878838405777326080
4103243746238298742411995907144156184576 2051663411494017649827026197542711853056
1025852474934442964224027220756672806912 512936622060938551767270671370994843648
256473503327327810711263866181826641920 128239347812093172769446198339077931008
64120971980261220091630231793621270528 32061135027237926899268682208851787776
16030892032172621876361124260446470144 8015608275363140151543953708233523200
4007885267319984682453672643121905664 2003983198479199644567684216063524864
1002011881649203473954266055283048448 501016082029403562812345001267167232
250513111617102694323778487446405120 125259091109751803620692237129613312
62630813205476130039747615268012032 31316040428038179134574555985608704

15658337126669146624637652168605696 7829327019659601840994013172203520
3914742737992315184834600130052096 1957410983077414724586096837001216
978725298579335928377446804488192 489372552809982247230922595237888
244691228165148265136560894115840 122348089962652703328830245306368
61175282921365637044690021777408 30588260430702461212482460450816
15294439700361051951309955006464 7647374592685436648189339893760
3823764667595173660361851142144 1911921019423814498314516168704 955979852525021083224053383168
477999597669067458645424340992 239004634537812187839410995200 119504735120545323178054909952
59753576486092276218202161152 29877392705955945423688433664 14938998584432876369137893376
7469650407943890013215784960 3734900761835670920931311616 1867488159849698417627365376
933762969390780687394537472 466890929428356082987696128 233450187080660911139061760
116727454723571890392137728 58364907953406662607372288 29183044272513690009337856
14591817284162024357494784 7296056216033601855160320 3648101894993095765786624
1824087840984695301996544 912062367236421360549888 456040406990247535050752
228024815181142194913280 114014713433580311150592 57008509638294762422272
28504831279899684634624 14252703870325994029056 7126496050351072870400 3563320082769574363136
1781696070181806145536 890866049489412554752 445442031943961018368 222725519571607879680
111365011585617625088 55683631692715655168 27842378796311248896 13921470873132335104
6960876174054522880 3480508455771439104 1740289412257808384 870162298314948608
435089945250496512 217549370671759360 108776884359135232 54389541691195392 27195320601411584
13597935178612736 6799105028259840 3399621233606656 1699844976541696 849939668140032
424978424004608 212493506969600 106248900968448 53125524226048 26563298983936 13281917927424
6641093181440 3320613699584 1660340404224 830186979328 415101878272 207555133440 103779663872
51890880512 25945964544 12973244416 6486753280 3243442176 1621753856 810893312 405454848
202731520 101367808 50684928 25342976 12671744 6336000 3168064 1584064 792048 396032 198020 99012
49507 24754 49509 99020 198044 396096 792208 1584448 3168960 6338048 12676352 25353216 50707456
101416960 202838016 405684224 811384832 1622802432 3245670400 6491471872 12983205888 25966936064
51934920704 103871938560 207748071424 415504531456 831025840128 1662085234688 3324237578240
6648609374208 13297487183872 26595511238656 53192096219136 106386339921920 212776974811136
425562539556864 851142258982912 1702318877704192 3404706474885120 6809550388723712
13619375655354368 27239301066522624 54479701644673024 108961602312601600 217927602671714304
435864001436450816 871745595058946048 1743526374489980928 3487123117724139520
6974386972936634368 13949055420849979392 27898673791653380096 55798473483213602816
111599198766240890880 223202901132109152256 446414809463473045504 892847633325455572992
1785731295447930109952 3571534648489898147840 7143213412167872151552 142867115054711896014848
28574006570176095453184 57149166061856797753344 114300637966722809200640
228605887619464045789184 457220998610964946354176 914460443966003602259968
1828957781420154623623168 3657989349816604085452800 7316126273585797847318528
14632547695076775047462912 29265685685963908800577536 58532551963548535012458496
117067465110338504847523840 234139652587159879340261376 468288749907285497970950144
936596389280502474522755072 1873230557492867906207219712 3746536672849461726737858560
7493224461426375282122555392 14986751154307654221538787328 29974106771525115757664927744
59949422468869846144504561664 119901262789378921547358535680 239807361282036301611415896064
479624393970629520256229441536 959268130754372874579254181888 1918574947134973417292098961408
3837227265522402170851379118080 7674609273549715014237120626688
15349528032109251373542966034432 30699675034238145437223381630976
61400588008515576254721662386176 122803651897109723269993123020800
245612255554376588061085842538496 491234414629067459164370878070784
982488636298763484413140142129152 1965016886678784100995077056233472
3930113001520082466327747656417280 7860384459365193461330682400735232
15721085831380443980011738977271808 31442805488061002074724226306146304
62886878626722232378849949315497984 125776292554644921216502892037406720

251557655711690755350611770887634944 503125452628183336536435515400912896
1006271187665970324743294978053111808 2012582940151147952827437850608795648
4025247009940710512336571490222735360 8050656279158250238036534558455758848
16101637076870158902799852272932093952 32203923190847634659053270857905340416
64409144455909903025013674339892985856 128820885060249073463841613927950581760
257646962417356681755311758352230383616 515304309428430433165880577697119207424
1030629388044295005642275277379555295232 2061300314463458289905578798729744351232
4122683705676653137053214085400756224000 8245533564852779388590541146684047491072
16491399436704505006149308245133165068288 32983463487406902470235068393796470308864
65968256202809589856343040594653220962304 131939170861610749544431888803427002613760
263883658635204638752355392835095126605824 527777951094375556831694016126672495968256
10555577169836683672317354493166309477449728 2111196874969232461942641908158547925925888
4222478820530195158501149659968953793904640 84445127782243850786234031007241623471915008
16890595846854622510931525389090678712041472 33781872258443086898789977527396220960505856
67565105646353857551433808553222168993857536 135132933551643082610575324103303792133406720
270271311621156900236566062200326492558196736 540553512278055270503962952388090801699160064
1081128802627593481069587560751057236563853312 2162301161398152842262498433451865739458772992
4324689435082237444771643490803234011579678720 8649553094736338410036580229405473088483622912
17299454638616403861059746954408956307615776768
34599606175520261804092666900013932876528615424
69200606147615431772131679782419906275651354624
138403999888380679872156051529623893596490956800
276813574963060992400097486988815949283358408704
553638300298721250111765741836768222747469807616
1107298901342641030846673019391809093856445595648
2214642404175679122939629110220163484435903152128
4429374011332152368371824363313417562317830225920
8858926428625892981728781012373016311527708295168
17718209669174962453427826596238394996839512276992
35437132962196277886796182335461514741247215927296
70875693172085261733473422956892478977630814601216
141754240839555935386708962485723856945534394695680
283514190669882694612942158115325511871614320377856
567039799321307036904932782518406619704319702728704
1134102434605697369167962497612324431330821529403392
2268250541137561329052118860375671246506007306698752
4536592426127455839536625451053387260700743109181440
9073367539959578041938026362710864056778943209930752
18147100455328488809605603646629907184312800402997248
36294931661475643070670309135676172510135428772265984
72591324824588617044258821956185061303290513477074944
145185572652451895894354051282035555172620338819235840
290376991311453115400380917303401975477319301368643584
580765674636004878024107464085465681218795850197630976
1161554733298207050494906187128254822965906195315949568
2323156234648808689883194892171156566988441380473274368
4646406005402406557553154820171606976090140740629299200
9292999083014391470679839712001801636406797440624099328
18586372310447939652506739567320778641266626799979200512
37173492909734192727307599421275908019439317437420404736
74348482397145012299203439415820517512690762549764816896
148699957949643278287583359978178437973005780449377648640

2974059022099930639535196822494316818412600715984513269761
5948237770413991426637452890850129754730171645962947133441
1189671499325624314840902427342325174527028371991373545472
2379390889136900688708628553029248796216044829580315328512
4758877559245105495470904502747694486756065830355767132160
9517946680432819227049103798873782762160084003101807214592
19036276484750854926312797184504353101616072690984160329728
38073319217272142797054773542522281357823954751529412460544
76148170930085151482967905432071713024831528242181008523264
152299406851252034743652527558197726668030293962606384250880
304604943684667533042738488504504054572795062881701502910464
609222147333661993196343843785225311619059075676380474638336
1218466881459597784061442142112288502818505605117871588691148
2436986669049263389672310309350638866263987902009341649092608
4874071417813142196231555529110153523157274033225030487244480
9748338995055515226236980974241505944206958005252645598527488
19497070308969492120021701685321962367564922407720570199212032
38994925255655907575138882844321825693431857609871698402738176
77991419786745661820468724635999453303467740808604512814104576
155985978124359016981319367166710510440143532794931257645465600
311978233350453420643402570122844228546703167945306979325444096
623969020904377614648332811824534872426238540601502886719913984
1247963150215696776019720966806762575518141490624783629577879552
2495976517245276645485552619928910812367611800093122971431862272
4992053468118319477863326612488592947397881237873357367415930880
9984307803492171329511095970238728540121077751120937583936274432
19969017341495407406591077431000542370892786052990320866081374208
39938838152012944308319925843047255323086833207477533128580399104
79879283242070147606915393648186851808776188617948849049996099584
159761780360228813194381871220558385942757421641885263685662801920
319529988472634662349865910289486136535924932095745658542666809344
639072832449623396621936156275711002372670041815441579428016029696
1278171375907954937088280983944899463346980438878783683541396881408
2556394173833326161865379310676753843897241588253368416453523406848
5112891191701484899108393306927417522201044597498338931648506101760
10225988071472634948972055985002654713215212036979882060779930779648
20452387519084600199454650712300948764056669757926172516525698711552
40905597790447861001930378909193176203365830883785161822983071727616
81812841085453043209902912787568909757236644503435957225829492064256
163628973180020728831890135513502934215483254478603181611385681346560
327264528378270742487948890903736097832986439900668897542224757129216
654542220793000054624235021560932654470012741688262863723356303130624
1309110769658917248545144522628786226548105207150375864724526184005632
2618274195463668775683638004271414288312369861848452004004679523500032
5236653703219006108553973926570512247057058618792304557120613357977600
10473518031021349331481343689196391834978755027775410212463735337910272
20947457311209372891709479050503518351686785635932422621372487919730688
41895757120752094240912541445228506066832122432628049635635010327281664
83793199238170885396812249578899950860581347186782508057050089630203904
167589768469675164623598832534685779174996899016617833685660317211688960
335186276926017116907146331823143313257662207319341302514440910325940224
167596508456341952283547499288457534082665308302723468828780593114251264

83799939224837673056760916332671705768249756472888143200170365532766208
41900812110752184985874041510557322247583429397207275992975217254203392
20950827304542766721683812427389395805520990278985240192932625871011840
10475624276854720475215302049750065243625132929683421194688821557460992
5237917450719028794794348942902716292244885359937111146455665089708032
2619011381505348675990523430465199981338602127516255847783459700342784
1309532018825591477291936194739520908277380787531978061169543427915776
654779173449254308294305337123220912942730255652914099223678502830080
327396168742856438971321288438340685873385058769919583931292645851136
163701375380542861897744954157535457637702494856691059125373020143616
8185233319482875215491463204795028616935623016421116314254985858680832
40926989349693036680478393508566421759930606450038398361206603644928
20463906050985848641749735496578850217591548908985607575569138974720
10232158713562589471630137119437244777608897296476007985267488063488
5116182200816127311192703245292532223211010069229606091375203319808
2558142522425479943285168965433221028808785745110604095058331303936
1279096972221448115486993154110087973006033227803202572214530473984
639561341615078129665700912751782715803836791525551548449947648000
319787098559716100793952624224260722552328484574750905396315029504
159896763155946568377527396036315043601369286693363018283828117504
79949988516017543179039239980249862963287165549675291934749360128
39975797727030901084657390971171102062944843876334537363792330752
19988300598026515289897580976108636322123052488915714380104990720
9994351166268790018733233233315860806386841519832080039156908032
4997276016762161196258837989288701725856078397603151444130660352
2498688225194963691575529680959736524259368017645131434341433344
1249369221004423392510820183637561092795348418244343573308768256
624697164705682469616937763397626961730506413833060714723409920
312354859454576621489232717488236688531669309271974821396217856
156180568278156004084998276638829948099042705813709642715365376
78091853414511848712690097266770775966125378495715937366310912
39046711344972847691440528107063288941364702042288526687469568
19523747991344885513268003790370594949833357418359542345891840
9762070155101673590407871763604772714492181907787410674024448
4881133157265452212090870816012123977033842553197525087551488
2440615618490033814488902875110930798410797076250672419045376
1220332329173670761466185171107899804152336437951291147157504
610178424551162307843959452330167104549637168888623042396160
305095342257744617477413159553192153511553059400800255606784
152550736119954040516423296470650377374143767178644495007744
76276900555517886147070006582352338996255502328444431106048
38139216525529376017964182464689744652719560533783307354112
19069991386649904481196680819101659903655684951672199577600
9535187255267560476705635202929223740475794818226372739072
4767689408605084356406464998153808764561873580308322844672
2383892594788194237230056197421502829442924875751729659904
1191970242636923148128439947883050638302456480674648948736
595997093939874588820925898527674930941725261736716533760
298004533280643801788815911556912271366111141568054296576
149005259795675154583584436924993538630679826133875163136
74504126475514204136380459035765470789152040741861588992
37252811526595415490484349804517086131482084208392798208

1862677990771686445638923504557571843419407402292740096

931357702606801058376814759444644690132355297082920140

465688204913879446967083883305251729277503446509735116

232848779262179182372880193444090556744414622239005081

116426728033709320631109222617777624425038760611571302

582145332181645250378891742567549852389351050518200320

291078512097371948801118686023105791326754148989272064

145542179051960636206395750381218328229416386359721984

72772551027617649006116078875441880397727849112403968

36387006264627489954517141280137298340373752522473472

18193868507723077702988121561276828240941790244372480

9097116941566205214358836241242503655848352113754112

4548649814635435788611805850923296595612904552660992

2274370579243884485022096790612670681650816524222464

1137208125585025537869145327881846532747590386057216

568615480774054416613621130228678862334886255001600

284313449377798032146334798258217229147988658487296

142159579184284427992929515701047513564267094736896

71081216839834919956345816136493206277269930115072

35541322043763812958113437211231327886203156430848

17771017833805082969026983177108026316885673902080

8885687322864129729498623874300194345334884794368

4442932864412858987241878080023187766113466318848

2221511033696826554867222111448139179779745120256

1110777817593611808056752591442342238251378540544

555400059169405169339947063580307443306442260480

277705604771002217325758915719721883743597625344

138855589978650924990772149824645022916987060224

69429188785900370659332420894714551981087653888

34715291291237639411639383438553296251840888832

17357994094762546746806278214874658256568975360 8679171271953136893896432355236334193608753152

4339672748262500207194862801517669629466509312 2169879930274215983720754712708586081064321024

1084961743208590931922039012329168673697693696 542491760640036935991850334152022153431613440

271251324837889203011340581069729985007190016 135628384677879969013377997531724446649286656

67815553468407668260542852264291950397489152 33908457298937676007198352881360838735167488

16954568931835758942062639815287851135795200 8477454607101339940263051594947641452003328

4238812374142400204747391641125678668054528 2119448722367065219681628742388768305053696

1059745628831465168494780832107348638040064 529883448239698863574373646510156561776640

264947041031832571450678438483319402266624 132476178971907855557085026855780261822464

66239418713949712694415417234950411255808 33120373970972748805144160521005345800192

16560519292485320631540306212267742986240 8280425799742133430254266082016406536192

4140295976620803272369189528949470789632 2070189526685269914805623008445369155584

1035115532530069096713325626208001458176 517568150858751618011919874096659169280

258789267726234343833588467544658804736 129397230011546439330608499020494012416

64699913079987853372211382134329311232 32350605577101243539559257379205808128

16175627307104280196506411845623480320 8087975912828969311616597500822028288

4044406908605289926248999453941615820 2022207510784565693458584516421065113

1011057836332432118963346529356611584 505539059371017885316885238303948800

252774600287909855576048605964795904 126389835445155384246827296388808704

63196185373177920352815144897609728 31598726511889074291108320800407552

15799680168594594202904534576005120 7899998540622325630127454375903232

3950078498473677079401320731901952 1975078863318095671869457137926144

9875592386996764020191269549506564937895228701524840517626704691202468997131952333835469809317314561234523324776952625340402641141766172740417888691664729503118131230864321059463101013784965152768154324700147413718519612073574407716389749875596598514966069248385827224619025363552466422988819291748087213544858959227125769646067471737910770147566551044823130449934524555407759769602411613582000046862870868131841205830969516415724018928189446029275740164040083012111564830146983163730007729647910912150737938133199075221176320007537048022387405589705654272376859956905742870917624627218843375634605773117498327049421876711962201344557711364711032803310758065183129602355563625320207729043701761177805424492518212747919365889145181624662804869939229446316203933672730001408147234532498720157178265607361874198888597535326208368101088642059360586956818405423366984442220380169202896150932958205706244601540309186847650611202300816271453608099184641150431194156896186531845752271262934941617356828761932775427011510272143812546180896574668807190771424232904589312359545776971049022259217977649136522640752648989004712246415196164494592428115755008002247341250054151208961123693143025212456965618578305116746547228093454475537154048140470087127452876807023645093860999168351189291567467724817559816422094274568780084132907581444390130027384012802195108994157117441097576487311114245487992387718348827440511694405632137205307251097606860402801508352343027012023091217151694198538248576018897960964288095348326402144090623836161072066786754565360441307955226802743410688134016401408006700954288128335054425292816753056808968376696176644188431974402094257930241047149936645235854540826179796992130901606406545211392327267123216363683848182005764091084802045583361022812165114163225571328127859206393088319660815983367991843996001998049990449953249774995599912199828399664799344159872031975046395136127905282558156851164160102330368204664832409337856818692096163741696032748994566549929984131001221122620076851252402585600104807268352209618731008419245850624838508478464167705051136033541681315846708470480896134172093972482683495566540853670985072640107344117628928214692530225152429393650384896858804480638976171764332101632034353553615093766870848161972224137419712018513922748449215951667254970083830661120109942366684577792219889131415666688439787058924355584879591710034755584175921860444160000

www.ingramcontent.com/pod-product-compliance
Lightning Source LLC
Chambersburg PA
CBHW081507220526
45467CB00010B/2815
* 9 7 9 8 3 8 8 1 0 9 5 3 8 *